简明中餐餐饮汉英双解辞典

主编 冯源

图书在版编目(CIP)数据

简明中餐餐饮汉英双解辞典/冯源主编. —北京:北京大学出版社,2009.3

ISBN 978-7-301-15006-1

Ⅰ.简… Ⅱ.冯… Ⅲ.饮食-文化-双解辞典-汉、英 Ⅳ.TS971-61

中国版本图书馆CIP数据核字(2009)第034329号

书　　　　名:简明中餐餐饮汉英双解辞典
著作责任者:冯　源　主编
责 任 编 辑:张建民
标 准 书 号:ISBN 978-7-301-15006-1/H·2219
出 版 发 行:北京大学出版社
地　　　　址:北京市海淀区成府路205号　100871
网　　　　址:http://www.pup.cn
电 子 邮 箱:zpup@pup.pku.edu.cn
电　　　　话:邮购部62752015　发行部62750672　编辑部62755217
　　　　　　　出版部62754962
印　刷　者:涿州市星河印刷有限公司
经　销　者:新华书店
　　　　　　　890毫米×1240毫米　A5　12.5印张　360千字
　　　　　　　2009年3月第1版　2009年3月第1次印刷
定　　　　价:38.00元

《简明中餐餐饮汉英双解辞典》
编撰委员会

主 编 撰　冯　源

副主编撰　纪俊超

汉语顾问　徐明旭

英语顾问　John F. Haynes

条目撰写　（按汉语拼音顺序排列）

陈　蔚　段纯淳　冯艳昌　冯　源　纪俊超
匡晓文　李枚珍　李文婷　毛春洲　王　琳
杨　红　张　华

编　　辑　（按汉语拼音顺序排列）

冯　源　金银星　凌定安　王　博　许　腾
张志文　朱　瀚

校　　对　（按汉语拼音顺序排列）

薄　隽　冯　源　郝帅帅　姜芳芳　金银星
王　博　温　强　许　腾　张　燕　张芳芳
张雪明　张志文　赵善梅　朱　瀚　朱千靖

前言

迄今为止，在我国乃至全世界都还没有一部汉英/英汉对照的中西餐饮辞典。随着我国与世界各国的交流日趋频繁，我国的旅游产业和旅游教育蓬勃发展，中外人士都需要汉英/英汉对照的中西餐饮辞典，以满足出国旅游、工作或者到中国旅游、工作的需要，以及学习的需要。编写一部简明实用、汉英/英汉对照的中西餐饮双解辞典，是时代的需要，也是我们"简明中西餐饮汉英/英汉双解辞典"项目组全体成员的愿望。

《简明中餐餐饮汉英双解辞典》是该项目的中餐分册。中华餐饮文化源远流长、丰富多彩、举世闻名，在人类文明史上占据重要地位。我们从国内出版的各种中餐餐饮书籍、各地旅游局网站、中国饭店协会网站、各地餐饮行业协会网站、中国烹饪协会网站、中华美食网等提供的数万条餐饮资料中筛选出 2777 个条目，一个条目为一个菜品、食品或饮品。我们为所有条目撰写了释文，并给释文配上英文。

本辞典是菜单型，不是菜谱，力求简明扼要、信息量大。它对菜品、食品、饮品的介绍以名称突出主题，以主料说明本质，以辅料与方法说明特点。

我国的餐饮史可以说是中华各民族相互交往与融合的镜子。祖国各地餐饮发展到今天，早已是你中有我，我中有你，甚至中西混合，日日更新，很难按传统的八大菜系分类。例如"糖醋里脊"这道菜就出现在鲁菜、苏浙菜、东北菜的菜谱里；"酒酿圆子"以大同小异的名称存在于多个菜系。因此，我们将条目按原产地或省、市、自治区标注。

本项目构思于2003年春，2004年春正式立项。我们组成了一支兼具应用语言学与旅游、餐饮专业背景，具备双语能力的队伍，开始艰难的跋涉。这个队伍来自五湖四海，其中一半是海归人员，对理解中西餐饮有独特优势。部分研究生和优秀本科生参加了一些辅助性工作，为本项目的完成作出了贡献。

编撰《简明中餐餐饮汉英双解辞典》是原创性工作，无前例可循。中外各国对很多物产和物品的命名并不一致，甚至对产地、源流众说纷纭。由于资料繁杂，考证艰难，资金短缺，我们的工作难度很大，受主客观条件限制，本辞典必然有许多缺失甚至错误，敬请广大读者批评指正。

冯　源

2009年2月8日

目录

一、条目

1. 收入的条目一律用粗体字顶格排印。

2. 所有条目按汉语拼音顺序排列。

3. 多音字按其在该条目里的发音排列。

4. 所有条目参照国家标准《汉语拼音正词法基本规则》和汉英双语《现代汉语词典》断词,注拼音,标四声。当以上《规则》和《词典》里无例可循或出现相互矛盾状况时,本辞典遵循汉语表述习惯,并根据自身的特殊性,处理词界。

5. 所有条目的释文均为汉英对照,先汉语后英语。

二、释文

1. 条目和汉语拼音用粗体字,拼音放在圆括弧(　)里,汉语和英语释文分段排列。

2. 三个或更多地名并列出现时,用简称。如:"江浙沪","闽粤港台"。

3. 释文内容顺序为:汉字条目,汉语拼音,产地,主料、辅料及制作方法简介,英语译名及其他汉语的英语译文。举两例:

例 1:

八珍扒大鸭 (bāzhēn pá dàyā)

福建菜。将仔鸭与干贝、萝卜、鱿鱼、香菇、玉兰片、火腿、鸡肉、虾仁等八种配料一起炖煮。

Stewed duck with eight treasures. Fujian dish. Young duck stewed with other eight ingredients, such as scallop, turnip, squid, fragrant mushroom, bamboo shoot, ham, chicken, and shrimp meat.

例 2:

董酒 (dǒngjiǔ)

白酒,产于贵州遵义。用高粱为主料,用小麦和多种中药制曲酿造,酒精含量为 30-60 度,属混合香型酒。

Dongjiu, liquor produced in Zunyi, Guizhou. White spirit made with sorghum and yeast of wheat and several herbs. Contains 30-60% alcohol, and has a mixed aroma.

三、译文

1. 英文语法以《新韦氏语法指南》为准。

2. 一般情况下,菜品和食品释文的英译采用省略句型,饮品采用省略句+陈述句型。见上述两例。

3. 单复数问题作如下处理:

   1) 概念明确、可数的东西一般用复数。如:两只鸭 two ducks,八宝 eight treasures,藕 lotus roots。

   2) 若没有指明数目,可数的大件食材一般用单数。如:鸡 chicken,鸭 duck。

   3) such as 后列举的东西一律用单数。

   4) 中草药、动物内脏、调料用单数。特别指明数目的除外。

4. 一些目前未被收入英语辞典,但已在世界各地广泛出现,独具特色的中国食品或特殊食材,采用拼音音译,用斜体标出。如:饺子 *jiaozi*、粽子 *zongzi*、馕 *nang*、糌粑 *zanba*、芷草 *zhicao*。

5. 所有酒品和茶品的拼音均用小写,其英文译名用斜体字,如碧螺春(bìluóchūn)*biluochun*,西凤酒 (xīfèngjiǔ)*xifengjiu*。乌龙(wūlóng) oolong 为唯一例外。

6. 为了避免歧义,统一部分英汉对应名词。如:田螺 field snail,鸡精 chicken essence,玉兰片 soaked bamboo shoot,泡椒 pickled pepper,猪里脊肉

pork tenderloin，猪排骨（根据上下文）pork rib，pork sparerib 或 pork steak，猪下水 hog offal，猪蹄 hog feet，鸡腿 drumstick，木耳 wood ear，鳜鱼 Chinese perch。

7. 各地使用的芡粉有所不同，很多菜谱并未对此作具体规定。因此，“芡粉”一律译为 starch。有具体说明的除外。

8. “味精”一词使用频繁，根据英文习惯，将其英译 monosodium glutamate 缩写为 MSG。

9. 为中华文化独有，英文里缺乏对应词或没有统一译名的词条，参照中国名家翻译并在中国得到广泛认可的译文，或者意译，力求信达。如干茶叶、刚摘的鲜茶叶以及茶汤分别译为 dry tea leaves，fresh tea leaves 以及 the tea。

四、术语

为了节约篇幅，这里对释文中经常出现的术语定义如下：

“调料”指盐、酱油、葱、姜、味精等五种普适性佐料，不包括其他佐料，如糖、醋、辣椒、花椒、蚝油、料酒、麻油等。

“卤汁”的基本内容与调料相同，加水或汤，或因各地习惯不同增减调料。

“清汤”指加淡盐和少量油的开水。

“高汤”在上海菜中指酱油汤，在其他地方指加了调料的肉汤。

“白汤”为不加调料的肉汤。

“鲜汤”指加了调味品的清汤。

“上汤”为长时间熬煮、加佐料的肉汤或鸡鸭汤。

“膏汤”与“上汤”意义相同。

“勾芡”指在少量水或汤里调入芡粉，煮开，然后与烹制的菜肴拌和。

阿婆铁蛋 (āpótiědàn)

台湾点心。先将鹌鹑蛋煮熟,去壳后入卤汤焖煮并浸泡多次,然后自然风干,直到蛋白缩成脆薄的一层。

Dried marinated quail eggs. Taiwan snack. Shelled coddled quail eggs boiled and marinated in gravy several times. Air-dried till egg white becomes thin and crisp.

阿宗面线 (āzōngmiànxiàn)

台湾点心。先将面线煮熟,倒入鱼头汤中,然后加入卤过的大肠。因首创人阿宗得名。

Noodles with hog intestine. Taiwan snack. Boiled noodles put in fish head soup, then combined with cooked hog large intestine. It is named after its creator A Zong.

安多面片 (ānduōmiànpiàn)

西藏食品。先将面团用手拉成小片,然后用羊肉汤煮熟。

Noodles in mutton soup. Tibetan food. Flour dough hand-made into small pieces, then boiled in mutton soup.

安溪色种 (ānxīsèzhǒng)

乌龙茶。产于福建安溪一带的乌龙茶的总称。除铁观音外还有六种,即本山、黄旦、毛蟹、乌龙、梅占、奇兰,各有特色。

Anxi *wulong* or oolong varieties. It is a general name for all kinds of oolong teas produced in Anxi, Fujian. Besides *tieguanyin*, there are *benshan*, *huangdan*, *maoxie*, oolong, *meizhan*, and *qilan*, each having a different aroma and taste.

鹌鹑蛋海参 (ānchúndàn hǎishēn)

山东曲阜孔府菜。将海参煮熟,配熟鹌鹑蛋和萝卜片装盘。

Quail eggs and sea cucumbers. Specialty of the Kong Family in Qufu, Shandong. Sea cucumbers boiled, then served with cooked quail eggs and sliced turnips.

鹌鹑蛋奶露 (ānchúndàn nǎilù)

浙江菜。先将鲜奶加白糖烧滚,用藕粉勾芡,然后倒在去壳的熟鹌鹑蛋上。

Quail eggs in milk soup. Zhejiang dish. Sugared fresh milk boiled, thickened with lotus root starch, then poured over shelled boiled quail eggs.

鹌鹑炖参竹 (ānchún dùn shēn zhú)

广东菜。鹌鹑、玉竹、北沙参加入盐、味精等调料炖煮。

Stewed quails with herb. Guangdong

dish. Quails stewed with herbs, such as solomonseal root and coastal glehnia root, then flavored with salt and MSG.

熬黄花鱼 (áohuánghuāyú)

浙江菜。将黄花鱼、五花肉、青菜加调料熬炖。

Stewed yellow croaker. Zhejiang dish. Yellow croaker stewed with streaky pork, Chinese cabbage, and seasonings.

鏊锅油鸡 (àoguō yóujī)

江苏菜。将仔鸡加桂皮、八角、小茴香等焖熟，斩块，浇上酱油和麻油。

Stewed hen with sauce. Jiangsu dish. Young hen stewed with cinnamon, aniseed, and fennel, chopped, then washed with sesame oil and soy sauce.

B

八宝鹌鹑 (bābǎo ānchún)

浙江菜。先将肉丁、竹笋、枸杞、莲子、调料等填入鹌鹑腹内，蒸熟，然后炸至金黄。

Fried quails with eight treasures. Zhejiang dish. Quails stuffed with seasoned eight ingredients, such as diced pork, bamboo shoot, Chinese wolfberry, and lotus seed, steamed, then deep-fried.

八宝冰 (bābǎobīng)

台湾点心。将大米、红枣、莲子、龙眼、栗子、薏米等八种原料熬成粥，拌入炼乳，冻成冰。

Frozen eight-treasure porridge. Taiwan snack. Porridge cooked with eight ingredients, such as rice, red date, husked lotus seed, longan, chestnut, and seed of Job's-tears, mixed evenly with condensed milk, then frozen.

八宝菠菜 (bābǎo bōcài)

东北菜。先将菠菜焯水，加入火腿、海米、鲜冬笋、干冬菇、黄瓜、胡萝卜等，再加调料拌匀。

Spinach with eight treasures. Northeastern dish. Spinach quick-boiled, combined with other eight ingredients, such as ham, dried shrimp, winter bamboo shoot, dried winter mushroom, cucumber, and carrot, then seasoned.

八宝刀鱼 (bābǎo dāoyú)

江苏名菜。刀鱼腹内装入腌渍过的猪腿肉、火腿、香菇等八种馅料，用猪网油包裹，蒸熟。

Anchovy with eight treasures. Famous Jiangsu dish. Long-tailed anchovy stuffed with seasoned eight ingredients, such as hog trotter ham, and dried mushroom, wrapped in web lard, then steamed.

八宝冬瓜盅 (bābǎo dōngguāzhōng)

广东菜。以半个焯水冬瓜为容器，放入瘦肉、火鸡肉、火腿、蟹肉、田鸡肉、鲜蚝，鲜汤等，蒸熟。

Meat soup in wax gourd. Guangdong dish. Halved wax gourd scalded, stuffed with eight ingredients, such as lean pork, turkey, ham, crab meat, frog meat, oyster, and seasoned soup, then steamed.

八宝豆腐 (bābǎo dòufu)

各地家常菜。嫩豆腐加猪肉、鸡肉、虾仁、火腿等翻炒，加盐、酱油等调料。

Sautéed tender bean curd and meat. Home-style dish in many places.

Tender bean curd sautéed with eight ingredients, such as ground pork, minced chicken, shrimp meat, and ham, then seasoned with salt, soy sauce, and other flavorings.

八宝炖鱼头 (bābǎo dùn yútóu)

湖南菜。先将鱼头腌渍,油煎,然后加入枸杞、海参、河虾、野笋、红枣、火腿、鸡腿菇等熬炖。

Fish head with eight treasures. Hunan dish. Fish head pickled, deep-fried, then stewed with eight ingredients, such as Chinese wolfberry, sea cucumber, shrimp, wild bamboo shoot, red date, ham, and drumstick mushroom.

八宝饭 (bābǎofàn)

各地传统甜点。先把糯米煮熟,再加入猪油、核桃仁、芝麻、甜豆沙、玫瑰膏、青梅干、杏干、葡萄干、松仁等,做成半圆形糕状。

Sticky rice and eight-treasure pastry. Traditional sweet pastry in many places. Sticky rice cooked, combined with various ingredients, such as lard, walnut, sesame, sweet bean paste, rose jam, dried green plum, dried apricot, raisin, and pine nut, then arranged into a dome-shape cake.

八宝鸽子 (bābǎo gēzi)

北京菜。先将鸽子腹内填入加调料的鸡脯肉、火腿、冬笋、红枣等八种馅料,炸至金黄,再放入汤中焖煮,然后蒸至酥烂。

Pigeon with eight treasures. Beijing dish. Pigeons stuffed with eight seasoned ingredients, such as chicken breast, ham, winter bamboo shoot, and red date, deep-fried, simmered, then steamed.

八宝鳜鱼 (bābǎo guìyú)

江苏菜。先将腌渍过的鸡脯肉、火腿、冬笋、虾仁等八种馅料塞入鳜鱼肚内,用蛋浆封口,然后将鱼身裹满蛋浆和面包屑,油炸。

Fried Chinese perch with eight treasures. Jiangsu dish. Chinese perch fish stuffed with seasoned eight ingredients, such as chicken breast, ham, winter bamboo shoot, and shrimp meat, sealed with egg paste, coated with egg paste and bread crumbs, then deep-fried.

八宝海参 (bābǎo hǎishēn)

湖北菜。将烩好的火腿、蹄筋、鸡肉、冬笋、虾米、香菇、莲子和荸荠倒在海参上,蒸熟并勾芡。

Sea cucumbers with eight treasures. Hubei dish. Sea cucumbers steamed with cooked eight treasures, such as ham, pork tendon, chicken, winter bamboo shoot, dried shrimp meat, dried mushroom, lotus seed, and water chestnut, then thickened with starch.

八宝黑米粥 (bābǎo hēimǐzhōu)

各地传统点心。用黑糯米与桂圆、红豆、绿豆、红枣、核桃、莲子、杏干、芝麻等八种配料熬成粥。

Black sticky rice porridge with eight treasures. Traditional snack in many places. Black sticky rice cooked into porridge with other eight ingredients, such as longan, red bean, mung bean, red date, walnut, lotus seed, dried apricot, and sesame.

八宝桔盅 (bābǎo júzhōng)

湖北菜。先将香菇、马蹄、火腿、鸡脯肉、冬笋、莲子、桃仁、青豆等拌匀,蒸熟,然后装入桔盅。

Eight treasures in orange cups. Hubei dish. Hollowed oranges filled with steamed mixture of dried mushrooms, water chestnuts, ham, chicken breast, winter bamboo shoots, lotus seeds, walnuts, and green soybean.

八宝辣酱 (bābǎo làjiàng)

上海菜。将虾仁、花生米、猪肉丁、鸭肫、竹笋、鸡丁等加辣酱拌炒。

Spicy eight treasures with sauce. Shanghai dish. Shrimp meat, peanuts, diced pork, duck gizzard, bamboo shoots, and diced chicken stir-fried, then flavored with chili sauce.

八宝梨罐 (bābǎo líguàn)

山东点心。雕梨为罐,放入糯米、梨肉、桔饼、桂圆、山楂等八种原料,蒸熟。

Eight treasures in pear cups. Shandong snack. Hollowed pears filled with eight ingredients, such as sticky rice, pear pulp, candied tangerine, longan meat, and haw, then steamed.

八宝酿 (bābǎoniàng)

新疆点心。先将当地香梨去核,填入葡萄干、瓜干、红枣,蒙上桑皮纸,然后蒸熟,浇蜜汁。

Steamed dried fruits in pear cups. Xinjiang snack. Cored local pears filled with raisins, dried melons, and red dates, covered with mulberry bark paper, steamed, then flavored with honey.

八宝酿凉瓜 (bābǎo niàng liángguā)

江苏菜。将虾仁、火腿、香菇、花生米、猪肉、鸭肫、竹笋、鸡肉等填入凉瓜,加入清汤,蒸熟。

Steamed bitter gourd with eight treasures. Jiangsu dish. Shrimp meat, ham, dried mushrooms, peanuts, pork, duck gizzard, bamboo shoots, and diced chicken filled in bitter gourds, then steamed in clear soup.

八宝糯米鸡 (bābǎo nuòmǐ jī)

浙江菜。先将鸡腹内填入加调料的江米饭、豌豆、火腿、莲子、薏仁、芡实、虾仁等八种馅料,蒸熟,然后油炸。

Chicken with eight treasures and sticky rice. Zhejiang dish. Whole chicken stuffed with seasoned eight ingredients, such as cooked sticky rice, pea, ham, lotus seed, seed of Job's tears, gorgon euryale seed, and shrimp meat, steamed, then deep-fried.

八宝青蟹饭 (bābǎo qīngxiè fàn)

福建菜。将糯米饭与白果、花生、冬菇、冬笋、火腿、调料等拌和,覆上青蟹片,蒸熟。

Crab and eight treasures with sticky rice. Fujian dish. Sticky rice mixed with gingkoes, peanuts, winter mushrooms, winter bamboo shoots, ham, and seasonings, covered with crab chunks, then steamed.

八宝碎扣鸭 (bābǎo suìkòuyā)

浙江菜。将生鸭焯过,再与猪肉、香菇、百合、薏米、莲子、白果、栗子、火腿等八种配料加调料蒸熟。

Duck with eight treasures. Zhejiang dish. Scalded whole duck seasoned, then steamed with other eight ingredients, such as pork, dried mushroom, lily bulb, seed of Job's tears, lotus seed, gingko, chestnut, and ham.

八宝鸭 (bābǎoyā)

山东菜。先将莲子、鸡丁、虾仁、青豆、香菇、干贝、调料等填入肥鸭,然后蒸熟,炸至金黄。

Duck with eight treasures.

Shandong dish. Fat duck stuffed with seasoned eight ingredients, such as husked lotus seed, diced chicken, shelled shrimp, green bean, dried mushroom, and dried scallop, steamed, then deep-fried.

八宝鸭羹 (bābǎo yāgēng)

浙江菜。将鸭肉、胡萝卜、火腿、香菇、豌豆、莲子等八种原料用鸡汤烩熟,勾芡。

Duck soup with eight treasures. Zhejiang dish. Duck braised in chicken soup with eight ingredients, such as carrot, ham, dried mushroom, pea, and lotus seed, then thickened with starch.

八宝芋巢 (bābǎo yùcháo)

台湾菜。先将芋头泥与葡萄干、冬瓜糖、桔饼等八种果脯蒸熟,然后勾芡。

Steamed taros with dried fruits. Taiwan dish. Mashed taros steamed with various dried fruits, such as raisin, candied melon, and candied tangerine, then thickened with starch.

八宝原壳鲜贝 (bābǎo yuánké xiānbèi)

山东菜。先将鲜鲍鱼肉与大虾肉、鸡肉、冬菇等八种原料煸炒,然后装入贝壳蒸熟,浇芡汁。

Abalone with eight treasures.

Shandong dish. Abalone meat stir-fried with eight ingredients, such as prawn meat, chicken, and winter mushroom. Put back in abalone shells, steamed, then washed with starched soup.

八宝鼋鱼 (bābǎo yuányú)

江苏菜。鼋鱼即甲鱼。将甲鱼肉和糯米、鲜笋、香菇、火腿等八种原料焖熟,加黄酒、盐、姜、葱。

Turtle with eight treasures. Jiangsu dish. Soft-shell turtle meat braised with eight ingredients, such as sticky rice, fresh bamboo shoot, dried mushroom, and ham, then seasoned with yellow wine, salt, ginger, and scallion.

八宝粥 (bābǎozhōu)

各地传统点心。将糯米或粳米加红枣、红豆、黑豆、绿豆、桂圆肉、百合、莲子、糖等煮成粥。

Eight-treasure porridge. Traditional snack in many places. Sticky rice or round grain rice boiled into porridge with red dates, red, black and green beans, dried longans, lily bulb cloves, lotus seeds, and sugar.

八卦汤 (bāguàtāng)

江苏菜。先将带肠乌龟肉用麻油煎透,再入砂锅煨烂,加盐、黄酒等调料。

Stewed turtle. Jiangsu dish. Turtle meat with intestine deep-fried in sesame oil, then stewed in a clay pot. Seasoned with salt, yellow wine, and other condiments.

八块鸡 (bākuàijī)

南方菜。先将整只母鸡切成八块,用盐、花椒、黄酒腌渍,然后油炸。

Fried chicken. Southern dish. Whole hen cut into eight chunks, pickled with salt, Chinese prickly ash, and yellow wine, then deep-fried.

八生火锅 (bāshēng huǒguō)

福建菜。先将火锅汤加调料煮开,然后放入鸡肫、牛百叶、海蛎、鱿鱼、生鱼、猪腰、青菜、豆腐涮熟。

Eight treasures in fire pot. Fujian dish. Seasoned soup boiled in a fire pot. Chicken gizzard, ox tripe, oyster, squid, fish fillet, hog kidney, green vegetable, and bean curd are added for quick-boiling.

八味酿笋 (bāwèi niàngsǔn)

浙江菜。先把竹笋塞满用香菇、蘑菇、干笋尖、榨菜、烤麸等做成的馅,再油炸,然后焖熟。

Eight-taste bamboo shoots. Zhejiang dish. Bamboo shoots stuffed with eight ingredients, such as dried mushroom, dried bamboo shoot tip, pickled potherb mustard tuber, and gluten, deep-fried, then simmered.

八仙豆腐 (bāxiān dòufu)

台湾菜。将豆腐泥与爆香的青豆、香菇、草菇、黄豆芽等拌匀,蒸熟。

Steamed bean curd with vegetables. Taiwan dish. Mashed bean curd mixed with quick-fried green beans, dried mushrooms, meadow mushrooms, and soybean sprouts, then steamed.

八仙过海闹罗汉 (bāxiān-guòhǎi nào luóhàn)

山东曲阜孔府宴菜。将鸡、虾、鱼翅、海参、鲍鱼、鱼骨、鱼肚等加调料蒸熟,摆八仙与罗汉状于罐中。

Chicken with seafood varieties. Feast dish of the Kong Family in Qufu, Shandong. Chicken, shrimp, shark fin, sea cucumbers, abalone, soft fish bones, and fish maw seasoned, steamed, then arranged in a pot in shapes of Chinese legendary Eight Immortals and Monks.

八鲜大补汤 (bāxiān dàbǔtāng)

江苏菜。先将干贝、海米、火腿、冬笋、猪肉、鸡肉炖至酥烂,再加入海参和人参酒煮熟，然后放入人参片焖煮，加调料。

Eight treasures with ginseng.

Jiangsu dish. Dried scallops, dried shrimp meat, ham, winter bamboo shoots, pork, and chicken stewed with sea cucumbers and ginseng wine, then simmered with sliced ginseng and seasonings.

八珍炒面（bāzhēn chǎomiàn）

广东食品。先把猪腰、猪肚、叉烧、虾球等炒熟，再放入焯熟的面条、小白菜、香菇等炒匀。

Stir-fried noodles with eight treasures. Guangdong food. Quick-boiled noodles stir-fried with hog kidney, hog tripe, barbecued pork, shrimp meat, baby Chinese cabbage, and dried mushrooms.

八珍冬瓜盅（bāzhēn dōngguāzhōng）

四川菜。将冬瓜去瓤成盅，放入干贝、萝卜、鱿鱼、香菇、玉兰片、火腿、鸡肉、虾仁等八种配料，加调料蒸熟。

Steamed eight treasures in wax gourd. Sichuan dish. Wax gourd hollowed, filled with eight ingredients, such as scallop, turnip, squid, dried mushroom, bamboo shoot, ham, chicken, and shrimp meat, then steamed with seasonings.

八珍豆腐（bāzhēn dòufu）

四川菜。先将豆腐块掏空，填入炒好的海带、鱿鱼、鸡脯、香菇，然后蒸熟，浇上高汤，配氽过的菜心装盘。

Bean curd with eight treasures. Sichuan dish. Hollowed cubed bean curd stuffed with cooked kelp, squid, chicken breast, and dried mushrooms, steamed, washed with broth, then served with quick-boiled Chinese baby cabbage.

八珍鲩鱼（bāzhēn huànyú）

广东菜。将鲩鱼煮熟，沥干，淋上滚猪油，佐咸味卤酱或酸甜酱食用。

Grass carp with special sauce. Guangdong dish. Grass carp boiled, drained, then washed with boiling lard. Served with salty sauce or sweet and sour sauce.

八珍扒大鸭（bāzhēn pá dàyā）

福建菜。将仔鸭与干贝、萝卜、鱿鱼、香菇、玉兰片、火腿、鸡肉、虾仁等八种配料炖煮，加盐，黄酒等调料。

Stewed duck with eight treasures. Fujian dish. Young duck stewed with eight ingredients, such as scallop, turnip, squid, dried mushroom, bamboo shoot, ham, chicken, and shrimp meat. Seasoned with salt, yellow wine, and other condiments.

八爪鱼煲（bāzhǎoyúbāo）

广东菜。八爪鱼即乌贼。将八爪鱼与珍珠笋、甘笋、青豆等煲煮，加入盐、蒜茸、麻油、胡椒粉等调料。

Braised cuttlefish and vegetables. Guangdong dish. Cuttlefish braised with baby corn, carrots, and green peas, then seasoned with salt, mashed garlic, sesame oil, and ground black pepper.

巴盟烩菜（bāménghuìcài）

内蒙巴盟菜。将豆角、土豆、肉和粉

条等先炒再烩煮，加盐、酸菜、辣椒等调味。

Braised vegetables with meat. Specialty in Bameng, Inner Mongolia. Stir-fried green string beans, potatoes, pork, and bean noodles stewed, then flavored with salt, sour pickled Chinese cabbage, and chili pepper.

巴蜀九味鸡 (bāshǔ jiǔwèijī)

四川凉菜。将卤熟的仔鸡切块，加入白糖、醋、麻油、辣椒等拌匀。

Marinated nine-taste chicken. Sichuan cold dish. Marinated chicken chopped and mixed with white sugar, vinegar, sesame oil, chili pepper, and other spices.

鲃肺汤 (bāfèitāng)

江苏菜。将鲃鱼片、鲃鱼肝、火腿片、笋片等入鸡汤迅速煮熟。

Barbel soup. Jiangsu dish. Barbel fillets, barbel liver, sliced ham, and sliced bamboo shoots quick-boiled in chicken soup.

拔丝白果 (básī báiguǒ)

东北家常菜。先将鸡蛋加淀粉调匀，摊成薄皮，切成块油炸，然后将糖熬至抽丝，放入炸好的蛋片拌匀。

Sugarcoated mini egg pieces.

Northeastern home-style dish. Sliced eggs-starch sheets deep-fried and quickly coated with heated melting sugar.

拔丝白薯球 (básī báishǔqiú)

山东菜。先将白薯球炸至金黄，再将白糖炒化，浇在薯球上，然后加桂花糖酱。

Sugarcoated sweet potato balls. Shandong dish. Sweet potato balls deep-fried, quickly coated with melting sugar, then sprinkled with osmanthus flower sauce.

拔丝脆皮豆腐 (básī cuìpí dòufu)

湖南菜。先将豆腐块裹芡粉油炸至皮脆，然后将糖熬至抽丝，倒入豆腐块迅速翻炒。

Sugarcoated bean curd. Hunan dish. Cubed bean curd coated with starch, deep-fried until surface turns crisp, then quickly fried with heated melting sugar.

拔丝红薯 (básī hóngshǔ)

南方菜。先将红薯块炸熟，再把白糖炒至起丝，然后放入红薯块迅速翻匀，撒上芝麻。

Sugarcoated fried yam. Southern dish. Sweet cubed potatoes deep-fried, quickly coated with heated melting sugar, then showered with sesame.

拔丝金枣 (básī jīnzǎo)

东北家常菜。先将枣裹上淀粉炸至金黄，再将糖熬至抽丝，放入炸好的枣搅匀。

Sugarcoated date. Northeastern home-style dish. Dates coated with starch, deep-fried, then quickly coated with heated melting sugar.

拔丝桔子 (básī júzi)

山东菜。先将桔瓣微炸，再把白糖炒至融化，然后倒入桔瓣迅速翻匀。

Sugarcoated Chinese orange.
Shandong dish. Chinese orange sections lightly fried and then coated with heated melting sugar.

拔丝空心小枣 (básī kōngxīn xiǎozǎo)

山东菜。先将空心小枣填入白糖水晶馅,再裹上淀粉,炸至金黄,然后将白糖炒至深黄起丝,倒入炸好的小枣拌匀。

Sugarcoated dates. Shandong dish. Chinese dates pitted, filled with lard and sugar, dusted with dry starch, deep-fried, then quickly coated with heated melting sugar.

拔丝莲籽 (básī liánzǐ)

山东菜。先将莲子蒸透,再炸至微黄,然后将白糖炒至金黄起丝,倒入莲子迅速拌匀。

Sugarcoated lotus seeds. Shandong dish. Lotus seeds steamed, fried, then quickly coated with heated melting sugar.

拔丝奶豆腐 (básī nǎidòufu)

内蒙点心。先将奶豆腐条裹蛋清面粉浆油炸,然后浇拔丝糖浆,撒芝麻。

Fried milk curd with toffee syrup. Inner Mongolian snack. Milk curd strips coated with flour and egg white paste, deep-fried, washed with toffee syrup, then sprinkled with sesame.

拔丝苹果 (básī píngguǒ)

北方家常点心。先将苹果块挂芡糊,炸至焦脆,然后放入炒化的糖液,迅速翻匀。

Sugarcoated apples. Home-style snack in northern areas. Apple sections coated with starch paste, deep-fried until crisp, then quickly mixed with heated melting sugar.

拔丝葡萄 (básī pútáo)

东北家常菜。先将葡萄去皮去籽,裹上面粉,挂鸡蛋淀粉糊,炸至浅黄,然后将糖熬至抽丝,放入炸好的葡萄拌匀。

Sugarcoated grapes. Northeastern home-style dish. Peeled and seeded grapes coated with flour and egg and starch paste, deep-fried, then quickly mixed with heated melting sugar.

拔丝肉段 (básī ròuduàn)

东北家常菜。先将猪肉切段,挂鸡蛋淀粉糊,炸至金黄,然后将糖熬至抽丝,放入炸好的肉段拌匀。

Sugarcoated pork. Northeastern home-style dish. Cubed pork coated with egg and starch paste, deep-fried, then quickly mixed with heated melting sugar.

拔丝山药 (básī shānyào)

河南菜。先将山药块挂蛋清淀粉糊油炸,然后与熬化的糖浆迅速翻炒。

Sugarcoated Chinese yam. Henan dish. Cubed Chinese yam coated with starch and egg white, deep-fried, then quickly coated with heated melting sugar.

拔丝土豆 (básī tǔdòu)

东北家常菜。先将土豆炸至金黄,再将糖熬至抽丝,然后放入炸好的土豆拌匀。

Sugarcoated potatoes. Northeastern home-style dish. Potatoes deep-fried, then quickly coated with heated melting sugar.

拔丝西瓜 (básī xīguā)

山东菜。先将西瓜块碾去水分,裹上蛋清淀粉糊,油炸,然后与熬化的糖浆迅速翻炒。

Sugarcoated watermelon. Shandong dish. Watermelon cubes squeezed, coated with starch and egg white, deep-fried, then quickly fried with heated melting sugar.

拔丝香蕉 (básī xiāngjiāo)

东北家常菜。先将香蕉去皮切块,挂上蛋糊,炸至金黄,再将糖熬至抽丝,放入炸好的香蕉块拌匀。

Sugarcoated bananas. Northeastern home-style dish. Peeled bananas cut into sections, coated with egg and starch paste, deep-fried, then quickly mixed with heated melting sugar.

拔丝香芋 (básī xiāngyù)

华南传统点心。先将香芋炸熟,再将白糖熬至起丝,倒入香芋迅速翻匀。

Sugarcoated taros. Traditional snack in southern areas. Taros deep-fried, then quickly coated with heated melting sugar.

拔丝湘莲 (básī xiānglián)

湖南菜。先将莲子蒸熟,再裹上蛋浆油炸,然后将糖熬至抽丝,倒入莲子快速翻炒。

Sugarcoated lotus seeds. Hunan dish. Steamed lotus seeds coated with egg paste, deep-fried, then quick-fried with heated melting sugar.

拔丝樱桃 (básī yīngtáo)

山东菜。先将樱桃稍炸,再把白糖炒至融化,然后放入樱桃迅速翻炒拌匀。

Sugarcoated cherry. Shandong dish. Cherries lightly fried and quickly coated with heated melting sugar.

拔丝鱼段 (básī yúduàn)

东北家常菜。先将鲤鱼段挂糊,炸至金黄,然后将糖熬至抽丝,放入炸好的鱼段拌匀。

Sugarcoated fish. Northeastern home-style dish. Carp chunks coated with starch paste, deep-fried, then quickly coated with heated melting sugar.

霸王别姬 (bàwáng-biéjī)

江苏菜。用甲鱼与鸡蒸炖,用盐、黄酒、姜、葱调味。

Stewed turtle and chicken. Jiangsu dish. Cultured turtle and chicken steamed in soup. Seasoned with salt, yellow wine, ginger, and scallion.

霸王花煲瘦肉 (bàwánghuā bāo shòuròu)

山东菜。先将霸王花、瘦肉、无花果、蜜枣、杏等加水煮开,然后加调料,用慢火炖熟。

Stewed zygophyllum flower and pork. Shandong dish. Zygophyllum flower and lean pork boiled with figs, honey dates, and dried apricots, then simmered with seasonings.

霸王全肘 (bàwáng quánzhǒu)

四川菜。先将猪肘氽过，放入辣椒、酱油、花椒等制成的卤汁煮熟，然后在原汁里浸泡。

Marinated hog knuckle. Sichuan dish. Scalded hog knuckle boiled and marinated in gravy seasoned with chili pepper, soy sauce, and Chinese prickly ash.

白扒猴头蘑 (báipá hóutóumó)

内蒙菜。当地猴头蘑加鸡汤和调料煨熟。

Braised mushrooms in chicken soup. Inner Mongolian dish. Local monkey-head mushrooms braised in seasoned chicken soup.

白扒裙边 (báipá qúnbiān)

山东菜。先将鼋鱼裙边和腌渍过的鸡块、猪肉片蒸烂，再用白油奶汤扒裙边至入味，勾芡。

Turtle's soft shells in milk soup. Shandong dish. Pickled turtle's soft shells steamed with pickled cubed chicken and sliced pork, braised in lard and milk soup, then thickened with starch.

白扒素海参 (báipá sùhǎishēn)

河南菜。先将淀粉与水、酱油调和，煮熟成棕色晶状，冷却后切成海参片形，然后与熟芝麻、蘑菇拌炒，勾芡，淋油。

Sautéed soy sea cucumbers. Henan dish. Starch mixed with water and soy sauce, boiled into crystal brown gruel, then cooled. Cut in the shape of sliced sea cucumbers, sautéed with prepared sesame and mushrooms, thickened with starch, then washed with sesame oil.

白扒通天鱼翅 (báipá tōngtiānyúchì)

山东菜。先将鱼翅用开水烫过，然后加上汤、黄酒、白糖等文火焖煮，勾芡。

Braised shark fin. Shandong dish. Shark fin quick-boiled, braised in broth over low heat, flavored with yellow wine and white sugar, then thickened with starch.

白扒鸭掌 (báipá yāzhǎng)

东北菜。先将鸭掌稍烫，再与火腿、冬笋、冬菇等加高汤用小火煨透，然后勾芡，淋麻油。

Stewed duck feet. Northeastern dish. Duck feet quick-boiled, stewed in broth with ham, winter bamboo shoots, and winter mushrooms, thickened with starch, then sprinkled with sesame oil.

白扒鱼翅 (báipá yúchì)

山东菜。将水发鱼翅和焯水的鸡块、鸭肉、猪肘等同蒸，然后用加调料的奶汤煨煮鱼翅，勾芡。

Braised shark fin in milk soup. Shandong dish. Shark fin soaked, steamed with quick-boiled chicken, duck, and pork joint, braised in seasoned milk soup, then thickened with starch.

白扒鱼肚 (báipá yúdǔ)

山东菜。先将鱼肚、冬菇、玉兰片、菜心等用白油、奶汤扒至入味，然后勾芡。

Fish maw in milk soup. Shandong dish. Fish maw braised in milk soup with winter mushrooms, soaked bamboo shoot slices, baby Chinese cabbage, and lard, then thickened with starch.

白拌黄螺 (báibàn huángluó)

福建菜。先将黄酒、白酱油、糖、葱调和,再将焯过的黄螺肉放入浸泡约二十分钟。

Marinated yellow snails. Fujian dish. Quick-boiled yellow snails marinated in mixture of yellow wine, white soy sauce, sugar, and scallion for about 20 minutes.

白菜串鸡 (báicài chuàn jī)

广东菜。先将整鸡去骨,鸡腹内涂上蛋清,塞入白菜和草菇,油炸,然后与汆过的五花肉及调料炖煮。

Braised chicken, pork, and cabbage. Guangdong dish. Boned whole chicken brushed inside with egg white, stuffed with Chinese cabbage and meadow mushrooms, deep-fried, then braised with scalded streaky pork and seasonings.

白菜炖全鸡 (báicài dùn quánjī)

山东家常菜。先将五花肉过油,然后与土鸡入高汤炖煮,加入白菜和调料。

Stewed chicken and cabbage. Home-style dish in Shandong. Local chicken stewed in broth with quick-fried streaky pork, then boiled with cabbage and flavorings.

白菜滚木棉鱼汤 (báicài gǔn mùmiányú tāng)

香港菜。先将木棉鱼稍煎,然后加入白菜、生姜煲煮,汤熬成乳白色时加入盐等调料。

Chinese cabbage and fish soup. Hong Kong dish. Kapok fish deep-fried, braised with Chinese cabbage and ginger until soup milky, then seasoned with salt and other flavorings.

白菜扣虾 (báicài kòu xiā)

江苏菜。先将大虾轻炒,然后与冬菇和白菜一道蒸熟,浇原汤。

Steamed prawns and Chinese cabbage. Jiangsu dish. Prawns lightly fried, steamed with mushrooms and Chinese cabbage, then washed with broth.

白菜牛肉卷 (báicài niúròu juǎn)

东北菜。先将白菜叶烫软,裹上牛里脊肉片煮熟,然后和熟胡萝卜、洋葱、花菜等放入盘中,浇少许芥末酱。

Cabbage-beef rolls. Northeastern dish. Sliced beef rolled in scalded cabbage leaves and boiled. Served with boiled carrots, onion, cauliflower, and a little mustard sauce.

白菜肉卷 (báicài ròu juǎn)

四川菜。将猪肉末与调料拌成馅,用白菜叶裹馅成卷,蒸熟。

Steamed pork and cabbage rolls. Sichuan dish. Seasoned ground pork rolled in Chinese cabbage leaves, then steamed.

白菜素鸭 (báicài sùyā)

各地家常菜。先将炸过的豆泡与白菜、草菇、胡萝卜等煸炒,然后焖煮,调味,勾芡。

Simmered cabbage and soy duck. Home-style dish in many places. Deep-fried bean curd balloons quick-fried with Chinese cabbage, meadow mushrooms, and carrots, simmered, then flavored with seasonings and starched soup.

白菜香蕈 (báicài xiāngxùn)

四川菜。将干蘑菇与白菜炖煮,加盐、味精、猪油。

Braised mushrooms and cabbage. Sichuan dish. Dried mushrooms braised with Chinese cabbage and flavored with salt, MSG, and lard.

白参藏猪肉夹 (báishēn zàngzhūròujiá)

西藏菜。将五花藏猪肉煮半熟,切片,夹上白萝卜片蒸熟。

Steamed turnip and pork. Tibetan dish. Half-cooked sliced Tibetan streaky pork put between sliver of white turnips, then steamed.

白茶 (báichá)

微发酵茶,产区主要在福建省,是中国茶类中的珍贵品种。制作时不揉不炒,成品茶满披白毫,因而得名。

Baicha, or white tea or light-fermented tea. It is a rare kind among Chinese teas, mainly produced in Fujian. It is processed without rubbing-twisting or baking, which helps retain the hairs of tea leaves.

白灼明螺 (báizhúo míngluó)

广东凉菜。将螺片用上汤焯熟,加入麻油、鱼露、胡椒粉拌匀,佐芥末酱或梅膏酱食用。

Boiled whelks with sauce. Guangdong cold dish. Sliced whelks quick-boiled in broth, then mixed with sesame oil, fish gravy, and ground black pepper. Served with mustard sauce or plum sauce.

白炒刀鱼丝 (báichǎo dāoyúsī)

江苏菜。先将刀鱼浆炸熟成丝,然后与冬菇、绿叶菜、火腿等加调料煸炒。

Saury with vegetables. Jiangsu dish. Minced saury meat fried into shreds, then sautéed with winter mushrooms, green vegetable, ham, and seasonings.

白炒目鱼卷 (báichǎo mùyújuǎn)

广东菜。目鱼即墨鱼。先将目鱼剞十字花刀,微炸,然后与花菇、青椒、冬笋、胡萝卜及调料同炒。

Sautéed ink fish and vegetables. Guangdong dish. Ink fish cross cut, lightly fried, then stir-fried with grained mushrooms, green pepper, bamboo shoots, carrots, and seasonings.

白炒虾球 (báichǎo xiāqiú)

广东菜。先将虾仁裹上蛋清淀粉浆稍炸,然后与冬笋、香菇、西红柿、青椒及调料同炒。

Sautéed shrimp meat and vegetables.

Guangdong dish. Shelled shrimp coated with starched egg white, lightly fried, then sautéed with bamboo shoots, mushrooms, tomatoes, green pepper, and seasonings.

白炒响螺 (báichǎo xiǎngluó)

福建菜。将葱段、冬笋、冬菇略炒，勾芡，然后将响螺片倒入速炒。

Sautéed whelks. Fujian dish. Winter bamboo shoots, dried mushrooms, and scallion lightly fried, thickened with starch, then sautéed with sliced whelks.

白葱辣牛肉 (báicōng làniúròu)

山东家常菜。将牛腿肉丁炸至略脆，投入白蔻、糖、盐、干辣椒，加汤，用文火煨酥。

Braised beef with white nutmeg. Home-style dish in Shandong. Diced rump beef fried to half-crisp, then braised in soup with white nutmeg, sugar, salt, and chilly pepper.

白肚 (báidǔ)

浙江菜。将猪肚加黄酒与盐烹熟，切条。

Boiled hog tripe. Zhejiang dish. Hog tripe boiled with yellow wine and salt, then cut into strips.

白干煮三丝 (báigān zhǔ sānsī)

江苏菜。将白豆腐干丝、鸡丝、火腿丝、鸡肫片、鸡肝片、笋片等放入鸡汤煮熟。

Bean curd and meat in chicken soup. Jiangsu dish. Shredded dried bean curd, chicken, and ham boiled in chicken soup with sliced chicken gizzard, chicken liver, and bamboo shoots.

白菇炖鸡 (báigū dùn jī)

山东菜。先将鸡块加汤、黄酒、盐等炖烂，再放入蘑菇，煮至汤呈乳白色。

Stewed chicken with mushrooms. Shandong dish. Cubed chicken stewed with mushrooms in soup flavored with yellow wine and salt.

白果煲鸡 (báiguǒ bāo jī)

广东菜。先将白果与鸡肉轻炒，再加入汤、葱、盐、味精等煲煮。

Braised chicken and gingkoes. Guangdong dish. Quick-fried chicken and gingkoes braised in soup flavored with scallion, salt, and MSG.

白果炒鸡丁 (báiguǒ chǎo jīdīng)

北京传统菜。先将鸡丁轻炒，再把调料炒出香味，然后倒入白果拌炒，勾芡，加入麻油。

Sautéed chicken with gingkoes. Traditional dish in Beijing. Quick-fried diced chicken and seasonings sautéed with gingkoes, thickened with starch, then flavored with sesame oil.

白果冬瓜汤 (báiguǒ dōngguātāng)

广东菜。将去皮切块的冬瓜、莲子、白果等放入锅中熬煮，加盐，姜、葱、麻油。

Wax gourd and gingko soup. Guangdong dish. Peeled chopped wax gourd stewed with lotus seeds and ginkgoes. Seasoned with salt, ginger, scallion, and sesame oil.

白果豆腐煎 (báiguǒ dòufu jiān)

南方菜。将白果和豆腐加调料炖煮。

Bean curd and gingko soup. Southern dish. Bean curd and gingkoes stewed with seasonings.

白果炖老鸭 (báiguǒ dùn lǎoyā)

广西菜。将白果与老鸭清炖。

Stewed duck and gingkoes. Guangxi dish. Mature duck stewed with gingkoes in clear soup.

白果桂花羹 (báiguǒ guìhuā gēng)

江苏点心。用白果和糖桂花煮汤,加糖,勾薄芡。

Gingko and osmanthus flower soup. Jiangsu snack. Gingkoes and sugared osmanthus flower boiled, flavored with sugar, then thickened with starch.

白果烧鸡 (báiguǒ shāojī)

南方菜。先将鸡加姜片和黄酒煮至半熟,再加入白果炆煮,用盐调味。

Stewed chicken with gingkoes. Southern dish. Chicken quick-boiled with ginger and yellow wine, stewed with gingkoes, then seasoned with salt.

白果鸭煲 (báiguǒ yā bāo)

广东菜。先将白果仁煮熟,油炸,然后与煮熟的鸭煲煮,放入豆芽及调料。

Braised duck and gingkoes. Guangdong dish. Gingkoes boiled, fried, then braised with boiled duck, bean sprouts, and seasonings.

白果鸭母卵 (báiguǒ yāmǔluǎn)

广东点心。将糯米粉加水做成鸭蛋状丸子,包入麻茸或莲茸馅,放入白果冰糖汤煮熟。

Sticky rice dumplings with gingkoes. Guangdong snack. Sticky rice flour and water made into balls like duck eggs, filled with sesame or lotus seed paste, then boiled in gingko and rock sugar soup.

白果芋泥 (báiguǒ yùní)

江苏菜。先把白果去皮去心,炸熟,再用糖腌渍,然后放在芋泥上。

Gingkoes and taros with sugar. Jiangsu dish. Peeled and cored gingkoes deep-fried, pickled with sugar, then put on the top of mashed taros.

白毫银针 (báiháoyínzhēn)

又名白毫、银针白毫。白茶,主产于福建福鼎、政和等地。外形芽头肥壮,遍披白毫。汤色清澈,呈浅黄。叶底银白。

Baihaoyinzhen, also called *baihao* or *yinzhenbaihao*. White tea produced in Fuding and Zhenghe, Fujian. Its dry leaves are stout, plump, and white-hairy. The tea is crystal and light yellow with infused leaves silvery white.

白花酿北菇 (báihuā niàng běigū)

广东菜。先将虾胶填入花菇,再把花菇抹上蛋清,贴上香菜叶,撒上火腿茸,蒸熟。

Steamed mushrooms with minced shrimp. Guangdong dish. Grained mushrooms filled with minced shrimp meat, sealed with egg white and coriander leaves, sprinkled with minced ham, then steamed.

白芨冰糖燕窝 (báijī bīngtáng yànwō)

浙江点心。将燕窝和白芨入冰糖水蒸熟,去渣。

Bird's nest and herb soup. Zhejiang snack. Edible bird's nests and bletilla striate steamed in rock sugar soup. Filtered before serving.

白椒爆牛肚 (báijiāo bào niúdǔ)

四川菜。将煮熟晒干褪色的辣椒切碎,与牛肚片爆炒,烹入葱、姜、酱油、料酒。

Stir-fried spicy ox tripe. Sichuan dish. Ox tripe stir-fried with discolored chili pepper and flavored with scallion, ginger, soy sauce, and cooking wine.

白椒炒肉丁 (báijiāo chǎo ròudīng)

四川菜。将煮熟晒干褪色的辣椒切碎,与肉丁爆炒,烹入葱、姜、酱油、料酒等。

Stir-fried spicy pork. Sichuan dish. Diced pork stir-fried with discolored chili pepper and flavored with scallion, ginger, soy sauce, and cooking wine.

白椒肚片 (báijiāo dǔpiàn)

川湘鄂家常菜。将煮熟晒干褪色的辣椒与肚片爆炒,加盐、料酒、酱油、味精等,勾芡。

Stir-fried hog tripe with pepper. Home-style dish in Sichuan, Hunan, and Hubei. Sliced hog tripe stir-fried with discolored chili pepper, flavored with salt, cooking wine, soy sauce, MSG, and starched soup.

白椒鸡蛋 (báijiāo jīdàn)

四川、湖北家常菜。先把鸡蛋煎好,葱煸香,再与煮熟晒干褪色的辣椒及调料煸炒。

Eggs with chili pepper. Home-style dish in Sichuan and Hubei. Fried eggs stir-fried with discolored chili pepper, scallion, and other condiments.

白椒焖牛肉 (báijiāo mèn niúròu)

四川菜。将煮熟晒干褪色的辣椒切碎,与牛肉焖熟,烹入葱、姜、酱油、料酒等。

Braised spicy beef. Sichuan dish. Beef braised with discolored chili pepper and flavored with scallion, ginger, soy sauce, and cooking wine.

白椒烧鹅 (báijiāo shāo é)

四川菜。将煮熟晒干褪色的辣椒切碎,与鹅肉红烧,烹入葱、姜、酱油、料酒、糖等调料。

Braised spicy goose. Sichuan dish. Goose braised with discolored chili pepper and flavored with scallion, ginger, soy sauce, cooking wine, and sugar.

白椒烧鸭 (báijiāo shāo yā)

四川菜。将煮熟晒干褪色的辣椒切碎,与鸭肉红烧,烹入葱、姜、酱油、料酒、糖等。

Braised spicy duck. Sichuan dish. Duck braised with discolored chili pepper and flavored with scallion, ginger, soy sauce, cooking wine, and sugar.

白椒猪肚 (báijiāo zhūdǔ)

四川菜。将煮熟晒干褪色的辣椒切碎,与猪肚片爆炒,烹入葱、姜、酱油、料酒等。

Stir-fried spicy hog tripe. Sichuan dish. Sliced hog tripe stir-fried with discolored chili pepper and flavored with scallion, ginger, soy sauce, and cooking wine.

白酒 (báijiǔ)

中国蒸馏酒的俗称。用高粱、玉米、稻米、甘薯等粮食或某些果品经过发酵、蒸馏制成。无色,酒精含量较高。

White spirit or Chinese liquor. Distilled from fermented sorghum, corn, rice, sweet potato, or fruits. It is colorless and high in alcohol content.

白菊牛柳 (báijú niúliǔ)

广东菜。先将牛柳腌渍,再加入蒜、葱爆炒,勾芡,然后撒上白菊花瓣。

Beef with white chrysanthemum. Guangdong dish. Pickled beef stirfried with garlic and scallion, thickened with starch, then garnished with white chrysanthemum petals.

福鼎白琳功夫茶 (fúdǐng báilín gōngfuchá)

红茶。产于福建福鼎太姥山的白琳、湖林等地。外形细长弯曲,茸毫呈粒状,色泽黄黑。汤色浅亮,叶底鲜红带黄。

Fuding Bailin *gongfucha*, black tea produced in Bailin and Hulin areas in Mount Tailao, Fujian. Its dry leaves are slim, crescent, dark yellow, and frizzy hairy. The tea is crystal and light yellow with infused leaves yellowish red.

白萝卜炖羊肉 (báiluóbo dùn yángròu)

西北菜。先将白糖熬至冒泡,再将羊肉放入拌炒,然后加入白萝卜和调料炖熟。

Mutton with white turnip.
Northwestern dish. Mutton stir-fried with heated melting sugar, then simmered with white turnips and seasonings.

白毛猴 (báimáohóu)

乌龙茶。产于福建政和县,外形粗壮卷曲,布满白毫。汤色清绿泛黄,叶底嫩绿。

Baimaohou oolong tea produced in Zhenghe, Fujian. Its dry leaves are stout, curly, and hairy. The tea is yellowish green with infused leaves tender green.

白牡丹 (báimǔdān)

白茶。产于福建政和、福鼎等地。外形肥嫩,叶背遍布白茸,色泽深灰绿。汤色浅黄。叶底浅灰,叶脉微红。

Baimudan white tea produced in Zhenghe and Fuding, Fujian. Its dry leaves are tender, plump, dark greyish green, and white hairy on the backs. The tea is light yellow with infused leaves light gray with concealed red veins.

白片羊肉 (báipiàn yángròu)

广东凉菜。将羊肉煮烂,撕去脂皮,置冷后切成薄片,佐葱和甜面酱食用。

Mutton with leek and sauce.

Guangdong cold dish. Mutton boiled, fat removed, chilled, and sliced. Served with leek and sweet fermented flour sauce.

白葡萄酒鲈鱼 (báipútáojiǔ lúyú)

山东菜。先将鲈鱼肉腌渍,略煎,加白葡萄酒烧入味,再浇上用奶油、面粉、鲜奶等炒成的面糊。

Braised bass with white wine.

Shandong dish. Pickled and lightly fried black spotted bass braised with white wine, then washed with dressing of butter, flour, and milk.

白切肉 (báiqiēròu)

东北菜。先将五花猪肉煮熟切片,然后浇上用蒜泥、麻油、味精等调好的汁。

Boiled pork with garlic sauce.

Northeastern dish. Boiled streaky pork sliced, then washed with dressing of garlic, sesame oil, and MSG.

白肉火锅 (báiròu huǒguō)

东北菜。先将熟五花猪肉、牛肉、羊肉切片,酸菜切丝,然后将肉和菜在火锅中涮,佐麻油和韭菜花调成的汁。

Fire pot with boiled meat.

Northeastern dish. Sliced streaky pork, beef, and mutton quick-boiled in fire pot soup with shredded sour pickled cabbage. Served with dressing of sesame oil and Chinese chive flower.

白肉血肠火锅 (báiròu xuècháng huǒguō)

东北菜。将熟白肉片和熟血肠片入火锅涮烫,加入海米、酸菜等,佐豆腐乳汁或韭菜花汁。

Prepared pork and blood sausage in fire pot. Northeastern dish. Sliced cooked pork and blood sausage quick-boiled in fire pot soup with dried shrimp meat, sour pickled cabbage, and other ingredients. Served with fermented bean curd juice or dressing of Chinese chive flower.

白烧四宝 (báishāo sìbǎo)

山东菜。把鸡腰、鸭舌、鸭掌、熟鸡皮、菜心、蘑菇等入汤加调料烧至汤近干,勾芡,浇鸡油。

Braised chicken and duck offal.

Shandong dish. Chicken kidney, duck tongue, duck feet, prepared chicken skins, baby vegetable, mushrooms, and seasonings braised until reduced. Thickened with starch, then washed with chicken oil.

白烧鱼翅 (báishāo yúchì)

福建菜。将熟鱼翅用鸡和蹄髈盖住,加清汤焖酥。

Shark fin with chicken and hog knuckle. Fujian dish. Prepared shark fin covered with chicken and hog knuckle, then stewed in clear soup.

白水羊肉 (báishuǐ yángròu)

西北菜。将带骨绵羯羊肉用小火翻煮熟透,佐醋、蒜汁。

Boiled mutton with sauce.

Northwestern snack. Mutton with bones from wether boiled thoroughly over low heat, then served with vinegar and garlic sauce.

白水羊头 (báishuǐ yángtóu)

北京菜。将羊头肉用清水煮熟,切片,撒上特制的椒盐。

Boiled lamb head with spicy salt.

Beijing snack. Lamb head meat boiled, then sliced. Served with spicy salt.

白松鸡 (báisōngjī)

山东菜。先将鸡脯肉蒸熟,抹上用鸡肉馅、猪油、蛋清、松仁粉等调成的糊再蒸,然后配氽过的油菜心装盘,浇上卤汁。

Steamed chicken breast. Shandong dish. Prepared chicken breast coated with mixture of minced chicken, lard, egg white, and minced pine nuts, steamed, then served with scalded baby cabbage and sauce.

白酥鸡 (báisūjī)

山东菜。先将鸡脯肉抹上用虾仁、猪肥膘等做成的酱,再与鸡肫、鸡肝一道蒸熟,然后用鸡汤加虾米、白酱油勾芡,浇在鸡肉上。

Chicken breast with meat varieties. Shandong dish. Chicken breast coated with shrimp meat and fat pork paste, steamed with chicken gizzard and chicken liver, then washed with chicken soup flavored with dried small shrimp, white soy sauce, and starch.

白酥里脊 (báisū lǐji)

东北家常菜。先将猪里脊肉饼挂糊,炸至金黄,切条,然后用胡萝卜和黄瓜加调料做成汁,浇在肉条上。

Fried pork tenderloin with vegetable sauce. Northeastern home-style dish. Ground pork tenderloin coated with egg and starch paste, deep-fried, cut into strips, then washed with sauce of carrots, cucumbers, and seasonings.

白汤鲫鱼 (báitāng jìyú)

江苏菜。先将鲫鱼略煎,再加水用文火焖煮至汤色变白,然后放入火腿、竹笋、香菇等煮熟,淋鸡油。

Crucian soup. Jiangsu dish. Crucian carp lightly fried, simmered until soup becomes white, then boiled with ham, bamboo shoots, and mushrooms. Washed with chicken oil.

白糖糕 (báitánggāo)

江西点心。先将米浆发酵,加入苏打和白糖,然后蒸熟。

Steamed sweet rice cakes. Jiangxi snack. Fermented rice paste mixed with soda and sugar, then steamed.

白兔冷盘 (báitù lěngpán)

浙江凉菜。把盐水鸡脯、盐水鸭肚、糖醋胡萝卜、黄瓜、蛋黄糕、蛋白糕、卤香菇等堆放成兔状。

Poultry and vegetable plate.

Zhejiang cold dish. Salted chicken breast, salted boned duck, sweet-sour pickled carrots, cucumbers, egg yolk cakes, egg white cakes, and marinated mushrooms arranged like a rabbit.

白煨脐门 (báiwēi qímén)

江苏菜。将鳝肉加调料煨熟,勾芡。

Simmered eel. Jiangsu dish. Eel meat simmered with seasonings and thickened with starch.

白炆猪耳 (báiwén zhūěr)

广东凉菜。将猪耳加八角炆熟,切片,配咸菜装盘,蘸米醋、盐、蒜茸、辣椒做的卤汁。

Hog ears with sauce. Guangdong cold dish. Hog ears simmered with aniseed, sliced, then served with preserved mustard turnips. Flavored with dressing of vinegar, salt, garlic, and chili pepper.

白鲞扣鸡 (báixiǎng kòu jī)

浙江菜。将熟鸡脯肉、鸡翅膀、咸鱼干蒸熟。

Dried salty fish with chicken.

Zhejiang dish. Prepared chicken breast, chicken wings, and dried salty fish steamed together.

白鲞玉脂 (báixiǎng yùzhī)

江苏菜。将豆腐和咸鱼干一道蒸熟,放上爆香的红椒丝和葱丝。

Dried salty fish with bean curd. Jiangsu dish. Bean curd and dried salty fish steamed, then garnished with quick-fried shredded red pepper and scallion.

白雪潭蟳 (báixuětánxún)

台湾菜。先将炒好的猪肉、香菇、竹笋等放在熟蛋白之间,浇热油,然后将备好的红蟳肉倒在蛋白上。

Mangrove crab meat with egg white. Taiwan dish. Pork, mushrooms, and bamboo shoots stir-fried, put between prepared egg white pieces, washed with hot oil, then covered with prepared mangrove crab meat.

白雪虾仁 (báixuě xiārén)

广东菜。先把虾仁与蛋清、生粉、盐、味精等拌和,冷冻,再炒成松散白雪状。

Sautéed egg white and shrimp meat. Guangdong dish. Shrimp meat mixed with egg white, starch, salt, and MSG, chilled, then sautéed into snow-like floss.

白油肝片 (báiyóu gānpiàn)

四川菜。先将猪肝片用盐和生粉腌渍,然后加泡椒、木耳、葱、菜心等爆炒。

Stir-fried hog liver. Sichuan dish. Sliced hog liver pickled with salt and starch, then stir-fried with pickled red pepper, wood ears, scallion, and Chinese baby cabbage.

白油肉片 (báiyóu ròupiàn)

东北家常菜。将猪肉片轻炒至发白,再与莴笋、木耳翻炒,烹入加调料的芡汁。

Sautéed pork, stem lettuce, and mushrooms. Home-style dish in northeastern areas. Sliced lean pork lightly fried until white, sautéed with stem lettuce and wood ears, then washed with seasoned starched sauce.

白玉干贝 (báiyù gānbèi)

广东菜。先将冬瓜块用开水焯过，然后与蒸熟的干贝和瘦猪肉炖煮，放入盐和葱。

Stewed scallops with wax gourd. Guangdong dish. Quick-boiled cubed wax gourd stewed with steamed dried scallops and lean pork, then seasoned with salt and scallion.

白玉鳝段 (báiyù shànduàn)

山东家常菜。先将葱、姜、蒜炒香，再放入鳝段焖煮，然后加入豆腐、火腿、冬笋、木耳及调料。

Sautéed eel with bean curd. Home-style dish in Shandong. Sectioned field eel sautéed with quick-fried scallion, ginger, and garlic, then simmered with bean curd, ham, winter bamboo shoots, wood ears, and seasonings.

白云边酒 (báiyúnbiānjiǔ)

白酒，产于湖北松滋。以高粱为原料，用小麦制曲，酒精含量 42-58 度，属兼香型酒。

Baiyunbian Jiu liquor produced in Songzi, Hubei. White spirit made with sorghum and wheat yeast. Contains 42-58% alcohol, and has a mixed aroma.

白云凤爪 (báiyúnfèngzhǎo)

广东点心。据传此菜创于广州白云山。冰冻鸡脚加白醋、姜、糖等烹煮，置凉。

Sweet and sour chicken feet.

Guangdong snack originated in Mount Baiyun in Guangzhou. Frozen chicken feet cooked with white vinegar, ginger, and sugar, then chilled.

白云猪手 (báiyúnzhūshǒu)

广东凉菜。据传此菜创于广州白云山。先将猪脚煮熟后置凉，再用黄酒、盐、味精等浸泡。

Marinated hog trotters. Guangdong cold dish originated in Mount Baiyun in Guangzhou. Hog trotters boiled, cooled, then marinated in gravy of yellow wine, salt, and MSG.

白斩鸡 (báizhǎnjī)

东南方菜。先将嫩鸡煮熟，迅速置冷使其表皮收缩，然后抹上花生油，斩成小块，佐以酱油、姜、葱、蒜、麻油等调成的卤汁。

Boiled chicken with dressing.

Southeastern dish. Young chicken boiled, quickly chilled to make its skins shrink, coated with peanut oil, then chopped. Served with dressing of soy sauce, ginger, scallion, garlic, and sesame oil.

白斩加积鸭 (báizhǎn jiājīyā)

海南凉菜。加积是海南的一个镇。将加积鸭煮熟，涂麻油，斩块，佐以蒜茸、姜茸等调成的酸桔汁。

Duck with dressing. Hainan cold dish. Jiaji is a town in Hainan. Local duck boiled, brushed with sesame oil, then chopped. Served with dressing of minced garlic, ginger, and sour tangerine juice.

白汁鲍鱼 (báizhī bàoyú)

四川菜。先将鲍鱼炖至汤变白，再

加入火腿、鸡肉等焖煮。

Stewed abalone. Sichuan dish.

Abalone stewed till soup milky, then simmered with ham and chicken.

白汁鲳鱼 (báizhī chāngyú)

广东菜。将鲳鱼加姜丝蒸熟，淋上用鱼露、胡椒粉、青红椒、京葱等炒制的卤汁。

Steamed pomfret. Guangdong dish. Pomfret steamed with shredded ginger, then washed with sauce of stir-fried green and red pepper, leek, fish gravy, and ground black pepper.

白汁甘笋 (báizhī gānsǔn)

浙江菜。将甘笋即胡萝卜加盐和牛油煮熟，放入焯熟的青豆和面粉牛奶汤，用盐、胡椒粉调味。

Carrot in milk soup. Zhejiang dish. Carrots boiled in soup seasoned with salt and butter, combined with scalded green beans and flour milk soup, then seasoned with salt and ground black pepper.

白汁鳜鱼 (báizhī guìyú)

江浙菜。先把鳜鱼蒸熟，再用蚕豆、火腿丁、虾仁、芡粉、水、盐、牛奶等做成卤汁，浇在鱼上。

Steamed Chinese perch with sauce. Jiangsu and Zhejiang dish. Chinese perch steamed, then washed with dressing of broad beans, diced ham, shrimp meat, starch, water, salt, and milk.

白汁猴头菇 (báizhī hóutóugū)

四川菜。先将牛奶、淀粉、盐等调成白汁，煮沸，然后放入猴头菇、猪肉、百合等炖煮。

Stewed mushroom in milk soup. Sichuan dish. Monkey-head mushrooms stewed with pork and lily bulbs in soup flavored with milk, starch, and salt.

白汁鮰鱼 (báizhī huíyú)

上海菜。先将鮰鱼块用沸水烫过，再用猪油稍煎，然后与竹笋闷烧，用黄酒、盐、姜、葱、胡椒粉调味。

Braised catfish with bamboo shoot. Shanghai dish. Longsnout catfish scalded, lightly fried with lard, braised with bamboo shoots, then flavored with yellow wine, salt, ginger, scallion, and ground black pepper.

白汁芦筋 (báizhī lújīn)

江苏菜。先将蹄筋用小火焖透，然后与火腿、鸡脯、冬菇、青豆等入鸡清汤烧沸，加调料，勾芡，淋鸡油。

Stewed pork tendon. Jiangsu dish. Pork tendons stewed. Boiled in clear chicken soup with seasonings, ham, chicken breast, winter mushrooms, and green soybean. Thickened with starch, then washed with chicken oil.

白汁酿鱼 (báizhī niàngyú)

山东菜。在黄鱼身上劈出相连的鱼片，卷上猪肉末，蒸熟，浇加调料的芡汁和鸡油。

Steamed yellow croaker with pork. Shandong dish. Yellow croaker carved into connected slices. Each piece rolled with seasoned ground pork. Steamed, then washed with seasoned starched soup and chicken oil.

白汁牛肉 (báizhī niúròu)

四川菜。先将牛奶和淀粉调成白汁,煮沸,然后放入牛肉、马铃薯、白果及调料炖煮。

Stewed beef in milk soup. Sichuan dish. Beef stewed with potatoes and ginkgoes in soup seasoned with starch and milk.

白汁裙边 (báizhī qúnbiān)

山东菜。先将鸡翅、猪肉、猪肋骨略炒,再加入奶汤、黄酒煮沸,然后放入鼋鱼裙边,用微火煨烂,加调料,勾芡。

Turtle's soft shells in milk soup. Shandong dish. Quick-fried chicken wings, pork, and pork ribs boiled in milk and yellow wine soup. Stewed with turtle's soft shells. Seasoned, then thickened with starch.

白汁乳狗 (báizhī rǔgǒu)

江苏菜。先将乳狗肉加黄酒和肉汤炖熟,再将火腿片、香菇片、青菜心等用原汤略烩,浇在狗肉上。

Stewed baby dog meat. Jiangsu dish. Baby dog meat stewed in pork and yellow wine soup, then covered with mixture of simmered sliced ham, mushrooms, baby cabbage, and broth.

白汁稀卤笋 (báizhī xīlǔ sǔn)

江苏菜。将虾仁、青鱼肉、猪腰等馅料填入春笋,加鸡汤蒸熟,勾芡。

Steamed stuffed bamboo shoot. Jiangsu dish. Spring bamboo shoots stuffed with shrimp meat, black carp meat, and hog kidney, steamed in chicken soup, then thickened with starch.

白汁鲜蘑 (báizhī xiānmó)

浙江菜。将鲜蘑放入牛奶中煮熟。

Mushroom in milk. Zhejiang dish. Fresh mushrooms boiled in milk.

白汁银鱼 (báizhī yínyú)

福建菜。先把银鱼氽过,再放入鸡汤煨煮,加入黄酒、盐、牛奶、菱角,然后勾芡。

Lake icefish in milk soup. Fujian dish. Quick-boiled lake icefish simmered in chicken soup, flavored with yellow wine, salt, milk, and water caltrop meat, then thickened with starch.

白汁鱼肚 (báizhī yúdǔ)

四川菜。先将鱼肚蒸熟,再将鲜笋、火腿、虾仁与盐、水、淀粉、牛奶等做成卤汁,浇在鱼肚上。

Steamed fish maw with milk sauce. Sichuan dish. Fish maw steamed, then marinated in gravy of fresh bamboo shoots, ham, shrimp meat, salt, water, starch, and milk.

白汁鱼丸 (báizhī yúwán)

四川菜。先将牛奶和淀粉调成白汁,煮沸,再放入鱼丸、虾仁等煮熟。

Boiled fish balls and shrimp meat. Sichuan dish. Fish balls and shelled shrimp boiled in soup enriched with starch and milk.

白灼蚶 (báizhuóhān)

闽粤琼家常菜。将蚶灼熟,加入葱、姜、红辣椒、花生油等拌匀。

Boiled ark shell. Home-style dish in Fujian, Guangdong, and Hainan. Ark shell boiled, then mixed with scallion, ginger, red pepper, and peanut oil.

白灼牛肉 (báizhuó niúròu)

广东菜。将牛肉灼熟,放在烫熟的豆芽上,撒上葱、姜、红椒丝、香菜,淋色拉油。

Boiled beef with bean sprouts. Guangdong dish. Beef boiled, put on a bed of quick-boiled mung bean sprouts, covered with shredded scallion, ginger, red pepper, and coriander, then washed with salad oil.

白灼排骨 (báizhuó páigǔ)

东北菜。先将猪排骨切段煮熟,然后浇上用葱、姜、盐、黄酒等制成的汁。

Boiled pork ribs with sauce. Northeastern dish. Chopped pork ribs boiled, then washed with dressing of scallion, ginger, salt, and yellow wine.

白灼虾 (báizhuóxiā)

闽粤琼家常菜。将基围虾灼熟,配葱、姜、酱油、麻油、辣椒等做成的卤汁。

Boiled shrimp with sauce. Home-style dish in Fujian, Guangdong, and Hainan. Greasy back shrimp boiled. Served with sauce of scallion, ginger, soy sauce, sesame oil, and hot pepper.

白灼响螺片 (báizhuó xiǎngluópiàn)

广东菜。先将海螺肉片入上汤氽熟,再与姜、葱或韭黄翻炒,用黄酒、盐、麻油调味。

Sautéed conch meat. Guangdong dish. Sliced conch quick-boiled in broth, sautéed with ginger and scallion or yellow Chinese chive, then flavored with yellow wine, salt, and sesame oil.

白灼象拔蚌 (báizhuó xiàngbábàng)

广东菜。将象拔蚌肉片灼熟,与豉油皇拌匀。

Boiled geoducks with fermented soybean sauce. Guangdong dish. Sliced geoducks boiled, then mixed with fermented soybean sauce.

白灼腰片 (báizhuó yāopiàn)

东北家常菜。先将猪腰切片煮熟,然后浇上用香菜梗、胡萝卜、酱油、味精等制成的卤汁。

Boiled hog kidney with sauce. Northeastern home-style dish. Boiled sliced hog kidney washed with dressing of coriander, carrot, soy sauce, and MSG.

白灼鱼片 (báizhuó yúpiàn)

东北菜。先将腌渍过的鱼片煮熟,再与焯水的青椒片、红椒片和葱丝摆盘,浇上热色拉油。

Boiled fish fillets with sauce. Northeastern dish. Pickled fish fillets boiled, garnished with quick-boiled sliced green and red pepper and shredded scallion, then sprinkled with hot salad oil.

白灼醉鲜虾 (báizhuó zuìxiānxiā)

广东菜。先将新鲜基围虾用白酒腌渍,然后焯水,配蒜茸、酱油、麻油等。

Boiled shrimp with sauce.

Guangdong dish. Fresh greasy back shrimp pickled in White spirit, then quickly boiled. Served with garlic soy sauce and sesame oil.

百搭肉片炒蘑菇 (bǎidā ròupiàn chǎo mógu)

南方家常菜。先用小火煸炒鸡脯肉,再加入蘑菇翻炒,然后加调料及高汤。

Sautéed chicken and mushrooms. Home-style dish in southern areas. Chicken breast and mushrooms sautéed over low heat, then flavored with broth and other condiments.

百合拌金针 (bǎihé bàn jīnzhēn)

广东凉菜。将汆过的金针、百合加橄榄油、盐等调料拌匀。

Quick-boiled lily buds and lily bulbs. Guangdong cold dish. Tiger lily buds and Chinese lily bulbs quick-boiled, then seasoned with olive oil, salt, and other flavorings.

百合参竹润肺汤 (bǎihé shēn zhú rùnfèitāng)

广东菜。用百合、红枣、沙参、玉竹、芦笋等煮成汤。

Lily and other herbs in soup. Guangdong dish. Lily bulbs boiled with red dates, straight ladybell roots, polygonatum, and asparagus.

百合炒肉片 (bǎihé chǎo ròupiàn)

南方菜。先把猪肉片用盐、料酒、蛋清、湿淀粉等腌渍,再微炸,然后与百合同炒。

Sautéed pork and lily bulb. Southern dish. Sliced pork pickled with salt, cooking wine, egg white, and wet starch, lightly fried, then sautéed with lily bulb cloves.

百合炖排骨 (bǎihé dùn páigǔ)

南方菜。将排骨同百合、蘑菇等熬炖,用盐、料酒、姜、葱等调味。

Stewed lily cloves and pork spareribs. Southern dish. Pork spareribs stewed with lily cloves and mushrooms. Seasoned with salt, cooking wine, ginger, and scallion.

百合炖银耳 (bǎihé dùn yíněr)

各地传统点心。将水发银耳与百合熬炖,加白糖。

Lily cloves and white wood ear soup. Traditional snack in many places. Water-soaked white wood ears stewed with lily cloves, then sweetened with sugar.

百合红枣汤 (bǎihé hóngzǎo tāng)

广东菜。用百合、红枣、赤小豆、薏仁等煮成汤。

Lily and date soup. Guangdong dish. Lily bulbs boiled with red dates, red beans, and seeds of Job's tears.

百合鸡丝 (bǎihé jīsī)

甘肃菜。先将鲜百合快炒,再加入鸡肉丝炒匀,加调料。

Sautéed lily bulb and chicken. Gansu

dish. Quick-fried Chinese lily bulbs sautéed with shredded chicken and seasonings.

百合牛肉 (bǎihé niúròu)

甘肃菜。先将牛肉用热油浸泡，再轻炒百合，然后将牛肉与百合翻炒，加入芡汁及调料。

Sautéed beef and lily bulb. Gansu dish. Beef steeped in hot oil, sautéed with quick-fried lily bulbs, then seasoned with starched soup and flavorings.

百合庆鸿运 (bǎihé qìng hóngyùn)

河南点心。将氽过的百合与红枣、龙眼干、汤圆同煮，勾芡。

Lily bulb and sticky rice *tangyuan*. Henan snack. Scalded Chinese lily bulbs boiled with Chinese red dates, dried longans, and sticky rice *tangyuan*, then thickened with starch.

百合烧肉 (bǎihé shāo ròu)

东北家常菜。先将五花猪肉过油，加入酱油、黄酒等焖至肉烂，然后放入百合，煮至汁浓。

Streaky pork with lily bulb. Home-style dish in northeastern areas. Streaky pork quick-fried, braised with soy sauce and yellow wine, then simmered with lily bulbs.

百合丝瓜炒鸡片 (bǎihé sīguā chǎo jīpiàn)

山东家常菜。先将丝瓜和腌过的鸡肉轻炒，再与百合及调料拌炒。

Sautéed chicken, lily bulb, and snake gourd. Home-style dish in Shandong. Snake gourd quick-fried with pickled chicken, then sautéed with Chinese lily bulbs and condiments.

百合田鸡 (bǎihé tiánjī)

广东菜。将田鸡、百合、胡萝卜入高汤炖煮，加盐、黄酒、姜、葱等。

Stewed frog with lily bulb. Guangdong dish. Frogs, lily bulbs, and carrots stewed in broth and seasoned with salt, yellow wine, ginger, and scallion.

百合西芹 (bǎihé xīqín)

南方菜。先将西芹氽过，再与百合轻炒，加调料。

Sautéed lily cloves and celery. Southern dish. Scalded American celery sautéed with lily cloves and flavorings.

百合玉米 (bǎihé yùmǐ)

南方菜。先将嫩玉米粒氽过，再与百合拌炒，用盐和味精调味，勾芡。

Sautéed lily cloves and corn. Southern dish. Quick-boiled tender corn sautéed with lily cloves and flavored with salt, MSG, and starched soup.

百合玉子虾球 (bǎihé yùzǐ xiāqiú)

南方家宴菜。先将玉子豆腐煎至金黄，炒熟虾仁，炒软豌豆角，然后将三者与鲜百合轻炒拌匀，加调料，勾芡。

Shrimp meat with bean curd. Family feast dish in southern areas. Deep-fried creamy bean curd, lightly fried shelled shrimp, and pea pods sautéed with fresh Chinese lily bulbs and seasonings, then thickened with starch.

百花北菇拼虾球 (bǎihuā běigū pīn xiāqiú)

广东菜。将加调料的虾茸填入北菇帽,抹上蛋清,撒火腿茸,配香菜、菜远等料蒸熟。

Steamed mushrooms with minced shrimp. Guangdong dish. Mushroom caps filled with seasoned minced shrimp meat, coated with egg white, showered with ham crumbs, then steamed with coriander and tender vegetable stalks.

百花鲳鱼 (bǎihuā chāngyú)

江苏菜。将腌渍过的鲳鱼片裹香菇丝、胡萝卜丝、火腿丝等成花束状,蒸熟。

Steamed pomfret. Jiangsu dish. Pickled pomfret fillets rolled into bouquets with shredded mushrooms, carrots, and ham, then steamed.

百花大虾 (bǎihuā dàxiā)

山东菜。将对虾肉面剖十字花刀,抹上腌渍过的鸡肉末和蛋清,蒸熟,然后浇上芡汁。

Steamed prawns. Shandong dish. Shelled prawns cross-carved, coated with pickled ground chicken and egg white, steamed, then washed with starched sauce.

百花冬瓜 (bǎihuā dōngguā)

南方家常菜。将切碎的冬瓜与蛋浆和盐调匀,蒸熟,浇麻油。

Steamed eggs with gourd. Home-style dish in southern areas. Chopped wax gourd mixed with egg paste and salt, steamed, then sprinkled with sesame oil.

百花冬瓜球 (bǎihuā dōngguāqiú)

江苏菜。先将冬瓜球开一小洞,填入虾胶,上嵌咸蛋黄粒,再裹上生粉蒸熟,浇芡汁,配煮熟的西蓝花装盘。

Steamed wax gourd and broccoli. Jiangsu dish. Wax gourd carved into balls with an opening on each, filled with shrimp paste, decorated with diced salty egg yolks, coated with corn starch, steamed, then washed with starched soup. Served with boiled broccoli.

百花豆腐 (bǎihuā dòufu)

东北家常菜。先将豆腐切菱形块,上面挖小孔,嵌入腌渍过的里脊肉末蒸熟,然后用甜椒和蛋清做成卤汁,浇在豆腐上。

Steamed bean curd with pork. Northeastern home-style dish. Bean curd cut into diamond cubes with a hole on each. Filled with pickled minced pork tenderloin, steamed, then washed with sauce of green pepper and egg white.

百花鸡 (bǎihuājī)

浙江菜。先将虾仁与盐、味精、黄酒、芡粉、胡椒粉、水等调成虾糊,再将虾糊抹在腌渍过的鸡脯肉上,然后将鸡肉炸至金黄。

Fried chicken breast. Zhejiang dish. Chicken breast pickled, coated with paste of shrimp meat, salt, MSG, yellow wine, starch, ground black pepper, and water, then deep-fried.

百花煎凤翼 (bǎihuā jiān fèngyì)

广东菜。将鸡翅腌渍,抹上虾泥煎熟,配芥蓝叶。

Fried chicken wings. Guangdong dish. Chicken wings pickled, coated with shrimp paste, fried, then garnished with mustard leaves.

百花煎酿竹荪 (bǎihuā jiān niàng zhúsūn)

广东菜。先将竹荪裹上虾茸,炸至金黄,再浇上特制的芡汁。

Long net stinkhorn with minced shrimp. Guangdong dish. Long net stinkhorn coated with minced shrimp meat, deep-fried, then washed with special sauce.

百花酒焖肉 (bǎihuājiǔ mèn ròu)

江苏菜。先将五花猪肉烤至皮焦,再加入百花酒、糖、酱油等调料,用砂锅焖至酥烂。

Braised streaky pork with liquor. Jiangsu dish. Streaky pork broiled until brown, seasoned with flower-scented liquor, sugar, and soy sauce, then braised in an earthen pot.

百花酿带子 (bǎihuā niàng dàizi)

广东菜。将虾茸用盐和味精调味,酿在带子上,撒上火腿末、香菜叶,蒸熟。

Steamed scallops with minced shrimp. Guangdong dish. Minced shrimp meat seasoned with salt and MSG, topped on scallops, garnished with ham crumbs and coriander leaves, then steamed.

百花酿黄瓜 (bǎihuā niàng huánggua)

南方菜。用黄瓜配腌渍过的猪肥肉、大虾肉、芥蓝、红尖椒、芹菜等蒸熟。

Steamed meat and vegetable varieties. Southern dish. Cucumbers steamed with pickled fat pork, shelled prawns, Chinese broccoli, red pepper, and celery.

百花酿青瓜 (bǎihuā niàng qīngguā)

安徽菜。先将虾胶丸嵌入去瓤的黄瓜段,用尖椒和芹菜点缀,然后蒸熟,勾芡。

Shrimp balls in cucumbers. Anhui dish. Hollowed cucumber sections filled with shrimp balls, garnished with pepper and celery, steamed, then thickened with starch.

百花酿油条 (bǎihuā niàng yóutiáo)

广东点心。将油条裹上鱼茸虾酱糊,炸至金黄。

Fried dough twist with seafood. Guangdong snack. Fried dough twists coated with fish and shrimp paste, then fried until golden.

百花酿鱼鳔 (bǎihuā niàng yúbiào)

福建、广东菜。先将鱼肚酿上虾胶,然后配火腿、芹菜、香菇、笋片蒸熟,浇芡汁。

Steamed fish maw with minced shrimp. Dish in Fujian and Guangdong. Fish maw laid with minced shrimp meat, steamed with ham, celery, dried mushrooms, and bamboo shoots, then washed with starched soup.

百花烧卖 (bǎihuā shāomài)

广东点心。将面粉皮包上用猪肉、麻油、虾米等做成的馅,做成袋形,蒸熟。

Steamed stuffed bun. Guangdong snack. Flour wrappers made into bag-shape buns filled with mixture of ground pork, sesame oil, and dried shrimp meat, then steamed.

百花蛇脯 (bǎihuā shépú)

广东菜。先将水律蛇肉片用盐、淀粉等腌渍,加葱、姜等蒸熟,再将虾胶涂在蛇肉片上,炸至金黄,然后加入蟹黄、蛋清等翻炒。

Snake meat with seafood flavorings. Guangdong dish. Sliced water snake meat pickled with starch and salt, steamed with scallion and ginger, coated with minced shrimp meat, deep-fried, then stir-fried with crab roe and egg white.

百花酥鸡 (bǎihuā sūjī)

北方家常菜。先将鸡肉末、猪肉末、火腿末、胡椒粉、鸡蛋清等调和,蒸熟,然后切条,挂糊,炸至金黄。

Fried ground pork and chicken. Home-style dish in northern areas. Minced chicken, pork, and ham mixed with ground black pepper and egg white, steamed, cut into strips, coated with flour paste, then deep-fried.

百花燕菜 (bǎihuā yàncài)

山东菜。先将燕菜用热清汤冲洗,再放入汤盘,四周围上蒸熟的鸽蛋。

Bird's nests with pigeon eggs. Shandong dish. Edible bird's nests washed with warm clear soup, then garnished with steamed pigeon eggs on a platter.

百花鱼肚 (bǎihuā yúdǔ)

广东菜。将鱼肚与青笋片、冬笋片、火腿片入鲜汤煮熟,加胡椒粉和料酒,勾芡。

Fish maw with bamboo shoots. Guangdong dish. Fish maw boiled in seasoned soup with lettuce stems, winter bamboo shoots, and sliced ham, flavored with ground black pepper and cooking wine, then thickened with starch.

百花芝士球 (bǎihuā zhīshìqiú)

广东点心。将奶酪球裹上鱼肉、猪肉、太白粉、调料等制成的浓浆,粘上面包屑,炸至金黄。

Fried cheese balls. Guangdong snack. Cheese balls coated with paste of fish, pork, starch, and seasonings, showered with bread crumbs, then deep-fried.

百花紫菜卷 (bǎihuā zǐcàijuǎn)

台湾菜。先将虾肉馅摊在紫菜皮上,上置煎蛋皮,再往蛋皮上抹虾肉馅,铺蟹柳,然后裹成卷,烘熟。

Stuffed laver roll. Taiwan dish. Laver sheets layered with shrimp meat, fried egg sheets, and crab meat, made into rolls, then baked.

百鸟朝凤 (bǎiniǎo cháofèng)

湖南菜。先将整鸡蒸熟,然后用熟

鸡蛋、香菇、菜心等围在鸡的周围。

Steamed chicken with eggs and vegetables. Hunan dish. Whole chicken steamed, then put on a platter surrounded with boiled eggs, dried mushrooms, and baby cabbage.

百泉春酒 (bǎiquánchūnjiǔ)

白酒，产于河南辉县。以高粱为原料，用百泉水酿造，酒精含量为39-68度，属浓香型酒。

Baiquanchun liquor produced in Hui County, Henan. White spirit made with sorghum and water from Baiquan Spring. Contains 39-68% alcohol, and has a thick aroma.

百叶包肉 (bǎiyè bāo ròu)

上海菜。用湿豆腐张或豆腐百叶裹调味肉末成卷状，煮熟。

Bean curd rolls with pork. Shanghai dish. Wet bean curd sheets made into rolls filled with seasoned minced pork, then boiled.

百叶瓜仔肉 (bǎiyè guāzǎi ròu)

湖南菜。将肉末、酱瓜、淀粉、调料等做成馅，填入牛胃囊，蒸熟，用刀切开装盘。

Stuffed ox tripe. Hunan dish. Ox tripe stuffed with mixture of minced pork, pickled cucumbers, starch, and seasonings, then steamed. Cut to serve.

扳指干贝 (bānzhǐ gānbèi)

福建菜。小萝卜掏去心做成环形，填入干贝，蒸烂，浇特制的卤汁。

Steamed scallops in radish. Fujian dish. Radishes cored, stuffed with dried scallops, steamed, then washed with special sauce.

班戟卷 (bānjǐjuǎn)

广东点心。将蛋浆、牛奶、奶油、糖、面粉搅拌均匀，小火煎熟，卷入切碎的鲜水果，撒上糖粉。

Fruit pancakes. Guangdong snack. Eggs mixed with milk, butter, sugar, and flour, then fried into pancakes. Pancakes made into rolls filled with chopped fresh fruits, then sprinkled with powdered sugar.

班腩烩饭 (bānnǎn huì fàn)

广东食品。先将鱼肉、胡萝卜、香菇、姜等煮熟，然后加调料，勾芡，浇在米饭上。

Fish and vegetable over rice. Guangdong food. Fish, carrots, dried mushrooms, and ginger boiled, seasoned, thickened with starch, then served over rice.

板筋拌黄瓜 (bǎnjīn bàn huángguā)

东北家常凉菜。先将熟牛板筋切条，再划成丝，加黄瓜条、辣椒油、醋、味精等拌匀。

Beef tendons with cucumbers. Northeastern home-style cold dish. Boiled beef tendons cut into strips, torn into shreds, then mixed with cucumbers strips, chili oil, vinegar, and MSG.

板栗山鸡 (bǎnlì shānjī)

东北菜。先将山鸡块腌制，油炸，再与油炸板栗入鲜汤焖煮，然后蒸透。

Pheasant with chestnuts. Northeastern dish. Cubed pheasant pickled, fried, simmered in seasoned soup with fried chestnuts, then steamed.

板栗烧肥鸭 (bǎnlì shāo féiyā)

广东菜。将肥鸭加调料焖熟，扣在蒸熟的板栗上，勾芡。

Braised duck and chestnuts.

Guangdong dish. Fat duck braised with seasonings, put on a bed of steamed chestnuts, then washed with starched soup.

板栗烧野鸡 (bǎnlì shāo yějī)

曹雪芹《红楼梦》菜谱。先将汆过的野鸡肉块与葱、姜、黄酒煸炒，再加入酱油、盐、糖、鸡精、鸡汤等煮熟，然后放入板栗、肥膘片、香菇、冬笋等焖烂，原汁勾芡，淋麻油。

Stewed pheasant and chestnut. Menu from *A Dream of Red Mansions* by Cao Xueqin. Quick-boiled pheasant stirred-fried with scallion, ginger, and yellow wine. Boiled in chicken soup seasoned with soy sauce, salt, sugar, and chicken essence. Stewed with chestnuts, sliced fat pork, dried mushrooms, and winter bamboo shoots. Washed with starched broth, then sprinkled with sesame oil.

板栗蒸仔鸭 (bǎnlì zhēng zǐyā)

湖南菜。先将鸭肉煸炒，然后加入熟板栗和调料蒸烂。

Steamed duck with chestnuts.

Hunan dish. Quick-fried duck steamed with prepared chestnuts and seasonings.

板鸭 (bǎnyā)

江苏、江西、四川菜。先将全鸭用热盐、茴香、五香粉腌过，再入卤水腌制，然后置阴凉处风干。食用前先用清水浸泡去盐，再入热水焖泡去腥，然后配冬笋、冬菇、萝卜、猪肉、鸡肉等做成各式菜肴。

Flattened salty duck. Jiangsu, Jiangxi, and Sichuan dish. Whole duck pickled with heated salt, fennel, and five-spice powder, steeped in gravy, then air-dried in shade. It must be soaked in cold water to dissolve salt and immersed in hot water to reduce smell before it is cooked with other ingredients such as bamboo shoot, mushroom, turnip, pork or chicken.

板鸭火锅 (bǎnyā huǒguō)

湖北菜。将板鸭片放入加麻油、姜、蒜、红辣椒等调料的高汤烫熟。

Salted duck in fire pot. Hubei dish. Flattened salty duck sliced, then quick-boiled in soup of sesame oil, ginger, garlic, and red pepper in a fire pot.

半煎煮银鱼 (bànjiān zhǔ yínyú)

福建菜。将银鱼煎至半熟，再与香芹、青蒜、红椒丝等同煮，佐豆酱。

Whitebait with vegetables. Fujian dish. Half-fried whitebait boiled with celery, garlic sprouts, and shredded red pepper. Served with fermented soybean sauce.

半月沉江 (bànyuè chén jiāng)

福建菜。先将油炸面筋泡用水浸软，切片，然后入清汤与冬菇、冬笋、当归烹煮，加盐、味精等。

Gluten soup. Fujian dish. Soaked fried gluten puffs sliced, boiled in clear soup with winter mushrooms, winter bamboo shoots, and Chinese angelica, then seasoned with salt, MSG, and other seasonings.

拌白菜黄瓜丝 (bàn báicài huángguāsī)

东北凉菜。将白菜丝、黄瓜丝、葱、蒜、干红辣椒丝等加调料拌匀。

Chinese cabbage with cucumbers. Northeastern cold dish. Sliced Chinese cabbage and cucumbers mixed with scallion, garlic, shredded chili pepper, and seasonings.

拌白菜水萝卜丝 (bàn báicài shuǐluóbosī)

东北凉菜。将白菜丝、水萝卜丝、葱丝加醋、味精、辣椒油等拌匀。

Chinese cabbage and turnips with dressing. Northeastern cold dish. Shredded Chinese cabbage and turnips mixed with scallion, vinegar, MSG, and chili oil.

拌菠菜豆腐皮 (bàn bōcài dòufupí)

河南家常菜。将豆腐皮和菠菜烫熟，加盐、酱油、醋、麻油拌匀。

Spinach with bean curd skin. Home-style dish in Henan. Bean curd skins and spinach quick-boiled and seasoned with salt, soy sauce, vinegar, and sesame oil.

拌葱头 (bàncōngtóu)

河南家常凉菜。将葱头切块、辣椒切丝，加精盐、酱油、陈醋、麻油等拌匀。

Onion and pepper with dressing. Home-style cold dish in Henan. Green onion heads sectioned, mixed with shredded hot pepper, then seasoned with salt, soy sauce, vinegar, and sesame oil.

拌脆鳝 (bàncuìshàn)

江苏菜。用酱油、麻油、味精等调成汁，浇在炸好的鳝鱼丝上。

Deep-fried eel. Jiangsu dish. Shredded eel deep-fried and washed with dressing of soy sauce, sesame oil, and MSG.

拌耳丝 (bàněrsī)

各地家常凉菜。先将猪耳朵煮熟切丝，然后与黄瓜丝和蛋丝拌和，放入麻油、辣油、酱油、醋等。

Hog ears with soy sauce. Home-style cold dish in many places. Boiled shredded hog ears mixed with shredded cucumbers and eggs, then seasoned with sesame oil, chili oil, soy sauce, and vinegar.

拌粉皮 (bànfěnpí)

陕西家常凉菜。将煮熟的芡粉浆冷却成冻，然后切条，撒上黄瓜丝，加入酱油、醋、葱花、麻油拌匀。

Starch jelly with sauce. Home-style cold dish in Shaanxi. Starch jelly strips garnished with shredded cucumbers and seasoned with soy sauce, vinegar, chopped scallion, and sesame oil.

拌凤爪 (bànfèngzhǎo)

四川凉菜。先将鸡爪煮熟,置凉,然后加入麻油、黄酒、味精、酱油等拌匀。

Chicken feet with sauce. Sichuan cold dish. Chicken feet boiled, chilled, then flavored with sesame oil, yellow wine, MSG, and soy sauce.

拌海蜇皮 (bànhǎizhépí)

江浙沪凉菜。先将海蜇用冷水泡,再用滚水烫,然后与腌过的萝卜丝拌匀,加调料。

Cool jellyfish. Cold dish in Jiangsu, Zhejiang, and Shanghai. Jellyfish soaked in cold water, then scalded. Mixed with shredded pickled turnips and seasonings.

拌合菜 (bànhécài)

山东家常菜。将炒熟的肉丝和焯过的豆芽、黄瓜、白菜等放在粉丝上,浇上特制的卤汁。

Pork, vegetables, and starch noodles. Home-style dish in Shandong. Stir-fried shredded pork, quick-boiled bean sprouts, cucumbers, and cabbage put on a bed of starch noodles, then washed with special dressing.

拌鸡冠肚皮 (bàn jīguān dǔpí)

江苏凉菜。将猪肚煮熟,切成鸡冠状片,加酱油、醋、麻油等拌和。

Hog tripe with sauce. Jiangsu cold dish. Hog tripe boiled, sliced like cockscomb, then seasoned with soy sauce, vinegar, and sesame oil.

拌鸡丝 (bànjīsī)

南方菜。将熟鸡脯肉丝与黄瓜拌和,浇上用酱油、麻油、醋兑成的卤汁。

Shredded chicken with sauce. Southern dish. Cooked shredded chicken breast mixed with cucum-bers, then washed with mixture of soy sauce, sesame oil, and vinegar.

拌鸡丝冻粉 (bàn jīsī dòngfěn)

东北家常凉菜。先将冻粉泡软,熟鸡肉切丝,然后加黄瓜丝、海米、酱油、麻油、味精等拌匀。

Chicken with agar. Northeastern home-style cold dish. Soaked agar and shredded prepared chicken mixed with shredded cucumbers, dried shrimp, soy sauce, sesame oil, and MSG.

拌鸡丝银芽 (bàn jīsī yínyá)

东北家常凉菜。先将熟鸡肉切丝,绿豆芽焯水,再加入盐、醋、花椒油、蒜泥。

Chicken with green bean sprouts. Northeastern home-style cold dish. Shredded prepared chicken and scalded green bean sprouts seasoned with salt, vinegar, Chinese prickly ash oil, and mashed garlic.

拌芥末肚 (bànjièmodǔ)

东北家常凉菜。先将熟猪肚切丝,然后加入冬笋丝、胡萝卜丝、盐、味精、芥末等拌匀。

Hog tripe with mustard. North-eastern home-style cold dish. Hog tripe boiled, shredded, mixed with shredded winter bamboo shoots and carrots, then seasoned with salt, MSG, and mustard.

拌两鸡丝 (bànliǎngjīsī)

东北家常凉菜。先将生鸡脯肉切丝后入油划散至熟，然后与熟鸡肉丝、酱油、辣椒油等拌匀。

Shredded chicken with sauce.

Northeastern home-style cold dish. Fried shredded chicken breast mixed with boiled shredded chicken, soy sauce, and chili oil.

拌绿豆芽 (bànlǜdòuyá)

各地家常凉菜。先将绿豆芽焯熟，然后与黄瓜丝拌匀，加入盐、葱、姜、醋、麻油等。

Mung bean sprouts with dressing. Home-style cold dish in many places. Scalded mung bean sprouts and shredded cucumbers mixed with scallion, ginger, salt, vinegar, and sesame oil.

拌明太鱼丝 (bàn míngtàiyúsī)

东北家常凉菜。先将明太鱼丝用水浸泡片刻，再用辣椒粉搓拌，加入糖、酱油、芝麻酱等，撒香菜叶。

Sweet and hot pollack. Northeastern home-style cold dish. Shredded pollack soaked in water briefly, rubbed with chili powder, mixed with sugar, soy sauce, and sesame sauce, then garnished with coriander.

拌三色 (bànsānsè)

广东凉菜。先将大白菜、甘笋、西芹汆熟，再与酱油、醋、盐、糖、麻油拌匀。

Three vegetables with sauce.

Guangdong cold dish. Chinese cabbage, carrots, and American celery quick-boiled and flavored with soy sauce, vinegar, salt, sugar, and sesame oil.

拌三色虾仁 (bàn sānsè xiārén)

东北家常凉菜。先将腌渍过的虾仁上浆，入油划散至熟，再将南荠丁、青豆焯水，然后将三样加卤汁拌匀。

Three-color shrimp meat.

Northeastern home-style cold dish. Pickled shrimp meat starched, fried, then mixed with scalded diced water chestnuts, green soybean, and dressing.

拌山药丝 (bànshānyàosī)

各地家常凉菜。将山药切丝，焯水，用酱油、盐、醋、葱、姜、麻油调味。

Boiled Chinese yam with sauce. Home-style cold dish in many places. Shredded Chinese yam quick-boiled, then flavored with soy sauce, salt, vinegar, scallion, ginger, and sesame oil.

拌兔丁 (bàntùdīng)

四川家常凉菜。先将整只嫩兔入沸水煮熟，迅速冷却，切块，然后用辣椒酱、麻油、芝麻酱、酱油等拌匀。

Rabbit with spicy sauce. Home-style cold dish in Sichuan. Whole young rabbit boiled, chilled, chopped, then mixed with chili sauce, sesame oil, sesame sauce, and soy sauce.

拌五彩干丝 (bàn wǔcǎi gānsī)

四川凉菜。将煮熟的胡萝卜、鸡肉、冬菇、辣椒、火腿切成丝,与豆腐干丝、酱油、麻油、辣油等拌匀。

Assorted vegetables with spicy sauce. Sichuan cold dish. Boiled shredded carrots, chicken, winter mushrooms, hot pepper, and ham mixed with shredded dried bean curd, soy sauce, sesame oil, and chili oil.

拌五香黄豆 (bàn wǔxiāng huángdòu)

云南凉菜。将黄豆煮熟,加入盐、五香粉、芝麻、味精等拌匀。

Five-spice soybean. Yunnan cold dish. Soybean boiled, then mixed with salt, five-spice powder, sesame, and MSG.

拌鸭掌 (bànyāzhǎng)

东北家常凉菜。先将熟鸭掌去骨切条,然后加黄瓜片、味精、醋、麻油等拌匀。

Duck feet with cucumbers.

Northeastern home-style cold dish. Cooked duck feet boned, slivered, then mixed with sliced cucumbers, MSG, vinegar, and sesame oil.

拌腌雪里蕻 (bàn yānxuělǐhóng)

东北家常凉菜。将雪里蕻切碎,装入坛内密封腌数日,启封后加白糖、麻油拌匀。

Seasoned preserved mustard.

Northeastern home-style cold dish. Chopped potherb mustard pickled, sealed in a pot for several days, then flavored with sugar and sesame oil.

拌银耳 (bànyíněr)

北京凉菜。先将银耳切成小朵,焯水,然后加白糖、香肠末、麻油等拌匀。

Sweet white wood ears. Beijing cold dish. White wood ears cut into small pieces, quick-boiled, then mixed with sugar, minced sausage, and sesame oil.

拌肘花 (bànzhǒuhuā)

东北凉菜。将熟猪肘片、黄瓜片、胡萝卜片用酱油和辣椒油拌匀。

Pork joint with vegetables. Northeastern cold dish. Sliced boiled pork joint mixed with sliced cucumbers and carrots, then flavored with soy sauce and chili oil.

瓣酥皮 (bànsūpí)

广东点心。先将面粉、猪油、鸡蛋、白糖混合,冷冻,然后放入模具烤制。

Flour, egg, and sugar cookie. Guangdong snack. Flour, lard, eggs, and white sugar mixed, chilled, then baked in a mould.

榜舍龟 (bǎngshèguī)

福建点心。先将糯米团包豆沙馅压成扁圆,再用模具做成龟形或寿字形,然后蒸熟。

Sticky rice and bean paste buns. Fujian snack. Sticky rice flour buns stuffed with sweetened bean paste, pressed flat, molded into turtles or Chinese characters of longevity, then steamed.

棒棒鸡 (bàngbàngjī)

四川凉菜。将煮熟的鸡肉用小木棒拍松,撕成鸡丝,浇上麻辣调味汁。

Chicken with spicy sauce. Sichuan cold dish. Boiled chicken pounded with a wooden stick to make the meat flossy, torn into shreds, then flavored with spicy sauce.

棒芽炒蚕豆 (bàngyá chǎo cándòu)

四川菜。先将椿芽氽过,切碎,然后与嫩蚕豆及调料拌炒。

Sautéed broad beans with toon leaves. Sichuan dish. Scalded chopped tender Chinese toon leaves sautéed with fresh tender broad beans and seasonings.

棒芽豌豆 (bàngyá wāndòu)

贵州菜。先将豌豆炸酥,再与氽过的椿芽、红椒、葱、盐等翻炒。

Sautéed toon leaves with fried peas. Guizhou dish. Scalded tender Chinese toon leaves sautéed with deep-fried peas, red pepper, scallion, and salt.

包谷粑 (bāogǔbā)

云贵川传统点心。把嫩玉米磨成稠浆,用新鲜玉米叶包上,蒸熟。

Steamed tender corn cakes.

Traditional snack in Yunnan, Guizhou, and Sichuan. Fresh and tender corn ground into thick paste, wrapped in fresh corn leaves, then steamed.

包面 (bāomiàn)

湖北点心。将精粉做成面皮,用猪肉末与麻油、盐、猪油渣等做成馅,包馅成团,或蒸或煮。

Steamed stuffed buns. Hubei snack. Fine flour dough made into buns filled with mixture of minced pork, sesame oil, salt, and lard dregs, then steamed or boiled.

包烧鱼 (bāoshāoyú)

四川菜。先将鲤鱼腌渍,鱼腹填入炒熟的泡椒、火腿、香菇,再用蛋清抹遍鱼身,烤熟。

Broiled carp. Sichuan dish. Pickled carp filled with stir-fried pickled pepper, ham, and dried mushrooms, coated with egg white, then broiled.

煲淋萝卜 (bāo lín luóbo)

湖南菜。白萝卜条加入红枣、木耳、腐竹及调料炖煮。

Stewed turnips. Hunan dish. Turnip strips stewed with red dates, wood ears, dried bean curd skins, and seasonings.

煲烧凤翅 (bāo shāo fèngchì)

浙江菜。先将鸡翅腌渍,再油炸、烹煮。

Braised chicken wings. Zhejiang dish. Chicken wings pickled, deep-fried, then braised.

煲仔鱼丸 (bāozǎi yúwán)

广东菜。先将鲮鱼茸与发菜、虾仁、香菜、葱、生菜等做成丸子,再与粉丝、生菜等煲煮,加调料。

Braised fish balls with vegetables. Guang-dong dish. Minced mud carp meat mixed with black moss, shrimp meat, coriander, scallion, and romaine lettuce, made into balls, then braised with vermicelli, romaine lettuce, and seasonings.

包子 (bāozi)

各地点心。将发酵面团做成各种馅料的包子,蒸熟。

Steamed stuffed buns. Snack in many places. Fermented flour dough made into dome-shape buns stuffed with a variety of fillings. Steamed.

宝丰酒 (bǎofēngjiǔ)

白酒,产于河南宝丰。以本地高粱为主料,用小麦、大麦、豌豆制曲酿造,酒精含量为 54-63 度,属清香型酒。

Baofeng liquor produced in Baofeng, Henan. White spirit made with local sorghum and yeast of wheat, barley, and peas. Contains 54-63% alcohol, and has a delicate aroma.

宝酿龙瓜 (bǎo niàng lóngguā)

浙江菜。将虾仁、鸡肉、干肉、笋、冬菇、火腿等加调料制成馅,填入丝瓜,蒸熟。

Steamed stuffed luffa. Zhejiang dish. Luffa stuffed with seasoned shrimp meat, chicken, dried preserved meat, bamboo shoots, winter mushrooms, and ham, then steamed.

宝糯米鸡 (bǎo nuòmǐ jī)

浙江菜。把江米饭、豌豆、火腿、莲子、薏米、芡实、虾米、口蘑、冬笋等加调料拌匀,装入鸡腹,先蒸熟,后油炸。

Fried and steamed stuffed chicken. Zhejiang dish. Chicken stuffed with seasoned sticky rice, peas, ham, lotus seeds, seeds of Job's tears, gorgon euryale seeds, dried shrimps, mushrooms, and winter bamboo shoots, steamed, then deep-fried.

宝箱豆腐 (bǎoxiāng dòufu)

江苏菜。先将豆腐块中间掏空,填入腌渍过的肉馅,面上嵌大虾仁,然后与香菇、青豆、调料等煨熟,勾芡。

Stuffed bean curd. Jiangsu dish. Cubed bean curd hollowed, stuffed with pickled minced pork, then topped with shelled prawns. Simmered with dried mushrooms, green soybean, and seasonings, then thickened with starch.

宝中藏宝 (bǎo zhōng cáng bǎo)

西藏菜。先将牦牛腿骨段盐腌,再将虫草放入骨内,然后蒸熟,浇上高汤,用鲜奶勾芡。

Yak bones with herb. Tibetan dish. Segmented yak leg bones pickled with salt, filled with Chinese caterpillar fungi, steamed, washed with broth, then thickened with starched fresh milk.

鲍贝烩豆瓣 (bàobèi huì dòubàn)

广东菜。将鲍贝与蚕豆瓣加调料入上汤烩煮,勾芡,淋麻油。

Simmered abalone and broad beans. Guangdong dish. Abalone simmered in broth with shelled broad beans and seasonings, thickened with starch, then washed with sesame oil.

鲍翅木瓜船 (bàochì mùguāchuán)

福建、广东菜。先将鱼翅和母鸡、猪

脚、排骨、猪皮等加调料炖烂，然后装入雕成船形的木瓜中稍蒸。

Shark fin and meat varieties in rock melon. Fujian and Guangdong dish. Shark fin stewed with hen, hog trotters, spareribs, hog skins, and seasonings, put in a rock melon carved like a ship, then steamed briefly.

鲍鱼干锅鸡 (bàoyú gānguō jī)

云南菜。将姜、蒜、干辣椒爆香，加入鸡丁和鲍鱼翻炒，加调料和水焖煮至汤近干，烹入香醋、黄酒。

Spicy chicken and abalone. Yunnan dish. Diced chicken and abalone stir-fried with quick-fried ginger, garlic, and chili pepper, simmered with seasonings until reduced, then flavored with vinegar and yellow wine.

鲍鱼鸡煲翅 (bàoyú jī bāo chì)

广东菜。先把氽过的去皮母鸡、猪小腿肉、鲍鱼、鱼翅一道放入煲内，加入山药、龙眼干、陈皮、姜片、大料、开水，用文火煲至酥熟；再分别用鸡汤焖煮鲍鱼，用上汤焖煮鱼翅；然后把鲍鱼切片装碗，鱼翅放在鲍鱼片上，浇汤汁。

Shark fin and abalone with chicken. Famous dish in Guangdong. Scalded skinned hen and pork hock stewed with abalone, shark fin, Chinese yam, dried longans, processed tangerine peels, ginger, aniseed, and cinnamon in an earthen pot. Abalone and shark fin simmered in chicken soup and broth respectively. Abalone sliced, covered with shark fin, then washed with gravy.

鲍鱼鸡翅 (bàoyú jīchì)

广东菜。先将鲍鱼加调料煮熟，再将鸡翅油炸，去骨，加调料焖酥，然后将鲍鱼和鸡翅同装一盘，用清炒的菜心装饰。

Braised abalone and chicken wings. Guangdong dish. Abalone braised with seasonings. Deep-fried chicken wings boned, then simmered with flavorings. Cooked abalone and chicken wings put on the same platter garnished with sautéed baby Chinese cabbage.

鲍鱼鸡片汤 (bàoyú jīpiàn tāng)

南方菜。将鲍鱼、鸡片、火腿片、笋片等放入烧沸的姜汤煮熟。

Abalone and chicken soup. Southern dish. Abalone and sliced chicken, ham, and bamboo shoots boiled in ginger soup.

鲍汁扒百灵菇 (bàozhī pá bǎilínggū)

广东菜。先将百灵菇与调料煲煮入味，切片，再蒸透，淋上鲍汁。

Mushrooms with abalone sauce. Guangdong dish. Lark mushrooms braised with seasonings, sliced, steamed, then washed with abalone sauce.

鲍汁扒鹅掌 (bàozhī pá ézhǎng)

广东菜。先将去骨鹅掌入原汤煨至酥烂，然后用鲍鱼汁、原汤等制芡汁，浇在鹅掌上。

Goose feet in abalone soup.

Guangdong dish. Boned goose feet braised in broth, then washed with starched sauce made with abalone soup and broth.

鲍汁扒辽参 (bàozhī pá liáoshēn)

广东菜。将辽参用冷水浸泡,蒸软,配鲍仔和西蓝花装盘,浇鲍汁。

Ginseng and broccoli in abalone gravy. Guangdong dish. Ginseng soaked in cold water and steamed. Served with prepared abalone and broccoli in abalone gravy.

爆炒鹅肠 (bàochǎo écháng)

湖南菜。先将红椒煸炒过,然后加入鹅肠、香菜杆、调料等翻炒。

Stir-fried goose intestine. Hunan dish. Goose intestine stir-fried with red pepper, coriander stalks, and seasonings.

爆炒肥肠 (bàochǎo féicháng)

湖南菜。先将肥肠煮至近熟,然后加入葱、姜、蒜、干辣椒等爆炒。

Stir-fried hog intestine. Hunan dish. Hog large intestine boiled, then stir-fried with scallion, ginger, garlic, and chili pepper.

爆炒鸡杂 (bàochǎo jīzá)

云贵川菜。将鸡杂切片,加入红椒丝、香菇、姜葱丝等爆炒,用盐、料酒、酱油等调味。

Stir-fried spicy chicken giblet. Yunnan, Guizhou, and Sichuan dish. Sliced chicken giblet stir-fried with shredded red pepper, dried mushrooms, ginger, and scallion, then flavored with salt, cooking wine, and soy sauce.

爆炒莲花白 (bàochǎo liánhuābái)

西北回族菜。将卷心菜加花椒和干红辣椒、盐翻炒。

Sautéed cabbage. Hui ethnic minority dish in northwestern areas. Cabbage sautéed with Chinese prickly ash, chili pepper, and salt.

爆炒面肺子 (bàochǎo miànfèizi)

新疆点心。先将羊肺灌入面粉浆煮熟,切块,然后与葱、姜等调料爆炒。

Stir-fried lamb lung. Xinjiang snack. Lamb lung filled with flour paste, boiled, sliced, then stir-fried with leek, ginger, and other seasonings.

爆炒墨鱼仔 (bàochǎo mòyúzǎi)

福建菜。先将姜葱煸香,再与墨鱼仔爆炒,然后加韭菜和佐料翻炒。

Stir-fried cuttlefish with chive. Fujian dish. Cuttlefish stir-fried with quick-fried ginger and scallion, then sautéed with Chinese chive and condiments.

爆炒牛肚 (bàochǎo niúdǔ)

湖南菜。先将肚丝焯好,然后加入红椒、青椒、葱、姜、蒜、爆炒,加盐、酱油、料酒。

Stir-fried ox tripe. Hunan dish. Scalded shredded ox tripe stir-fried with red and green pepper, scallion, and garlic. Seasoned with salt, soy sauce, and cooking wine.

爆炒肉片 (bàochǎo ròupiàn)

山东菜。先将挂浆肉片入油锅拨

散，然后与笋、木耳、黄瓜等爆炒，加调料，勾芡。

Stir-fried pork. Shandong dish. Sliced pork coated with starch paste, stir-fried with bamboo shoots, wood ears, and cucumbers, seasoned, then starched.

爆炒鳝片 (bàochǎo shànpiàn)

四川菜。将鳝鱼片配青椒、花椒、姜丝等爆炒，加辣椒、盐、酱油、料酒。

Stir-fried spicy eel. Sichuan dish. Sliced field eel stir-fried with green pepper, Chinese prickly ash, and shredded ginger. Seasoned with hot pepper, salt, soy sauce, and cooking wine.

爆炒双脆 (bàochǎo shuāngcuì)

福建菜。先将葱、姜、蒜末煸香，再与腌渍过的鸡肫和肚头爆炒，勾芡。

Stir-fried chicken gizzard and hog tripe. Fujian dish. Pickled chicken gizzard and hog tripe stir-fried with quick-fried scallion, ginger, and garlic, then thickened with starch.

爆炒田螺 (bàochǎo tiánluó)

西南家常菜。先将田螺肉腌渍，焯水，然后与干辣椒、花椒、辣椒酱、葱、姜、蒜苗等爆炒，加盐、糖、味精、麻油。

Stir-fried field snails. Home-style dish in southwestern areas. Pickled and scalded field snails stir-fried with chili pepper, Chinese prickly ash, hot pepper sauce, scallion, ginger, and garlic sprouts, then seasoned with salt, sugar, MSG, and sesame oil.

爆炒血鸭 (bàochǎo xuèyā)

湖南郴州特色菜。先将土鸭肉煸炒，再加入红椒和泉水焖煮，然后与鸭血爆炒，加调料。

Stir-fried duck and duck blood. Specialty of Chenzhou, Hunan. Quick-fried local duck simmered with red pepper and spring water, then stir-fried with duck blood and condiments.

爆炒鸭肠 (bàochǎo yācháng)

四川菜。先将鸭肠用沸水烫过，然后与红椒、姜、花椒、盐、料酒等爆炒。

Stir-fried spicy duck intestine. Sichuan dish. Scalded duck intestine stir-fried with red pepper, ginger, Chinese prickly ash, salt, cooking wine, and other condiments.

爆炒鸭丝 (bàochǎo yāsī)

北京菜。先将鸭脯肉切丝，芹菜切段，然后加葱、姜、蒜等爆炒。

Stir-fried duck breast. Beijing dish. Shredded duck breast stir-fried with celery, scallion, ginger, garlic, and other condiments.

爆炒羊肝 (bàochǎo yánggān)

四川菜。先将羊肝片用芡粉蛋清腌渍，然后与辣椒、蒜苗、姜、葱等爆炒，加入盐、料酒、酱油。

Stir-fried lamb liver. Sichuan dish. Sliced lamb liver pickled with starch and egg white paste, stir-fried with hot pepper, garlic sprouts, ginger, and scallion, then seasoned with salt, yellow wine, and soy sauce.

爆炒羊尾 (bàochǎo yángwěi)

山东菜。先将羊尾片挂浆，入油锅拨散，再加入黄酒、盐、清汤等煸炒。

Stir-fried lamb tail. Shandong dish. Sliced lamb tail coated with starch paste, lightly fried, then stir-fried with yellow wine, salt, and clear soup.

爆炒腰花 (bàochǎo yāohuā)

各地传统菜。先将腰花和木耳汆过，再与葱、姜、蒜、冬笋、料酒等爆炒，勾芡。

Stir-fried hog kidney. Traditional dish in many places. Scalded sliced hog kidney and wood ears stir-fried with scallion, ginger, garlic, winter bamboo shoots, cooking wine, and other condiments, then thickened with starch.

爆炒芷江鸭 (bàochǎo zhǐjiāngyā)

湘西特色菜。先将芷江鸭切成块，然后加入芷草、姜、大葱、红椒、花椒爆炒。

Stir-fried duck with herb. Specialty of western Hunan. Chopped Zhijiang duck stir-fried with local herb *zhicao*, ginger, leek, red pepper, and Chinese prickly ash.

爆炒猪肺 (bàochǎo zhūfèi)

四川菜。先将猪肺入沸水煮熟，切片，然后与辣椒、香菇、调料等爆炒。

Stir-fried spicy hog lung. Sichuan dish. Quick-boiled sliced hog lung stir-fried with chili pepper, dried mushrooms, and seasonings.

爆炒猪肝片 (bàochǎo zhūgānpiàn)

各地家常菜。将猪肝与汆过的竹笋、胡萝卜和芹菜爆炒，加盐、料酒、酱油等调料。

Stir-fried hog liver with vegetables. Home-style dish in many places. Hog liver stir-fried with quick-boiled bamboo shoots, carrots, and celery, then seasoned with salt, cooking wine, soy sauce, and other flavorings.

爆炒猪心 (bàochǎo zhūxīn)

云贵川菜。先将猪心入沸水汆过，切片，然后与辣椒、花椒、酱油、葱、姜等爆炒。

Stir-fried hog heart. Yunnan, Guizhou, Sichuan dish. Scalded sliced hog heart stir-fried with chili pepper, Chinese prickly ash, soy sauce, scallion, ginger, and other seasonings.

爆汆熏鱼 (bàocuān xūnyú)

江苏菜。先将青鱼块用黄酒和酱油腌渍，炸至深黄，然后入鸡清汤，与笋片、冬菇等焖煮。

Fried and simmered black carp. Jiangsu dish. Black carp pickled with yellow wine and soy sauce, deep-fried until dark yellow, then simmered with bamboo shoots and winter mushrooms in clear chicken soup.

爆脆海蜇皮 (bào cuì hǎizhépí)

福建菜。先将姜葱煸香，然后与海蜇丝爆炒，加酱油、黄酒、盐。

Stir-fried jellyfish. Fujian dish.

Shredded jellyfish stir-fried with ginger and garlic, then seasoned with soy sauce, yellow wine, and salt.

爆回锅牛肉 (bào huíguōniúròu)

湖南菜。先将牛柳炸熟,然后与椰菜、青红椒翻炒,加盐、料酒等调料。

Sautéed beef. Hunan dish. Fried beef strips sautéed with cauliflower and green and red pepper. Seasoned with soy sauce, salt, cooking wine, and other condiments.

爆两样 (bàoliǎngyàng)

山东菜。先将猪肝滑散,再放入熟猪肠、黄瓜、胡萝卜、木耳及调料拌炒。

Stir-fried hog liver and intestine. Shandong dish. Lightly fried hog liver stir-fried with prepared hog intestine, cucumbers, carrots, wood ears, and flavorings.

爆焖羊羔肉 (bàomèn yánggāoròu)

西藏菜。羊羔肉加调料爆炒。

Stir-fried lamb meat. Tibetan dish. Lamb meat stir-fried with seasonings.

爆目鱼花 (bàomùyúhuā)

江浙菜。把墨鱼切成荔枝花刀块,焯水沥干,再与玉兰片、木耳等爆炒,用盐、甜面酱、葱、姜等调味,勾芡。

Inkfish with wood ears and bamboo shoots. Jiangsu and Zhejiang dish. Inkfish sliced like litchi flowers and scalded. Stir-fried with soaked bamboo shoots and wood ears, flavored with salt, sweet fermented flour sauce, scallion, and ginger, then thickened with starch.

爆牛肚 (bàoniúdǔ)

山东菜。将卤牛肚片与葱、姜、料酒快速翻炒。

Stir-fried ox tripe. Shandong dish. Sliced marinated ox tripe stir-fried with scallion, ginger, and cooking wine.

爆三脆 (bàosāncuì)

山东菜。先将玉兰片、香菇、豌豆等汆过,然后放入海蜇、爆炒过的肚头、鸡胗拌炒,加盐、料酒等调料。

Jellyfish with hog tripe and chicken gizzard. Shandong dish. Jellyfish, prepared hog tripe, and chicken gizzard stir-fried with scalded bamboo shoots, dried mushrooms, and peas. Seasoned with salt, cooking wine, and other condiments.

爆三样 (bàosānyàng)

山东菜。先将腌渍过的猪心、猪腰、猪肉略炒,然后与大葱、番茄等翻炒,加黄酒、酱油、醋等调味。

Stir-fried pork, hog heart, and kidney. Shandong dish. Pickled pork, hog heart, and kidney lightly fried, sautéed with leek and mushrooms, then flavored with yellow wine, soy sauce, and vinegar.

爆烧冬笋 (bàoshāo dōngsǔn)

贵州、云南菜。先煸炒大料,再放入炸过的冬笋块和调料煨煮,然后加入柿子椒,勾芡。

Stir-fried bamboo shoots. Guizhou and Yunnan dish. Deep-fried winter bamboo shoots stewed with quick-fried aniseed and other seasonings, combined with sweet pepper, then thickened with starch.

爆双脆 (bàoshuāngcuì)

浙江菜。猪肚尖和鸭肫加佐料爆炒,烹入用黄酒、清汤、淀粉、盐等调成的卤汁。

Stir-fried hog tripe and duck gizzard. Zhejiang dish. Hog tripe and duck gizzard stir-fried with condiments, then flavored with sauce of yellow wine, clear soup, starch, and salt.

爆双丁 (bàoshuāngdīng)

东北家常菜。鸡肉丁和黄瓜丁爆炒,加盐、甜面酱、料酒等。

Stir-fried chicken and cucumbers. Northeastern home-style dish. Diced chicken and cucumbers stir-fried and flavored with salt, sweet fermented flour sauce, and cooking wine.

爆四丁 (bàosìdīng)

浙江菜。将虾仁、鸡脯丁、鸡肫丁、猪里脊丁分别炒熟,装入四只盘中,同时上桌。

Fried four kinds of meat. Zhejiang dish. Shrimp meat, diced chicken breast, chicken gizzard, and diced pork tenderloin stir-fried separately. Served on separate platters.

爆素鳝丝 (bàosùshànsī)

浙江菜。先将冬菇丝挂上淀粉糊油炸,再将豌豆、黄酒、酱油、白糖等入汤煮沸,勾芡,倒在冬菇丝上。

Fried mushrooms with sauce. Zhejiang dish. Shredded winter mushrooms coated with starch paste, fried, then washed with gravy of peas, yellow wine, soy sauce, sugar, and starch.

爆乌花 (bàowūhuā)

苏浙皖沪菜。先将鲜墨鱼片焯水,然后与玉兰片、木耳、青豆等加佐料爆炒,勾芡,撒香菜和胡椒粉。

Stir-fried squid. Jiangsu, Zhejiang, Anhui, and Shanghai dish. Scalded sliced fresh squid stir-fried with soaked bamboo shoots, wood ears, green soybean, and condiments. Thickened with starch, then sprinkled with coriander and ground black pepper.

爆鱿鱼卷 (bàoyóuyújuǎn)

东北菜。先将鱿鱼块轻炒至起卷,然后加入用鸡汤、盐、味精等调成的卤汁爆炒。

Stir-fried squid. Northeastern dish. Sliced squid lightly fried until rolled up, then stir-fried with sauce of chicken soup, salt, and MSG.

爆鱼丁 (bàoyúdīng)

山东菜。先将鱼肉丁用黄酒、盐、蛋清、湿淀粉等腌渍,入油锅拨散,然后加入冬笋、清汤、黄酒、芡汁等翻炒。

Stir-fried fish. Shandong dish. Diced fish pickled with yellow wine, salt, egg white, and starch, lightly fried, then stir-fried with winter bamboo shoots and sauce of clear soup, yellow wine, and starch.

爆糟排骨 (bàozāo páigǔ)

福建菜。将排骨用糖、盐、红糟腌渍,裹蛋白和菱粉糊,炸熟。

Fried sweet and salty ribs. Fujian dish. Pork ribs pickled with sugar, salt, and red fermented sticky rice juice, coated with egg white and water caltrop starch paste, then deep-fried.

北葱焖羊肉 (běicōng mèn yángròu)

广东菜。将羊肉与大葱焖煮,加料酒、盐、姜、草果、八角等。

Stewed mutton with leek. Guangdong dish. Mutton stewed with leek and flavored with cooking wine, salt, ginger, tsaoko amomum fruit, and aniseed.

北大仓酒 (běidàchāngjiǔ)

白酒,产于黑龙江齐齐哈尔。以当地高粱为原料,用大麦、小麦、大豆、玉米等制曲酿造,酒精含量 38-50 度,属酱香型酒。

Beidachang liquor produced in Qiqihaer, Heilongjiang. White spirit made with local sorghum and yeast of barley, wheat, peas, and corn. Contains 38-50% alcohol, and has a soy aroma.

北港毛尖 (běigǎngmáojiān)

黄茶。产于湖南岳阳北港。外形呈金黄色,毫尖显露。汤色金黄,叶底肥嫩。

Beigangmaojian, yellow tea produced in Beigang of Yueyang, Hunan. Its dry leaves are golden and hairy. The tea has an orange color with infused leaves tender and stout.

北菇扒菜胆 (běigū pá càidǎn)

东北家常菜。先将泡软的干北菇蒸熟装盘,然后将炒熟的芥菜嫩梗摆在盘边,浇调味卤汁。

Mushrooms with mustard tender. Northeastern home-style dish. Soaked northern forest mushrooms steamed, put on a platter with sautéed tender mustard stems, then washed with sauce.

北菇扒双蔬 (běigū pá shuāngshū)

广东菜。将北菇蒸熟装盘,四周围上汆过的芦笋和芥菜,淋鸡油、撒盐和胡椒粉。

Steamed mushrooms with greens. Guangdong dish. Steamed mushrooms surrounded with quick-boiled asparagus and mustard greens on a platter. Washed with chicken oil, then sprinkled with salt and ground black pepper.

北菇鹅掌 (běigū ézhǎng)

广东菜。先将鹅掌炸熟,再与香菇和笋花入上汤焖煮,勾芡,然后配火腿片装盘。

Simmered goose feet with mushrooms. Guangdong dish. Fried goose feet simmered in broth with dried mushrooms and dried bamboo shoots, thickened with starch, then served with sliced ham.

北菇海参煲 (běigū hǎishēn bāo)

东北家常菜。先将水发海参煮熟切片,北菇泡软,然后将海参和北菇加调料同煮入味。

Stewed mushrooms and sea cucumbers. Northeastern home-style dish. Sliced soaked sea cucumbers stewed with soaked northern forest mushrooms and condiments.

北菇扣鹅掌 (běigū kòu ézhǎng)

江苏菜。鹅掌裹淡酱油和淀粉油炸,然后加入香菇、猪脚浓汤、米酒、蚝油等烹煮。

Goose feet with mushrooms. Jiangsu dish. Goose feet coated with light soy sauce and starch, deep-fried, then braised with dried mushrooms in hog trotter soup seasoned with yellow wine and oyster sauce.

北菇蒸滑鸡 (běigū zhēng huájī)

江西菜。先将鸡翅和香菇切丝,再用葱、姜、盐、料酒腌渍,淋油,然后蒸熟。

Steamed chicken wings and mushrooms. Jiangxi dish. Shredded chicken wings and dried mushrooms pickled with scallion, ginger, salt, and cooking wine, washed with oil, then steamed.

北京爆肚 (běijīngbàodǔ)

北京菜。先将猪肚丝焯熟,放入垫有香菜的盘中,然后浇用麻油、芝麻酱、韭菜花等调成的卤汁。

Boiled hog tripe with sauce. Beijing dish. Shredded hog tripe quick-boiled, bedded on coriander, then served with dressing of sesame oil, sesame paste, and Chinese chive flower.

北京炒肝 (běijīngchǎogān)

北京菜。将猪肝片和猪大肠段入卤汤煮熟,勾芡。

Braised hog liver and large intestine. Beijing dish. Sliced hog liver and segmented hog large intestine braised in gravy, then thickened with starch.

北京炒疙瘩 (běijīng chǎogēda)

北京食品。先将面团搓成条,切成丁,煮熟,然后与胡萝卜丁及调料翻炒。

Stir-fried flour dough and carrots. Beijing food. Flour dough cut into dices, boiled, then stir-fried with diced carrots and seasonings.

北京烤鸭 (běijīngkǎoyā)

北京名菜。先将北京鸭用沸水焯烫,再入烤炉烤熟,然后配葱丝、黄瓜条、甜面酱、薄饼。

Broiled Beijing duck. Famous Beijing dish. Scalded Beijing duck broiled, then served with shredded leek stems, cucumbers strips, sweet fermented flour sauce, and thin pancakes.

北京烤羊肉 (beǐjīng kǎoyángròu)

北京菜。先将羊肉片用酱油、卤虾油、葱丝和香菜末等腌渍,然后放铁叉上翻烤。

Broiled mutton. Famous Beijing dish. Sliced mutton pickled with soy sauce, shrimp sauce, shredded leek, and chopped coriander, then broiled on iron bars.

北京涮羊肉 (běijīng shuànyángròu)

北京名菜。将羊肉片、白菜、粉丝等入火锅涮熟,佐芝麻酱、腌韭菜花汁、豆腐乳汁等。

Mutton in fire pot. Famous Beijing dish. Sliced mutton, Chinese cabbage, and starch vermicelli quick-boiled in fire pot soup and served with sesame sauce, preserved Chinese chive flower sauce, and fermented been curd juice.

北京酥鲫鱼 (běijīng sūjìyú)

北京菜。将鲫鱼加花椒、丁香、白糖、桂皮等用微火炖酥。

Stewed crucian. Beijing dish. Crucian stewed with Chinese prickly ash, clove, sugar, and cinnamon.

北芪党参凤肝汤 (běiqí dǎngshēn fènggān tāng)

广东汤。将北芪、党参、鸡肝加调料熬炖。

Chicken liver and herb soup. Guangdong soup. Milkvetch root stewed with pilose asiabell root, chicken liver, and seasonings.

北芪炖乳鸽 (běiqí dùn rǔgē)

广东菜。将乳鸽与北芪、淮山、红枣、生姜炖煮,加盐和黄酒。

Stewed pigeon with herbs. Guangdong dish. Young pigeon stewed with herbs, such as milkvetch root, Chinese yam, red date, and ginger, then seasoned with salt and yellow wine.

北芪杞子羊肉汤 (běiqí qǐzǐ yángròu tāng)

广东汤。将羊肉与北芪、枸杞子等加盐、黄酒熬炖。

Mutton and herb soup. Guangdong soup. Mutton stewed with milkvetch root, Chinese wolfberry, salt, and yellow wine.

北芪瘦肉 (běiqí shòuròu)

广东菜。先将瘦猪肉腌渍,然后加入北芪、天麻等炖煮。

Stewed pork with herbs. Guangdong dish. Pickled lean pork stewed with herbs such as milkvetch root and tall gastrodia tuber.

北杏燕窝汤 (běixìng yànwō tāng)

广东点心。用北杏、燕窝、冰糖等炖制。

Apricot and bird's nest soup. Guangdong snack. Apricots stewed with edible bird's nests and rock sugar.

贝母鸡块 (bèimǔ jīkuài)

西藏菜。先将鸡块用酥油炸熟,然后加入贝母、生姜、豆粉、调料等炖煮。

Chicken stewed with herb. Tibetan dish. Chopped chicken deep-fried with Tibetan butter, then stewed with fritillary, ginger, soybean powder, and seasonings.

贝母甲鱼 (bèimǔ jiǎyú)

浙江菜。先将甲鱼块腌渍,然后和川贝母入鸡汤蒸熟。

Steamed turtle with herb. Zhejiang dish. Pickled turtle chunks steamed with fritillary bulbs in chicken soup.

贝母牛肉 (bèimǔ niúròu)

新疆菜。将贝母和牛肉入肉汤炖

熟，加调料。

Stewed beef with herb. Xinjiang dish. Fritillary and beef stewed in seasoned broth.

贝丝扒菜胆 (bèisī pá càidǎn)

广东菜。先将干贝泡软，撕成丝，然后与芥菜心入膏汤煮熟，加盐、姜、葱、黄酒等，勾芡。

Scallops with baby Chinese broccoli. Guangdong dish. Hand-shredded soaked scallops braised in broth with baby Chinese broccoli, seasoned with salt, ginger, scallion, and yellow wine, then thickened with starch.

焙面鲤鱼 (bèimiàn lǐyú)

河南菜。将龙须面炸焦，盖在熘好的糖醋黄河鲤鱼上。

Sweet and sour carp with noodles. Henan dish. Yellow River carp cooked sweet and sour, then covered with deep-fried fine noodles.

焙子 (bèizǐ)

内蒙点心。小麦面饼，有白焙子、咸焙子和甜焙子等。

Baked flour cakes. Inner Mongolian snack. Flour dough made into plain, salty, and sweet cakes.

本山 (běnshān)

乌龙茶。产于福建安溪，属安溪色种。外形肥壮重实，梗鲜亮细瘦，色泽鲜润。汤色橙黄，叶底黄绿。

Benshan, one of the oolong varieties from Anxi, Fujian. Its dry leaves are stout, solid, sleek, and lustrous on slim stalks. The tea has an orange color with infused leaves yellowish green.

荸荠百合 (bíqí bǎihé)

南方点心。将荸荠与百合加冰糖煮熟，置凉。

Water chestnuts and lily cloves. Southern snack. Water chestnuts boiled with lily cloves and rock sugar, then chilled.

荸荠鸡片 (bíqí jīpiàn)

江苏家常菜。先将鸡脯肉熘熟，再与荸荠片翻炒，加入糖、盐、葱等。

Sautéed chicken and water chestnuts. Home-style dish in Jiangsu. Lightly fried chicken breast sautéed with sliced water chestnuts, then flavored with sugar, salt, and scallion.

碧螺春 (bìluóchūn)

绿茶。中国名茶，产于江苏太湖中的洞庭东、西两山。外形纤细，色泽青翠，布满茸毫，卷曲成螺。汤色清澈明亮，叶底嫩绿显翠。

Biluochun, famous green tea in East and West Dongting Hills in Taihu Lake, Jiangsu. Its dry leaves are tenuous, verdant and spirally hairy. The tea is crystal clear with infused leaves tender green.

碧绿鳜鱼卷 (bìlǜ guìyújuǎn)

广东菜。先用鳜鱼皮包火腿成卷，炸熟，然后与蒜茸、姜片、油菜心及调料拌炒。

Fish skin and ham rolls. Guangdong dish. Chinese perch skins made into rolls filled with ham, deep-fried, then stir-fried with garlic, ginger, baby Chinese broccoli, and flavorings.

碧绿花枝玉带 (bìlǜ huāzhī yùdài)

广东菜。用带子配上花枝片、甘笋、蒜茸、姜等翻炒。

Cuttlefish and scallops with vegetables. Guangdong dish. Scallops and cuttlefish stir-fried with carrot cuts, mashed garlic, ginger, and other condiments.

碧绿蹄筋 (bìlǜ tíjīn)

江苏菜。先用鸡汤把蹄筋和火腿片加调料煮透,再放入过油丝瓜条煮滚,然后勾芡,淋猪油。

Pork tendons with loofah. Jiangsu dish. Pork tendons and sliced ham boiled in chicken soup with lightly fried loofah and seasonings, thickened with starch, then washed with lard.

碧绿虾仁 (bìlǜ xiārén)

四川菜。先将虾仁腌渍,与嫩豌豆一道用油滑熟,然后用黄酒、葱、姜、芡汁翻炒,配轻炒豆苗装盘。

Shrimp meat with peas. Sichuan dish. Shrimp meat pickled, lightly fried with peas, sautéed with sauce of yellow wine, scallion, ginger, and starch, then garnished with quick-fried pea sprouts.

碧绿鲜带子 (bìlǜ xiāndàizi)

广东菜。将带子与蒜茸、虾米、姜、葱、蚝油等翻炒,配氽过的芥蓝装盘。

Sautéed scallops with Chinese broccoli. Guangdong dish. Scallops sautéed with mashed garlic, dried shrimp meat, ginger, scallion, and oyster sauce, then served with quick-boiled Chinese broccoli.

碧绿鸭条 (bìlǜ yātiáo)

广东菜。鸭条与嫩菜茎、腊鸭丝、蒜茸等同炒,用黄酒、胡椒粉、酱油、麻油等调味。

Stir-fried duck with vegetable stalks. Guangdong dish. Duck strips stir-fried with tender vegetable stalks, smoked shredded duck, and mashed garlic, then flavored with yellow wine, ground black pepper, soy sauce, and sesame oil.

碧螺春水烧豆腐 (bìluóchūnshuǐ shāo dòufu)

江苏菜。将豆腐放入碧螺春茶汤,加入茶叶和枸杞煮沸,勾芡。

Bean curd in tea soup. Jiangsu dish. Bean curd boiled in *biluochun* tea soup with tea leaves and Chinese wolfberry, then thickened with starch.

碧桃鸡 (bìtáojī)

山东菜。先将鸡脯肉片挂浆,内卷核桃仁,外裹切碎的火腿、冬菇、黄瓜皮,然后炸熟,浇卤汁。

Chicken rolls with sauce. Shandong dish. Starched sliced chicken breast made into rolls filled with walnut meat, and coated with chopped ham, dried mushrooms, and cucumber peels. Deep-fried, then marinated in sauce of yellow wine, salt, and starch.

碧玉卷 (bìyùjuǎn)

福建点心。将大米和韭菜磨成浆，烙成薄饼，卷入香菇、猪肉、笋丝、辣椒粉等做成的馅。

Rice and chive rolls. Fujian snack. Rice and Chinese chive ground into paste, fried into green thin pancakes, then made into rolls filled with dried mushrooms, pork, shredded bamboo shoots, and chili powder.

碧玉牛筋 (bìyù niújīn)

上海菜。先将牛筋焖酥，再加胡萝卜和白萝卜煮熟，然后加花生酱、酱油、XO酱烹煮，冷却后切片。

Beef tendons with carrots and turnips. Shanghai dish. Stewed beef tendons boiled with carrots and white turnips, then simmered with peanut sauce, soy sauce, and XO sauce. Cooled, then sliced.

碧玉豌豆仁 (bìyù wāndòurén)

福建菜。先将豌豆仁与素火腿炒熟，然后放在剪成小勺样的生白菜叶上。

Sautéed peas and soy ham. Fujian dish. Fresh peas and soy ham sautéed, then put in small spoon-shape Chinese cabbage leaves.

避风塘炒蟹 (bìfēngtáng chǎoxiè)

台湾菜。将海蟹块裹上淀粉糊油炸，然后加干辣椒、蒜茸、姜、葱翻炒。

Stir-fried sea crab. Taiwan dish. Sea crab chunks coated with starch paste, deep-fried, then stir-fried with chili pepper, minced garlic, ginger, and scallion.

扁大枯酥 (biǎndàkūsū)

江苏菜。用猪肋条肉粒、生荸荠粒、蛋黄、米粉做成圆饼煎熟，将菊花叶炒至翠绿做配饰，将芡汁浇在饼上。

Pork cake with chrysanthemum leaves. Jiangsu dish. Pancakes made with diced streaky pork, diced water chestnuts, egg yolks, and rice flour, garnished with lightly fried fresh chrysanthemum leaves, then washed with starched soup.

扁豆肉丝 (biǎndòu ròusī)

河南家常菜。先将葱、姜丝炝锅，再放入挂浆猪肉丝和汆过的扁豆丝煸炒，然后加盐，用高汤勾芡。

Sautéed pork and common beans. Home-style dish in Henan. Starched shredded pork sautéed with scalded shredded common beans, then flavored with quick-fried scallion, ginger, salt, and starched broth.

扁豆烧鱼 (biǎndòu shāo yú)

家常菜。先将草鱼腌渍，煎至两面变色，然后加清汤、扁豆、木耳煨熟，勾芡。

Braised grass carp with beans. Home-style dish in many places. Pickled grass carp fried, simmered with kidney beans and wood ears in clear soup, then thickened with starch.

扁豆薏米炖鸡脚 (biǎndòu yìmǐ dùn jījiǎo)

广东菜。将鸡脚、扁豆、薏米、姜片等炖熟，用盐调味。

Chicken feet with seeds of Job's tears. Guangdong dish. Chicken feet stewed with white album beans, seeds of Job's tears, and ginger, then seasoned with salt.

汴京焖炉烤鸭 (biànjīng mènlú kǎoyā)

河南开封菜。先将京冬菜加调料填入鸭腹,然后将蜂蜜涂遍鸭身,焖烤至柿红色。

Broiled duck with preserved cabbage. Specialty in Kaifeng, Henan. Duck filled with seasoned preserved Chinese cabbage, coated with honey, then broiled until redgolden.

遍地锦装鳖 (biàndì jǐnzhuāngbiē)

山东菜。先将甲鱼肉煮半熟,再与口蘑和玉兰片及调料煸炒,焖至汁浓,然后配鸭蛋、火腿等蒸熟。

Turtle with duck eggs and ham. Shandong dish. Turtle meat half-boiled, stir-fried with Mongolian mushrooms, soaked bamboo shoots, and condiments, braised until reduced, then steamed with duck eggs and ham.

缤纷花枝片 (bīnfēn huāzhīpiàn)

广东菜。将洋葱、胡萝卜、木耳、荷兰豆及调料翻炒,勾芡。

Sautéed vegetables. Guangdong dish. Onion and carrots sautéed with wood ears, snow peas, and flavorings, then thickened with starch.

冰白玉水果色拉 (bīng báiyù shuǐguǒ sèlā)

广东家常凉菜。将白玉豆腐、苹果、西瓜切块,加牛奶和糖拌匀,冰镇。

Bean curd and fruit salad. Home-style cold dish in Guangdong. Diced creamy bean curd, apples, and watermelon mixed with sugared milk, then chilled with ice.

冰粉 (bīngfěn)

云贵川点心。将冰粉即薜荔果粉用凉开水溶释,放置数小时成透明胶冻,加入凉红糖水、炒过的碎花生或芝麻。

Ficus pumila jelly with sugar. Yunnan, Guizhou, and Sichuan snack. Ficus pumila seed powder melted in cold water, crystallized, then served with cold brown sugar soup, roasted crumbed peanuts, or sesame.

冰凉糕 (bīngliánggāo)

湖北点心。米浆与白糖、苏打、米酒调和,蒸熟,切成菱形块。

Cool rice cakes. Hubei snack. Rice paste mixed with white sugar, soda, and sweet fermented sticky rice juice, steamed, then cut into diamond cubes.

冰梅子姜炆鸡翼 (bīngméi zijiāng wén jīyì)

广东菜。先将腌鸡翅爆煎至金黄,再与煸过的子姜片同炒,炆煮,勾芡。

Chicken wings with tender ginger. Guangdong dish. Pickled chicken wings fried until golden over high heat, stir-fried with sliced baby ginger, simmered, then thickened with starch.

冰皮盐水鸡 (bīngpí yánshuǐjī)

广东菜。将整鸡与月饼、糕粉、煮甜甜、桂皮、八角等烹煮。

Braised chicken and moon cakes. Guangdong dish. Whole chicken braised with moon cakes, cake powder, sweetener, cinnamon, and aniseed.

冰霜丸子 (bīngshuāng wánzi)

东北菜。先将猪肉丸子油炸,再将糖熬至抽丝,放入丸子翻匀,撒上白糖。

Sugarcoated pork meatballs. Northeastern dish. Pork meatballs deep-fried, quickly coated with heated melting sugar, then showered with granulated sugar.

冰糖百合 (bīngtáng bǎihé)

南方点心。将百合入冰糖水煮熟。

Lily cloves soup. Southern snack. Lily cloves boiled in rock sugar soup.

冰糖甲鱼 (bīngtáng jiǎyú)

上海菜。甲鱼块加冰糖、猪油等炖熟。

Stewed turtle. Shanghai dish. Turtle chunks stewed and flavored with rock sugar and lard.

冰糖莲子粥 (bīngtáng liánzǐ zhōu)

东南方点心。将糯米和莲子加冰糖熬煮。

Sticky rice and lotus seed porridge. Southeastern snack. Sticky rice and lotus seeds cooked into porridge, then flavored with rock sugar.

冰糖排马面 (bīngtáng pái mǎmiàn)

江苏菜。先将猪头肉去骨切块,加醋、黄酒、冰糖等焖至酥烂,然后浇上卤汁,用绿叶菜缀边。

Stewed hog head. Jiangsu dish. Hog head boned, chopped, stewed with vinegar, yellow wine, and rock sugar, washed with gravy, then decorated with vegetables greens.

冰糖湘莲 (bīngtáng xiānglián)

湖南传统点心。先将莲子蒸熟,然后加桂圆肉、菠萝、冰糖、清水,用中火炖煮。

Stewed sweet lotus seeds. Hunan snack. Steamed lotus seeds stewed in water over medium heat with logan pulp, pineapple, and rock sugar.

冰糖燕窝 (bīngtáng yànwō)

各地点心。将水发燕窝放入冰糖水,蒸熟,撒上樱桃等果脯。

Bird's nest soup with sugar. Snack in many places. Water-soaked edible bird's nests steamed in rock sugar soup, then garnished with candied cherries or other preserved fruits.

冰糖燕窝炖乳鸽 (bīngtáng yànwō dùn rǔgē)

广东菜。将乳鸽、燕窝、冰糖等用文火炖煮。

Pigeon with bird's nests. Guangdong dish. Young pigeon stewed with edible bird's nests and rock sugar.

冰糖银耳 (bīngtáng yínĕr)

各地家常点心。将水发银耳用冰糖水熬煮。

White wood ear and sugar soup. Home-style snack in many places. Soaked white wood ears boiled in rock sugar soup.

冰糖鼋鱼 (bīngtáng yuányú)

东北名菜。先将鼋鱼焯水,过油,然后与鸡肉、猪肉、冰糖等炖至汁浓。

Stewed sweet soft-shell turtle. Famous northeastern dish. Scalded soft-shell turtle lightly fried, then stewed with chicken, pork, and rock sugar.

冰糖蒸鸡胆 (bīngtáng zhēng jīdăn)

广东菜。将鸡胆和冰糖蒸熟,加盐。

Steamed chicken gallbladder with rock sugar. Guangdong dish. Chicken gallbladder steamed with rock sugar, then flavored with salt.

冰糖肘子 (bīngtáng zhǒuzi)

各地传统菜。将蹄髈先煮后炖,加入冰糖、葱、姜、盐、料酒。

Sweetened hog knuckle. Traditional dish in many places. Scalded hog knuckle stewed with rock sugar, scallion, ginger, salt, and cooking wine.

冰糟鳕鱼冻 (bīngzāo xuěyúdòng)

上海菜。先把鳕鱼片煮熟,再加入香糟鱼汤,冷却结冻。

Cod jelly. Shanghai dish. Sliced cod fish boiled in gravy of yellow wine, fish soup, and sweet fermented sticky rice juice, then chilled into jelly.

冰糟肘子 (bīngzāo zhǒuzi)

山东菜。先将猪肘煮至半熟,再抹上糖浆,炸至发红,然后切块,放入冰糖,加清汤蒸至酥烂。

Steamed sugared hog knuckle. Shandong dish. Hog knuckle half-boiled, coated with syrup, fried until red, cut into big pieces, then steamed in clear soup seasoned with rock sugar.

冰醉鸡 (bīngzuìjī)

各地家常凉菜。先将鸡腿去骨腌渍,蒸熟,再用加胡椒粉的黄酒浸泡,然后冷却。

Drumsticks in yellow wine. Home-style cold dish in many places. Boned and pickled drumsticks steamed, then marinated in yellow wine flavored with ground black pepper. Served chilled.

菠菜鸡煲 (bōcài jī bāo)

广东菜。先将腌渍过的鸡块与冬菇、葱、姜爆炒,然后加菠菜煲煮,加调料。

Stewed chicken with spinach. Guangdong dish. Pickled chicken stir-fried with winter mushrooms, scallion, and ginger, then stewed with spinach. Seasoned.

钵钵鸡 (bōbōjī)

四川菜。先将仔鸡入加了姜、葱、黄酒的汤烫熟,然后迅速冷却,切块,用竹签串上,浸入用花生酱、芝麻酱、蚝油、酱油、黄酒、盐、花椒油、红油等调成的卤汁。

Marinated spicy chicken. Sichuan dish. Young hen quick-boiled in clear soup seasoned with ginger, scallion, and yellow wine. Chilled, chopped, strung on bamboo sticks, then marinated in gravy of peanut sauce, sesame sauce, oyster sauce, soy sauce, yellow wine, salt, Chinese prickly ash oil, and chili oil.

饽饽四品 (bōbo sìpǐn)

北京点心。由豆黄、芝麻卷、金糕、枣泥糕组成。

Four kinds of vegetable cakes. Beijing snack. A set of 4 kinds of cakes, including black soybean cakes, sesame rolls, haw cakes, and mashed date cakes.

菠菜炒粉丝 (bōcài chǎo fěnsī)

东北家常菜。先将菠菜切段,粉丝煮熟,然后加葱、盐、味精等同炒。

Sautéed spinach with vermicelli. Northeastern home-style dish. Prepared spinach sautéed with prepared starch vermicelli, scallion, salt, and MSG.

菠菜蛋汤 (bōcài dàntāng)

各地家常菜。先将姜、葱、盐等加入清汤煮开,然后放入菠菜和煎好的蛋皮煮熟,加麻油。

Spinach and egg soup. Home-style dish in many places. Spinach and fried egg sheets boiled in clear soup seasoned with salt, ginger, and scallion, then flavored with sesame oil.

菠菜豆腐汤 (bōcài dòufu tāng)

各地家常菜。将菠菜和豆腐煮汤,加盐、葱、姜、麻油等。

Spinach and bean curd soup. Home-style dish in many places. Spinach boiled with bean curd and seasoned with salt, scallion, ginger, and sesame oil.

菠菜炖鸡 (bōcài dùn jī)

四川菜。先将母鸡氽过,再与菠菜、火腿、笋片等炖煮,加调料。

Stewed hen with spinach. Sichuan dish. Scalded hen stewed with spinach, ham, sliced bamboo shoots, and seasonings.

菠菜粉皮 (bōcài fěnpí)

东北家常凉菜。先将菠菜切段焯透,粉皮切条,然后加醋、酱油、麻油等拌匀。

Spinach with bean jelly. Northeastern home-style cold dish. Scalded spinach mixed with bean jelly strips and flavored with vinegar, soy sauce, and sesame oil.

菠菜鸡煲 (bōcài jī bāo)

南方菜。先将腌渍过的鸡块与冬菇爆炒,然后与煮熟的甘笋入菠菜煲烧开。

Braised chicken with spinach. Southern dish. Pickled chopped chicken stir-fried with winter mushrooms, then boiled in an earthen pot with prepared carrots and spinach.

菠菜油豆腐 (bōcài yóudòufu)

东北家常菜。将菠菜和油豆腐加调

料翻炒。

Spinach and fried bean curd. Northeastern home-style dish. Spinach sautéed with deep-fried bean curd and seasoned.

菠菜鱼丸汤 (bōcài yúwán tāng)

浙江菜。海鳗鱼茸加姜汁、盐等做成鱼丸,煮熟,加入菠菜烫熟。

Eel ball and spinach soup. Zhejiang dish. Minced sea eel meat seasoned with ginger juice and salt, made into balls, then boiled in spinach soup.

菠饺鱼肚 (bōjiǎo yúdǔ)

四川菜。先用菠菜汁同面粉和成面团,擀成饺皮,包入肉馅,然后将饺子入鱼肚火腿汤煮熟。

Spinach *jiaozi* with fish maw. Sichuan dish. Spinach juice and flour mixed, made into *jiaozi* stuffed with prepared meat, then boiled in fish maw and ham soup.

钵酒焗乳鸽 (bō jiǔ jú rǔgē)

广东菜。将炸过的乳鸽放入用黄酒、白糖、盐调味的高汤,慢火焗熟。

Braised young pigeon. Guangdong dish. Deep-fried young pigeon braised in broth flavored with yellow wine, sugar, and salt.

玻璃白菜 (bōli báicài)

广东菜。将油煎过的白菜杆放入高汤,加入草菇、猪肉、火腿等蒸熟。

Steamed Chinese cabbage with pork. Guangdong dish. Deep-fried Chinese cabbage stems steamed with meadow mushrooms, pork, and ham in broth.

玻璃皮蛋 (bōli pídàn)

各地家常凉菜。先将皮蛋压成泥,摊在玻璃纸上,蒸熟,置凉后切条,然后浇上酱油、醋、白糖。

Steamed preserved eggs. Home-style cold dish in many places. Mashed preserved eggs laid on a piece of glassine, steamed, chilled, and cut into strips. Served with soy sauce, vinegar, and sugar.

玻璃烧卖 (bōli shāomài)

四川点心。用面皮包上白菜肉馅,蒸熟。

Steamed pork and cabbage buns. Sichuan snack. Flour wrappers made into buns filled with prepared Chinese cabbage and pork, then steamed.

玻璃酥虾 (bōli sūxiā)

广东潮州菜。先将去壳大虾剖开,腌渍,酿入用马蹄、韭黄、香菇、火腿等做成的馅,然后裹上淀粉浆,炸至金黄。

Fried stuffed prawns. Specialty in Chaozhou, Guangdong. Shelled prawns cut on the backs, pickled, stuffed with mixture of water chestnuts, yellow Chinese chive, dried mushrooms, and ham, coated with starch paste, then deep-fried until golden.

玻璃桃仁 (bōli táorén)

山东菜。将白糖炒成流质状,放入炸至微黄的核桃仁,迅速翻匀。

Sugarcoated walnuts. Shandong dish. Walnuts lightly fried, then quick-fried with heated melting sugar.

玻璃虾球 (bōli xiāqiú)

广东菜。先将鲜虾仁剖成两半，挂蛋清芡粉浆，慢火煎熟，然后加调料翻炒、勾芡。

Stir-fried shrimp meat. Guangdong dish. Shelled fresh shrimp halved, coated with starched egg white, fried over low heat, stir-fried with seasonings, then starched.

玻璃鲜墨 (bōli xiānmò)

湖南菜。将新鲜墨鱼片用鸡汤烹煮，加入姜汁、芡汁、鸡油。

Braised cuttlefish. Hunan dish. Sliced fresh cuttlefish braised in chicken soup flavored with ginger juice, starched soup, and chicken oil.

玻璃鱿鱼 (bōli yóuyú)

四川菜。将水发鱿鱼片入清汤焖煮，加入胡椒粉、盐、味精、料酒，配焯过的菠菜装盘。

Simmered squid with spinach.

Sichuan dish. Water-soaked sliced squid boiled in clear soup seasoned with ground black pepper, salt, MSG, and cooking wine, then served with scalded spinach.

菠萝炒牛肉 (bōluó chǎo niúròu)

台湾家常菜。先将腌渍过的牛肉片滑熟，然后与菠萝块快炒。

Quick-fried beef and pineapple. Home-style dish in Taiwan. Pickled sliced beef lightly fried, then quick-fried with diced pineapple.

菠萝干贝 (bōluó gānbèi)

南方菜。将蒸熟的干贝与菠萝和胡萝卜拌炒，用煸过的蒜茸和盐调味，勾芡。

Sautéed pineapple and scallops.

Southern dish. Steamed dried scallops sautéed with pineapple and carrots, seasoned with quick-fried minced garlic and salt, then thickened with starch.

菠萝咕噜虾球 (bōluó gūlū xiāqiú)

南方菜。先将虾球炸至金黄，再与菠萝丁翻炒，加盐、味精、料酒，然后勾芡。

Shrimp with pineapple. Southern dish. Deep-fried shrimp balls stir-fried with diced pineapple, seasoned with salt, MSG, and cooking wine, then thickened with starch.

菠萝古老肉 (bōluó gǔlǎoròu)

山东菜。先将肥肉丁挂浆，炸至金黄，再将番茄酱稍炒，加姜汁、白醋、酱油、糖等勾芡，然后放入菠萝和肉丁拌炒。

Sautéed fat pork and pineapple. Shandong dish. Diced fat pork coated with starch paste, deep-fried, sautéed with diced pineapple, then flavored with sauce of quick-fried tomato sauce, ginger juice, white vinegar, soy sauce, sugar, and starch.

菠萝蜜炒牛肉 (bōluómì chǎo niúròu)

广东菜。先将牛肉用酱油、鱼露、胡椒粉、蒜茸等腌渍，再与菠萝蜜、青椒等同炒。

Sautéed beef and jackfruit.

Guangdong dish. Beef pickled with soy sauce, fish gravy, ground black pepper, and minced garlic, then sautéed with jackfruit and green pepper.

菠萝田鸡腿 (bōluó tiánjītuǐ)

山东菜。先将田鸡腌渍,炸熟,再同菠萝炒匀,然后盛入菠萝盅内。

Sautéed frog and pineapple.

Shandong dish. Frogs pickled, fried, and sautéed with pineapple. Served in a halved hollowed pineapple.

菠萝虾球 (bōluó xiāqiú)

山东菜。先将虾仁挂浆,炸至金黄,再与用油烫过的菠萝、蛋黄酱等拌匀。

Prawn meat with pineapple.

Shandong dish. Shelled prawns coated with starch paste, deep-fried until golden, then mixed with quick-fried pineapple and mayonnaise.

菠萝鱼 (bōluóyú)

湖南菜。先将鳜鱼油炸,然后浇上用糖、菠萝、盐熬成的汁。

Fried Chinese perch with sweet sauce. Hunan dish. Chinese perch deep-fried and then marinated in sauce of sugar, pineapple, and salt.

拨鱼子 (bōyúzǐ)

山西家常食品。先将面粉加水调成糊,再用特制三棱筷拨面糊成鱼肚形入锅煮熟,然后浇上荤或素卤菜。

Hand-made noodles with special dressing. Home-style food in Shanxi. Flour and water paste shaped into fish maw with special triangular chopsticks, boiled, then covered with marinated dishes of meat or vegetables.

钵仔焗鱼肠 (bōzǎi jú yúcháng)

广东菜。将鱼肠、油条、粉丝、陈皮等拌匀,盛入瓦钵,加入高汤蒸熟。

Steamed fish intestine. Guangdong dish. Fish intestine, fried dough twists, starch noodles, and dried tangerine peels mixed, put in an earthen container, then steamed in broth.

博饼 (bóchēng)

广东点心。将糯米粉与油和匀,包上特制的甜馅或咸馅,煎黄。

Stuffed sticky rice buns. Guangdong snack. Sticky rice flour mixed with oil, made into buns stuffed with sweet or salty filling, then deep-fried.

博山豆腐箱 (bóshān dòufuxiāng)

山东博山菜。将海米、虾仁、肉末加调料拌和,填入掏空的豆腐块,蒸熟,浇上汤。

Steamed stuffed bean curd. Specialty in Boshan, Shandong. Hollowed cubed bean curd filled with seasoned mixture of dried shrimp meat, fresh shrimp meat, and ground pork, steamed, then washed with broth.

薄荷茶香骨 (bòhe cháxiānggǔ)

南方家常菜。先将猪排骨用薄荷水和糯米香茶浸泡,再用酱油、辣椒水、盐、糖腌渍,然后炸熟,加入葱、姜、辣椒等。

Peppermint and tea scented pork ribs. Home-style dish in southern areas. Pork ribs steeped in peppermint tea and fragrant sticky rice tea, pickled with soy sauce, hot pepper juice, salt, and sugar, deep-fried, then flavored with hot pepper, scallion, and ginger.

簸箕饭 (bòjifàn)

福建传统食品。把米浆倒入平底容器,蒸熟成薄皮,卷入炒熟的猪肉、香菇、莴笋、豆芽等作馅,裹成筒状,装竹盘分食。

Steamed stuffed rice rolls. Traditional Fujian food. Rice paste steamed into thin cakes in flat containers, made into rolls filled with cooked pork, dried mushrooms, stem lettuce, and bean sprouts, then served on bamboo plates.

布袋豆腐 (bùdài dòufu)

山东菜。先将玉脂豆腐块炸熟,掏空,再填入木耳丝、胡萝卜丝、虾仁丝,用韭黄扎口呈袋状,蒸熟,淋上鲍汁。

Steamed stuffed bean curd. Shandong dish. Cubed tender bean curd deep-fried, hollowed, stuffed with shredded wood ears, carrots, and shrimp meat, then tied into small bags with yellow Chinese chive. Steamed, then washed with abalone sauce.

布袋鸡 (bùdàijī)

山东菜。将炒熟的猪肉、竹笋、蘑菇等填进鸡腹,先炸后蒸。

Steamed stuffed chicken. Shandong dish. Chicken stuffed with quick-fried pork, bamboo shoots, and mushrooms, fried, then steamed.

才鱼火锅 (cáiyú huǒguō)

湖北菜。将才鱼片裹蛋浆焯水,然后放在鱼骨上,加盐、姜、葱、色拉油等煮熟。

Dark sleeper in fire pot. Hubei dish. Dark sleeper fillets coated with egg paste, scalded, put on a bed of fish bones, then boiled in fire pot soup seasoned with salt, ginger, scallion, and salad oil.

C

彩烩鸡肝 (cǎihuì jīgān)

河南菜。先将鸡肝烫熟，再与多种颜色的蔬菜拌炒，加盐、料酒等，勾芡。

Sautéed chicken liver with vegetables. Henan dish. Poached chicken liver sautéed with colorful vegetables, seasoned with salt and cooking wine, then thickened with starch.

彩烩鱼丝 (cǎihuì yúsī)

东北菜。将鱼肉丝与冬笋丝、胡萝卜丝、黄瓜丝、香菇丝同炒，加盐、黄酒，勾芡。

Colorful fish and vegetables.

Northeastern dish. Shredded fish stir-fried with shredded winter bamboo shoots, carrots, cucumbers, and dried mushrooms, seasoned with salt and yellow wine, then thickened with starch.

彩椒熘肥肠 (cǎijiāo liū féicháng)

东北菜。将熟猪大肠加青椒、红椒及盐、黄酒熘炒，勾芡。

Hog large intestine with pepper. Northeastern dish. Hog large intestine stir-fried with green and red pepper, salt, and yellow wine, then thickened with starch.

彩椒山药 (cǎijiāo shānyào)

东北家常菜。将山药泥加调料填入掏空的青椒，蒸熟装盘，浇卤汁。

Steamed Chinese yam in pepper cup. Northeastern home-style dish. Green pepper cups stuffed with mashed and seasoned Chinese yam, steamed, then washed with special sauce.

彩色里脊肉 (cǎisè lǐjiròu)

陕西菜。先将里脊肉丝腌渍，滑散，再与微炸过的土豆丝、香菇、青椒、胡萝卜翻炒，烹入料酒、盐、味精，勾芡。

Sautéed pork with colorful vegetables. Shaanxi dish. Pickled shredded pork tenderloin lightly fried, sautéed with lightly fried shredded potatoes, dried mush-rooms, green pepper, and carrots, then seasoned with cooking wine, salt, MSG, and starched soup.

彩色肉丝 (cǎisè ròusī)

福建菜。先将猪里脊肉丝用盐、黄酒、鸡蛋清及淀粉腌渍，然后与辣椒丝滑熟，勾芡。

Sautéed pork with pepper. Fujian dish. Shredded pork tenderloin pickled with salt, yellow wine, egg white, and starch, sautéed with shredded pepper over low heat, then thickened with starch.

彩丝金花 (cǎisī jīnhuā)

山西家常菜。将炸粉丝、炒金菇、汆银芽装盘,用炸珍珠叶、甘笋花和芫荽点缀。

Fried starch noodles with mushrooms. Home-style dish in Shanxi. Fried starch noodles, sautéed golden mushrooms, and scalded mung bean sprouts served on the same platter garnished with fried pearl leaves, carrot cuts, and coriander.

彩丝龙虾 (cǎisī lóngxiā)

广东菜。先将龙虾汆过,去壳,再裹上淀粉糊油炸,然后与红椒、芹菜、肥肉及调料合炒。

Stir-fried lobster with vegetables. Guangdong dish. Scalded lobster shelled, coated with starch paste, deep-fried, then stir-fried with red pepper, celery, fat pork, and seasonings.

彩塘滑豆腐 (cǎitáng huá dòufu)

湖南菜。将豆腐、鲜贝、草虾仁、香菇、胡萝卜、芦笋等入高汤炖煮,加料酒、盐,勾芡。

Stewed bean curd with vegetables. Hunan dish. Bean curd stewed with fresh scallops, shrimp meat, dried mushrooms, carrots, and asparagus in broth, flavored with cooking wine and salt, then thickened with starch.

彩圆缤纷 (cǎiyuán bīnfēn)

台湾菜。将糯米粉分别用菠菜汁、胡萝卜汁、水和面,做成绿色、红色、白色汤圆,煮熟,放入红豆山药汤中。

Colored sticky rice *tangyuan*. Taiwan dish. White, green, and red *tangyuan* made from sticky rice flour with spinach juice, carrot juice, and water, respectively. Boiled, then served in soup of red bean and Chinese yam.

彩云鱼肚 (cǎiyún yúdǔ)

山东菜。先往鱼肚上铺面粉、蛋清、鸡肉馅、火腿、冬菇、黄瓜皮,然后蒸熟,浇芡汁。

Steamed fish maw. Shandong dish. Fish maw layered with flour, ground chicken, egg white, starch, ham, winter mushrooms, and cucumber peels, steamed, then marinated in starched soup.

菜胆灵芝鲍 (càidǎn língzhī bào)

广东菜。先将鲍鱼、母鸡、排骨、火腿等煨熟,再与灵芝同蒸,配焯水的菜心装盘,浇勾芡的上汤。

Abalone with lucid ganoderma fungus. Guangdong dish. Abalone stewed with hen, pork spareribs, and ham, steamed with lucid ganoderma fungus, garnished with baby cabbage, then washed with starched broth.

菜胆奶油鸡 (càidǎn nǎiyóu jī)

广东菜。先将整鸡腌渍,鸡腹填入葱、姜,蒸熟,然后配灼熟的菜胆装盘,浇上用鲜奶、鸡汤、生粉、盐做成的卤汁。

Steamed chicken with baby cabbage. Guangdong dish. Pickled chicken

stuffed with scallion and ginger, steamed, garnished with boiled baby Chinese cabbage on a platter, then washed with sauce of fresh milk, chicken soup, starch, and salt.

菜豆腐 (càidòufu)

西南家常菜。将水发黄豆磨成浆，煮开后放入小青菜，点少量酸卤，煮成菜豆腐。佐各式辣酱食用。

Jellied vegetable bean curd. Home-style dish in southwestern areas. Soaked soybean ground into paste and boiled. Baby greens and a little sour bittern added, then made into jellied vegetable bean curd. Served with a variety of spicy sauces.

菜饭 (càifàn)

上海传统家常食品。先将青菜略炒，黄豆炖烂，咸猪肉丁煮熟，然后放入生米，加水和盐煮熟。

Rice cooked with vegetable and meat. Traditional home-style food in Shanghai. Rice cooked with quick-fried vegetables, stewed soybean, prepared diced salty pork, and salt.

菜泡饭 (càipàofàn)

东南沿海家常食品。将炒青菜与大米饭加水烩煮。

Boiled rice and vegetables. Home-style food in southeastern coastal areas. Cooked rice boiled with sautéed vegetables.

菜脯拌豆腐 (càipú bàn dòufu)

南方菜。将碎萝卜干加盐、葱、辣椒炒熟，浇在蒸熟的豆腐丁上。

Bean curd with cabbage. Southern dish. Steamed diced bean curd covered with sautéed chopped preserved turnips, scallion, and hot pepper.

菜脯蛋 (càipúdàn)

台湾菜。先将切碎的萝卜干加葱花炒香，然后与蛋浆调匀，煎熟。

Fried preserved turnips and eggs. Taiwan dish. Chopped preserved turnips lightly fried with scallion, mixed with egg paste, then fried.

菜脯鲈鱼 (càipú lúyú)

上海菜。将鲈鱼肉腌渍，煎至金黄，斩块，配氽过的小唐菜装盘。

Fried perch. Shanghai dish. Perch pickled, fried until golden, then chopped. Served with scalded baby Shanghai greens.

菜干扣肉 (càigān kòuròu)

福建菜。将菜干和炸过的瘦肉加蒜、黄酒、酱油、糖等拌匀，蒸熟。

Steamed pork with preserved vegetable. Fujian dish. Dried vegetable and fried lean pork mixed with garlic, yellow wine, soy sauce, and sugar, then steamed.

菜椒牛柳 (càijiāo niúliǔ)

南方家常菜。先将牛肉片腌渍，然后与大蒜、菜椒、酱油等翻炒。

Stir-fried beef with sweet pepper. Home-style dish in southern areas. Pickled sliced beef stir-fried with garlic, sweet pepper, and soy sauce.

菜苔炒腊肉 (càitái chǎo làròu)

湖北菜。先将腊肉片氽过，然后与菜苔及调料同炒。

Stir-fried rape blossoms with smoked

cured pork. Hubei dish. Sliced smoked cured pork quick-boiled, then stir-fried with rape blossoms and flavorings.

菜头萝卜炖牛肉 (càitóu luóbo dùn niúròu)

四川菜。将菜头、萝卜、牛肉等炖煮,用盐、黄酒、葱、姜调味。

Stewed beef and vegetables. Sichuan dish. Mustard tubers, turnips, and beef stewed, then flavored with salt, yellow wine, scallion, and ginger.

菜头鱼汤 (càitóu yútāng)

四川家常菜。先将炸透的鲫鱼放入加调料的汤中煮沸,然后加入菜头、香菇、豆腐。

Crucian and vegetable soup. Home-style dish in Sichuan. Deep-fried crucian boiled in seasoned soup with mustard tubers, dried mushrooms, and bean curd.

菜茎斑球 (càiyuǎn bānqiú)

广东菜。将菜茎与鱼丸、笋片翻炒,加蒜茸、姜、黄酒、盐、高汤。

Sautéed vegetable and fish balls. Guangdong dish. Vegetable stalks, fish balls, and sliced bamboo shoots sautéed with mashed garlic, ginger, yellow wine, salt, and broth.

菜远炒鸡球 (càiyuǎn chǎo jīqiú)

东北家常菜。菜远即青菜心。先将鸡块腌渍,油炸,然后与冬菇及调料翻炒,配焯过的菜心装盘。

Chicken with baby greens and mushrooms. Northeastern home-style dish. Pick-led chopped chicken fried, then stir-fried with winter mushrooms and condiments. Served with scalded baby greens.

菜远炒牛肉 (càiyuǎn chǎo niúròu)

东北家常菜。先将牛肉片腌渍,再过油,然后与汆过的青菜心同炒。

Beef stir-fried with baby greens. Northeastern home-style dish. Pick-led sliced beef lightly fried, then stir-fried with scalded baby greens.

菜远炒排骨 (càiyuǎn chǎo páigǔ)

东北家常菜。先将排骨块腌渍,油炸,然后与汆过的青菜心及调料同炒。

Pork spareribs with baby greens. Northeastern home-style dish. Pick-led pork spareribs deep-fried, then stir-fried with scalded baby greens and seasonings.

菜远炒虾球 (càiyuǎn chǎo xiāqiú)

东北家常菜。先将虾仁腌渍,油炸,然后与汆过的青菜心及调料同炒。

Shrimp meat with baby greens. Northeastern home-style dish. Fried pickled shrimp meat stir-fried with scalded baby greens and seasonings.

蚕豆炒韭菜 (cándòu chǎo jiǔcài)

安徽、河南家常菜。将蚕豆和韭菜拌炒,用盐、味精调味。

Sautéed broad beans with Chinese chive. Home-style dish in Anhui and Henan. Fresh broad beans sautéed with Chinese chive, then seasoned with salt and MSG.

蚕豆炒西芹 (cándòu chǎo xīqín)

四川菜。先将蚕豆氽过,再与西芹、辣椒、葱等翻炒,加盐和味精。

Sautéed broad beans and celery. Sichuan dish. Scalded broad beans sautéed with American celery, hot pepper, and scallion, then flavored with salt and MSG.

蚕豆炒虾仁 (cándòu chǎo xiārén)

江苏家常菜。将挂芡的鲜虾仁与氽过的嫩蚕豆快炒,烹入盐、姜、葱、黄酒。

Broad beans with shrimp meat. Home-style dish in Jiangsu. Starched fresh shrimp meat quick-fried with scalded fresh broad beans, then seasoned with salt, ginger, scallion, and yellow wine.

蚕豆烩鲜贝 (cándòu huì xiānbèi)

东南方菜。先将蚕豆煮熟,再与鲜贝、虾仁、香菇及调料等烩煮。

Braised scallops and broad beans. Southeastern dish. Boiled broad beans braised with scallops, shrimp meat, dried mushrooms, and flavorings.

蚕豆素虾仁 (cándòu sùxiārén)

东北家常菜。先将鸡蛋清、水、糯米粉、盐调和,做成虾仁状,煮熟,然后与熟蚕豆及调料同炒。

Soy shrimp with broad beans. Northeastern home-style dish. Egg white, water, and sticky rice flour mixed, made into shrimp-shape pieces, boiled, then sautéed with prepared broad beans and seasonings.

蚕茧葡萄虾 (cánjiǎn pútáoxiā)

福建菜。葡萄裹上虾泥,用威化纸包上炸熟。

Fried shrimp with grapes. Fujian dish. Grapes coated with minced shrimp meat, rolled in wafer, then fried.

曹雪芹家酒 (cáoxuěqín jiājiǔ)

产于河北唐山,以高粱、大麦为主要原料,酒精含量为35-46度,属浓香型酒。

Caoxueqin jiajiu liquor produced in Tangshan, Hebei. White spirit made with sorghum and barley. Contains 35-46% alcohol, and has a thick aroma.

漕溜鱼片 (cáoliū yúpiàn)

山东菜。先将石斑鱼片腌渍,过油,再与葱段、木耳等拌炒,勾芡。

Sautéed grouper. Shandong dish. Pickled and lightly fried grouper fillets stir-fried with scallion sections and wood ears, then thickened with starch.

草菇菜心 (cǎogū càixīn)

东北家常凉菜。先将草菇和油菜心焯水,然后加盐、味精等拌匀。

Mushrooms with rape greens. Northeastern home-style cold dish. Meadow mushrooms and baby rape greens quick-boiled and flavored with salt and MSG.

草菇氽虾丸 (cǎogū cuān xiāwán)

广东菜。将虾丸、鱼丸、草菇等入鸡汤煮熟,加调料。

Fish and shrimp balls in soup.

Guangdong dish. Shrimp balls, fish balls, and meadow mushrooms boiled in chicken soup, then seasoned.

草菇烩菜苔 (cǎogū huì càitái)

四川菜。将草菇、菜苔、虾仁、香菇等加调料烩煮。

Sautéed mushrooms and cabbage blossoms. Sichuan dish. Meadow mushrooms sautéed with cabbage blossoms, shrimp· meat, dried mushrooms, and seasonings.

草菇鸡团 (cǎogū jītuán)

广东菜。将草菇氽过,撒上粟粉,填入鸡肉茸成团状,入上汤煮熟,配熟芥菜胆装盘。

Steamed mushrooms with chicken. Guangdong dish. Scalded meadow mushrooms coated with corn starch, filled with minced chicken, arranged into balls, then boiled in broth. Served with prepared baby Chinese broccoli.

草菇烧笋 (cǎogū shāo sǔn)

广东菜。先将冬笋炸熟,再与草菇、猪肚等入上汤焖煮,勾芡,用火腿片点缀。

Braised mushrooms and bamboo shoots. Guangdong dish. Winter bamboo shoots deep-fried, braised with meadow mushrooms and hog tripe in broth, thickened with starch, then garnished with shredded ham.

草菇丝瓜 (cǎogū sīguā)

四川菜。将草菇、丝瓜、红椒、葱等焖烧,用麻油、盐、味精等调味。

Simmered mushrooms and luffa. Sichuan dish. Meadow mushrooms simmered with luffa, red pepper, and scallion, then seasoned with sesame oil, salt, and MSG.

草菇田螺 (cǎogū tiánluó)

四川菜。先将田螺煮熟,再与草菇、红椒、洋葱及调料焖煮。

Simmered mushrooms and field snails. Sichuan dish. Boiled field snails simmered with meadow mushrooms, red pepper, onion, and seasonings.

草原八珍 (cǎoyuán bāzhēn)

内蒙名菜。先将驼掌、驴冲、鹿鞭、驼峰、猴头蘑、牛鞭蒸熟,然后扣上发菜饼,浇麻油。

Steamed prairie eight treasures. Famous Inner Mongolian dish. Camel palm, donkey tail, deer penis, camel hump, monkey-head mushrooms, and beef penis steamed together, covered with black moss cakes, then washed with sesame oil.

叉烤鳜鱼 (chākǎo guìyú)

江苏菜。先将炒过的肉丝、京冬菜、笋丝等填入鳜鱼腹,然后用猪网油把鱼包好,烤熟。

Broiled Chinese perch. Jiangsu dish. Chinese perch stuffed with cooked shredded pork, preserved baby Chinese cabbage, and shredded bamboo shoots, wrapped in web

lard, then broiled.

叉烤酥方 (chākǎo sūfāng)

江苏菜。将带肋骨的方肉涂上叉烧酱,别在铁叉上烤熟。

Grilled streaky pork. Jiangsu dish. Cubed streaky pork with ribs coated with barbecue sauce, then grilled on forks.

叉烧包 (chāshāobāo)

广东点心。用发面团包入叉烧、葱、姜等制成的馅,蒸熟。

Buns stuffed with broiled pork. Guangdong snack. Fermented dough made into buns filled with mixture of barbecued pork, scallion, and ginger, then steamed.

叉烧酱汁排骨 (chāshāo jiàngzhī páigǔ)

安徽菜。先将排骨裹上用叉烧酱、蒜粉、生粉、蚝油、香草等调制的浓浆,然后用中火焗熟。

Broiled pork ribs with sauce. Anhui dish. Pork ribs coated with thick paste of barbecue sauce, garlic powder, starch, oyster sauce, and vanilla powder, then broiled over medium heat.

叉烧肉 (chāshāoròu)

南方菜。将猪肉用盐、糖、叉烧酱腌渍,再慢火炸至金红,然后加叉烧酱和红曲焖至汁干。

Broiled pork. Southern dish. Pork pickled with salt, sugar, and barbecue sauce, fried until golden brown over low heat, then simmered with barbecue sauce and fermented red yellow wine until reduced.

叉烧鸭 (chāshāoyā)

浙江菜。全鸭肚内填入干荷叶、花椒等香料,外涂叉烧酱,用铁叉叉上,在炭火上烤熟。

Grilled duck. Zhejiang dish. Duck stuffed with fragrant condiments, such as dry lotus leaf and Chinese prickly ash, coated with barbecue sauce, put on an iron fork, then grilled over charcoal.

叉烧羊排 (chāshāo yángpái)

四川菜。用铁叉串上羊排,裹上麻辣调料,放在烤炉上烤熟。

Grilled lamb ribs. Sichuan dish. Lamb ribs strung on iron forks, coated with spicy seasonings, then grilled.

叉烧鱼 (chāshāoyú)

四川菜。先将鲤鱼腌渍,鱼腹内填入煸炒过的榨菜肉丝泡椒馅,再将鱼身裹上猪网油和豆粉蛋清浆,然后用铁叉别上,放在木炭火上烤熟。

Grilled fish. Sichuan dish. Pickled carp stuffed with quick-fried pork, preserved mustard tubers, and pickled pepper, wrapped in web lard, coated with bean starch and egg paste, then grilled on an iron fork over charcoal.

茶梅凉拌莲藕 (chá méi liángbàn liánǒu)

安徽、江苏家常点心。将梅肉与梅汁拌匀,加入汆过的藕片。

Lotus roots with plum sauce. Home-style snack in Anhui and Jiangsu.

Plum meat and plum soup mixed, then combined with quick-boiled sliced lotus roots.

茶树菇炒五花肉 (cháshùgū chǎo wǔhuāròu)

湖南家常菜。先将猪五花肉煎熟，然后与茶树菇、葱头、蒜、剁椒、酱油等翻炒。

Stir-fried mushroom and pork.

Home-style dish in southern areas. Fried streaky pork stir-fried with tea bush mushrooms, onion heads, garlic, pickled pepper, and soy sauce.

茶树菇乡螺鸭 (cháshùgū xiāngluó yā)

湖南菜。用茶树菇与田螺、口水鸭、辣椒及调料翻炒。

Mushrooms with duck and field snails. Hunan dish. Chopped duck stir-fried with tea bush mushrooms, field snails, hot pepper, and other condiments.

茶香骨 (cháxiānggǔ)

浙江菜。先将猪排骨用葱姜及盐腌渍，再用浓茶和黄酒烹煮，然后炸至金黄。

Fried tea-scented pork ribs. Zhejiang dish. Pork ribs pickled with scallion, ginger, and salt, boiled in strong tea and yellow wine soup, then deep-fried.

茶香鸡 (cháxiāngjī)

广东菜。先将鸡腌渍，再放入用茶叶、八角、草果、花椒、丁香等熬成的卤水炆煮，然后用黄酒、茶叶、黄糖焖熏。

Tea-scented chicken. Guang-dong dish. Pickled chicken braised in gravy of tea, aniseed, tsaoko amomun fruit, Chinese prickly ash, and clove over low heat, then smoked with yellow wine, tea leaves, and yellow sugar in a covered wok.

茶香焗乳鸽 (cháxiāng jú rǔgē)

广东菜。用炒香的茶叶、黄糖等焗乳鸽，佐卤汁。

Broiled pigeon with tea leaves. Guangdong dish. Baby pigeons broiled with prepared tea leaves and yellow sugar. Served with special sauce.

茶香虾 (cháxiāngxiā)

各地家常菜。将虾仁加鲜茶蕊、黄酒、白糖、盐翻炒，勾芡，淋猪油。

Sautéed shrimp meat with tea leaves. Home-style dish in many places. Shrimp meat and fresh tea leaves sautéed, flavored with yellow wine, sugar, and salt, thickened with starch, then washed with lard.

茶香腰花 (cháxiāng yāohuā)

湖南菜。将猪腰花与青红椒及调料煸炒，放入少许龙井茶叶。

Hog kidney with green tea. Hunan dish. Hog kidney stir-fried with green and red pepper and other seasonings. Flavored with *longjingcha* tea leaves.

茶香鱼片 (cháxiāng yúpiàn)

安徽、江苏菜。先将鳜鱼片腌渍，放在铁锅里，用茶叶、糖、米、葱等焖熏至熟，然后抹上麻油。

Tea-scented fish fillets. Anhui and Jiangsu dish. Pickled Chinese perch fillets smoked with tea leaves, sugar, rice, and scallion in a covered wok, then brushed with sesame oil.

茶叶蛋 (cháyèdàn)

各地家常菜。先将鸡蛋用清水煮至蛋清凝固,再轻敲击至蛋壳开裂,然后放入五香盐茶汤炆煮。

Eggs boiled in tea. Home-style dish in many places. Eggs soft-boiled, shells cracked, then braised in salted five-spice and tea soup.

茶叶虾 (cháyèxiā)

台湾菜。将河虾加铁观音茶叶和秘制酱料爆炒。

Stir-fried shrimp with tea leaves. Taiwan dish. River shrimp stir-fried with special sauce and *tieguanyin* tea leaves.

茶叶熏鸡 (cháyè xūnjī)

湖南菜。仔鸡与茶叶、饭锅巴、菜叶、麻油、红糖、烧酒等一起蒸熟。

Steamed chicken with tea leaves. Hunan dish. Young chicken steamed with tea leaves, rice crust, vegetable leaves, sesame oil, brown sugar, and Chinese liquor.

拆冻鲫鱼 (chāidòng jìyú)

江苏菜。先将鲫鱼煎至两面呈黄色,再煮熟,然后拆去鱼骨,倒入烹鱼的汤汁,置凉凝冻。

Crucian carp jelly. Jiangsu dish. Crucian carp fried until golden on both sides, boiled, boned, marinated in broth, then chilled into jelly.

拆骨肉炖白菜粉 (chāigǔròu dùn báicàifěn)

东北菜。将熟猪骨肉拆下,与白菜和粉条加调料炖煮。

Pork with cabbage and starch noodles. Northeastern dish. Cooked pork stripped from bones, then stewed with Chinese cabbage, starch noodles, and seasonings.

拆骨掌翅 (chāigǔ zhǎngchì)

江苏菜。将鸡脚与鸡翅加调料焖熟,去骨,切段装盘。

Chicken feet and wings. Jiangsu dish. Chicken feet and wings braised with seasonings, boned, then chopped in sections.

拆烩鲢鱼头 (chāihuì liányútóu)

江苏菜。先将鲢鱼头煮熟,去骨,再与笋片、香菇、鸡片、肫片、鱼片等入鸡汤焖煮,用糖、盐、姜、葱、料酒调味,勾芡,然后加醋、胡椒粉、猪油。

Braised carp head. Jiangsu dish. Silver carp head boiled, boned, braised in chicken soup with sliced bamboo shoots, mushrooms, chicken, chicken gizzard, and fish fillets. Seasoned with sugar, salt, ginger, scallion, cooking wine, starch, vinegar, ground black pepper, and lard.

长白山葡萄酒 (chángbáishān pútáojiǔ)

葡萄酒,产于吉林。以当地野生葡萄为主要原料酿造,酒精含量为7-12度,属甜型葡萄酒。

Changbaishan wine produced in

Jilin. Wine made with local wild grapes. Contains 7-12% alcohol, and has a sweet aroma.

长乐烧酒 (chánglèshāojiǔ)

白酒，产于广东五华。以糙米为原料，用药曲酿造。酒精含量在45-58度之间，属米香型酒。

Changleshaojiu, liquor produced in Wuhua, Guangdong. White spirit made with brown rice and herb yeast. Contains 45-58% alcohol, and has a rice aroma.

长寿果蒸鸡 (chángshòuguǒ zhēng jī)

四川菜。先将全鸡氽过，腌渍，蒸熟，再去骨切块，入鸡汤蒸透，然后浇上用花生、胡萝卜、青笋、猪油、味精、芡粉等做成的卤汁。

Steamed chicken with peanut sauce. Sichuan dish. Scalded and pickled whole chicken steamed, boned, chopped, re-steamed in chicken soup, then marinated in sauce of peanuts, carrots, stem lettuce, lard, MSG, and starch.

长寿鱼 (chángshòuyú)

河南菜。先将黄河鲤鱼腌渍，油煎，然后加入枸杞子烹煮。

Simmered carp with Chinese wolfberry. Henan dish. Pickled and fried Yellow River carp simmered with Chinese wolfberry.

肠粉 (chángfěn)

广东点心。将米粉加水揉成面团，擀成粉皮，做成卷状蒸熟，配酱油、麻油等调料食用。

Rice flour rolls with sauce. Guangdong snack. Rice flour mixed with water, made into sheets, rolled, then steamed. Served with soy sauce and sesame oil.

朝鲜生拌鱼 (cháoxiǎn shēngbàn yú)

朝鲜族凉菜。将鲤鱼肉切丝，加麻油、白糖、辣椒酱拌匀。

Pickled raw carp. Korean ethnic minority cold dish. Shredded raw carp pickled with sesame oil, sugar, and chili sauce.

潮汕功夫茶 (cháoshàn gōngfuchá)

源于潮汕，现流行于闽、粤、台和海外华埠的一种品茶艺术。采用乌龙茶叶、紫砂茶具，讲究沏泡技艺、巡茶礼仪、品茶评茶，是一种独特的茶艺。

Tea ceremony. It is an art of tea appreciation originated in Chaozhou and Shantou, Guangdong, now widely performed in Fujian, Guangdong, Taiwan, and many overseas Chinese communities. It uses quality Oolong tea and delicate red stoneware tea set, paying special attention to the procedures of tea making, etiquette of tea serving, and tea appraisal.

潮汕蚝烙 (cháoshànháolào)

广东菜。先把鲜蚝肉与葱、味精、鱼露、芡粉水调和，再用猪油煎熟，然后把鸭蛋浆浇在上面，撒上辣椒酱，两面煎脆。

Fried oyster and eggs. Guangdong dish. Fresh oyster meat mixed with

scallion, MSG, fish gravy, and starched water, fried with lard, washed with duck egg paste and chili pepper sauce, then fried until both sides become crisp.

潮汕鱼丸 (cháoshànyúwán)

广东菜。先将鱼丸蒸熟,再与冬菇、笋花、紫菜入上汤烹煮,用鱼露和胡椒粉调味。

Fish ball soup. Guangdong dish. Steamed fish balls boiled in broth with winter mushrooms, bamboo shoots, and laver, then flavored with fish gravy and ground black pepper.

潮式拌海蜇 (cháoshì bàn hǎizhé)

广东潮州家常菜。将浸泡过的海蜇控干水分,加入芹菜、红尖椒、蚝油拌匀,撒上熟芝麻。

Jellyfish with flavorings. Home-style dish in Chaozhou, Guangdong. Steeped jellyfish drained, mixed with celery, red pepper, and oyster sauce, then topped with prepared sesame.

潮式手撕大龙虾 (cháoshì shǒusī dàlóngxiā)

潮州菜。将龙虾用上汤煮熟, 撕成条状,伴生菜、桔油。

Lobster with tangerine oil. Specialty in Chaozhou, Guangdong. Lobster boiled in broth, torn into strips, then served with shredded romaine lettuce and tangerine oil.

潮州冻花蟹 (cháozhōudònghuāxiè)

广东菜。花蟹加姜末蒸熟,配香菜、香醋。

Steamed crab with vinegar. Guangdong dish. Spotted crab steamed with mashed ginger. Served with coriander and vinegar.

潮州冻肉 (cháozhōudòngròu)

广东菜。将猪肉、猪蹄、猪皮等煮熟,冷冻后切块,佐鱼露、香菜。

Pork jelly with fish sauce. Guangdong dish. Pork, pork trotters, and hog skins boiled, chilled into jelly, then cut into pieces. Served with fish sauce and coriander.

潮州烧雁鹅 (cháozhōu shāo yàné)

广东菜。先将整鹅卤熟,切块,再裹上生粉,炸至皮脆,然后浇胡椒油,撒芫荽叶,配腌萝卜、梅膏酱。

Marinated goose with sauce. Guangdong dish. Marinated and chopped goose coated with starch, fried until crisp, washed with pepper oil, then decorated with coriander leaves. Served with pickled turnips and plum sauce.

潮州生淋鱼 (cháozhōu shēnglínyú)

广东潮州家常菜。将鲩鱼用开水焯熟,淋上烧滚的猪油,佐咸酱或酸甜酱。

Grass carp with sauce. Home-style dish in Chaozhou, Guangdong. Scalded grass carp washed with boiling lard. Served with special salty sauce or sweet-sour sauce.

潮州油果 (cháozhōuyóuguǒ)

广东潮州点心。将糯米粉加水做成面皮,包入番薯茸、花生、芝麻,炸至金黄。

Fried stuffed sticky rice balls. Snack in Chaozhou, Guangdong. Sticky rice flour and water made into buns, stuffed with mixture of mashed sweet potatoes, peanuts, and sesame, then deep-fried until golden.

炒鹌鹑松 (chǎoānchúnsōng)

广东菜。将鹌鹑肉末和猪瘦肉末用蛋黄和生粉拌匀,与冬菇粒和笋粒拌炒,加盐、黄酒、味精、姜末。

Quail and pork floss. Guangdong dish. Ground quail meat and lean pork mixed with egg yolks and starch, then stir-fried with diced winter mushrooms and bamboo shoots. Seasoned with salt, yellow wine, MSG, and minced ginger.

炒白鸽松 (chǎobáigēsōng)

广东菜。先将白鸽肉末和猪瘦肉末与盐、蛋清、黄酒等调和,用麻油炒至半熟,然后与碎韭黄和火腿末同炒,配熟薄饼皮、生菜叶、香醋。

White pigeon floss. Guangdong dish. Ground white pigeon meat and lean pork mixed with salt, egg white, and yellow wine, half-fried with sesame oil, then sautéed with chopped yellow Chinese chive and diced ham. Served with prepared thin flour cakes, romaine lettuce, and vinegar.

炒白果 (chǎobáiguǒ)

福建菜。将氽过的白果与虾仁、香芹及调料拌炒、勾芡。

Sautéed ginkgoes with celery. Fujian dish. Quick-boiled ginkgoes sautéed with shrimp, celery, and condiments, then thickened with starch.

炒碧桃里脊 (chǎo bìtáo lǐji)

山东菜。先将里脊肉片挂蛋清,卷入桃仁,然后入油滑透,加卤汁翻炒。

Sautéed pork with walnuts. Shandong dish. Pork tenderloin fillets coated with egg white, made into rolls filled with walnuts, lightly fried, then sautéed with seasoned sauce.

炒豆芽 (chǎodòuyá)

东北家常菜。将黄豆芽和苹果丝加盐、醋、糖等翻炒,淋猪油。

Sautéed soy bean sprouts. Northeastern home-style dish. Soybean sprouts and shredded apples sautéed with salt, vinegar, and sugar, then washed with lard.

炒芙蓉蟹黄 (chǎo fúróng xièhuáng)

南方菜。将熟蟹黄片放入用蛋清、芡粉、料酒、高汤等调成的卤汁,低温加热,待蟹黄呈芙蓉片状浮起,装盘并撒上火腿末。

Sautéed crab yolks. Southern dish. Prepared crab roe pieces put into gravy of egg white, starch, cooking wine, and broth, boiled over low heat until yolk pieces float, then topped with ham crumbs.

炒腐皮丝 (chǎofǔpísī)

四川菜。先将豆腐皮用水泡发,切丝,然后与黄豆芽、红椒丝等爆炒,用盐、酱油、花椒等调味。

Bean curd skins and soybean

sprouts. Sichuan dish. Shredded soaked bean curd skins sautéed with soybean sprouts and shredded red pepper, then seasoned with salt, soy sauce, and Chinese prickly ash.

炒鸽松 (chǎogēsōng)

广东潮州菜。先将乳鸽茸和猪瘦肉末加芡粉水调和,用文火油炸,然后与马蹄末、韭黄末、香菇末、火腿末等拌炒,配薄饼和生菜叶。

Pigeon floss with vegetables. Specialty in Chaozhou, Guangdong. Minced young pigeon meat and lean pork mixed with starched water, fried over low heat, then stir-fried with chopped water chestnuts, yellow Chinese chive, dried mushrooms, and ham. Served with small thin pancakes and lettuce leaves.

炒桂花翅 (chǎoguìhuāchì)

浙江菜。先将水发鱼翅、姜、葱、盐、老鸡、猪脚一起炖煮,再与香菇丝、鸡蛋、肉末、葱、辣椒同炒,撒火腿末,配香醋、薄饼皮、生菜叶。

Stewed shark fin sautéed with mushrooms. Zhejiang dish. Soaked shark fin stewed with ginger, scallion, mature chicken, and hog trotters. Sautéed with shredded dried mushrooms, eggs, minced pork, scallion, and pepper, then sprinkled with minced ham. Served with vinegar, thin flour cakes, and romaine lettuce.

炒桂花鱼肚 (chǎo guìhuā yúdǔ)

广东菜。将鱼肚煮熟切碎,与鸡蛋烹炒,用鸡粉、姜片、香菜调味。

Stir-fried fish maw. Guangdong dish. Diced boiled fish maw stir-fried with egg paste, then flavored with chicken powder, ginger, and coriander.

炒桂竹笋 (chǎoguìzhúsǔn)

台湾菜。将桂竹笋丝同肉丝、辣椒加调料爆炒。

Bamboo shoots with pork and pepper. Taiwan dish. Shredded bamboo shoots stir-fried with shredded pork, hot pepper, and seasonings.

炒海瓜子 (chǎohǎiguāzi)

福建菜。先将葱、姜煸出香味,再与海瓜子和青辣椒翻炒。

Stir-fried clam. Fujian dish. Tiny clam stir-fried with green pepper and quick-fried scallion and ginger.

炒合菜 (chǎohécài)

山东菜。先将猪五花肉丝加甜酱煸炒,再与芹菜、粉皮、豆腐皮、炸豆腐、胡萝卜等翻炒。

Sautéed pork with vegetables. Shandong dish. Shredded streaky pork lightly fried with sweet fermented flour sauce, then sautéed with celery, rice jelly, bean curd skins, fried bean curd, and carrots.

炒鲎片 (chǎohòupiàn)

福建菜。先将炸过的鲎肉片与香菇拌炒、勾芡,然后倒在过油的冬笋片上。

Horseshoe crab with mushrooms. Fujian dish. Deep-fried horseshoe crab meat stir-fried with dried mushrooms, thickened with starch, then served over lightly fried winter bamboo shoots.

炒胡饽子 (chǎohúbōzi)

宁夏菜。将烙饼条入羊肉汤焖煮,加盐、葱、辣椒。

Simmered mutton and pancakes. Ningxia dish. Flour pancake strips simmered in mutton soup with salt, leek, and chili pepper.

炒胡萝卜酱 (chǎo húluóbojiàng)

浙江菜。先把胡萝卜和豆腐干炸透,再与肉丁煸炒,用黄酱、姜末、海米、料酒、酱油调味。

Stir-fried carrots with sauce.

Zhejiang dish. Deep-fried carrots and dried bean curd stir-fried with diced pork, then flavored with fermented soybean and flour sauce, minced ginger, dried shrimp meat, cooking wine, and soy sauce.

炒鸡米 (chǎojīmǐ)

山东菜。先将鸡脯肉末腌渍,再入油锅轻炒,然后加入汆过的冬笋、荸荠、菠菜梗翻炒。

Sautéed chicken with vegetables. Shandong dish. Pickled minced chicken breast lightly fried and sautéed with scalded winter bamboo shoots, water chestnuts, and spinach stalks.

炒姜丝肉 (chǎojiāngsīròu)

东北菜。将猪肉丝和姜丝加葱、盐、花椒水等速炒,勾芡,淋猪油。

Stir-fried pork with ginger.

Northeastern dish. Shredded pork and ginger quick-fried with scallion, salt, and Chinese prickly ash soup, thickened with starch, then washed with lard.

炒茭白绿蚕豆 (chǎo jiāobái lǜcándòu)

江浙皖家常菜。先将茭白片焯水,再与绿蚕豆、红辣椒片、胡椒粉、盐等同炒。

Sautéed wild rice stems and broad beans. Home-style dish in Jiangsu, Zhejiang, and Shanghai. Scalded sliced wild rice stems sautéed with fresh broad beans, red pepper, ground black pepper, and salt.

炒卷心菜 (chǎojuǎnxīncài)

各地家常菜。将卷心菜丝加盐、酱油、醋等爆炒。

Sautéed cabbage. Home-style dish in many places. Shredded cabbage sautéed and seasoned with salt, soy sauce, and vinegar.

炒腊野鸭条 (chǎolàyěyātiáo)

湖南洞庭湖风味菜。将腊制野鸭切条,与水芹爆炒。

Stir-fried smoked duck. Specialty in Dongting Lake area, Hunan. Smoked cured wild duck strips quick-fried with cress.

炒鲈鱼片 (chǎolúyúpiàn)

各地家常菜。先将鸡腿菇片和菜心炒熟,再将鲈鱼片加调料炒熟,勾芡,倒在鸡腿菇和菜心上。

Stir-fried perch. Home-style dish in many places. Sliced perch stir-fried, thickened with starch, then served

over quick-fried drumstick mushrooms and baby greens.

炒麦穗花鱿 (chǎo màisuì huāyóu)

广东菜。鱿鱼、笋、香菇用猪油、红辣椒、鱼露等烹炒。

Sautéed squid with bamboo shoots and mushrooms. Guangdong dish. Squid, bamboo shoots, and dried mushrooms sautéed with lard, red pepper, and fish sauce.

炒毛蟹 (chǎomáoxiè)

江苏菜。先将河蟹油煎,然后加入糖、酱油、毛豆等烩煮。

Simmered river crab. Jiangsu dish. River crab fried, then simmered with sugar, soy sauce, and green soybean.

炒米 (chǎomǐ)

内蒙食品。糜子经蒸、炒、碾等工序做成,可拌黄油、奶皮、白糖或奶茶食用。

Roasted millet. Inner Mongolian food. Millet steamed, roasted, and ground. Served with butter, milk skin, sugar, or milk tea.

炒面 1 (chǎomiàn)

各地传统食品。将糯米、大米、大麦、小米、玉米、高粱等用文火炒熟,磨成粉,用糖水、盐水或醪糟汤调和。

Roasted grain powder. Traditional food in many places. Sticky rice, round grain rice, millet, barley, corn, or sorghum roasted over low heat, ground into powder, then mixed with sugared or salted water, or sweet fermented sticky rice soup.

炒面 2 (chǎomiàn)

各地传统食品。先将面条煮熟,过凉水,然后与肉丝、青红椒丝、菜心及调料翻炒。

Stir-fried noodles with vegetables. Traditional food in many places. Boiled noodles quickly cooled in cold water, then stir-fried with shredded pork, red and green pepper, baby cabbage, and seasonings.

炒墨鱼花 (chǎomòyúhuā)

各地家常菜。先将墨鱼过油,再将煸过的葱蒜与笋片入清水烧开,加盐、黄酒,然后将墨鱼入汤烩煮,勾薄芡。

Braised cuttlefish with bamboo shoots. Home-style dish in many places. Lightly fried cuttlefish braised in soup of sliced bamboo shoots, quick-fried scallion and garlic, salt, and yellow wine, then lightly starched.

炒木樨肉 (chǎomùxīròu)

各地家常菜。将炒肉丝与炒鸡蛋、黄花菜、菠菜、木耳及调料翻炒。

Sautéed pork, eggs, and vegetables. Home-style dish in many places. Shredded pork stir-fried with fried eggs, day-lily buds, spinach, wood ears, and seasonings.

炒年糕 (chǎoniángāo)

南方点心。将年糕片与轻炒过的肉丝、虾仁、菜心等翻炒,烹入酱油、盐、糖。

Stir-fried rice cakes. Southern snack. Sliced rice cakes stir-fried with quick-fried shredded pork, shrimp meat, and baby vegetable, then flavored with soy sauce, salt, and sugar.

炒青菜 (chǎoqīngcài)

各地家常菜。青菜旺火快炒,加盐、葱、蒜。

Sautéed green vegetable. Home-style dish in many places. Green vegetable quickly sautéed over high heat and seasoned with salt, scallion, and garlic.

炒全蟹 (chǎoquánxiè)

山东菜。先将螃蟹蒸熟,剔出蟹黄和蟹肉,然后将蟹黄和蟹肉与醋、黄酒、酱油、姜汁等翻炒。

Stir-fried crab. Shandong dish. Steamed crab meat and roe stir-fried with vinegar, yellow wine, soy sauce, and ginger juice.

炒肉拉皮 (chǎo ròu lāpí)

东北凉菜。拉皮即凉粉。先将腌渍过的猪肉丝炒熟,再与拉皮、海蜇、黄瓜丝等拌和,加盐、麻油等拌匀。

Pork with bean jelly. Northeastern cold dish. Stir-fried pickled shredded pork mixed with bean jelly strips, jellyfish, and shredded cucumbers, then flavored with salt and sesame oil.

炒肉木耳白菜 (chǎo ròu mùěr báicài)

东北菜。将猪肉片、白菜片、水发木耳翻炒,用盐、酱油、姜、葱等调味。

Pork with wood ears and cabbage. Northeastern dish. Sliced pork and Chinese cabbage stir-fried with wood ears and flavored with salt, soy sauce, scallion, and ginger.

炒三泥 (chǎosānní)

河南菜。先将煮熟的红枣、山药、蚕豆捣成泥,分别用猪油和糖翻炒,然后成品字形装盘,撒青梅和糖桂花。

Date, yam, and broad bean paste. Henan dish. Boiled and mashed red dates, Chinese yam, and broad beans sautéed with lard and sugar separately, arranged into a triangle on a platter, then garnished with candied green plums and sugared osmanthus flower.

炒三丝 (chǎosānsī)

东北菜。将土豆、胡萝卜、青椒切丝,然后加盐、味精、花椒水等炒熟。

Sautéed potato, carrot, and green pepper. Northeastern dish. Shredded potatoes, carrots, and green pepper sautéed with salt, MSG, and Chinese prickly ash water.

炒三鲜 (chǎosānxiān)

浙江菜。将草菇片、茭白丝、猪肉丝加调料翻炒。

Sautéed three freshes. Zhejiang dish. Sliced meadow mushrooms, shredded wild rice stems, and shredded pork sautéed with seasonings.

炒沙茶牛肉 (chǎo shāchá niúròu)

福建菜。先将芥蓝菜叶用高汤、辣酱、沙茶酱翻炒,勾薄芡,然后与过油牛肉片速炒。

Stir-fried beef with mustard greens.

Fujian dish. Mustard greens sautéed with broth, chili sauce, barbecue sauce, and a little starch, then quick-fried with lightly fried sliced beef.

炒山药泥 (chǎoshānyàoní)

浙江菜。先把山药蒸熟,碾成泥,然后加花生油、白糖等炒透,用金糕点缀。

Stir-fried Chinese yam. Zhejiang dish. Steamed and mashed Chinese yam stir-fried with peanut oil and white sugar, then garnished with haw cakes.

炒生鸡丝 (chǎoshēngjīsī)

山东菜。先将挂浆鸡丝入油锅滑透,然后与煸炒过的冬笋丝翻炒,加黄酒、鸡汤。

Sautéed chicken with bamboo shoots. Shandong dish. Starched shredded chicken lightly fried, sautéed with quick-fried shredded winter bamboo shoots, then flavored with yellow wine and chicken soup.

炒双冬 (chǎoshuāngdōng)

上海菜。冬菇和冬笋先煸炒,再加调料焖煮,然后勾芡。

Sautéed mushrooms and bamboo shoots. Shanghai dish. Winter mushrooms and winter bamboo shoots sautéed, simmered with seasonings, then thickened with starch.

炒四宝 (chǎosìbǎo)

浙江菜。先将腌渍过的肚尖、鸡片、鸡肫、虾仁炒散,再将鸭掌、火腿、香菇、笋片入清汤煮沸,加黄酒、盐、味精,勾芡,然后把四宝与汤菜拌炒,淋猪油。

Stir-fried four meat treasures. Zhejiang dish. Hog tripe, sliced chicken, chicken gizzard, and shrimp meat pickled and lightly fried. Duck feet, ham, mushrooms, and sliced bamboo shoots boiled in clear soup, seasoned with yellow wine, salt, and MSG, then thickened with starch. The prepared meat treasures sautéed with the boiled, then washed with lard.

炒素什锦 (chǎosùshíjǐn)

湖南菜。先将鲜蘑、香菇、黄瓜、胡萝卜、西红柿、西蓝花、玉米笋、马蹄焯水,然后加入鸡汤翻炒。

Sautéed vegetables. Hunan dish. Fresh and dried mushrooms, cucumbers, carrots, tomatoes, broccoli, baby corn, and water chestnuts scalded, sautéed, then flavored with chicken soup.

炒笋菇 (chǎosǔngū)

各地家常菜。先将竹笋和蘑菇翻炒,再放入焯水西芹与调料焖熟,勾芡。

Sautéed bamboo shoots and mushrooms. Home-style dish. Bamboo shoots and mushrooms sautéed, simmered with scalded American celery and seasonings, then thickened with starch.

炒铁雀头脯 (chǎo tiěquètóupú)

江苏菜。先将铁雀脯肉和雀头裹鸡蛋清和湿淀粉油炸,再与青豆、栗子片、白果片、青菜心及调料煸炒。

Buntings with vegetables. Jiangsu dish. Yellow-breast bunting breasts and heads coated with egg white and starch, deep-fried, then stir-fried with green soybean, sliced chestnuts, sliced gingkoes, baby greens, and flavorings.

炒乌鱼片 (chǎowūyúpiàn)

广东菜。先将乌鱼片用胡椒粉、麻油、盐、料酒等腌渍,油煎,然后与菜心同炒,勾芡。

Sautéed cuttlefish with baby cabbage. Guangdong dish. Sliced cuttlefish pickled with ground black pepper, sesame oil, salt, and cooking wine. Fried, sautéed with baby Chinese cabbage, then thickened with starch.

炒西施舌 (chǎoxīshīshé)

福建菜。先将芥菜杆、香菇、冬笋片翻炒,再倒入卤汁煮滚并勾芡,然后与汆过的西施舌同炒。

Clam meat with vegetables. Fujian dish. Mustard stalks, mushrooms, and winter bamboo shoots quick-fried, boiled in gravy, starched, then stir-fried with scalded clam.

炒虾仁 (chǎoxiārén)

各地传统菜。先将挂浆虾仁入白油锅滑散,然后与汆过的青豆翻炒,烹入黄酒、盐、清汤。

Sautéed shrimp meat. Traditional dish in many places. Starched shrimp meat lightly fried with lard, sautéed with green soybean, then seasoned with yellow wine, salt, and clear soup.

炒虾仁腰果 (chǎo xiārén yāoguǒ)

福建菜。先将腌过的虾仁速冻,再把虾仁放入热油锅泡熟,然后与炸熟的腰果拌匀。

Shrimp meat and cashew nuts. Fujian dish. Pickled and quick-chilled shrimp meat steeped in heated oil, then mixed with fried cashew nuts.

炒虾蟹 (chǎoxiāxiè)

江苏菜。先将去壳的大虾挂浆,过油,再将蟹黄和蟹肉加盐和黄酒炒熟,然后将大虾、蟹黄、蟹肉速炒,撒上胡椒粉和香菜。

Stir-fried prawns and crab. Jiangsu dish. Shelled prawns pickled with egg white, starch, and salt, then lightly fried. Crab roe and crab meat sautéed with salt and yellow wine. Prepared prawns and crab roe and meat quick-fried, then sprinkled with ground black pepper and coriander.

炒鲜花菇 (chǎoxiānhuāgū)

福建菜。先将葱、蒜煸香,然后倒入花菇及盐翻炒。

Sautéed grained mushrooms. Fujian dish. Lightly fried scallion and garlic sautéed with fresh grain mushrooms and salt.

炒苋菜 (chǎoxiàncài)

各地家常菜。将苋菜加盐、花生油、蒜泥等炒熟。

Sautéed amaranth. Home-style dish

in many places. Amaranth sautéed with salt, peanut oil, and mashed garlic.

炒香螺 (chǎoxiāngluó)

广东菜。先将角螺肉油炸，再和香菇、咸菜、红椒、笋花、葱翻炒，勾芡。

Sautéed horn snail with vegetables. Guangdong dish. Horn snail meat fried, sautéed with dried mushrooms, preserved vegetable, red pepper, bamboo shoots, and scallion, then thickened with starch.

炒蟹肉 (chǎoxièròu)

四川菜。将蟹肉、虾仁、洋葱、红椒及调料爆炒。

Stir-fried crab and shrimp meat. Sichuan dish. Crab meat stir-fried with shrimp meat, onion, red pepper, and seasonings.

炒血糯 (chǎoxuènuò)

江苏菜。先把红糯米、白糯米、青梅干、红绿瓜丝、蜜枣、桂圆肉、熟猪油、甜桂花等一同蒸熟，然后用猪油和糖翻炒。

Sticky rice with candied fruits. Jiangsu dish. Red and white sticky rice steamed with candied green plums, shredded candied red and green gourd, sweet dates, dried longan meat, lard, and sugared osmanthus flower, then stir-fried with lard and sugar.

炒鸭肠 (chǎoyācháng)

广东菜。将鸭肠用开水烫过，与蒜、葱、青椒同炒，加料酒、酱油、醋、辣椒油、麻油。

Stir-fried duck intestine. Guangdong dish. Scalded duck intestine stir-fried with garlic, scallion, and green pepper, then flavored with yellow wine, soy sauce, vinegar, chili oil, and sesame oil.

炒羊肚 (chǎoyángdǔ)

各地家常菜。将羊肚丝、红辣椒、笋丝、韭黄爆炒，加卤汁炒匀。

Stir-fried lamb tripe. Home-style dish in many places. Shredded lamb tripe stir-fried with red pepper, bamboo shoots, and yellow Chinese chive, then drenched with seasoned soup.

炒羊肉片 (chǎoyángròupiàn)

北方家常菜。羊肉片加姜、葱、蒜、豆瓣酱爆炒，放入红椒、香菜，浇芡汁、麻油。

Stir-fried mutton. Home-style dish in northern areas. Sliced mutton stir-fried with ginger, scallion, garlic, and fermented broad bean sauce, combined with red pepper and coriander, then washed with starched soup and sesame oil.

炒羊杂 (chǎoyángzá)

东北菜。羊杂即下水，包括羊心、羊肚、羊肝、羊肺、羊肠、羊腰等。将煮熟的羊杂与尖椒、洋葱及调料爆炒，勾芡。

Stir-fried lamb offal. Northeastern dish. Boiled lamb offal, including heart, tripe, liver, lung, intestine, and kidney, stir-fried with hot pepper, onion, and seasonings, then thickened with starch.

炒莜面 (chǎoyóumiàn)

内蒙食品。先将莜麦面条煮熟,再与蔬菜翻炒,然后加入羊肉汤。

Oat noodles and vegetables. Inner Mongolian food. Oat noodles boiled, stir-fried with vegetables, then marinated in mutton soup.

炒孜然鸡心 (chǎo zīrán jīxīn)

东北菜。先将鸡心切开,用蚝油、盐、孜然腌制,然后加葱姜爆炒。

Chicken heart with cumin. Northeastern dish. Chicken heart cut into halves, pickled with oyster sauce, salt, and cumin, then stir-fried with ginger and scallion.

车前子油焖虾 (chēqiánzǐ yóumènxiā)

广东菜。先将虾炸熟,再用高汤焖煮,然后同车前子翻炒。

Sautéed shrimp with herb.

Guangdong dish. Shrimp fried, simmered in broth, then sautéed with semen plantago seeds.

陈醋果仁菠菜 (chéncù guǒrén bōcài)

东北凉菜。先将菠菜焯水,然后将熟花生仁砸碎,加陈醋、蒜等拌匀。

Spinach with peanuts. Northeastern cold dish. Quick-boiled spinach mixed with mashed prepared peanuts, vinegar, and garlic.

陈醋全家福 (chéncù quánjiāfú)

山西菜。鱿鱼和海蜇丝焯水,放入花生米、松花蛋、生黄瓜、生西芹,加陈醋拌匀。

Seafood, eggs, and vegetables with vinegar. Specialty in Shanxi. Squid and shredded jelly fish scalded, combined with peanuts, preserved eggs, cucumbers, and American celery, then flavored with vinegar sauce.

陈皮扒鸭 (chénpí pá yā)

东北菜。先将鸭稍烫,风干,炸至金黄,然后加鸡汤、陈皮等蒸烂,用原汤勾芡。

Steamed duck in chicken soup. Northeastern dish. Scalded duck air-dried, deep-fried, steamed with processed tangerine peels in chicken soup, then flavored with starched broth.

陈皮贡菜 (chénpí gòngcài)

广东凉菜。先将贡菜用开水烫熟,加入陈皮和话梅,再用糖、盐、葱油等拌匀。

Mustard stems with tangerine peels. Guangdong cold dish. Preserved mustard stems scalded, mixed with processed tangerine peels and candied plums, then flavored with sugar, salt, and scallion oil.

陈皮鸡 (chénpíjī)

四川菜。先将鸡块炸熟,再与煸炒过的辣椒、花椒、陈皮、糖醋汁等烹煮。

Spicy chicken with tangerine peels. Sichuan dish. Fried chicken braised with quick-fried chili pepper, Chinese prickly ash, processed tangerine peels, and sweet-sour sauce.

陈皮牛肉 (chénpí niúròu)

四川菜。先将牛肉丁微炸，炖至酥烂，然后入原汤，与煸炒过的陈皮、花椒、辣椒、酱油、葱、姜、蒜、酒酿等焖煮收汁。

Stewed beef with tangerine peels. Sichuan dish. Lightly fried diced beef stewed, then simmered with quick-fried processed tangerine peels, Chinese prickly ash, chili pepper, soy sauce, scallion, ginger, garlic, sweet fermented sticky rice juice, and broth until reduced.

蛏干炖薯仔 (chēnggān dùn shǔzǎi)

福建菜。先将水发蛏干蒸熟，然后放入甜薯仔，加肉汤炖熟。

Braised razor clam with sweet potatoes. Fujian dish. Soaked dried razor clam meat steamed, then simmered with small sweet potatoes in pork broth.

蛏干烧肉 (chēnggān shāoròu)

湖南菜。先将蛏干蒸熟，然后加入带皮五花肉、熟笋片、香葱、姜、酱油、白糖、料酒等炖煮。

Stewed razor clam meat with pork. Hunan dish. Steamed dried razor clam meat stewed with streaky pork, sliced bamboo shoots, scallion, ginger, soy sauce, white sugar, and cooking wine.

蛏溜奇 (chēngliūqí)

福建菜。把鸡蛋浆、肥猪肉粒、调料和氽过的蛏肉拌匀，炒熟。

Stir-fried razor clam with eggs. Fujian dish. Scalded razor clam meat mixed with egg paste, diced fat pork, and flavorings, then stir-fried.

蛏子豆腐荠菜汤 (chēngzi dòufu jìcài tāng)

上海菜。先将蛏子过油，然后与嫩豆腐、荠菜入清汤煮沸，加盐、葱、少许麻油。

Clam, bean curd, and shepherd's purse soup. Quick-fried razor clam, tender bean curd, and shepherd's purse boiled in clear soup and seasoned with salt, scallion, and sesame oil.

成都素烩 (chéngdūsùhuì)

四川成都菜。先将白萝卜、胡萝卜、马铃薯、黄瓜等用开水氽过，再与蘑菇及调料爆炒，加入上汤、腐竹、芦笋、玉米笋等烩煮，勾芡。

Simmered assorted vegetables. Specialty in Chengdu, Sichuan. Scalded turnips, carrots, potatoes, and cucumbers stir-fried with mushrooms and flavorings. Simmered in broth with bean curd skins, asparagus, and baby corn, then thickened with starch.

成吉思汗铁板烧 (chéngjísīhán tiěbǎnshāo)

内蒙菜。将羊肉片加入桂圆肉、枸杞等腌渍，在铁板上烤熟，佐辣椒汁等食用。因受元太祖成吉思汗喜爱而得名。

Broiled mutton with chili sauce. Inner Mongolian dish. Sliced mutton pickled with longan meat and

Chinese wolfberry, broiled on an iron board, then served with chili sauce. It is named after Chengjisihan, the first emperor of the Yuan Dynasty, who favored this dish.

成珠小凤饼 (chéngzhū xiǎofèngbǐng)

广东点心,又称鸡仔饼。面粉皮包入菜心馅,做成小饼,烤熟。

Vegetable cakes. Guangdong snack. Small flour cakes filled with prepared baby cabbage, then baked.

豉姜泥鳅 (chǐjiāng níqiū)

浙江菜。先将泥鳅焯水洗净,加猪油、姜、蒜、豆豉等烹煮。

Braised loach with fermented soybean. Zhejiang dish. Loach scalded, cleaned, then braised with lard, ginger, garlic, and fermented soybean.

豉椒炒鳝片 (chǐjiāo chǎo shànpiàn)

广东菜。鳝片、豆豉、青椒、红椒与胡椒粉、酱油等翻炒。

Eel with fermented beans and pepper. Guangdong dish. Sliced eel stir-fried with fermented soybean, green and red pepper, ground black pepper, and soy sauce.

豉椒蛤蜊肉 (chǐjiāo gélíròu)

四川菜。将蛤蜊肉与豆豉、红辣椒、木耳等爆炒,用盐、黄酒、味精等调味。

Stir-fried clam meat with fermented soybean. Sichuan dish. Clam meat stir-fried with fermented soybean, red pepper, and wood ears, then flavored with salt, yellow wine, and MSG.

豉椒划水 (chǐjiāo huáshuǐ)

湖南菜。先将鱼尾即划水腌渍,挂蛋清芡粉浆油炸,再加入豆豉、干辣椒、笋片、青蒜等翻炒,然后入鸡汤微焖,勾芡。

Fish tails with fermented soybean and pepper. Hunan dish. Pickled carp tails coated with egg white and starch, deep-fried, stir-fried with fermented soybean, chili pepper, sliced bamboo shoots, and garlic sprouts, then simmered in starched chicken soup.

豉椒牛肉 (chǐjiāo niúròu)

湖南家常菜。先将牛肉片腌渍,再与豆豉和蒜茸爆炒,加入青椒和红椒。

Beef with fermented soybean and pepper. Home-style dish in Human. Pickled sliced beef stir-fried with fermented soybean, mashed garlic, and green and red pepper.

豉椒蒸腊肉 (chǐjiāo zhēng làròu)

湖南菜。将熟腊肉片做底,上放炒香的豆豉和红辣椒,蒸透。

Smoked pork with fermented soybean and pepper. Hunan dish. Sliced smoked cured pork covered with quick-fried fermented soybean and red pepper, then steamed.

豉椒蒸鱼 (chǐjiāo zhēng yú)

湖南菜。先将草鱼腌渍,再加豆豉、姜、蒜翻炒,然后用急火蒸。

Steamed carp with fermented

soybean and pepper. Hunan dish. Pickled grass carp stir-fried with fermented soybean, ginger, and garlic, then steamed over high heat.

豉油皇蒸生鱼 (chǐyóuhuáng zhēng shēngyú)

广东菜。生鱼加葱、姜及煸炒过的豆豉蒸熟,浇上用酱油和热油调制的卤汁。

Steamed fish with fermented soybean sauce. Guangdong dish. Snakehead fish steamed with ginger, scallion, and quick-fried fermented soybean, then washed with dressing of soy sauce and heated oil.

豉油鲤鱼段 (chǐyóu lǐyúduàn)

湖南菜。先将鲤鱼段裹上蛋浆油炸,再加入特制的豉油炖煮。

Carp with fermented soybean sauce. Hunan dish. Carp chunks coated with egg and starch paste, deep-fried, then braised with special fermented soybean sauce.

豉油蒸鲩鱼 (chǐyóu zhēng huànyú)

南方菜。先将鲩鱼盐渍,然后放入豆豉、葱、姜、猪油蒸熟,撒胡椒粉、葱丝,浇卤汁。

Grass carp with fermented soybean. Southern dish. Salt-pickled grass carp steamed with fermented soybean, scallion, ginger, and lard, then flavored with ground black pepper, shredded scallion, and starched sauce.

豉汁凉瓜炆蟹 (chǐzhī liángguā wén xiè)

广东菜。先将蟹肉、凉瓜、豆豉与生粉、胡椒粉、盐、上汤调成的卤汁翻炒,然后焖熟。

Crab and gourds with fermented soybean. Guangdong dish. Crab meat, bitter gourds, and fermented soybean sautéed with sauce of starch, ground black pepper, salt, and broth, then simmered.

豉汁梅子蟠龙鳝 (chǐzhī méizi pánlóngshàn)

广东、福建菜。将白鳝切成连而不断的小段,蟠于盘中,配豉汁、杨梅等蒸熟。

Steamed eel with sauce. Guangdong and Fujian dish. White eel cut into connected segments, coiled on a platter, then steamed with fermented soybean sauce and prepared red bay berries.

豉汁排骨 (chǐzhī páigǔ)

广东菜。将排骨用黄酒和淀粉略腌,淋豉汁,蒸熟,撒葱花。

Steamed pork ribs with sauce. Guangdong dish. Chopped pork ribs pickled with yellow wine and starch, topped with fermented soybean sauce, steamed, then sprinkled with chopped scallion.

豉汁平鱼 (chǐzhī píngyú)

四川菜。在平鱼腹中塞入豆豉、辣椒、姜、葱等料,蒸熟。

Butterfish with fermented soybean. Sichuan dish. Butterfish stuffed with fermented soybean, hot pepper, ginger, and scallion, then steamed.

豉汁鱼云 (chǐzhī yúyún)

广东菜。将大头鱼与盐、胡椒粉、生粉、豆豉、蒜茸、姜、葱、辣椒拌和,用锡纸包好,蒸熟或烤熟。

Fish with fermented soybean. Guangdong dish. Big head fish mixed with salt, ground black pepper, starch, fermented soybean, minced garlic, ginger, scallion, and hot pepper, wrapped in foil, then steamed or broiled.

豉汁蒸凤爪 (chǐzhī zhēng fèngzhǎo)

广东点心。先将鸡爪炸至金黄色,然后用热水浸至酥软,加豉汁、胡椒粉等蒸透。

Steamed chicken feet with sauce. Guangdong snack. Chicken feet deep-fried, soaked in warm water until soft, then steamed with fermented soybean sauce and ground black pepper.

豉汁蒸黄骨鱼 (chǐzhī zhēng huánggǔyú)

广东菜。将豆豉铺在新鲜黄骨鱼上,加豉汁和酱油,蒸熟。

Steamed yellow-bone fish with sauce. Guangdong dish. Yellow-bone fish covered with fermented soybean, topped with fermented soybean juice and soy sauce, then steamed.

豉汁蒸鲫鱼 (chǐzhī zhēng jìyú)

四川菜。先将鲫鱼用豆豉蒜茸汁腌渍,蒸熟,然后撒上香菜和葱花,淋猪油。

Steamed crucian with fermented soybean. Sichuan dish. Crucian pickled with fermented soybean juice and minced garlic, steamed, then sprinkled with coriander, scallion, and lard.

豉汁蒸排骨 (chǐzhī zhēng páigǔ)

广东菜。先将排骨腌渍,然后放入煸炒过的蒜和豆豉,蒸熟。

Steamed pork steak with fermented soybean. Guangdong dish. Pork steaks pickled, mixed with quick-fried garlic and fermented soybean, then steamed.

豉汁蒸鱼片 (chǐzhī zhēng yúpiàn)

广东菜。先将鱼片用豆豉汁、姜末、红椒末略腌,蒸熟,然后撒上葱和胡椒粉,淋麻油。

Fish fillets with fermented soybean juice. Guangdong dish. Fish fillets pickled with fermented soybean juice, minced ginger, and red pepper, steamed, then flavored with chopped scallion, ground black pepper, and sesame oil.

赤豆糕 (chìdòugāo)

上海点心。将煮烂的赤豆与糯米粉、生粉、发酵粉、糖调和,蒸熟成糕,冷却后切块。

Red bean cakes. Shanghai snack. Boiled red beans, sticky rice flour, starch, baking powder, and sugar mixed, then steamed into cakes. Cooled, then cut into cubes.

赤豆酿鲤鱼 (chìdòu niàng lǐyú)

湖南菜。将赤小豆和陈皮填入鲤鱼肚,然后将鱼蒸熟,淋麻油。

Steamed stuffed carp. Hunan dish. Carp filled with red beans and processed tangerine peels, steamed, then flavored with sesame oil.

虫草杜仲炖海虾 (chóngcǎo dùzhòng dùn hǎixiā)

南方药膳。将海虾与冬虫夏草、生杜仲等炖煮。

Stewed prawns with herbs. Folk medicinal dish in southern areas. Prawns stewed with Chinese caterpillar fungus and raw eucommia bark.

虫草鹅掌 (chóngcǎo ézhǎng)

四川菜。将鹅掌汆过,加入虫草、香菇等熬炖。

Stewed goose feet with herb. Sichuan dish. Scalded goose feet stewed with Chinese caterpillar fungus and dried mushrooms.

虫草峰蘑菇 (chóngcǎo fēng mógu)

西藏菜。将雪鸡与藏北虫草、蘑菇、黄芪炖煮。

Snow cock stewed with herbs. Tibetan dish. Tibetan snow cock stewed with northern Tibetan caterpillar fungus, mushrooms, and milkvetch roots.

虫草甲鱼 (chóngcǎo jiǎyú)

山东菜。先将甲鱼用开水烫过,然后装入土罐,加虫草、清汤蒸熟。

Steamed turtle with Chinese caterpillar fungus. Shandong dish. Scalded soft-shell turtle steamed in clear soup with Chinese caterpillar fungus in an earthen pot.

虫草葵花鲍 (chóngcǎo kuíhuābào)

广东菜。将鲍鱼、虫草、火腿、香菇及调料入高汤焖熟,用菜椒片和黄瓜皮摆成葵花形饰盘。

Braised abalone with herb. Guangdong dish. Abalone, Chinese caterpillar fungus, ham, and dried mushrooms braised in seasoned broth. Served with sliced pepper and cucumber rinds arranged like a sunflower.

虫草牛舌 (chóngcǎo niúshé)

西藏菜。先将牛舌加藏葱、藏茴香等煮熟,去皮,切片,然后加入蝉和虫草蒸熟,浇卤汁。

Steamed ox tongue. Tibetan dish. Ox tongue boiled with Tibetan leek and Tibetan fennel, skinned, sliced, steamed with cicadae and Chinese caterpillar fungus, then marinated in sauce.

虫草全鸭 (chóngcǎo quányā)

各地传统药膳。先将虫草、姜、葱填入鸭腹,然后将鸭放入加调料的清汤蒸熟。

Steamed duck with herb. Traditional folk medicinal dish in many places. Whole duck stuffed with Chinese caterpillar fungus, ginger, and scallion, then steamed in seasoned clear soup.

虫草水鱼裙腿 (chóngcǎo shuǐyú-qúntuǐ)

药膳。水鱼即甲鱼。将水鱼、虫草、枸杞等装入竹盅,加高汤,蒸至酥烂。

Steamed turtle with herbs. Folk medicinal dish. Cultured turtle, Chinese caterpillar fungus, and Chinese wolfberry steamed in broth in a bamboo cup.

虫草松茸鸡 (chóngcǎo sōngróng jī)

西藏菜。将藏土鸡与虫草、松茸等煮熟。

Chicken with matsutake and mushrooms. Tibetan dish. Tibetan chicken boiled with Chinese caterpillar fungus and pine mushrooms.

臭豆腐 (chòudòufu)

江浙沪皖点心。豆腐发酵后油炸至发脆,佐辣酱食用。

Fried fermented bean curd. Snack in Jiangsu, Zhejiang, Shanghai, and Anhui. Bean curd fermented, deep-fried, then served with hot pepper sauce.

臭豆腐焖排骨 (chòudòufu mèn páigǔ)

湖南菜。先将猪排骨油煎,然后加入臭豆腐、辣椒及调料煸炒、焖煮。

Pork steaks with fermented bean curd. Hunan dish. Fried pork steaks stir-fried with fermented bean curd, hot pepper, and seasonings, then braised.

臭豆腐猪手煲 (chòudòufu zhūshǒu bāo)

湖南菜。先将煮熟的猪蹄与臭豆腐、香菇、花生仁及调料煸炒,然后放入沙锅炖煮。

Hog trotters with fermented bean curd. Hunan dish. Hog trotters stir-fried with fermented bean curd, dried mushrooms, peanuts, and flavorings, then stewed in a clay pot.

出汁豆腐 (chūzhī dòufu)

浙江菜。先将豆腐块、香菇、茄块裹面粉油炸,然后浇上卤汁和白萝卜泥。

Fried bean curd and vegetables. Zhejiang dish. Bean curd, mushrooms, and cubed eggplants coated with flour, deep-fried, then washed with special sauce and mashed white turnips.

川贝炖鹧鸪 (chuānbèi dùn zhègū)

广东菜。将鹧鸪与川贝、北杏、姜片同炖,用盐、味精调味。

Partridge with herbs. Guangdong dish. Partridge stewed with herbs, such as fritillary bulb, apricot kernel, and ginger, then flavored with salt and MSG.

川贝酿梨 (chuānbèi niàng lí)

浙江点心。先将雪梨去皮挖核,再将糯米饭、冬瓜条、川贝母、冰糖装入雪梨,蒸至酥烂,浇上冰糖汁。

Steamed pear with herb. Zhejiang snack. Pear peeled, cored, filled with cooked sticky rice, candied wax gourd, fritillary bulb, and rock sugar, steamed, then washed with rock sugar soup.

川红功夫茶 (chuānhóng gōngfuchá)
红茶。产于四川宜宾等地。外形肥壮圆紧,显金毫,乌黑油润。汤色浓亮,叶底厚软红匀。
Sichuan gongfu tea, of Sichuan black tea, produced in Yibin, Sichuan. Its dry leaves are stout, round-tight, golden-hairy, and glittery black. The tea is thick red with infused leaves flat and soft red.

川椒煸乳鸽 (chuānjiāo biān rǔgē)
四川菜。先将乳鸽腌渍,然后与辣椒、姜、葱等爆炒。
Stir-fried spicy pigeon. Sichuan dish. Pickled young pigeon stir-fried with hot pepper, ginger, and scallion.

川椒生炒鸡 (chuānjiāo shēngchǎo jī)
四川菜。先将鸡块腌渍,再与红椒、花生、香菇等爆炒,用盐、黄酒、花椒、酱油等调味。
Stir-fried spicy chicken. Sichuan dish. Pickled chicken stir-fried with red pepper, peanuts, and dried mushrooms, then flavored with salt, yellow wine, Chinese prickly ash, and soy sauce.

川南汤火锅 (chuānnán tānghuǒguō)
四川菜,源于川南。先将牛肉汤加入各种火锅调料煮沸,再放入鳝鱼、笋片、豆腐、白菜等烫熟。
Beef soup fire pot. Sichuan dish originated in southern Sichuan. Eel, bamboo shoots, bean curd, and cabbage quick-boiled in beef soup seasoned with spicy condiments.

川式炒鸭掌 (chuānshì chǎoyāzhǎng)
四川菜。先将鸭掌汆过,然后与干辣椒、葱、姜、豆瓣酱等爆炒。
Stir-fried duck feet. Sichuan dish. Scalded duck feet stir-fried with chili pepper, scallion, ginger, and fermented broad bean sauce.

川味腊肉 (chuānwèi làròu)
川黔滇菜。先将猪肉用盐、花椒、八角等腌渍,再用松柏叶、桂枝等熏至棕红色,然后加入各种调料或蒸、或煮、或炒。
Smoked cured pork. Sichuan, Guizhou, and Yunnan dish. Pork pickled with salt, Chinese prickly ash, and aniseed, then smoked with pine, cypress, or bay leaves until red brown. It may be steamed, boiled, or stir-fried alone or with any other ingredients.

川味牛排 (chuānwèi niúpái)
四川菜。先将牛排略腌,过油,然后与红辣椒、豆瓣酱、花椒等焖煮收汁。
Braised spicy beef steak. Sichuan dish. Beef steak pickled briefly, lightly fried, then braised with red pepper, fermented broad bean sauce, and Chinese prickly ash until reduced.

川味泡菜 (chuānwèi pàocài)
四川传统小菜。将萝卜、菜头、辣椒、姜、葱头、蒜、莲花白、黄瓜等各种蔬菜洗净,控干水分,放入用盐、糖、白酒、花椒等调制的泡菜卤水腌制。
Pickled vegetables. Traditional side dish in Sichuan. Fresh vegetables, such as turnip, mustard tuber, chili pepper, ginger, green onion head, garlic, cabbage, and cucumbers, washed, drained, then pickled for several days in brine seasoned with salt, sugar, Chinese liquor, and Chinese prickly ash.

川味烧鲶鱼 (chuānwèi shāoniányú)
四川菜。先将鲶鱼腌渍,过油,然后与豆瓣酱、红辣椒、葱、花椒等焖熟。
Simmered spicy catfish. Sichuan dish. Pickled and lightly fried catfish simmered with fermented broad bean sauce, red pepper, scallion, and Chinese prickly ash.

川味香肠 (chuānwèi xiāngcháng)
四川菜。先将碎猪肉与盐、辣椒粉、花椒等拌匀,灌入猪小肠,再用松柏叶、桂枝等熏至棕红色,然后或蒸、或煮、或炒。
Smoked spicy sausage. Sichuan dish. Chopped pork mixed with salt, chili ground black pepper, and Chinese prickly ash, filled in hog casing, then smoked with pine, cypress, or bay leaves until red brown. Steamed, boiled, or stir-fried.

川西肉豆腐 (chuānxi ròudòufu)
四川菜。用豆腐皮包裹肉末、嫩豆腐、香菇及调料,蒸熟。
Steamed stuffed bean curd buns. Sichuan dish. Mixture of ground pork, tender bean curd, dried mushrooms, and seasonings wrapped into buns with bean curd skins, then steamed.

串烤大虾 (chuànkǎo dàxiā)
东南沿海家常菜。先将虾用葱、盐等腌制,然后用扦子串上,烤熟。
Grilled prawns. Home-style dish in southeastern coastal areas. Prawns pickled with onion and salt, strung on iron bars, then grilled.

串烧藏茹 (chuànshāo zàngrú)
西藏菜。将藏菇用竹签串好,炸熟,浇芡汁。
Fried Tibetan mushrooms. Tibetan dish. Tibetan mushrooms strung on bamboo prods, deep-fried, then washed with starched soup.

炊饼绿豆糕 (chuībǐng lǜdòugāo)
福建点心。将糯米粉、白糖或红糖加桂花做成软饼,剖开,与绿豆糕做成夹饼,伴茶食用。
Sticky rice and mung bean cakes. Fujian snack. Sticky rice flour, white or brown sugar, and osmanthus flower made into soft cakes. Each rice cake cut into halves, then sandwiched with a mung bean cake. Served with tea.

炊莲花鸡 (chuīliánhuājī)
广东菜。先将鸡肉裹淀粉浆炸熟,再与香菇和笋尖入上汤焖煮,加入黄酒和盐,然后倒在面皮做的莲花盘里。
Chicken, mushrooms, and bamboo shoots. Guangdong dish. Chicken

coated with starch paste, fried, braised in broth with dried mushrooms and tender bamboo shoots, then seasoned with yellow wine and salt. It is usually served in a flour cake made like a lotus flower.

炊麒麟鱼 (chuīqílínyú)

广东菜。麒麟鱼即鲈鱼。先将鲈鱼用蛋清、黄酒、盐、味精、胡椒粉腌渍,再与香菇、火腿同蒸,浇芡汁。

Steamed perch. Guangdong dish. Perch pickled with egg white, yellow wine, salt, MSG, and ground black pepper, steamed with dried mushrooms and ham, then washed with starched soup.

炊水晶鸡 (chuīshuǐjīngjī)

广东菜。将鸡茸、虾肉、蛋清、火腿末拌成馅,用鸡皮包馅成饺,蒸熟,浇芡汁。

Steamed chicken and shrimp dumplings. Guangdong dish. Ground chicken mixed with shrimp meat, egg white, and diced ham, wrapped into *jiaozi* with chicken skins, steamed, then washed with starched soup.

炊鱼翅盒 (chuīyúchìhé)

福建菜。将腌过的鱼翅用老母鸡、排骨、猪脚盖上炖熟,再把鱼翅裹上虾胶,放在装有蛋白的小碟里,面上置火腿、芫荽、芹菜、竹笋,蒸熟,然后浇上原汤芡汁。

Stewed shark fin. Fujian dish. Pickled shark fin covered with mature hen, pork ribs, and hog trotter, then stewed. Prepared shark fin coated with shrimp paste, put in small dishes with whipped egg white, topped with ham, coriander, celery, and bamboo shoots, re-steamed, then washed with starched broth.

捶烩鸡片 (chuíhuì jīpiàn)

山东菜。先将鸡脯肉捶成薄片,入油锅滑透,再将煸炒过的玉兰片、冬菇、火腿片入清汤煮沸,然后放入鸡片,勾芡。

Chicken slices in soup. Shandong dish. Chicken breast pounded into thin slices, lightly fried, boiled in clear soup with quick-fried soaked bamboo shoots, winter mushrooms, and ham, then thickened with starch.

锤肉炖鲫鱼 (chuíròu dùn jìyú)

山东菜。先将鲫鱼用黄酒腌渍,再焯水,然后配肉丸入原汤蒸熟。

Steamed crucian carp and meatballs. Shandong dish. Crucian carp pickled with yellow wine, scalded, then steamed in broth with meat balls.

春饼卷菜 (chūnbǐng juǎn cài)

北京食品。白菜水氽后与火腿片及调料拌匀,用烙饼裹成卷状。

Cabbage and ham rolls. Beijing food. Scalded cabbage and sliced ham mixed with seasonings, then rolled in flour pancakes.

春卷 (chūnjuǎn)

南方点心。用面皮包菜肉馅成卷,

炸至金黄,佐醋或辣酱。

Fried vegetable rolls or spring rolls. Southern snack. Ground pork and chopped vegetables rolled in flour wrappers, then deep-fried. Served with vinegar or hot pepper sauce.

春梅红烧参片 (chūnméi hóngshāo shēnpiàn)

广西菜。先将水发海参片加调料煨熟,再把虾胶瓤进冬菇,用芫荽和火腿末点缀,蒸熟,然后将烧好的海参和冬菇同装一盘。

Sea cucumbers with mushrooms. Guangxi dish. Sliced soaked sea cucumbers braised with seasonings. Winter mushrooms filled with minced shrimp meat, decorated with coriander and diced ham, then steamed. Cooked sea cucumbers and mushrooms served on the same platter.

春日合菜 (chūnrì hécài)

山东菜。先将豆芽煸炒,再加入粉丝、菠菜、炒熟的鸡蛋及调料拌炒。

Sautéed vegetables with eggs. Shandong dish. Quick-fried bean sprouts sautéed with starch noodles, spinach, fried eggs, and flavorings.

春笋白拌鸡 (chūnsǔn báibànjī)

江苏凉菜。将熟笋片用酱油、醋、糖等调成的卤汁拌和,放上煮熟的鸡片,浇上卤汁。

Bamboo shoots with chicken. Jiangsu cold dish. Sliced cooked bamboo shoots seasoned with mixture of soy sauce, vinegar, and sugar, covered with sliced boiled chicken, then washed with special sauce.

春笋拌豆腐 (chūnsǔn bàn dòufu)

四川凉菜。先将笋丝和豆腐分别汆过,然后加辣椒油、酱油等拌匀。

Spicy bamboo shoots and bean curd. Sichuan cold dish. Shredded bamboo shoots and bean curd scalded separately, then mixed together with chili oil and soy sauce.

春笋炒步鱼 (chūnsǔn chǎo bùyú)

浙江菜。将步鱼块和油炸笋块拌炒,用酱油、白糖、黄酒等调味,勾芡。

Black sleeper with bamboo shoots. Zhejiang dish. Chinese black sleeper chunks and deep-fried bamboo shoots sautéed, flavored with soy sauce, sugar, and yellow wine, then thickened with starch.

春笋烧仔鸭 (chūnsǔn shāo zǐyā)

四川菜。将腌渍好的鸭块与鲜笋、辣椒、香菇等焖煮,加酱油、料酒、糖、姜、葱。

Braised bamboo shoots and duck. Sichuan dish. Pickled chopped duck braised with fresh bamboo shoots, hot pepper, and dried mushrooms, then seasoned with soy sauce, yellow wine, sugar, ginger, and scallion.

春笋豌豆 (chūnsǔn wāndòu)

浙江菜。豌豆和春笋丁入沸水汆过,加调料炒熟,勾芡。

Sautéed bamboo shoots and peas. Zhejiang dish. Fresh peas and diced spring bamboo shoots quick-boiled, sautéed with seasonings, then thickened with starch.

椿芽烘蛋 (chūnyá hōng dàn)

各地家常菜。将鸡蛋浆、豆粉、水、盐、椿芽末调匀,微火煎熟。

Fried eggs with toon leaves. Home-style dish in many places. Egg paste mixed with bean powder, water, salt, and chopped tender Chinese toon leaves, then fried over low heat.

椿芽山椒蛋 (chūnyá shānjiāo dàn)

四川菜。将剁碎的椿芽、山椒与鸡蛋浆、盐等调匀,煎熟。

Stir-fried toon leaves and eggs. Sichuan dish. Chopped tender Chinese toon leaves and wild hot pepper mixed with egg paste and salt, then fried.

椿芽酸椒鱼 (chūnyá suānjiāo yú)

四川菜。先将刀鱼油炸,再与椿芽、泡辣椒、香菇等加盐、酱油、姜、葱焖烧。

Simmered spicy toon and carp. Sichuan dish. Fried knife fish simmered with tender Chinese toon leaves, pickled pepper, and dried mushrooms, then flavored with salt, soy sauce, ginger, and scallion.

纯正莲包 (chúnzhèng liánbāo)

广东点心。用发酵面粉制皮,用莲蓉作馅,包成半开口的包子,蒸熟。

Bun filled with mashed lotus seeds. Guangdong snack. Fermented dough made into half-open buns stuffed with sweetened mashed lotus seeds. Steamed.

莼菜汆塘鱼片 (chúncài cuān táng-yúpiàn)

江苏菜。将鱼片入猪肉汤烧沸,加入火腿丝,倒入莼菜汤。

Fish and water shield soup. Jiangsu dish. Fish fillets boiled in pork soup with shredded ham, then combined with water shield soup.

鹑蛋烧卖 (chúndàn shāomài)

广东点心。用面皮包上肉末、香菇、青椒、鹌鹑蛋等制成的馅,顶上摺边,蒸熟。

Buns filled with pork, quail eggs, and vegetables. Guangdong snack. Buns made with flour wrappers, filled with dried mushrooms, green pepper, and quail eggs, frilled on the top, then steamed.

糍粑 (cíbā)

南方点心。将蒸熟的糯米捣烂,捏成饼,加热或油煎后裹白糖、熟黄豆粉、花生粉或芝麻粉。

Sticky rice cakes. Southern snack. Steamed sticky rice pestled thoroughly, made into cakes, then heated or fried. Served with sugar, roasted soybean, peanut, or sesame powder.

糍粑排骨 (cíbā páigǔ)

湖南农村菜。先将排骨用糍粑裹好,然后油炸。

Fried pork ribs in sticky rice. Hunan countryside dish. Pork ribs wrapped in cooked sticky rice, then deep-fried.

糍粑鱼 (cíbāyú)

湖北菜。先将腌过的鱼晒干,煎熟,然后加豆瓣酱、酱油、料酒、辣椒焖熟。

Spicy dried fish. Hubei dish. Pickled fish sun-dried, deep-fried, then simmered with fermented broad-bean sauce, soy sauce, cooking wine, and chili pepper.

葱爆羊肉 (cōngbào yángròu)

北京菜。将羊腿肉片和大葱用麻油爆炒,加入蒜茸、醋、盐、酱油等。

Stir-fried mutton and leek. Beijing dish. Sliced gigot meat and leek stir-fried with sesame oil, then flavored with minced garlic, vinegar, salt, soy sauce, and sesame oil.

葱姜炒花蟹 (cōngjiāng chǎo huāxiè)

沿海家常菜。先将腌过的花蟹切成两半,炸至金黄,然后加葱、姜等翻炒,勾芡。

Crab with ginger and scallion. Home-style dish in coastal areas. Pickled crab cut into halves, deep-fried, stir-fried with scallion and ginger, then thickened with starch.

葱烤鲫鱼 (cōngkǎo jìyú)

各地家常菜。将腌过的鲫鱼炸酥,与煸过的葱、姜、酱油等烹煮收汁。

Braised crucian carp. Home-style dish in many places. Crucian carp pickled with salt, deep-fried, then braised with quick-fried scallion, ginger, and soy sauce until reduced.

葱烧鲤鱼 (cōngshāo lǐyú)

东北菜。先将鲤鱼炸至金黄,然后加葱、胡萝卜稍煮,勾芡。

Carp with scallion. Northeastern dish. Carp deep-fried, simmered with scallion and carrots, then thickened with starch.

葱烧蹄筋 (cōngshāo tíjīn)

福建菜。先将猪肉片煸炒,然后加入肉汤,放入煮熟的蹄筋、青菜心、胡萝卜、木耳等烩煮,勾芡。

Braised pork tendons with vegetables. Fujian dish. Sliced pork lightly fried, braised in broth with boiled pork tendons, baby greens, carrots, and wood ears, then thickened with starch.

葱烧乌参 (cōngshāo wūshēn)

浙江菜。将发好的乌参加入煸香的葱、料酒、糖、酱油等煮透,勾芡,配焯过的菜心装盘。

Braised sea cucumbers with green onion. Zhejiang dish. Prepared black sea cucumbers braised with quick-fried green onion, cooking wine, sugar, and soy sauce, thickened with starch, then served with scalded baby greens.

葱烧武昌鱼 (cōngshāo wǔchāngyú)

湖北菜。先将武昌鱼腌渍晒干,再油煎,然后加上葱等调料蒸制。

Steamed dried carp with scallion. Hubei dish. Pickled *wuchang* carp sun-dried, fried, then steamed with scallion and other condiments.

葱头牛肉丝 (cōngtóu niúròusī)

河南家常菜。先将牛肉丝挂芡,芹菜焯水,葱头炸香,然后将牛肉丝炒散,与葱头和芹菜及调料急炒。

Stir-fried beef with onion. Home-style dish in Henan. Starched shredded beef fried over low heat, then stir-fried with scalded celery, lightly fried onion, and seasonings.

葱油饼 (cōngyóubǐng)

北方食品。将面粉、花生油加水和盐调匀,放入葱花,做成圆饼,用油烙熟。

Scallion-flavored pancakes. Northern food. Flour mixed with peanut oil, water, salt, and chopped scallion, then pan fried.

葱油鸡 (cōngyóujī)

各地家常菜。将鸡入葱姜水煮熟,切块,佐葱油汁。

Chicken with scallion oil. Home-style dish in many places. Chicken boiled with scallion and ginger, chopped, then served with dressing of scallion oil.

葱油烤鱼 (cōngyóu kǎo yú)

台湾菜。将鲷鱼去骨,腹内填入葱、胡椒粉、辣椒等佐料,烤熟,浇麻油。

Grilled porgy. Taiwan dish. Porgy boned, stuffed with green onion, ground black pepper, and hot pepper, grilled, then washed with sesame oil.

葱蒸干贝 (cōngzhēng gānbèi)

江苏菜。将炒过的笋丝、香菇丝和炸黄的葱白段、干贝一起蒸熟。

Steamed scallops. Jiangsu dish. Dried scallops steamed with stir-fried shredded bamboo shoots, dried mushrooms, and fried scallion stems.

丛台酒 (cóngtáijiǔ)

白酒,产于河北邯郸。以华北红高粱为原料,用小麦制曲酿造,酒精含量为40-53度,属浓香型酒。

Congtaijiu liquor produced in Handan, Hebei. White spirit made with red sorghum grown in northern China and wheat yeast. Contains 40-53% alcohol, and has a thick aroma.

醋熘白菜 (cùliū báicài)

东北家常菜。将白菜片加花椒油、葱、姜、醋、糖等速炒,勾芡,淋猪油。

Sautéed cabbage with vinegar. Northeastern home-style dish. Sliced Chinese cabbage sautéed with Chinese prickly ash oil, scallion, ginger, vinegar, and sugar, thickened with starch, then washed with lard.

醋熘黄瓜 (cùliū huángguā)

四川菜。将黄瓜片与红辣椒、姜等翻炒,放入盐、醋、糖、酱油等。

Sautéed cucumbers with vinegar. Sichuan dish. Sliced cucumbers sautéed with red pepper and ginger, then flavored with salt, vinegar, sugar, and soy sauce.

醋熘鸡 (cùliūjī)

四川菜。先将鸡块油炸,再与香菇、辣椒、姜等翻炒,放入醋和酱油,勾芡。

Spicy chicken with vinegar. Sichuan dish. Deep-fried cubed chicken stir-fried with dried mushrooms, red pepper, and ginger, flavored with vinegar and soy sauce, then thickened with starch.

醋熘里脊 (cùliū lǐji)

四川菜。先将里脊肉用酱油腌渍,然后与红辣椒、花椒等爆炒,加入醋、酱油,勾芡。

Stir-fried pork tenderloin with vinegar. Sichuan dish. Pork tenderloin pickled with soy sauce, stir-fried with red pepper and Chinese prickly ash, flavored with vinegar and soy sauce, then thickened with starch.

醋焖鸡三件 (cù mèn jī-sānjiàn)

湖南菜。将鸡肫、鸡爪、鸡翅加辣椒、酱油、醋焖煮。

Sour chicken varieties. Hunan dish. Chicken gizzard, feet, and wings braised with hot pepper, soy sauce, and vinegar.

醋酥鲫鱼 (cùsū jìyú)

山东菜。将鲫鱼和肉皮卷加调料小火煨至酥烂,浇醋。

Braised crucian and hog skins. Shandong dish. Crucian carp braised with hog skin rolls and seasonings, then flavored with vine and broth.

氽白菜汤 (cuānbáicàitāng)

湖南菜。将白菜心放入高汤,加入火腿和冬菇,煮熟。

Chinese cabbage soup. Hunan dish. Baby Chinese cabbage boiled in broth with ham and winter mushrooms.

氽白肉 (cuānbáiròu)

东北菜。将熟五花猪肉片、酸菜丝、粉条放入肉汤炖熟。

Streaky pork with pickled cabbage. Northeastern dish. Boiled sliced streaky pork stewed in broth with pickled Chinese cabbage and starch noodles.

氽飞龙汤 (cuānfēilóngtāng)

内蒙菜。用大兴安岭林区产鹧鸪与姜汁、香菜、麻油等煮汤。

Francolin soup. Inner Mongolian dish. Forest francolins cooked in soup seasoned with ginger juice, coriander, and sesame oil.

氽鸡蓉豌豆 (cuān jīróng wāndòu)

江苏菜。将鸡肉做成丸子,用清水煮熟,配土豆泥和新鲜豌豆,浇黄油。

Chicken meatballs with potatoes and peas. Jiangsu dish. Chicken meatballs boiled, washed with butter, then served with mashed potatoes and fresh peas.

氽肉丝雪里蕻 (cuān ròusī xuělǐhóng)

东北菜。先将猪肉丝煮熟,撒上熟雪里蕻末,再将汤加调料烧开,浇在肉丝上。

Pork with mustard greens.

Northeastern dish. Boiled shredded pork covered with prepared chopped potherb mustard greens, then washed with boiled and seasoned broth.

汆丸子汤 (cuānwánzitāng)

各地家常菜。将肉末与鸡蛋、淀粉、姜末、葱花、盐等调匀，做成丸子，用清汤煮熟，加入菜心。

Boiled meatballs. Home-style dish in many places. Ground pork mixed with egg paste, starch, minced ginger, chopped scallion, and salt, then made into balls. Boiled in clear soup with baby greens.

汆猪肝汤 (cuān zhūgāntāng)

各地家常菜。将猪肝片用盐、芡粉腌渍，入清汤烫熟，用葱、姜、酱油，猪油调味，加入菜心。

Hog liver soup. Home-style dish in many places. Sliced hog liver pickled with salt and starch, quick-boiled in clear soup with baby greens, then seasoned with scallion, ginger, soy sauce, and lard.

脆皮八宝鸭 (cuìpí bābǎoyā)

北京菜。鸭腹内填入用江米、莲子、冬笋、冬菇、红枣等八种料制成的馅，蒸熟，然后裹玉米粉炸至金黄。

Crisp duck with eight treasures. Beijing dish. Duck stuffed with eight ingredients such as glutinous rice, lotus seeds, winter bamboo shoot, winter mushroom, and red date. Steamed, coated with corn flour, then deep-fried.

脆皮大肠 (cuìpí dàcháng)

广东菜。先将猪大肠入卤汤煮熟，沥干，然后挂生粉，炸至外皮焦脆。

Crisp hog large intestine.

Guangdong dish. Hog large intestine boiled in gravy, drained, coated with starch paste, then deep-fried until crisp.

脆皮黄瓜 (cuìpí huángguā)

东北家常凉菜。先将黄瓜条腌渍，然后加入酱油、花椒油、辣椒油拌匀。

Cucumbers with chili oil.

Northeastern home-style cold dish. Pickled cucumber strips flavored with soy sauce, Chinese prickly ash oil, and chili oil.

脆皮鸡 (cuìpíjī)

广东菜。先将嫩母鸡汆过，炸至半熟，挂糖浆，风干，然后炸至棕红。

Crisp chicken. Guangdong dish. Scalded young hen half-fried, coated with syrup, air-dried, then deep-fried until red-brown.

脆皮卷 (cuìpíjuǎn)

江苏菜。先将猪大肠加调料煮熟，再用酱油和糖腌渍，然后炸脆，淋醋、撒椒盐。

Crisp hog large intestine. Jiangsu dish. Hog large intestine boiled with spices, marinated in sugared soy sauce, deep-fried until crisp, then sprinkled with vinegar and spicy salt.

脆皮烤乳猪 (cuìpí kǎorǔzhū)

广东菜。先将乳猪用盐、五香粉、白酒、腐乳等腌渍，涂上糖、醋及花生油，然后置炭火上烤至猪皮呈大红色。

Broiled baby hog. Guangdong dish. Baby hog pickled with salt, five-spice powder, White spirit, and fermented bean curd, coated with sugar, vinegar, and peanut oil, then broiled over charcoal until red and crisp.

脆皮糯米鸭 (cuìpí nuòmǐyā)

南方菜。先将板鸭肉夹在两层糯米饭之间压实,然后慢火煎至金黄。

Fried duck with sticky rice. Southern dish. Flattened salty duck meat pressed in between prepared sticky rice, then fried over low heat until golden-brown.

脆皮肉丸 (cuìpí ròuwán)

四川菜。将肉末与蛋浆、淀粉、盐、花椒粉等调和,做成肉丸,炸至皮脆肉酥。

Fried spicy meatballs. Sichuan dish. Ground pork mixed with egg paste, starch, salt, and ground Chinese prickly ash, made into balls, then deep-fried until crisp.

脆皮乳鸽 (cuìpí rǔgē)

广东菜。先将乳鸽放入用桂皮、甘草、八角、鸡汤制成的卤水中烹煮,然后涂上用饴糖和白醋调成的糊,炸至皮脆。

Fried pigeon. Guangdong dish. Young pigeons boiled in chicken soup flavored with cinnamon, licorice root, and aniseed, coated with maltose and vinegar paste, then deep-fried.

脆皮三丝 (cuìpí sānsī)

四川菜。以笋丝、黄瓜丝、肉丝为馅,用豆腐皮包馅,炸至脆黄。

Fried vegetable and meat buns. Sichuan dish. Shredded bamboo shoots, cucum3bers, and pork wrapped into buns with bean curd skins, then fried until crisp.

脆皮石鱼卷 (cuìpí shíyújuǎn)

江西菜。先将庐山石鱼腌渍,然后炸脆,撒上椒盐。

Crisp rockfish with spicy salt. Jiangxi dish. Rockfish from Mount Lushan pickled, deep-fried until crisp, then flavored with spicy salt.

脆皮鸭 (cuìpíyā)

浙江菜。先将鸭用花椒、八角、黄酒、盐等腌渍,蒸熟,然后油炸。

Fried duck. Zhejiang dish. Duck pickled with prickly ash, aniseed, yellow wine, and salt. Steamed, then deep-fried.

脆皮鱼丝 (cuìpí yúsī)

上海凉菜。把煮熟的鱼皮丝、火腿丝、香菇丝、西蓝花杆丝加调料拌匀。

Crisp fishskin. Shanghai cold dish. Boiled shredded fish skins mixed with shredded prepared ham, dried mushrooms, and broccoli stems, then seasoned.

脆皮炸鸡 (cuìpí zhájī)

广东菜。先用盐、八角、桂皮、陈皮等制成的卤水将鸡煮熟,然后炸至金黄,撒椒盐。

Crisp chicken with spicy salt. Guangdong dish. Chicken boiled in gravy of salt, aniseed, cinnamon, and processed tangerine peels, then deep-fried. Flavored with spicy salt.

脆皮纸包鸡 (cuìpí zhǐbāojī)

广东菜。先将鸡氽熟,去骨切块,同甜酱、胡椒粉、糖、盐、香菜、葱、氽核桃仁拌和,然后用米纸包成块,裹蛋清淀粉浆,炸至金黄。

Fried tasty chicken. Guangdong dish. Scalded chicken boned, chopped, mixed with sweet flour sauce, ground black pepper, sugar, salt, coriander, scallion, and fried walnuts, wrapped in edible rice paper, coated with starched egg white, then deep-fried.

脆哨炒豆豉 (cuìshào chǎo dòuchǐ)

湘鄂川黔滇菜。将炸脆的猪肉丁即哨子与豆豉、干辣椒、葱、蒜、盐等爆炒。

Stir-fried crisp pork and fermented soybean. Hunan, Hubei, Sichuan, Guizhou, and Yunnan dish. Diced pork deep-fried crisp, then stir-fried with fermented soybean, chili pepper, scallion, garlic, and salt.

脆薯凤尾虾 (cuì shǔ fèngwěixiā)

广东菜。将虾裹上薯茸和蛋清淀粉浆,炸至金黄。

Crispy shrimp. Guangdong dish. Shrimp coated with mashed sweet potatoes, washed with starched egg white, then deep-fried.

脆炸草虾 (cuìzhá cǎoxiā)

广东菜。将鲜草虾裹上用盐、生粉、鸡蛋调成的浆,炸至金黄。

Fried prawns. Guangdong dish. Fresh giant tiger prawns coated with mixture of salt, starch, and egg paste, then deep-fried.

脆炸牛奶 (cuìzhá niúnǎi)

广东点心。将牛奶、蛋清、糖、生粉等做成奶糕,裹上用面粉、油等调成的面浆,炸至金黄。

Fried milk cakes. Guangdong snack. Milk, egg white, sugar, and starch made into cakes, coated with flour and oil paste, then deep-fried.

脆炸生蚝 (cuìzhá shēngháo)

广东菜。将鲜蚝肉裹上用盐、生粉、鸡蛋调成的浆,炸至金黄。

Fried oyster. Guangdong dish. Fresh oyster meat coated with salt, starch, and egg paste, then deep-fried.

脆炸双菇 (cuìzhá shuānggū)

东北菜。先将香菇条和平菇条焯水,然后挂蛋粉糊,炸至金黄。

Fried mushrooms. Northeastern dish. Scalded dried mushroom strips and oyster mushrooms coated with egg and starch paste, then deep-fried.

脆炸网油卷 (cuìzhá wǎngyóujuǎn)

山东菜。将猪网油抹上蛋糊,放上用猪肉、鸡肉、对虾、冬笋、口蘑等拌成的馅,裹成卷,先蒸后炸。

Fried stuffed web lard rolls. Shandong dish. Web lard coated with flour and egg paste, made into rolls filled with mixture of pork, chicken, prawn meat, winter bamboo shoots, and Mongolian mushrooms. Steamed, then deep-fried.

脆炸响铃 (cuìzhá xiǎnglíng)

浙江、福建菜。将豆腐皮裹猪肉末成卷,切段,油炸。

Crisp bean curd skin rolls. Zhejiang and Fujian dish. Fried bean curd skin rolls filled with ground pork, sectioned, then deep-fried.

翠包赛螃蟹 (cuìbāo sàipángxiè)

福建菜。先将腌过的鱼肉丁炒熟,烹醋,然后用生菜包裹成卷。

Sautéed fish with lettuce. Fujian dish. Pickled diced fish sautéed, flavored with vinegar, then wrapped in lettuce.

翠贝献金龙 (cuìbèi xiàn jīnlóng)

香港菜。在元贝上插上火腿条或胡萝卜条,蒸熟。

Steamed scallops with carrots. Hong Kong dish. Scallops topped with ham or carrot strips, then steamed.

翠豆虾仁 (cuìdòu xiārén)

四川菜。先将嫩豌豆角用开水烫过,然后与虾仁翻炒,加盐和味精。

Sautéed pea pods and shrimp meat. Sichuan dish. Scalded green pea pods sautéed with shrimp meat and seasoned with salt and MSG.

翠绿蟹黄鸡翼球 (cuìlǜ xièhuáng jīyìqiú)

香港菜。先将鸡翅去骨切块,用盐、酱油、糖、芡粉、黄酒等腌渍,然后与姜、葱、蟹黄、菜茎等煸炒,原汤勾芡。

Stir-fried chicken wings. Hong Kong dish. Boned and chopped chicken wings pickled with salt, soy sauce, sugar, starch, and yellow wine, stir-fried with ginger, scallion, crab roe, and vegetable stalks, then washed with starched broth.

翠绿牙鲆鱼卷 (cuìlǜ yápíngyújuǎn)

广东菜。先把葱、姜、蒜炒香,再放入炸好的鱼片与菜茎拌炒,勾芡。

Stir-fried fish and vegetables. Guangdong dish. Deep-fried fish fillets stir-fried with vegetable stalks, mixed with quick-fried scallion, ginger, and garlic, then thickened with starch.

翠竹粉蒸鱼 (cuìzhú fěnzhēngyú)

湖南菜。先将鱼块拌入熟米粉、猪油及调料,然后放入竹筒蒸熟。

Steamed fish in bamboo tube. Hunan dish. Fish chunks mixed with prepared rice flour, lard, and seasonings, then steamed in a bamboo tube.

达里湖酥鱼 (dálǐhú sūyú)

内蒙菜。将内蒙达里湖华子鱼油炸至酥,配辣酱或椒盐。

Fried fish with chili sauce. Inner Mongolian dish. *Huazi* fish from Dahri Lake in Inner Mongolia deep-fried. Served with chili sauce or spicy salt.

大拌菜 (dàbàncài)

东北凉菜。将紫甘蓝和生菜切块,与白糖、油炸花生、白芝麻拌匀。

Vegetables with flavorings. Northeastern cold dish. Sliced purple cabbage and lettuce mixed with white sugar, fried peanuts, and white sesame.

大棒骨 (dàbànggǔ)

东北菜。将猪骨放入用姜、葱、八角、茴香、黄酒、盐、酱油等制成的卤汤煮熟。

Marinated hog bones. Northeastern dish. Hog bones marinated in gravy of ginger, scallion, aniseed, fennel, yellow wine, salt, and soy sauce.

大丰收 (dàfēngshōu)

东北凉菜。把生菜、萝卜条、黄瓜条、大葱条、西红柿块等同装一篮,佐大酱。

Assorted fresh vegetables with sauce. Northeastern cold dish. Lettuce, turnips, cucumbers, leek, and tomatoes put in a wicker basket, then served with sweet fermented flour sauce.

大黄瓜镶肉 (dàhuángguā xiāng ròu)

浙江菜。先将大黄瓜挖空,填入肉馅,蒸熟,然后加调料焖煮。

Simmered cucumbers stuffed with pork. Zhejiang dish. Large cucumbers hollowed, filled with seasoned ground pork, steamed, then simmered with flavorings.

大酱炖鲫鱼 (dàjiàng dùn jìyú)

东北菜。将鲫鱼放入清汤,加入大酱、盐、葱、姜,炖至汁浓。

Stewed crucian. Northeastern dish. Crucian stewed with fermented soybean sauce, salt, leek, and ginger.

大卷 (dàjuǎn)

福建菜。将萝卜丁、瘦肉、笋、豆腐末、地瓜粉加水拌匀,蒸熟,然后切块。

Big vegetable and meat rolls. Fujian dish. Diced turnips mixed with lean pork, bamboo shoots, mashed bean curd, and sweet potato flour, steamed, then cut to serve.

大良炒牛奶 (dàliáng chǎo niúnǎi)

广东顺德菜。顺德古称大良。将牛奶蛋清浆与虾仁和烤鸭丝拌炒,撒上炸榄仁。

Stir-fried milk and meat. Specialty of Shunde, Guangdong. Shunde was called Daliang in ancient times. Milk and egg white mixture stir-fried with shrimp meat and shredded broiled duck, then topped with fried olives.

大良肉卷 (dàliángròujuǎn)

广东顺德菜。将腌过的猪瘦肉片裹火腿条成圆筒形,用蛋清淀粉浆封口,炸至金黄。

Fried pork and ham rolls. Specialty in Shunde, Guangdong. Ham strips rolled in pickled sliced lean pork, sealed with starched egg white, then deep-fried.

大良野鸡卷 (dàliáng yějījuǎn)

广东顺德菜。将猪肥膘薄片裹鸡肉、火腿成圆柱形,用蛋清淀粉浆封口,先蒸熟,再油炸。

Pork, ham, and chicken rolls. Specialty in Shunde, Guangdong. Sliced fat pork made into rolls filled with chicken and ham, sealed with starched egg white, steamed, then fried.

大麦茶 (dàmàichá)

各地传统清凉饮料。用文火将大麦炒至焦黄,与茶叶及配料一道碾碎,筛去粗皮,用开水冲泡。

Barley tea. Traditional summer beverage in many places. Barley baked over low heat until yellow brown, mixed with tea leaves and other ingredients, then ground. Coarse bran sifted out, then infused in hot water to make tea.

大盘鸡 (dàpánjī)

新疆菜。先把鸡块用酱油、盐、料酒等腌渍,再与姜、蒜、干辣椒等翻炒,然后加入糖,与蘑菇、土豆等焖煮。

Spicy chicken with mushrooms and potatoes. Xinjiang dish. Chicken pickled with soy sauce, salt, and cooking wine, stir-fried with ginger, garlic, and chili pepper, flavored with sugar, then stewed with mushrooms and potatoes.

大蒜仔蹄筋 (dàsuànzǎi tíjīn)

四川菜。将大蒜仔与猪蹄筋炖煮,用黄酒、花椒、姜、葱、酱油、盐调味。

Stewed pork tendons with garlic. Sichuan dish. Garlic cloves stewed with pork tendons and seasoned with yellow wine, Chinese prickly ash, ginger, scallion, soy sauce, and salt.

大蹄扒海参 (dàtí pá hǎishēn)

山东菜。先将猪蹄煮熟,炸至金黄,再入鸡汤煮开,用黄酒、盐等调味,然后加入海参煨炖,勾芡。

Stewed hog trotters and sea cucumbers. Shandong dish. Boiled hog trotters deep-fried, stewed with sea cucumbers in chicken soup, flavored with yellow wine and salt, then thickened with starch.

大碗茶 (dàwǎnchá)

北方民间茶俗。用大壶冲泡,大桶装茶,大碗畅饮。

Tea in big bowl. Folk tea drinking custom in northern areas. Tea infused in a big pot or a barrel, then served in big bowls.

大围山蒸方肉 (dàwéishān zhēng fāngròu)

湖南浏阳菜。将五花肉拌上特制调料蒸熟,用小白菜饰盘。

Steamed pork with baby Chinese cabbage. Specialty of Liuyang, Hunan. Streaky pork mixed with special condiments, steamed, then served with baby Chinese cabbage.

大叶苦丁茶 (dàyè kǔdīngchá)

南方传统清凉饮料,产于华南和西南地区,常用作保健茶。将老苦丁茶叶经过萎凋、杀青、揉捻、干燥四个过程制成。外形条索粗壮,无茸毫,汤色透明深绿,叶底摊张,呈深绿,味先苦后甘,耐冲泡。

Broad-leaf *kuding* tea. Traditional summer beverage in southern areas. Produced in southern and south-western areas, also used as a tonic drink. Made with full-grown leaves of broadleaf holly trees through withering, gentle-baking, rubbing-twisting, and drying. Its dry leaves are stout, tight, and hairless. The tea is crystal dark green with a biting bitter taste then a sweet aftertaste. The infused leaves are flat. Good for repeated brewing.

大枣百合汤 (dàzǎo bǎihé tāng)

浙江点心。将去核大枣与百合入砂锅煮烂,加入冰糖。

Red date and lily bulb soup. Zhejiang snack. Dates cored, boiled with lily bulb cloves in clay pot, then sweetened with rock sugar.

大炸羊 (dàzháyáng)

内蒙菜。将羊肉裹上蛋浆、淀粉、麻油、盐等调成的糊,炸熟。

Fried mutton. Inner Mongolian dish. Mutton coated with mixture of egg paste, starch, sesame oil, and salt, then deep-fried.

大煮干丝 (dàzhǔ gānsī)

江浙家常菜。把豆腐干丝、熟鸡丝、熟鸡肝、笋、虾仁等放入鸡汤煮沸。

Dried bean curd in soup. Home-style dish in Jiangsu and Zhejiang. Shredded dried bean curd boiled with prepared shredded chicken, chicken liver, bamboo shoots, and shrimp meat.

带皮蛇火锅 (dàipíshé huǒguō)

湖南菜。将带皮蛇段加入辣酱、香菜、青红椒、葱、姜,用大火煨熟。

Snake in fire pot. Hunan dish. Snake sections with skins braised over high heat with chili sauce, coriander, green and red pepper, scallion, and ginger.

带鱼扒白菜 (dàiyú pá báicài)

东北家常菜。先将带鱼炸至金黄,然后加白菜和调料焖至收汁,勾芡。

Ribbonfish with cabbage. North-eastern home-style dish. Ribbonfish deep-fried, simmered with Chinese cabbage and seasonings, then thickened with starch.

带子上朝 (dàizǐ shàngcháo)

山东菜。先将家鸭和野鸭炸至深红,再加入桂皮、花椒等炖熟,浇卤汁,然后将熟野鸭摆放在熟家鸭的怀里装盘。

Stewed duck and widgeon. Shandong dish. Duck and widgeon deep-fried until golden brown, stewed with cinnamon and Chinese prickly ash, washed with special sauce, then served with the widgeon set in the duck.

丹阳封缸酒 (dānyáng fēnggāngjiǔ)

黄酒,产于江苏丹阳,亦称曲阿酒。将糯米发酵,然后兑入 50 度小曲米酒,封缸 2-3 年酿造而成。

Danyang fenggangjiu, yellow wine from Danyang, Jiangsu, also called Qu'e Wine. It is a drink made from fermented glutinous rice and yellow wine of 50% alcohol content. Sealed in earthen jars for 2-3 years.

担担面 (dàndanmiàn)

四川点心。现煮现卖的辣味热汤面。

Noodles in hot soup. Sichuan snack. Noodles quick-boiled and served in hot and spicy soup.

淡菜炒笋尖 (dàncài chǎo sǔnjiān)

江苏菜,用蒸熟的扁尖笋、淡菜与原汤、鸡汤等炒至汤干。

Mussels with bamboo shoots.

Jiangsu dish. Steamed baby bamboo shoots and dried mussel sautéed with broth and chicken soup.

淡菜皱纹肉 (dàncài zhòuwénròu)

山东菜。先将带皮猪肉用沸水烫煮,炸透,再加调料煮至肉皮起皱,然后切成长条,加入淡菜蒸透。

Steamed streaky pork and dried mussels. Shandong dish. Scalded pork with skins deep-fried, boiled with seasonings till skins wrinkled, then cut into strips. Steamed with dried mussels.

蛋插人参果 (dàn chā rénshēnguǒ)

西藏点心。先将蛋清和蛋黄分别加入水和豆粉调和,蒸熟,切块,然后同装一盘,插上西芷人参果,淋酥油。

Egg cake with Tibetan silverweed root. Tibetan snack. Separated eggs mixed with water and soybean powder, whipped separately, and steamed into cakes. Cakes cut into cubes, decorated with Tibetan silverweed roots, then washed with melted Tibetan butter.

蛋黄里脊卷 (dànhuáng lǐji juǎn)

广东菜。将猪里脊肉片涂上用葱、姜、料酒、盐、淀粉调成的糊,裹熟蛋黄成卷,蒸熟。

Fried pork tenderloin. Guangdong dish. Sliced pork tenderloin coated with thick paste of scallion, ginger, cooking wine, and starch, made into rolls filled with prepared egg yolks, then steamed.

蛋饺 (dànjiǎo)

上海菜。以蛋浆做成饺皮,以猪肉末及调料为馅做成饺子,蒸熟。可与蔬菜煮汤。

Egg *jiaozi*. Shanghai dish. Egg paste fried into wrapping sheets, filled with seasoned minced pork, made into *jiaozi*, then steamed. Good also for soup with vegetables.

蛋梅鸡 (dànméijī)

江苏菜。将整鸡加调料焖烂,置盘中,与梅子形鸡蛋虾茸烧卖装盘,配焯过的青菜心。

Braised chicken with dumplings. Jiangsu dish. Whole chicken braised with seasonings. Put on the same platter with plum-shape dumplings made with eggs and shrimp, then served with scalded baby greens.

蛋泡银鱼 (dànpào yínyú)

东北菜。先将银鱼腌渍,然后挂芡粉蛋糊,炸至金黄。

Fried icefish. Northeastern dish. Icefish pickled, coated with egg and starch paste, then deep-fried.

蛋馓 (dànsǎn)

广东点心。将鸡蛋、白糖、炼奶等制成馅料,卷入馓皮,烤熟。

Sweet egg rolls. Guangdong snack. Prepared flour dough sheets made into rolls filled with mixture of eggs, sugar, and condensed milk, then baked.

蛋松 (dànsōng)

浙江菜。先把蛋浆煎成丝状,除油,压干,再把蛋丝搓成蛋松。

Egg floss. Zhejiang dish. Egg paste fried into threads, pressed until oil drained, then rubbed into floss.

蛋酥花生仁 (dànsū huāshēngrén)

北京家常菜。将花生仁裹上鸡蛋豆粉糊,炸至金黄。

Fried peanuts with egg paste. Home-style dish in Beijing. Peanuts coated with egg and soybean powder paste, then fried until golden-brown.

淡糟鲜竹蛏 (dànzāo xiānzhúchēng)

福建菜。先将香糟、黄酒、香菇及过油冬笋片煮沸,然后与氽过的蛏肉加调料翻炒。

Razor clam with mushrooms and bamboo shoots. Fujian dish. Dried mushrooms and lightly fried winter bamboo shoots boiled with yellow wine and sweet fermented rice juice, then stir-fried with scalded razor clam meat and flavorings.

淡糟香螺片 (dànzāo xiāngluópiàn)

广东菜。先将螺肉片用黄酒略腌,再与冬笋、花菇翻炒,放入蒜、红糟,然后浇上特制的卤汁。

Whelks with fermented sticky rice juice. Guangdong dish. Sliced whelks pickled with yellow wine, sautéed with winter bamboo shoots and grained mushrooms, flavored with garlic and sweet fermented sticky rice juice, then washed with special sauce.

当归牛腩 (dāngguī niúnǎn)

福建菜。先将煸炒过的牛腩、冬笋倒入猪骨汤烧沸,然后加入当归炖煮。

Stewed beef flank with herb. Fujian dish. Beef flank and winter bamboo shoots quick-fried, then simmered with Chinese angelica in hog bone soup.

当红脆皮鸽 (dānghóng cuìpígē)

上海菜。先将乳鸽煮熟,再裹上白醋和麦芽糖风干,然后油炸,切块。

Crispy sweet and sour pigeon. Shanghai dish. Young pigeon boiled, coated with white vinegar and maltose, air-dried, then deep-fried. Chopped before serving.

党参贡米粥 (dǎngshēn gòngmǐ zhōu)

四川点心。将黑米与党参熬成粥,加红糖。

Black rice and herb porridge.

Sichuan snack. Black rice and pilose asiabell root boiled into porridge, then sweetened with brown sugar.

刀切酥 (dāoqiēsū)

内蒙点心。先将面粉、饴糖、清水和成皮面,另将面粉与糖粉、油和成酥面,再用皮面包酥面,擀成条状,切片,烤熟。

Sliced flour pastry. Inner Mongolian snack. Flour, maltose, and water made into wrappers packed with pastry of flour, sugar, and vegetable oil, kneaded into rolls, sliced, then baked.

稻花香酒 (dàohuāxiāngjiǔ)

白酒,产于湖北宜昌。用红高粱、小麦、大米、糯米、玉米等五种粮食为原料酿造,酒精含量为28-52度,属浓香型酒。

Daohuaxiangjiu, liquor produced in Yichang, Hubei. White spirit made with red sorghum, wheat, rice, glutinous rice, and corn. Contains 28-52% alcohol, and has a thick aroma.

德州扒鸡 (dézhōupájī)

山东德州菜。先将全鸡抹上饴糖浆,炸至金黄,再放入丁香、砂仁、小茴香、肉桂等十六种调料焖煮。

Marinated whole chicken. Dish originated in Dezhou, Shandong. Whole chicken coated with syrup, deep-fried until golden, then marinated in gravy of sixteen condiments such as clove, amomi fruit, fennel, and cinnamon.

灯影牛肉 (dēngyǐngniúròu)

四川名菜。先将牛肉切成菲薄的大片,用盐腌至鲜红色,再用木炭火烘干,蒸熟,切成小片,然后用文火炸透,加入黄酒、辣椒粉、花椒粉、白糖、味精、五香粉、芝麻等翻匀,晾凉,淋上麻油。

Dried multi-taste beef slices.

Famous Sichuan dish. Beef sliced into broad thin pieces, pickled with salt until bright red, dried over charcoal, then steamed. Cut into small pieces, deep-fried over low heat, then mixed with yellow wine, chili powder, ground Chinese prickly ash, sugar, MSG, five-spice powder, and sesame. Cooled, then sprinkled with sesame oil.

地黄花粥 (dìhuánghuāzhōu)

药膳。用粟米煮粥,加入地黄花末。

Corn and herb porridge. Folk-medicinal dish. Ground corn cooked into porridge, then flavored with minced dried Chinese foxglove flower.

地三鲜 (dìsānxiān)

河南传统菜。将炸土豆、炸茄子、炸青椒与葱、蒜等调料拌炒,勾芡。

Sautéed three vegetables.

Traditional dish in Henan. Fried potatoes, eggplants, and green pepper sautéed with scallion, garlic, and other condiments, then washed with starched soup.

地四鲜 (dìsìxiān)

河南传统菜。将煸炒过的猪肉与炸好的土豆、茄子、青椒翻炒,浇卤汁。

Sautéed pork with vegetables.

Traditional dish in Henan. Lightly fried pork sautéed with fried potatoes, eggplants, and green pepper, then flavored with special sauce.

滇红功夫红茶 (diānhóng gōngfū hóngchá)

红茶。产于滇西、滇南两地。外形条索紧结,色泽乌润,金毫显露。汤色红浓艳亮,叶底红匀嫩亮,香高味浓。

Yunnan black tea produced in the western and southern areas of Yunnan. Its dry leaves are tight, straight, off-black, and golden-hairy. The tea is strong, aromatic, and bright red with infused leaves even and tender red.

吊烧鸡 (diàoshāojī)

广东菜。先把鸡用蒜、葱、姜、花生酱、芝麻酱等腌渍,吊烧,然后炸至金黄。

Grilled and fried chicken. Guangdong dish. Chicken pickled with garlic, scallion, ginger, peanut sauce, and sesame sauce, broiled, then deep-fried.

丁香排骨 (dīngxiāng páigǔ)

山东菜。将排骨段用开水氽熟,浇上用红曲米、丁香、糖、芡粉等调制的卤汁。

Pork ribs with clove. Shandong dish. Sectioned pork ribs quick-boiled, then marinated in gravy of red yeast rice, clove, sugar, and starch.

顶汤菜胆炖金钩翅 (dǐngtāng càidǎn dùn jīngōuchì)

广东菜。把菜胆、瘦肉丁及火腿片放入炖盅,将金钩翅放在面上,加入姜、葱、盐、汤等炖煮。

Steamed shark fin with vegetables. Guangdong dish. Shark fin put on a bed of vegetable stalks, diced lean pork, and sliced ham, then steamed in soup with ginger, scallion, and salt.

顶香肉丝 (dǐngxiāng ròusī)

河南家常菜。先用姜、葱炝锅,煸炒豆芽和香菇丝,然后放入腌渍过的里脊肉丝翻炒,勾芡。

Stir-fried pork and bean sprouts. Home-style dish in Henan. Pickled shredded pork tenderloin stir-fried with bean sprouts and shredded mushrooms, flavored with quick-fried ginger and scallion, then thickened with starch.

鼎边糊 (dǐngbiānhú)

福建食品。先把米浆泼在铁锅内缘,烘干至熟,然后放入用蛏、蚬、蚝、香菇、虾米、黄花菜等熬煮至沸腾的上汤。

Assorted seafood chowder. Fujian food. Rice paste poured onto the inner edge of a wok and baked. Immersed in boiling rich soup of razor clam, corbicula clam, oysters, dried mushrooms, dried shrimp meat, and dried day lily.

东安鸡 (dōngānjī)

湖南东安菜。先将仔鸡煮至半熟,切块,加米醋、花椒末等翻炒,然后入肉汤焖煮,勾芡。

Simmered young chicken. Specialty of Dong'an, Hunan. Half-boiled and chopped young hen stir-fried, flavored with rice vinegar and Chinese prickly ash, simmered in broth, then thickened with starch.

东北大拉皮 (dōngběi dàlāpí)

东北凉菜。拉皮即凉粉。先将拉皮切条,然后加黄瓜丝、萝卜丝、焯水菠菜及调料拌匀。

Bean jelly with vegetables.

Northeastern cold dish. Bean jelly strips mixed with shredded cucumbers and turnips, scalded spinach, and seasonings.

东北风味手抓骨 (dōngběi fēngwèi shǒuzhuāgǔ)

东北菜。先将煮熟的猪骨油炸,然后加辣椒粉、孜然、芝麻等炒入味。可手拿进食。

Spicy hog bones. Northeastern dish. Hog bones boiled, deep-fried, then stir-fried with chili powder, cumin, and sesame. Often served as finger food.

东北家常泡菜 (dōngběi jiācháng pàocài)

东北凉菜。将胡萝卜、白萝卜、白菜、黄瓜等放入加干辣椒、胡椒粉、白糖的盐水中,装坛密封腌制10天左右。

Pickled vegetables. Northeastern cold dish. Red and white turnips, Chinese cabbage, and cucumbers pickled in brine of salt, chili pepper, ground black pepper, and sugar, then sealed in pot for about ten days.

东北老虎菜 (dōngběi lǎohǔcài)

东北凉菜。先将青辣椒、红辣椒、豆腐皮、大葱等切丝,然后加香菜、盐、麻油等拌匀。

Spicy vegetables. Northeastern cold dish. Shredded green and red pepper, dried bean curd, and leek mixed, then seasoned with coriander, salt, and sesame oil.

东北熘三样 (dōngběi liūsānyàng)

东北菜。先将煮熟的猪大肠、猪肚、

猪肝切片，然后加胡萝卜片及调料翻炒，勾芡。

Stir-fried hog intestine, tripe, and liver. Northeastern dish. Sliced boiled hog large intestine, tripe, and liver stir-fried with sliced carrots and condiments, then thickened with starch.

东北乱炖 (dōngběiluàndùn)

东北菜。先将猪肉块、猪排骨、土豆块、豆腐块油炸，再与卷心菜、西红柿、老虎豆角、辣椒、洋葱、姜等炖煮，加盐、酱油、味精、胡椒粉、猪油等。

Stewed pork and assorted vegetables. Northeastern dish. Deep-fried pork, pork ribs, potatoes, and bean curd stewed with cabbage, tomatoes, tiger beans, hot pepper, onion, and ginger, then seasoned with salt, soy sauce, MSG, ground black pepper, and lard.

东北一锅出 (dōngběi yīguōchū)

东北菜。将猪排骨段、土豆和豆角等炖煮，同时将小馒头放入锅中，加盖焖熟。

Pork ribs with vegetables and steamed bread. Northeastern dish. Chopped pork ribs stewed with potatoes, string beans, and small steamed bread in a covered pot.

东壁龙珠 (dōngbìlóngzhū)

福建菜。先将五花肉末、虾茸、香菇丁及调料做成丸子，蒸熟，再嵌入龙眼肉炸酥，然后与芥蓝叶煸炒。

Shrimp and meat balls with longan. Fujian dish. Ground streaky pork, minced shrimp meat, and diced dried mushrooms mixed, made into balls, filled with dried longan pulp, then deep-fried. Sautéed with mustard greens.

东华鲊 (dōnghuázhà)

河南菜。先将焖熟的鲊鱼片裹上调料糊，然后用荷叶包好蒸熟。据传创于河南开封东华饭店。

Carp in lotus leaves. Henan dish. Pickled carp fillets simmered, coated with paste of seasonings, wrapped in lotus leaves, then steamed. It was created in Donghua Restaurant in Kaifeng, Henan.

东江豆腐煲 (dōngjiāng dòufubāo)

广东东江菜。在豆腐块上挖一洞，填入加调料的猪肉、鱼肉、虾米，炸至金黄，然后入高汤焖烧。

Braised bean curd. Specialty in Dongjiang, Guangdong. Cubed bean curd hollowed, stuffed with seasoned pork, fish, and shrimp meat, deep-fried, then braised in broth.

东坡肉 (dōngpōròu)

杭州菜。先将猪五花肉方块氽过，然后加糖、黄酒、酱油、姜、葱等焖至酥烂。据传由北宋诗人苏东坡创制。

Braised sweet and salty pork. Hangzhou dish. Scalded cubed streaky pork braised with sugar, yellow wine, soy sauce, ginger, and scallion. It was created ley Su Dongpo, a famous poet in the

Northern Song Dynasty.

东山羊 (dōngshānyáng)

海南菜。先将海南东山山羊肉用沸腾的姜葱水氽过,再稍炸,然后加葱、蒜、八角、桂皮、盐、黄酒等焖煮。

Braised chevon. Specialty in Hainan. Chevon from Dongshan Hill, Hainan scalded in soup of ginger and scallion, lightly fried, then braised with scallion, garlic, aniseed, cinnamon, salt, and yellow wine.

冬菜扣肉 (dōngcài kòuròu)

河南菜。将煎过的五花肉铺列碗底,放辣椒、冬菜、酱油、糖,蒸熟。

Preserved cabbage and pork. Henan dish. Deep-fried streaky pork covered with hot pepper, preserved Chinese cabbage, soy sauce, and sugar, then steamed.

冬虫夏草烩西红柿 (dōngchóng-xiàcǎo huì xīhóngshì)

广东菜。先将虫草和豌豆用开水烫过,再和西红柿焖煮,加盐和黄酒。

Tomato and herb soup. Guangdong dish. Chinese caterpillar fungus and peas scalded, braised with tomatoes, then flavored with salt and yellow wine.

冬菇扒笋胆 (dōnggū pá sǔn dǎn)

各地家常菜。先将冬菇、玉米笋、油菜焯水,然后浇上用盐、味精、麻油等调制的卤汁。

Mushrooms and greens. Home-style dish in many places. Winter mushrooms, baby corn, and rape greens boiled, then washed with dressing of salt, MSG, and sesame oil.

冬菇笋炖老豆腐 (dōnggū sǔn dùn lǎodòufu)

江苏菜。将鸡肉、火腿、鸡骨、猪骨、老豆腐、冬笋片、冬菇加调料用砂锅炖煮。

Mushroom and bean curd soup. Jiangsu dish. Chicken, ham, chicken bones, and hog bones stewed with winter bamboo shoots, dried mushrooms, tough bean curd, and seasonings.

冬菇蒸鸡 (dōnggū zhēng jī)

广东菜。将嫩鸡加入冬菇、红枣、腊肠、姜丝、蚝油等蒸熟。

Steamed chicken with mushrooms. Guangdong dish. Young chicken steamed with winter mushrooms, red dates, smoked sausage, ginger, and oyster sauce.

冬菇猪蹄 (dōnggū zhūtí)

河南菜。将氽过的猪蹄配冬菇入清水炖熟,加盐、姜、葱、味精。

Stewed hog trotters and mushrooms. Henan dish. Scalded hog trotters stewed with winter mushrooms and seasoned with salt, ginger, scallion, and MSG.

冬瓜球 (dōngguāqiú)

上海菜。先将冬瓜球掏空,填入虾米猪肉馅,用蛋清鸡茸封口,然后蒸熟,浇特制的卤汁。

Stuffed wax gourd balls. Shanghai dish. Wax gourd balls hollowed, filled with seasoned ground pork and dried shrimp meat, sealed with minced chicken and egg white, steamed, then washed with special sauce.

冬瓜四灵 (dōngguā sìlíng)

江苏名菜。将冬瓜、火腿、鸡肉、鱼、甲鱼肉及调料煮熟。

Wax gourds with four meat varieties. Famous Jiangsu dish. Wax gourd, ham, chicken, fish, and turtle meat stewed with seasonings.

冬瓜薏米煲鸭 (dōngguā yìmǐ bāo yā)

广东菜。先将鸭块稍煎,然后与冬瓜、薏米煲煮,放入姜汁、米酒、盐、陈皮。

Duck, gourd, and seeds of Job's tears. Guangdong dish. Quick-fried cubed duck braised with wax gourd and seeds of Job's tears, then seasoned with ginger juice, fermented sticky rice juice, salt, and processed tangerine peels.

冬笋烧腐竹 (dōngsǔn shāo fǔzhú)

四川菜。先将冬笋和腐竹炸黄,再入鲜汤加调料焖煮。

Simmered bamboo shoots and bean curd skins. Sichuan dish. Winter bamboo shoots and bean curd skins deep-fried, then simmered in seasoned soup.

董酒 (dǒngjiǔ)

白酒,产于贵州遵义。用高粱为主料,用小麦和多种中药制曲酿造,酒精含量为 30-60 度,属混合香型酒。

Dongjiu, liquor produced in Zunyi, Guizhou. White spirit made with sorghum and yeast of wheat and several herbs. Contains 30-60% alcohol, and has a mixed aroma.

冻顶乌龙 (dòngdǐngwūlóng)

乌龙茶。中国名茶,产于台湾南投县鹿谷乡,被誉为台湾茶中之圣。外形紧结弯曲,色泽新鲜墨绿。汤色澄清金黄。叶底呈淡绿,带红边。

Dongding, famous oolong tea produced in Lugu Village in Nantou, Taiwan, reputed to be the best tea in the province. Its dry leaves are tight, crescent, freshened, and dark green. The tea is crystal golden with infused leaves light green with red fringes.

冻米糖 (dòngmǐtáng)

江西点心,亦称米花糖。先将湿糯米蒸熟,晒干或冻干,用油炸成米花,然后用糖浆拌和,入木框轻压成形,切块。

Fried sticky rice cookie. Jiangxi snack. Sticky rice soaked, steamed, dried, then fried into popped rice. Mixed with syrup, put in a wooden case, gently pressed, then cut into cubes.

洞天乳酒 (dòngtiānrǔjiǔ)

果酒,产于四川青城山。当地新鲜猕猴桃汁经发酵、加醪糟汁、冰糖、少量曲酒等制成,色如碧玉,浓似乳汁,香甜醇厚。酒精含量为 12-18%。

Dongtianrujiu, kiwi wine produced in Qingchengshan, Sichuan. Local fresh kiwi juice fermented, then combined with fermented sticky rice juice, rock sugar, and a little liquor. It is jade green, thick, sweet, and aromatic. Contains 12-18% alcohol.

洞庭金龟 (dòngtíngjīnguī)

湖南菜。先将龟肉和五花肉煸炒，然后加调料，用小火煨煮。

Stewed turtle and pork. Hunan dish. Turtle meat and streaky pork quick-fried, then stewed with seasonings.

豆瓣鲳鱼 (dòubàn chāngyú)

四川菜。先将鲳鱼腌渍，煎透，再与煸炒过的豆瓣酱、红椒、鲜笋等焖煮。

Simmered butterfish with sauce. Sichuan dish. Butterfish pickled, fried, then simmered with fermented broad bean sauce, red pepper, and bamboo shoots.

豆瓣海参 (dòubàn hǎishēn)

四川菜。先将水发海参蒸熟，再用开水汆过，然后放入加了豆瓣酱的鸡汤炖煮，加酱油、糖、胡椒粉、料酒，勾芡，撒芹菜心。

Stewed sea cucumbers with sauce. Sichuan dish. Scalded soaked sea cucumbers stewed in chicken soup, seasoned with fermented broad bean sauce, soy sauce, sugar, ground black pepper, and cooking wine, thickened with starch, then garnished with baby celery.

豆瓣里脊 (dòubàn lǐji)

四川菜。先将里脊肉片腌渍，再与豆瓣酱、辣椒等爆炒。

Stir-fried pork tenderloin with sauce. Sichuan dish. Pickled sliced pork tenderloin stir-fried with fermented broad bean sauce and chili pepper.

豆瓣牛肉 (dòubàn niúròu)

四川菜。将腌渍过的牛肉片与豆瓣酱、西芹、辣椒等爆炒。

Stir-fried beef with sauce. Sichuan dish. Pickled sliced beef stir-fried with fermented broad bean sauce, American celery, and chili pepper.

豆瓣蹄髈(dòubàn típǎng)

四川菜。先将猪蹄髈煮熟，切块，再与豆瓣酱、红辣椒、葱、姜、蒜等爆炒。

Stir-fried hog knuckle with sauce. Sichuan dish. Boiled hog knuckle cut into pieces, then stir-fried with fermented broad bean sauce, red pepper, scallion, ginger, and garlic.

豆瓣鲜鱿 (dòubàn xiānyóu)

浙江菜。先把鲜鱿鱼用开水烫熟，然后与豆瓣酱及调料炒匀。

Sautéed squid with sauce. Zhejiang dish. Scalded fresh squid sautéed with fermented broad bean sauce and seasonings.

豆瓣鲜鱼 (dòubàn xiānyú)

四川菜。先将鲫鱼炸至金黄，再与豆瓣酱、红辣椒、香菇等焖煮。

Simmered crucian with sauce. Sichuan dish. Deep-fried crucian simmered with fermented broad bean sauce, red pepper, and dried mushrooms.

豆豉炒腊肉 (dòuchǐ chǎo làròu)

南方传统菜。先将肥腊肉片煸炒,再与豆豉和茭白片翻炒。

Stir-fried fermented soybean and preserved pork. Traditional dish in southern areas. Quick-fried sliced smoked cured pork stir-fried with quick-fried fermented soybean and sliced wild rice stems.

豆豉肥肠 (dòuchǐ féicháng)

四川菜。先将猪大肠油炸,再与豆豉、红辣椒、火腿片等翻炒。

Hog large intestine with fermented beans. Sichuan dish. Hog large intestine deep-fried, then stir-fried with fermented soybean, red pepper, and sliced ham.

豆豉蚵 (dòuchǐkē)

琼台闽粤家常菜。先将蚵即牡蛎微烫,然后加豆豉、葱、蒜等略炒。

Oyster with fermented soybean. Home-style dish in Hainan, Taiwan, Fujian, and Guangdong. Scalded oysters stir-fried or steamed with fermented soybean, scallion, and garlic.

豆豉划水 (dòuchǐ huáshuǐ)

福建菜。将煎好的鱼尾加豆豉等调料焖熟。

Fish tails with fermented soybean. Fujian dish. Fish tails deep-fried, then simmered with fermented soybean and other condiments.

豆豉回锅鱼 (dòuchǐ huíguōyú)

贵州菜。先将鱼片油炸,然后与煸过的豆豉焖烧,加入黄酒、盐、葱、姜、蒜。

Braised fish with fermented soybean. Guizhou dish. Fried fish fillets braised with quick-fried fermented soybean and flavored with yellow wine, salt, scallion, ginger, and garlic.

豆豉鸡 (dòuchǐjī)

贵州菜。先将鸡块过油,然后与煸过的豆豉焖烧,加入黄酒、盐、葱、姜、蒜。

Braised chicken with fermented soybean. Guizhou dish. Lightly fried cubed chicken braised with quick-fried fermented soybean, then flavored with yellow wine, salt, scallion, ginger, and garlic.

豆豉蒸带鱼 (dòuchǐ zhēng dàiyú)

南方家常菜。将带鱼用煸炒过的豆豉盖住,蒸熟。

Steamed hairtail fish with fermented soybean. Home-style dish in southern areas. Hairtail fish covered with quick-fried fermented soybean and steamed.

豆豆面 (dòudoumiàn)

西北回族食品。将手擀面与白豆在羊肉西红柿汤中煮熟,撒花椒叶。

Noodles with white beans. Hui ethnic minority food in northwestern areas. Hand-made noodles boiled with white beans in mutton and tomato soup, then flavored with Chinese prickly ash leaves.

豆腐煲 (dòufǔbāo)

南方家常菜。先将煸炒过的猪肉、

豆腐、鲜贝加调料焖熟，然后放入虾仁和芦笋略烩，勾芡，撒胡椒粉，淋猪油。

Braised bean curd with meat varieties. Home-style dish in southern areas. Quick-fried pork, bean curd, and fresh shellfish braised with seasonings. Simmered with shelled shrimp and asparagus, then thickened with starch. Flavored with ground black pepper and lard.

豆腐蛋花汤 (dòufu dànhuā tāng)

河南菜。先将豆腐入汤煮开，加酱油、葱，勾薄芡，然后倒入蛋浆烫熟，加醋、猪油。

Bean curd and egg soup. Henan dish. Bean curd soup flavored with soy sauce and scallion, then lightly starched. Beaten eggs added, then flavored with vinegar and lard.

豆腐皮包子 (dòufupí bāozi)

曹雪芹《红楼梦》菜谱。用精白面粉制皮，用豆腐皮、金针、木耳、香菇、青菜加调料制馅，做成包子，蒸熟。

Steamed vegetable buns. Menu from *A Dream of Red Mansions* by Cao Xueqin. Fine flour dough made into buns stuffed with seasoned mixture of bean curd skins, tiger lily buds, wood ears, dried mushrooms, and green vegetable, then steamed.

豆皮素菜卷 (dòupí sùcài juǎn)

南方家常菜。将香菇、木耳、莲子、芡粉、盐等做成馅，用豆皮裹馅成卷，用葱扎好，蒸熟，淋芡汁。

Steamed vegetable rolls. Home-style dish in southern areas. Bean curd skin rolls stuffed with mixture of dried mushrooms, wood ears, husked lotus seeds, starch, and salt. Tied with scallion, steamed, then washed with starched soup.

豆茸酿枇杷 (dòuróng niàng pípá)

江苏菜。先将枇杷去皮去核，酿入糖渍猪油丁和甜豆沙，用松仁和红樱桃末点缀，蒸熟，然后浇白糖桂花芡汁。

Steamed loquat with bean paste. Jiangsu dish. Loquats peeled, pitted, filled with sugared diced lard and sweet bean paste, decorated with pine nuts and chopped cherries, then steamed. Washed with starched soup flavored with sugar and osmanthus flower.

豆沙芝麻球 (dòushā zhīmaqiú)

广东点心。将糯米粉、黄油、砂糖与水调和，做成球状，包入豆沙馅，裹上芝麻，炸至金黄。

Fried sticky rice balls. Guangdong snack. Sticky rice flour mixed with butter, sugar, and water, made into balls filled with sweetened bean paste, coated with sesame, then deep-fried.

豆沙粽子 (dòushā zòngzi)

上海点心。将糯米和甜赤豆沙用竹叶或芦苇叶包成粽子，煮熟。

Sticky rice and red bean *zongzi*. Shanghai snack. Sticky rice and sweet red bean paste wrapped into *zongzi* with bamboo or reed leaves, then boiled.

独咸茄 (dúxiánqié)

江苏家常菜。将茄子和泡涨的黄豆用清水煮熟,加酱油、盐、花椒油。

Salty eggplants and soybean. Home-style dish in Jiangsu. Eggplants and soaked soybean boiled and seasoned with soy sauce, salt, and Chinese prickly ash oil.

杜家鸡 (dùjiājī)

湖北菜。先将熟鸡块摆成鸡形,放上葱姜丝,然后浇上加调料的鸡汤和热麻油。

Boiled chicken with sesame oil. Hubei dish. Chicken chunks boiled, assembled back to its original shape, added scallion and ginger, then washed with seasoned chicken soup and heated sesame oil.

杜康酒 (dùkāngjiǔ)

白酒,中国传统四大名酒之一,产于河南洛阳。以小麦、糯米、高粱为原料酿造,酒精含量为 52-55 度,属浓香型酒。据传,杜康为中酿酒鼻祖。

Dukangjiu liquor, one of the top 4 traditional alcoholic drinks in China. White spirit produced in Luoyang, Henan. It is made with wheat, glutinous rice, and sorghum. Contains 52-55% alcohol, and has a thick aroma. Du Kang was the legendary creator of white spirit in China's history.

炖菜核 (dùncàihé)

江苏菜。先将菜心滑油,放入锅底,再将火腿片、冬笋、鸡脯片、虾仁等放在菜心上,然后加调料,小火炖熟。

Stewed baby cabbage. Jiangsu dish. Baby cabbage lightly fried, put in a pot, covered with sliced ham, bamboo shoots, chicken breast, and shrimp meat. Seasoned, then stewed.

炖刀鱼五花肉菜心 (dùn dāoyú wǔhuāròu càixīn)

东北菜。先将刀鱼段和五花肉片加番茄酱、盐等入汤炖熟,然后放入焯水的油菜心。

Stewed saury and pork. Northeastern dish. Pacific saury and sliced streaky pork stewed in soup flavored with tomato sauce and salt, then garnished with scalded baby rape greens.

炖橄榄螺头汤 (dùn gǎnlǎn luótóu tāng)

福建菜。先将螺头、拍裂的橄榄、烫熟的瘦肉等用鸡汤拌匀,蒸熟,然后浇加调料的鸡汤。

Conch and olive soup. Fujian dish. Conches, cracked olives, and scalded lean pork mixed in chicken soup, steamed, then washed with seasoned chicken soup.

炖家野 (dùnjiāyě)

江苏菜。先将家鸡和野鸡加调料同锅炖酥,然后放入冬笋、冬菇、火腿、豌豆苗等煮沸。

Stewed chicken and pheasant.

Jiangsu dish. Domestic chicken and pheasant stewed together, then boiled with bamboo shoots, dried mushrooms, ham, and pea sprouts.

炖口蘑龙凤珠 (dùn kǒumó lóngfèng zhū)

山东菜。先将鸡肉块稍炸，再加口蘑、奶汤及调料小火炖烂，然后放入对虾煮熟。

Simmered prawns, chicken, and mushrooms. Shandong dish. Lightly fried cubed chicken stewed with Mongolian mushrooms in seasoned milk soup, then boiled with prawns.

炖文武鸭 (dùnwénwǔyā)

江苏菜。先将鸭头和鸭颈置砂锅底，再将烧鸭、白鸭各半只肚腹对齐放入锅中，加黄酒、鸭清汤炖熟。

Stewed white and red duck. Jiangsu dish. One half of a broiled duck and one half of a boiled duck put on a bed of duck head and neck in a casserole, then stewed in rice wine and clear duck soup.

炖羊肉鲜蘑香菜 (dùn yángròu xiānmó xiāngcài)

东北菜。先将羊肉加胡椒、红葡萄酒等炖熟，然后加香菜和鲜蘑菇稍焖。

Mutton with mushrooms. Northeastern dish. Mutton stewed with pepper and red wine, then simmered with coriander and fresh mushrooms.

炖猪蹄 (dùnzhūtí)

各地家常菜。将猪蹄配调料先用武火煮熟，再用文火炖至酥烂。

Stewed hog trotters. Home-style dish in many places. Hog trotters and scallion boiled, then stewed.

多味鸡片 (duōwèi jīpiàn)

四川凉菜。先将整只嫩鸡用沸水氽熟，用凉水迅速冷却，然后去骨切片，放入麻油、葱花、辣椒油、酱油、醋等拌匀。

Chicken with multi-taste sauce. Sichuan cold dish. Young chicken quick-boiled, then chilled in cold water. Boned, sliced, then mixed with sesame oil, chopped scallion, chili oil, soy sauce, and vinegar.

多味牛肉丸 (duōwèi niúròuwán)

北方菜。牛肉丸煎至棕色，浇上多味卤汁。

Multi-taste beef balls. Northern dish. Beef meatballs deep-fried, then marinated in multi-taste gravy.

多味茄泥 (duōwèi qiéní)

东北家常菜。将茄子煮烂，加米醋、白糖、酱油、香菜、辣椒油等拌匀。

Multi-taste eggplants. Northeastern home-style dish. Eggplants boiled thoroughly, then flavored with vinegar, sugar, soy sauce, coriander, and chili oil.

剁椒鱼头 (duòjiāo yútóu)

湖南菜。先将鱼头腌渍，劈开，然后加入大量剁椒和其他调料蒸熟。

Fish head with pickled pepper. Hunan dish. Pickled fish head steamed with pickled red pepper and other condiments.

鹅肝酱片 (égānjiàngpiàn)

上海菜。将熟鹅肝搅成糜,放入模具成型,冷却凝固,切片。

Goose liver cakes. Shanghai dish. Goose liver cooked, ground, then put into a mould to take shape. Frozen solid, then sliced.

鹅肉炖宽粉 (éròu dùn kuānfěn)

东北菜。先将鹅肉焯水,然后加宽粉条、黄酒、盐等炖熟。

Goose with starch noodles. Northeastern dish. Scalded cubed goose stewed with starch noodles, then flavored with salt and yellow wine.

鹅肉炖土豆 (éròu dùn tǔdòu)

东北菜。先将鹅肉块焯水,然后加土豆及调料炖熟。

Stewed goose and potatoes. Northeastern dish. Scalded cubed goose stewed with potatoes and seasonings.

额河烤鱼 (èhékǎoyú)

新疆烧烤。将额河大红鱼、狗鱼、花翅子等与牛羊腰、烤馕裹上孜然、辣椒粉等调料烧烤。

Grilled fish. Xinjiang barbecue. Red fish, pike fish and Huachizi arctic fish from Erqisi River, beef and lamb kidney, and dry flour cakes or *nang* coated with cumin, chili powder, and other condiments, then grilled.

二锅头 (èrguōtóu)

北京传统白酒,以小麦、高粱、玉米为原料,用麸曲发酵,酒精含量为45-65度,属清香型酒。

Erguotou, traditional white spirit in Beijing made with wheat, sorghum, corn, and bran yeast. Contains 45-65% alcohol, and has a delicate aroma.

发财好市 (fácáihǎoshì)

香港菜。用生蚝、发菜、冬菇入高汤炖煮,加蚝油、豉油等调料。

Oysters, black moss and mush-rooms. Hong Kong dish. Oysters, black moss and winter mush-rooms stewed in broth with oyster sauce and fermented soybean sauce.

发菜火腩焖蚝豉 (fàcài huǒnǎn mèn háochǐ)

广东菜。将烧腩、发菜、胡萝卜、蚝豉焖煮,加入酱油、盐、糖,勾芡。

Pork and black moss with fermented soybean. Guangdong dish. Grilled streaky pork braised with prepared black moss carrots, and oyster-flavored fermented soybean. Season-ed with soy sauce, salt, and sugar, then thickened with starch.

发菜瑶柱脯 (fàcài yáozhùpú)

广东菜。将发菜和干贝入高汤蒸熟,加蒜、黄酒、蚝油、麻油。

Scallops and black moss Guangdong dish. Scallops and black moss steamed in broth and flavored with garlic, yellow wine, oyster sauce, and sesame oil.

番茄焖明虾 (fānqié mèn míngxiā)

浙江菜。将煸炒过的芹菜、青椒、番茄、胡椒末、干辣椒等入鲜汤煮沸,放入明虾焖熟。

Prawns with tomatoes. Zhejiang dish. Prawns simmered in seasoned clear soup with quick-fried celery, green pepper, tomatoes, ground black pepper, and chili pepper.

翻沙白肉 (fānshā báiròu)

东北菜。先将猪肥膘片挂糊炸熟,然后将糖熬至翻沙,放入肉片快炒,挂匀糖霜。

Sugarcoated fat pork. Northeastern dish. Sliced fat pork coated with starch paste, deep-fried, then quick-fried with heated melting sugar.

翻沙里脊 (fānshā lǐji)

东北菜。先将猪里脊肉条挂糊,炸至外皮焦脆,然后将糖熬至翻沙,放入肉条快炒,挂匀糖霜。

Sugarcoated pork tenderloin. Northeastern dish. Pork tenderloin strips coated with starch paste, deep-fried, then quick-fried with heated melting sugar.

番茄煳 (fānqiéhú)

南方菜。先将蒜末炒香,再与洋葱丁和番茄泥拌炒,放入月桂叶、鼠尾草、兰姆酒、胡椒粉。

Sautéed tomatoes with onion. Southern dish. Tomato potage made with fried garlic, onion, bay leaves, mouse-tail grass, rum, and ground black pepper.

方鱼蒸鸡 (fāngyú zhēng jī)

江苏菜。先将鸡块腌渍，蒸熟，再加入炸脆舂碎的方鱼茸和芹菜稍蒸。

Steamed chicken with dried flatfish. Jiangsu dish. Pickled chopped chicken steamed, then re-steamed with fried minced dried flatfish and celery.

坊子酒 (fāngzijiǔ)

白酒，产于山东潍坊。以薯干为主要原料制成，酒精含量 62 度。

Fangzijiu, liquor produced in Weifang, Shandong. White spirit made with dried sweet potatoes. Contains about 60% alcohol.

肥脆牛肉 (féicuì niúròu)

东北菜。先将熟牛肉、熟猪肥肉分别切片，然后用两片牛肉夹一片猪肉，裹芡粉蛋浆，炸至金黄。

Fried beef and fat pork. Northeastern dish. Sliced boiled fat pork sandwiched between sliced boiled beef, coated with egg and starch paste, then deep-fried.

肥鱼火锅 (féiyú huǒguō)

湖北菜。将鱼、虾仁、冬瓜、菜心、粉条、泡辣椒等放入加调料的火锅汤涮熟。

Fish and vegetables in fire pot. Hubei dish. Fish, shrimp meat, white gourd, baby Chinese cabbage, starch noodles, and pickled pepper quick-boiled in seasoned soup in a fire pot.

翡翠豆腐 (fěicuì dòufu)

南方菜。将莴苣和豆腐拌炒，放入姜、盐、味精。

Sautéed stem lettuce and bean curd. Southern dish. Stem lettuce sautéed with bean curd and seasoned with ginger, salt, and MSG. .

翡翠干贝冬瓜羹 (fěicuì gānbèi dōngguā gēng)

浙江菜。先把冬瓜蒸烂，捣成茸，再与干贝入鸡汤加调料煮沸，勾芡，然后加入青菜汁煮熟。

Wax gourd, scallop, and vegetable soup. Zhejiang dish. Minced steamed wax gourd boiled with dried scallops in seasoned chicken soup, thickened with starch, then boiled again with green vegetable juice.

翡翠金钱蛋 (fěicuì jīnqiándàn)

北方菜。先将干辣椒煸香，再与煮熟的鸡蛋片翻炒，加盐，配汆过的菜心装盘。

Sautéed spicy eggs. Northern dish. Boiled eggs cut into pieces, sautéed with chili pepper, seasoned with salt, then garnished with scalded baby greens.

翡翠裙边 (fěicuì qúnbiān)

浙江菜。先将鼋鱼裙边加调料煨至酥熟，再加入绿菜茸搅匀，勾芡。

Turtle's soft shell with vegetables. Zhejiang dish. Turtle's soft shell stewed with seasonings. Mixed with chopped green vegetable, then thickened with starch.

翡翠蹄筋 (fěicuì tíjīn)

江苏菜。将炸过的猪蹄筋入鸡清汤焖煮至软糯,加盐、虾米、丝瓜条,勾芡。

Braised pork tendons. Jiangsu dish. Deep-fried pork tendons braised soft. Salt, dried shrimp meat, and slivered luffa added. Thickened with starch.

翡翠鲜虾饺 (fěicuì xiānxiājiǎo)

广东点心。以虾仁和肥肉及调料制馅,太白粉制皮,包成饺子,蒸熟。

Shrimp and pork dumplings. Guangdong snack. *Jiaozi* made of potato starch, stuffed with mixture of shrimp meat, fatty pork, and seasonings. Steamed.

肺汤 (fèitāng)

各地传统药膳,又名补肺汤。先将沙参、红枣、核桃仁煮熟,然后加入山药熬炖。

Stewed Chinese yam with herb. Traditional folk-medicinal dish in many places, also called nutritious soup for lung. Chinese yam stewed with boiled straight ladybell root, red dates, and walnuts.

汾酒 (fénjiǔ)

白酒,中国传统四大名酒之一,原产于山西汾阳杏花村,距今已有1500多年历史。用当地高粱、大麦、豌豆酿造,酒精含量在38-65度之间,属清香型酒。

Fenjiu liquor, one of the top 4 traditional alcoholic drinks in China. White spirit produced in Xinghua Village of Fenyang, Shanxi with a history of over 1500 years. Made from local sorghum, barley, and peas. Contains 38-65% alcohol, and has a delicate aroma.

粉果 (fěnguǒ)

广东点心。以菠菜汁、太白粉制皮,包入用蘑菇加调料制成的馅,蒸熟。

Steamed vegetable buns. Guangdong snack. Potato starch and spinach juice made into buns, stuffed with seasoned mushrooms, then steamed.

粉皮鱼头 (fěnpí yútóu)

广东潮汕家常菜。先将鱼头煎透,加调料,用小火烧至汤汁浓稠,然后放入粉皮,撒青蒜丝。

Fish head with starch noodles. Home-style dish in Chaozhou and Shantou, Guangdong. Deep-fried fish head simmered with seasonings until thickened. Starch noodles added. Garnished with shredded garlic leaves.

粉丝鸭肠煲 (fěnsī yācháng bāo)

四川菜。将粉丝与鸭肠放入沙锅煲煮,用花椒、料酒、盐、葱、姜等调味。

Stewed vermicelli and duck intestine. Sichuan dish. Vermicelli and duck intestine stewed in a casserole, then seasoned with Chinese prickly ash, cooking wine, salt, scallion, and ginger.

粉丝椰汁鳝片窝 (fěnsī yēzhī shànpiàn wō)

广东菜。先将黄鳝段用盐和黄酒腌渍,过油,然后与绿豆粉丝煲煮,加入盐、葱、姜、椰汁。

Eel and vermicelli in coconut milk. Guangdong dish. Field eel sections pickled with salt and yellow wine, lightly fried, braised with mung bean vermicelli, then flavored with salt, scallion, ginger, and coconut milk.

粉蒸鮰鱼 (fěnzhēng huíyú)

江苏菜。先将鮰鱼块用豆瓣酱、黄酒、酱油、糖等调成的汁腌渍,然后用炒米粉拌和,淋麻油,蒸熟。

Steamed long-snout catfish Jiangsu dish. long-snout catfish chunks pickled with mixture of fermented broad bean sauce, yellow wine, soy sauce, and sugar, coated with ground roasted rice, washed with sesame oil, then steamed.

粉蒸里脊 (fěnzhēng lǐji)

四川菜。先将猪里脊肉稍腌,然后裹上糯米粉、醪糟、花椒粉等蒸熟。

Steamed pork tenderloin. Sichuan dish. Pickled pork tenderloin coated with ground sticky rice, sweet fermented sticky rice juice, and ground Chinese prickly ash, then steamed.

粉蒸牛肉 (fěnzhēng niúròu)

四川菜。先将牛肉片稍腌,然后裹上糯米粉、醪糟、花椒粉等蒸熟。

Steamed beef. Sichuan dish. Pickled sliced beef coated with sticky rice flour, sweet fermented sticky rice juice, and ground Chinese prickly ash, then steamed.

粉蒸排骨 (fěnzhēng páigǔ)

各地传统菜。将排骨裹上炒米粉、盐、白胡椒粉,用豆瓣酱拌匀,蒸熟。

Steamed pork steaks in ground rice. Traditional dish in many places. Pork steaks coated with roasted ground rice, salt, and ground white pepper, mixed with fermented broad bean sauce, then steamed.

粉蒸肉 (fěnzhēngròu)

四川菜。先将猪肉用红糖、醪糟、豆腐乳汁等腌渍,然后加入米粉拌匀,用嫩豌豆盖上,蒸至酥烂。

Steamed pork in ground rice. Sichuan dish. Pork pickled with brown sugar, sweet fermented sticky rice juice, and fermented bean curd juice, mixed with ground rice, covered with tender peas, then steamed.

风干牦牛肉 (fēnggān máoniúròu)

西藏传统菜。将生牦牛肉切条,撒上盐和其他调料,风干。

Air-dried yak meat. Tibetan traditional dish. Yak meat strips cored with salt and other condiments, then air-dried.

风鸡斩肉 (fēngjī zhǎnròu)

江苏菜。用猪五花肉做成肉圆,先油煎,再放入风鸡砂锅加调料炖熟。

Stewed chicken and meatballs. Jiangsu dish. Streaky pork meatballs deep-fried. Stewed with air-dried chicken and flavorings in a casserole.

风味老鸭煲 (fēngwèi lǎoyābāo)

各地家常菜。将酸萝卜丝和火腿丝填入鸭腹,放入高汤,加香菇及调料煲熟。

Stewed duck. Home-style dish in many places. Duck stuffed with shredded ham and pickled turnips, then stewed with dried mushrooms in seasoned broth.

风味石榴鸡 (fēngwèi shíliujī)

山东菜。先将鸡脯丁、火腿丁、鲜鱿鱼丁及调料炒成馅,再用面粉鸡蛋皮包裹,用韭菜扎好,蒸熟。

Steamed assorted meat. Shandong dish. Diced chicken breast, ham, and fresh squid stir-fried with condiments, wrapped in egg-flour sheet, fastened with Chinese chive, then steamed.

风鸭糊 (fēngyāhú)

福建点心。将鸭内外抹调料,风干,切丁,与冬笋、香菇、碎肉煮成糊。

Duck chowder. Fujian snack. Pickled duck air-dried, diced, then cooked into chowder with winter bamboo shoots, dried mushrooms, and minced pork.

枫亭糕 (fēngtínggāo)

福建点心。糯米粉、白砂糖、花生仁、芝麻等加水拌匀,蒸熟,切块。

Sticky rice, peanut, and sesame cakes. Fujian snack. Sticky rice flour, granulated sugar, peanuts, sesame, and water mixed into paste, then steamed. Cut into cubes to serve.

封羊肉 (fēngyángròu)

内蒙传统菜。先将羊肉用油煸过,再用特制的面饼盖住焖煮。

Braised mutton with cakes. Inner Mongolian traditional dish. Quick-fried mutton braised in a pot covered with a special pancake as lid.

蜂巢芋角 (fēngcháo yùjiǎo)

广东点心。用芋泥和糯米粉制皮,用猪肉、虾米、糖、香菇、蚝油等制馅,包成角状,炸至金黄。

Fried taro buns. Guangdong snack. Mashed taros and sticky rice flour kneaded into wrappers, made into horn-shape buns stuffed with mixture of pork, dried shrimp meat, sugar, dried mushrooms, and oyster sauce, then deep-fried.

蜂蜜萝卜膏 (fēngmì luóbo gāo)

广东点心。将白萝卜和蜂蜜炖至膏状。

Turnip and honey jam. Guangdong snack. White turnips and honey boiled into jam.

凤冠鲍脯 (fèngguān bào pú)

广东菜。先将鸡脯肉切成斜纹片,裹上蛋清淀粉浆,然后与鲍鱼翻炒,烹入盐、姜、葱、黄酒,勾芡。

Stir-fried chicken and abalone. Guangdong dish. Chicken breast carved into diagonal pieces, coated with starched egg white, stir-fried with abalone, flavored with salt, ginger, scallion, and yellow wine, then thickened with starch.

凤凰花开 (fènghuáng huākāi)

南方菜。将面包剪成花形,用加调料的肉末酿在上面,中央置半只熟鹌鹑蛋,炸至金黄。

Fried quail eggs and pork on bread. Southern dish. Bread cut into a flower, spread with seasoned minced meat, garnished with a boiled half quail egg at the center, then deep-fried.

凤凰球 (fènghuángqiú)

广东点心。将面粉与蛋浆、黄油、白糖、果酱等拌和,制成球状,炸至金黄。

Fried sweet flour balls. Guangdong snack. Flour mixed with egg paste, butter, sugar, and fruit jam, made into balls, then deep-fried.

凤凰水仙 (fènghuángshuǐxian)

乌龙茶。产于广东潮安凤凰山地区。外形挺直肥大,色泽黄褐,油润有光。汤色橙黄清澈,带浓醇的天然花香。叶底肥厚柔软,边缘朱红。

Fenghuangshuixian, oolong tea produced in Mount Fenghuang in Chaoan, Guangdong. Its dry leaves are plump, straight, yellowish brown, and lustrous. The tea is crystal, having an orange color and a rich natural flower aroma. The infused leaves are fat and soft with red fringes.

凤凰粟米羹 (fènghuáng sùmǐ gēng)

广东点心。将粟米加淀粉、白糖、鸡蛋、水等煮成羹。

Sweet corn porridge. Guangdong snack. Corn boiled into porridge with starch, sugar, and eggs.

凤凰胎 (fènghuángtāi)

宫廷菜。鸡腹中未生的鸡蛋与鱼白加盐和料酒快炒。

Sautéed eggs and fish sperm. Imperial palace dish. Unborn eggs quick-fried with fish sperm and seasoned with salt and cooking wine.

凤凰鱼翅 (fènghuáng yúchì)

山东菜。先将嫩鸡炸至金黄,蒸至酥烂,再将鱼翅和冬笋用鸡汤扒透,浇上原汤芡汁。

Steamed chicken with shark fin. Shandong dish. Deep-fried and steamed young chicken braised with shark fin and winter bamboo shoots in chicken soup, then washed with starched broth.

凤姜仔鸭 (fèngjiāng zǐyā)

湖北菜。将汆过的鸭抹上麦芽糖炸熟,然后把冬菇、冬笋、火腿丁及调料填入鸭腹,蒸熟,勾芡。

Tasty stuffed duck. Hubei dish. Scalded duck coated with maltose and deep-fried. Stuffed with winter mushrooms, winter bamboo shoots, ham, and condiments, steamed, then washed with starched soup.

凤梨咖喱鸡 (fènglí gālíjī)

闽粤台菜。先将腌渍过的鸡丁加咖喱粉炒熟,然后加入菠萝丁稍煮,装入菠萝盅。

Chicken with curry. Guangdong, Fujian, and Taiwan dish. Diced chicken pickled, stir-fried with curry, quick-boiled with diced pineapple, then served in a hollowed pineapple.

凤丝仙人掌 (fèngsī xiānrénzhǎng)

广东菜。先将鸡肉丝腌渍,然后与仙人掌、姜、韭黄加盐和黄酒同炒,勾芡。

Stir-fried chicken and cactus. Guangdong dish. Pickled shredded chicken stir-fried with cactus, ginger, yellow Chinese chive, salt, and yellow wine, then thickened with starch.

凤尾虾 (fèngwěixiā)

南方菜。先将大虾去壳留尾,稍腌,然后用竹签串成虾排,炸至金黄酥脆。

Fried prawns. Southern dish. Prawns shelled save the tails, pickled, strung on bamboo sticks, then fried until golden and crisp.

凤尾腰花 (fèngwěi yāohuā)

湖南菜。先将猪腰切成羽状片,然后加入红椒、冬笋、香菇及调料煸炒。

Stir-fried hog kidney. Hunan dish. Hog kidney shredded into feather-like pieces, then stir-fried with red pepper, bamboo shoots, dried mushrooms, and condiments.

奉化芋艿头 (fènghuà yùnǎitou)

上海菜。先把红芋艿头切成莲花瓣形,蒸熟装盘,然后把海参、鱼肚、香菇、火腿、鸡脯肉、胡萝卜丁加调料翻炒,勾芡,倒入莲花形芋艿片中间。

Red taros with assorted meat. Shanghai dish. Red taros sectioned like lotus petals, steamed, and arranged on a platter like a lotus flower. Diced sea cucumbers, fish maw, mushrooms, ham, chicken breast, and carrots stir-fried, thickened with starch, then poured in the middle of the taros.

佛手三丝 (fóshǒu sānsī)

浙江凉菜。将佛手瓜丝、荷兰豆丝、泡椒丝用开水烫过,加调料凉拌。

Three vegetables with sauce. Zhejiang cold dish. Shredded chayotes, snow peas, and pickled red pepper scalded, then mixed with seasonings.

夫妻肺片 (fūqī fèipiàn)

四川凉菜。先将牛肉、牛杂氽过,加入花椒、八角、桂皮、盐、料酒等煮熟,置凉,切片,然后浇上用油炸花生、味精、辣椒油、酱油、花椒粉、芝麻酱等调制的卤汁。

Beef varieties with spicy sauce. Sichuan cold dish. Scalded beef and ox offal boiled with Chinese prickly ash, aniseed, cinnamon, salt, and cooking wine. Cooled, sliced, then washed with dressing of fried peanuts, MSG, chili oil, soy sauce, ground Chinese prickly ash, and sesame sauce.

芙蓉蛋 (fúróngdàn)

各地家常菜。鸡蛋浆加水、盐、火腿末、虾仁等调匀,蒸熟,撒葱花、浇麻油。

Steamed eggs. Home-style dish in many places. Egg paste mixed with water, salt, minced ham, and shrimp meat, steamed, then flavored with chopped scallion and sesame oil.

芙蓉豆腐 (fúróng dòufu)

南方菜。先将嫩豆腐、蛋清、牛奶、清汤、盐等调和,蒸熟,切块,然后放入氽过的火腿片、香菇片、青豆,浇上勾芡的鸡汤。

Steamed bean curd with egg white. Southern dish. Mixture of tender bean curd, egg white, milk, clear soup, and salt steamed, then cut into pieces. Quick-boiled ham, dried mushrooms, and green beans added. Washed with starched chicken soup.

芙蓉鸡片 (fúróng jīpiàn)

四川菜。先将鸡蛋清同鸡肉茸调成糊,用文火烙成花瓣状片,然后入鲜汤煮沸,放入火腿片、冬笋片、豌豆苗、盐、味精、胡椒粉,勾芡,淋鸡油。

Chicken and egg pieces in soup. Sichuan dish. Mixture of egg white and ground chicken breast fried into petal-like pieces over low heat, boiled in seasoned soup with sliced ham, winter bamboo shoots, pea sprouts, salt, MSG, and ground black pepper. Thickened with starch, then sprinkled with chicken oil.

芙蓉蟹 (fúróngxiè)

南方菜。先将蟹粉加姜、葱、料酒、酱油、米醋、糖等炒熟,然后调入蛋清,蒸熟,浇芡汁。

Steamed crab meat and egg white. Southern dish. Crab meat quick-fried with ginger, scallion, cooking wine, soy sauce, vinegar, and sugar. Mixed with egg white, steamed, then washed with starched soup.

芙蓉燕菜 (fúróng yàncài)

四川菜。先将鹌鹑蛋蒸熟,然后配烫过的燕菜装盘。

Bird's nest with quail eggs. Sichuan dish. Scalded edible bird's nests served with boiled quail eggs.

芙蓉银鱼 (fúróng yínyú)

江苏菜。将银鱼与香菇、火腿、青菜丝、溜过的蛋白加调料煮熟,勾芡。

Lake icefish with egg white and vegetables. Jiangsu dish. Lake icefish boiled with dried mushrooms, ham, shredded green vegetable, sautéed egg white, and seasonings, then thickened with starch.

符离集烧鸡 (fúlíjí shāojī)

安徽菜。先将整鸡用饴糖腌渍,再炸至金黄,然后入老卤汤煮透。

Fried and marinated chicken. Anhui dish. Whole chicken pickled with syrup, deep-fried until golden, then marinated in aged gravy.

福建老酒 (fújiànlǎojiǔ)

黄酒,产于福建。以糯米为主要原料酿制,贮存期在三年以上,酒精含量15度左右。

Fujian aged yellow wine. Made with glutinous rice and stored for over 3 years. Contains about 15% alcohol.

福建雪芽 (fújiànxuěyá)

白茶。产于福建福安。外形芽壮、毫多、叶绿。汤色黄绿清亮,叶底黄绿嫩匀。

Fujianxueya, white tea produced in Fuan, Fujian. Its dry leaves are plump, stout, hairy, and green. The tea is crystal and yellowish green with infused leaves yellowish green and even.

福山咖啡 (fúshānkāfēi)

海南咖啡。以海南福山咖啡豆为原料焙制。口味浓而不苦,香而不烈,清香沁人。

Coffee made with roasted coffee beans grown in Fushan, Hainan. It tastes heavy but not bitter, and has a light refreshing aroma.

甫里鸭羹 (fǔlǐyāgēng)

江苏菜。将干贝、虾米、鸭块、笋片、香菇、火腿、鱼圆等放在砂锅里加调料焖熟。

Braised duck. Jiangsu dish. Dried scallops, dried shrimp, cubed duck, sliced bamboo shoots, dried mushrooms, sliced ham, and fish balls braised with seasonings in a casserole.

腐皮包黄鱼 (fǔpí bāo huángyú)

浙江菜。将黄鱼肉加蛋清、淀粉拌成馅,裹上豆腐皮,炸至金黄。

Fried yellow croaker rolls. Zhejiang dish. Yellow croaker meat mixed with egg white and starch, wrapped in bean curd skins, then deep-fried.

腐皮肉卷 (fǔpí ròujuǎn)

江苏菜。用豆腐皮卷猪肉馅,用面糊封口,油炸。

Stuffed bean curd skin rolls. Jiangsu dish. Ground pork wrapped in dried bean curd skins, sealed with flour paste, then deep-fried.

腐乳滑鸡 (fǔrǔ huájī)

山东菜。先将鸡翅用腐乳酱、料酒等腌渍,炒至半熟,再加少量水焖酥,然后加入胡萝卜片略炒,最后撒葱花、淋麻油。

Chicken wings with fermented bean curd. Shandong dish. Chicken wings pickled with fermented bean curd and yellow wine, half-fried, then simmered with a little water. Quick-fried with carrot cuts, then sprinkled with chopped scallion and sesame oil.

腐乳扣肉 (fǔrǔ kòuròu)

上海菜。先将氽过的五花肉用热油泡至棕红,切片,加姜、葱、腐乳汁、鲜汤、盐、糖等炒香,然后蒸熟,浇原汁。

Steamed pork with fermented bean curd. Shanghai dish. Quick-boiled streaky pork fried until red brown, sliced, then sautéed with ginger, scallion, fermented bean curd juice, seasoned soup, salt, and sugar. Steamed, then washed with broth.

腐乳肉 (fǔrǔròu)

上海菜。把猪肋条肉块先烤后煮,再将红乳腐卤与黄酒浆抹在猪肉上,蒸烂。

Pork with fermented bean curd.

Shanghai dish. Cubed streaky pork roasted, boiled, coated with red fermented bean curd and yellow wine, then steamed.

腐竹焖肉 (fǔzhú mèn ròu)

四川菜。先将肉片用湿淀粉、盐、料酒腌渍,再爆炒,然后与腐竹、辣椒、花椒等焖煮。

Simmered bean curd skins and pork. Sichuan dish. Sliced pork pickled with wet starch, salt, and cooking wine, stir-fried, then simmered with bean curd skins, chili pepper, and Chinese prickly ash.

复元汤 (fùyuántāng)

各地传统药膳。将山药、肉苁蓉、菟丝子、核桃仁用布袋装好,同粳米、羊肉、羊脊骨等用武火烧沸,转文火炖烂。

Mutton and herb soup. Traditional folk-medicinal dish in many places. Chinese yam, desert cistanche, dodder seeds, and walnuts put in a cloth bag, boiled with round grain rice, mutton, and lamb backbones, then stewed.

富春鸡 (fùchūnjī)

江苏菜。先将整鸡和鸡蛋煮熟,炸至金黄,再放入砂锅加调料焖煮,加香菇和笋片。

Fried and braised chicken and eggs. Jiangsu dish. Chicken and eggs boiled, deep-fried, then braised with seasonings, mushrooms, and sliced bamboo shoots.

G

盖浇酸奶米饭 (gàijiāo suānnǎi mǐfàn)

西藏点心。在米饭上浇酸奶和熔化的酥油,撒白糖。

Rice with yogurt. Tibetan snack. Cooked rice washed with yogurt and melted Tibetan butter, then showered with sugar.

干贝四宝 (gānbèi sìbǎo)

山东菜。将干贝、火腿、冬笋、冬菇蒸熟,浇原汤。

Steamed scallops with other three treasures. Shandong dish. Dried scallops, ham, winter bamboo shoots, and winter mushrooms steamed, then washed with broth.

干贝莴笋 (gānbèi wōsǔn)

上海菜。先将干贝蒸熟,撕成丝,再与焯过的莴笋入葱味高汤煮开,勾芡。

Scallops with stem lettuce. Shanghai dish. Steamed dried scallops hand-shredded, boiled in clear soup with scalded stem lettuce, flavored with soy sauce and scallion, then thickened with starch.

干贝绣球 (gānbèi xiùqiú)

江苏菜。用虾仁和鱼肉做成丸子,裹上干贝丝、火腿丝、香菇丝蒸熟,浇上卤汁。

Fish and shrimp balls. Jiangsu dish. Fish and shrimp balls coated with shredded dried scallops, ham, and dried mushrooms. Steamed, then washed with special sauce.

干煸扁豆 (gānbiān biǎndòu)

上海菜。将肉末、虾米、榨菜末与炸过的四季豆加调料煸炒。

Sautéed long beans. Shanghai dish. Fried long beans sautéed with ground pork, dried shrimp meat, chopped preserved mustard tubers, and flavorings.

干煸豆角 (gānbiān dòujiǎo)

南方家常菜。先将豇豆炒至水分收干,然后加黄酒、豆瓣酱煸炒,淋猪油。

Sautéed cowpeas. Home-style dish in southern areas. Cowpeas sautéed, flavored with yellow wine and fermented broad bean sauce, then washed with lard.

干煸黄豆芽 (gānbiān huángdòuyá)

川黔滇菜。将黄豆芽同干辣椒及调料翻炒。

Sautéed soybean sprouts. Sichuan, Guizhou, and Yunnan dish. Soybean sprouts sautéed with chili pepper and seasonings.

干煸辣子鸡块 (gānbiān làzijīkuài)

西南家常菜。先将鸡块腌渍,稍炒,再炸至金黄,然后与煸炒过的干红辣椒及调料翻炒。

Stir-fried hot chicken. Home-style dish in southwestern areas. Quick-fried pickled chicken deep-fried, then stir-fried with prepared chili pepper and other condiments.

干煸鱿鱼丝 (gānbiān yóuyúsī)

四川菜。先将水发鱿鱼丝汆过,沥干,然后与干辣椒、笋丝及调料翻炒。

Stir-fried squid. Sichuan dish. Soaked shredded squid quick-boiled, drained, then stir-fried with chili pepper, shredded bamboo shoots, and seasonings.

干菜焖肉 (gāncài mèn ròu)

广东菜。先将猪五花肉加桂皮、茴香、黄酒等煮熟,再放入干菜焖煮,用盐和黄酒调味。

Stewed pork with preserved vegetable. Guangdong dish. Streaky pork boiled with cinnamon, fennel, and yellow wine, then stewed with preserved vegetable, salt, and yellow wine.

干炒鸡牛柳 (gānchǎo jī niúliǔ)

四川菜。先将鸡肉和牛肉条腌过,过油,然后放入辣椒、姜、葱、花椒,加盐、酱油、料酒翻炒。

Stir-fried chicken and beef. Sichuan dish. Lightly fried pickled chicken and beef strips stir-fried with chili pepper, ginger, scallion, and Chinese prickly ash, then seasoned with salt, soy sauce, and cooking wine.

干炒牛河 (gānchǎo niúhé)

广东点心。将河粉与牛肉片、胡萝卜、葱、盐、麻油等翻炒。

Stir-fried rice noodles with beef. Guangdong snack. Rice noodles stir-fried with sliced beef, carrots, scallion, salt, and sesame oil.

干锅带皮牛肉 (gānguō dàipíniúròu)

湖南菜。先将带皮牛肉、尖红椒和白萝卜及调料煸炒,然后盛入干锅。

Stir-fried chili beef. Hunan dish. Beef with skins stir-fried with red pepper and white turnips, then served on a bed of spices in a wok.

干锅红油手撕鸡 (gānguō hóngyóu shǒusījī)

湖南农家菜。先将腌渍过的鸡撕成条状,然后加红辣椒、红油及调料爆炒,然后盛入干锅。

Stir-fried chili chicken. Hunan dish. Pickled chicken hand-torn into strips, stir-fried with red pepper and chili oil, then served on a bed of spices in a wok.

干锅土鸭 (gānguō tǔyā)

湘西特色菜。将土鸭条加入黄豆芽、芹菜、蒜苗、泡椒、姜、香菜及调料爆炒,然后盛入干锅。

Stir-fried duck with vegetables. Specialty of western Hunan. Local duck cut into strips, stir-fried with soybean sprouts, celery, garlic shoots, pickled pepper, ginger, coriander, and other condiments. Served on a bed of spices in a wok.

干锅香辣鸡翅 (gānguō xiānglà jīchì)

西南菜。先将辣椒、花椒、蒜、葱炒香,然后与炸过的鸡翅加调料翻炒,然后盛入干锅。

Stir-fried hot chicken wings. Southwestern dish. Fried chicken wings stir-fried with quick-fried hot pepper, Chinese prickly ash, garlic, scallion, and other condiments. Served on a bed of spices in a wok.

干煎大虾 (gānjiān dàxiā)

江浙沪菜。先将大虾去须足,切成两段,盐渍,炸透,然后烹入料酒、糖、酱油。

Fried prawns with sauce. Jiangsu, Zhejiang, and Shanghai dish. Prawns cut into 2 sections without feelers, pickled with salt, deep-fried, then stir-fried with cooking wine, sugar, and soy sauce.

干煎黄鱼 (gānjiān huángyú)

浙江菜。先将黄鱼身打斜刀,用料酒、盐、姜、葱腌渍,再挂蛋浆炸至金黄,然后淋麻油、卤汁。

Fried yellow croaker with sauce. Zhejiang dish. Yellow croaker cross-carved on its body, pickled with cooking wine, salt, ginger, and scallion, coated with egg paste, deep-fried, then washed with sesame oil and special sauce.

干煎牛排 (gānjiān niúpái)

东北菜。先将牛里脊片腌渍,煎至金黄,然后切成条,浇上用洋葱、胡萝卜、盐等做的卤汁。

Fried beef with sauce. Northeastern dish. Pickled beef tenderloin fried until golden, cut into strips, then washed with dressing of onion, carrots, and salt.

干椒红麻童子鸡 (gānjiāo hóngmá tóngzǐjī)

四川菜。先将嫩鸡块腌渍,裹生粉,炸至金黄,然后加干辣椒、花椒等翻炒。

Fried hot chicken. Sichuan dish. Pickled cubed young chicken coated with corn starch, deep-fried, then stir-fried with chili pepper and Chinese prickly ash.

干焗蟹塔 (gānjú xiètǎ)

广东菜。将蟹肉、虾肉、胡椒粉、鸡蛋清等拌和,堆在蟹壳上,焗熟。

Baked crab and shrimp meat. Guangdong dish. Crab meat, shrimp meat, ground black pepper, and egg white mixed, piled on crab shells, then baked.

干烧陈皮肉丁 (gānshāo chénpí ròudīng)

四川菜。先将肉丁用陈皮、盐、酱油、黄酒、葱、姜等腌渍,再油炸,然后与葱、姜、辣椒等炒匀,烹入黄酒、酱油、醪糟汁。

Spicy pork with tangerine peels. Sichuan dish. Diced pork pickled with processed tangerine peels, salt, soy sauce, yellow wine, scallion, and ginger, then fried. Stir-fried with scallion, ginger, chili pepper, yellow wine, soy sauce, and sweet fermented sticky rice juice.

干烧凤翼翅 (gānshāo fèngyìchì)

广东菜。先将汆过的鸡翅去骨,然后塞入腌过的竹笋、香菇、火腿、芹菜,炸熟。

Fried chicken wings with vegetables. Guangdong dish. Chicken wings quick-boiled, boned, filled with pickled bamboo shoots, dried mushrooms, ham, and celery, then deep-fried.

干烧鳜鱼 (gānshāo guìyú)

四川菜。先将鳜鱼炸透,再加入辣椒、姜、葱、盐、酱油、黄酒等焖烧。

Simmered spicy Chinese perch. Sichuan dish. Chinese perch deep-fried, then simmered with hot pepper, ginger, scallion, salt, soy sauce, and yellow wine.

干笋炒腊肉 (gānsǔn chǎo làròu)

湖北菜。将干红椒、姜、青蒜煸香,放入腊肉片与浸软的干笋加调料翻炒。

Stir-fried preserved pork and bamboo shoots. Hubei dish. Smoked cured pork and soaked bamboo shoots stir-fried with quick-fried chili pepper, ginger, garlic sprouts, and seasonings.

干笋炒肉 (gānsǔn chǎo ròu)

南方菜。将焯过的带皮五花肉切片,然后与发好煮熟的干笋加青蒜、葱、姜、辣椒、酱油爆炒。

Stir-fried pork and bamboo shoots. Southern dish. Sliced quick-boiled streaky pork with skin stir-fried with boiled soaked bamboo shoots. Flavored with garlic sprouts, scallion, ginger, pepper, and soy sauce.

干炸带鱼 (gānzhá dàiyú)

上海菜。将带鱼段用姜末、花椒粉、大料粉、盐、酱油腌渍,然后用植物油文火炸透。

Fried ribbonfish. Shanghai dish. Sectioned ribbonfish pickled with minced ginger, ground Chinese prickly ash, aniseed, salt, and soy sauce, then deep-fried in vegetable oil over low heat.

干炸肝花 (gānzhá gānhuā)

广东菜。先将猪肝片和白膘肉片腌渍,再用豆腐皮裹成卷,先蒸后炸。

Fried fat pork and hog liver. Guangdong dish. Pickled sliced hog liver and fat pork rolled in bean curd sheets, steamed, then fried.

干炸鳅鱼 (gānzhá qiūyú)

湖南菜。先将鳅鱼用米酒腌渍,油炸,然后加入辣椒、姜、葱、盐等翻炒。

Stir-fried river eel. Hunan dish. River eel pickled with sweet fermented sticky rice juice, deep-fried, then stir-fried with chili pepper, ginger, scallion, and salt.

干蒸烧卖 (gānzhēng shāomài)

广东点心。薄面皮包入用糯米、猪肉、虾米等制成的馅,顶上张开并摺边,蒸熟。

Steamed buns stuffed with sticky

rice. Guangdong snack. Buns made with flour dough wrappers, stuffed with mixture of sticky rice, pork, and dried shrimp meat, frilled on open tops, then steamed.

干抓肉 (gānzhuāròu)

西藏菜。将连骨的羊肉与羊头、羊内脏等煮熟,手抓进食。

Boiled mutton and lamb offal. Tibetan dish. Mutton boiled with lamb head and offal. Served as finger food.

甘麦红枣汤 (gān mài hóngzǎo tāng)

台湾菜。将猪排骨加红枣、小麦、甘草等炖煮。

Sparerib and herb soup. Taiwan dish. Pork spareribs stewed with red dates, wheat, and liquorice root.

橄榄菜肉末蒸虾 (gǎnlǎncài ròumò zhēng xiā)

南方菜。将大虾从背部剖开,填入炒好的猪肉末,配橄榄菜蒸透。

Steamed pork in shrimp. Southern dish. Prawns cut open on the backs, filled with sautéed minced pork, then steamed with preserved salty vegetable.

赣南小炒鱼 (gànnán xiǎochǎoyú)

江西菜。先将草鱼块腌渍,裹生粉油炸,然后与青红椒、葱、姜、蒜翻炒,用米酒、醋、糖、盐、酱油调味。

Stir-fried spicy grass carp. Jiangxi dish. Pickled grass carp chunks coated with starch and fried. Stir-fried with red and green pepper, scallion, ginger, and garlic, then flavored with sweet fermented sticky rice juice, vinegar, sugar, salt, and soy sauce.

高沟捆蹄 (gāogōukǔntí)

江苏菜。用猪小肠衣裹住猪蹄髈肉条,捆紧,入原汤煮熟。

Boiled hog knuckle. Jiangsu dish. Hog knuckle strips wrapped in hog small intestine, tied hard, then boiled in broth.

高桥松饼 (gāoqiáosōngbǐng)

上海点心,原产于上海高桥镇。用精白粉、熟猪油、绵白糖、赤豆、桂花等制成饼,烤熟。

Sweet flour cakes. Shanghai snack, originated in Gaoqiao Town, Shanghai. Cakes made with fine flour, lard, soft white sugar, red beans, and osmanthus flower.

高邮咸蛋 (gāoyóuxiándàn)

江苏菜。高邮水乡鸭蛋用盐腌制后煮熟,因蛋黄大、油多而著名。

Preserved salty duck eggs. Jiangsu dish. Duck eggs from Gaoyou, a region of rivers and lakes, preserved with salt, then boiled. It is known for its big size and oily yolk.

格萨尔烤羊腿 (gésàěr kǎoyángtuǐ)

西藏名菜,将藏羊前腿腌制后烘烤,配生菜和红心萝卜。

Broiled lamb forelegs. Famous Tibetan dish. Tibetan lamb forelegs pickled, broiled, then served with lettuce and pink turnips.

蛤蚧炖鹰龟 (géjiè dùn yīngguī)

广西菜。先将当地鹰嘴龟和蛤蚧氽过,再加入姜和黄酒焖熟,然后与淮山、枸杞子、冬虫夏草同炖。

Stewed tortoise and geckos. Guangxi dish. Local hawk-mouth tortoise and geckos quick-boiled, simmered with ginger and yellow wine, then stewed with Chinese yam, Chinese wolfberry, and Chinese caterpillar fungus.

蛤蜊氽鲫鱼 (gélí cuān jìyú)

江苏菜。先把鲫鱼用开水烫过,油煎,加调料焖熟,再把蛤蜊用开水烫熟,放在鱼的两边。

Quick-boiled carp and clam. Jiangsu dish. Crucian carp scalded, fried, simmered with seasonings, then put on a platter with quick-boiled clam on both sides.

功德豆腐 (gōngdédòufu)

广州潮汕家常菜。先将豆腐炸至金黄,再与香菇、冬笋、青椒、红椒入素鲜汤烧煮,然后分层装盘,顶上缀樱桃。

Fried bean curd with vegetables. Home-style dish in Chaozhou and Shantou, Guangdong. Bean curd deep-fried, then boiled with mushrooms, bamboo shoots, green and red pepper in seasoned soup. Layered on a platter, then garnished with cherries on the top.

宫保鸡丁 (gōngbǎojīdīng)

四川菜。先将鸡肉丁用米酒、盐、蛋清、芡粉腌渍,再用热油快炸,然后同炸好的花生和干辣椒爆炒,加入姜、葱、蒜,勾芡。

Stir-fried spicy chicken. Sichuan dish. Diced chicken pickled with sweet fermented sticky rice juice, salt, egg white, and starch, then quickly fried over high heat. Stir-fried with fried peanuts, chili pepper, ginger, scallion, and garlic, then thickened with starch.

共阿馍馍 (gòngāmómo)

西藏点心。用面粉、鸡蛋、水和成面皮,肉为馅,做成包子,或蒸或炸。

Stuffed buns. Tibetan snack. Flour, egg paste, and water mixed, made into buns stuffed with meat, then steamed or fried.

共和凉菜卷 (gònghé liángcàijuǎn)

山东菜。先将芹菜、胡萝卜、白菜梗、香菇等切丝,水烫,加芥末酱拌匀,然后用熟薄面皮裹成卷。

Cold vegetable rolls. Shandong dish. Shredded celery, carrots, Chinese cabbage stalks, and dried mushrooms quick-boiled, then flavored with mustard sauce. Wrapped into rolls with thin pancakes.

贡眉 (gòngméi)

白茶,又称寿眉。产于福建建阳、建瓯等地。外形毫多肥壮,色泽翠绿。汤色透明橙黄,叶底柔软鲜亮。

Gongmei, also called *shouei*, white tea produced in Jianyang and Jianzhen, Fujian. Its dry leaves are plump, stout, jade green, and hairy. The tea is crystal yellowish with infused leaves soft and lustrous.

沟帮子熏鸡 (gōubāngzi xūnjī)

东北菜。先将小公鸡入料汤略泡，再入卤汤慢火煮酥，然后刷麻油，入熏锅用糖闷熏。

Smoked cockerel. Northeastern dish. Cockerel immersed in seasoned soup for a while, marinated in gravy over low heat, washed with sesame oil, then smoked with sugar in a covered wok.

狗肉炖干白菜粉 (gǒuròu dùn gānbáicài fěn)

东北菜。先将狗肉加调料煮熟，拆骨，再将干白菜焯软，与粉条、狗肉入原汤炖煮。

Dog meat with cabbage and noodles. Northeastern dish. Dog meat boiled with seasonings, boned, then stewed with dried Chinese cabbage and starch noodles in broth.

枸杞蛤蜊蒸蛋 (gǒuqǐ gélí zhēng dàn)

湖南菜。将枸杞、蛤蜊、蛋浆调和，加盐蒸熟。

Steamed clam and eggs with Chinese wolfberry. Hunan dish. Clam, Chinese wolfberry, and egg paste mixed, flavored with salt, then steamed.

枸杞菊花煲排骨 (gǒuqǐ júhuā bāo páigǔ)

广东菜。将排骨与枸杞、姜、菊花加盐煲煮。

Pork ribs with herbs. Guangdong dish. Pork ribs stewed with Chinese wolfberry, ginger, and chrysanthemum flower, then seasoned with salt.

枸杞银耳汤 (gǒuqǐ yíněr tāng)

广东菜。将银耳、枸杞、冰糖、桂花等一起熬煮。

Chinese wolfberry and white wood ear soup. Guangdong dish. White wood ears stewed with Chinese wolfberry, rock sugar, and osmanthus flower.

咕咾肉 (gūlǎoròu)

各地家常菜。先将猪肉腌渍，油煎，再与菠萝、青椒、红椒等翻炒，加入番茄酱、糖、醋。

Pork with pineapple and pepper. Home-style dish in many places. Pickled and fried pork sautéed with pineapple, green and red pepper, then flavored with tomato sauce, sugar, and vinegar.

古井贡酒 (gǔjǐnggòngjiǔ)

白酒，产于安徽亳县。用高粱为主料，用大麦、小麦、豌豆制曲酿造，酒精含量为 30-60 度，属浓香型酒。

Gujinggongjiu, liquor produced in Haoxian, Anhui. White spirit made with sorghum and yeast of barley, wheat, and peas. Contains 30-60% alcohol, and has a thick aroma.

古田干片 (gǔtiángānpiàn)

福建福安点心。将薄豆腐干片与多种调料烹煮，染成黄色，晾干。

Savoury dried bean curd. Snack in Fu'an, Fujian. Thin bean curd

pieces boiled with assorted seasonings, died with yellow color, then air-dried.

古突 (gǔtū)

藏族年夜饭,即面团肉粥。用羊肉、麦粒、杏干、豌豆、辣椒、羊毛、瓷片、线团、木炭等九种材料熬煮而成。

Tibetan chowder. Tibetan chowder prepared for the eve of the Tibetan New Year. It is boiled with nine ingredients such as mutton, Tibetan oat, dried apricot, peas, hot pepper, wool, porcelain pieces, wool coils, and charcoal.

骨头炖白菜粉条 (gǔtou dùn báicài fěntiáo)

东北菜。先将猪骨头焯水,然后加白菜、粉条及盐、黄酒、姜、葱炖熟。

Hog bone with cabbage and noodles. Northeastern dish. Quick-boiled hog bones stewed with Chinese cabbage and starch noodles. Seasoned with salt, yellow wine, ginger, and scallion.

骨头炖酸菜冻豆腐粉条 (gǔtou dùn suāncài dòngdòufu fěntiáo)

东北菜。先将猪骨头焯水,然后加酸菜、冻豆腐、粉条及调料炖熟。

Hog bone with vegetables. Northeastern dish. Quick-boiled hog bones stewed with pickled Chinese cabbage, frozen bean curd, and starch noodles, then seasoned.

顾渚紫笋 (gùzhúzǐsǔn)

绿茶。又称湖州紫笋、长兴紫笋。产于浙江长兴顾渚村。外形细嫩,银毫明显,色泽带紫。汤色清澈,叶底细嫩成朵。

Guzhuzisun, also called *huzhouzisun* or *changxingzisun*, green tea produced in Guzhu Village of Changxing, Zhejiang. Its dry leaves are tenuous, hairy and purplish. The tea is crystal clear with infused leaves like flowers.

挂霜丸子 (guàshuāng wánzi)

山东菜。先将肉丸炸透,再将糖炒至出霜,倒入肉丸翻匀。

Sugarcoated meatballs. Shandong dish. Meatballs deep-fried, then quickly fried with heated melting sugar.

怪味蚕豆 (guàiwèi cándòu)

四川传统点心。将水发蚕豆炸酥,加入盐、花椒粉、辣椒粉、醋、糖、姜末、蒜茸、甜面酱、芝麻酱、辣油等翻炒至干。

Savoury broad beans. Sichuan traditional snack. Soaked broad beans deep-fried, then stir-fried with salt, ground Chinese prickly ash, chili powder, vinegar, sugar, minced ginger, garlic, sweet fermented flour sauce, sesame sauce, and chili oil until dried.

怪味鸡 (guàiwèijī)

四川菜。先将鸡煮熟,切块,然后加入盐、酱油、花椒粉、辣椒粉、糖、芝麻、辣油、麻油等拌匀。

Savoury chicken. Sichuan dish. Boiled and chopped chicken mixed with salt, soy sauce, ground Chinese prickly ash, chili powder, sugar, sesame, chili oil, and sesame oil.

怪味牛肉卷 (guàiwèi niúròujuǎn)

四川菜。将牛肉末与花椒粉、盐、糖等调成馅,用面皮裹馅成卷,油炸。

Fried savoury beef rolls. Sichuan dish. Ground beef mixed with ground Chinese prickly ash, salt, and sugar, rolled in flour sheets, then deep-fried.

怪味烟熏排骨 (guàiwèi yānxūn páigǔ)

四川菜。先将排骨用五香粉、盐、糖、姜、葱、料酒腌渍,稍蒸,再入卤汤煮透,油炸至干,然后用木炭烟熏至暗红,淋麻油。

Smoked savoury pork steaks.

Sichuan dish. Pork steaks pickled with five-spice powder, salt, sugar, ginger, scallion, and cooking wine. Quickly steamed, marinated in gravy, deep-fried until dried, smoked over charcoal, then sprinkled with sesame oil.

关东煮鸡 (guāndōngzhǔjī)

东北菜。将鸡加调料煮熟,用原汤浸泡两天,捞出风干,放入缸中储存。食用前入汤加调料煨炖。

Stewed preserved chicken. North-eastern dish. Chicken boiled with condiments, marinated in broth for 2 days, air-dried, then stored in a jug. Stewed in soup with seasonings before serving.

棺材板 (guāncaibǎn)

台湾点心。先将面包外面煎黄,再把中间挖空,填入用鸡肝、牛奶、虾仁等制成的馅,然后盖上面包片。

Fried stuffed bread. Taiwan snack. Bread fried until crisp, hollowed, filled with prepared chicken liver, milk, and shrimp meat, then covered with a piece of fried bread.

灌肠 (guàncháng)

西藏传统菜。将半熟米、肉末、羊血、茴香等灌入羊肠,煮熟后切段,可与其他菜烹炒或单独食用。

Rice, meat, and blood sausage. Tibetan traditional dish. Lamb intestine stuffed with half-cooked rice, ground meat, lamb blood, and fennel, boiled, then cut into segments. It may be cooked with other ingredients or served alone.

灌蟹鱼圆 (guànxiè yúyuán)

江苏菜。将鱼肉与肥猪肉末及蛋清做成鱼圆,塞入蟹粉馅,煮熟。

Fish balls with crab meat. Jiangsu dish. Fish meat, ground fat pork, and egg white mixed, made into balls, stuffed with miced crab meat, then boiled.

罐焖核桃鸡块 (guànmèn hétao jīkuài)

广东菜。将鸡块和核桃仁放入瓦罐,加鸡汤焖煮,用盐、黄酒、姜、葱调味。

Braised chicken with walnuts. Guangdong dish. Chopped chicken and walnuts braised in chicken soup in an earthen pot, then seasoned with salt, yellow wine, ginger, and scallion.

广东大叶青 (guǎngdōng dàyèqīng)

黄茶。产于广东韶关、湛江等地。外形条索肥壮显毫,色泽青润显黄。汤色橙黄明亮,汤味浓醇回甘。叶底淡黄,叶张完整。

Guangdong dayeqing, yellow tea produced in Shaoguan and Zhanjiang, Guangdong. Its dry leaves are plump, stout, hairy, and yellowish green. The tea is crystal yellow with a rich sweet aftertaste. The infused leaves are light yellow and in good shape.

广东粉果 (guǎngdōngfěnguǒ)

广东点心。将澄面、生粉加水和油做成面皮,包入用猪肉、冬菇、盐等做成的馅,蒸熟。

Steamed stuffed buns. Guangdong snack. Wheat starch, corn starch, water, and oil mixed, made into wrappers, packed with mixture of seasoned pork, winter mushrooms, and salt, then steamed.

广东裹蒸粽 (guǎngdōng guǒzhēngzòng)

广东点心。将糯米、五花肉、鸡肉、香菇、叉烧、绿豆、蛋黄、栗子、烤鸭肉等用竹叶或芦叶包裹成粽子,蒸熟。

Steamed savoury *zongzi*.

Guangdong snack. Sticky rice, streaky pork, chicken, dried mushrooms, barbecued pork, mung beans, egg yolks, chestnuts, and broiled duck mixed, wrapped into *zongzi* with bamboo or reed leaves, then steamed.

广东卤猪肉 (guǎngdōng lǔzhūròu)

广东菜。先将猪肉氽过,然后放入用八角、甘草、桂皮、沙姜、丁香、花椒等制成的卤汤中慢火煮熟。

Marinated pork. Guangdong dish. Scalded pork marinated over low heat in gravy of aniseed, licorice root, cinnamon, kaempferia rootstock clove, and Chinese prickly ash.

广式烧填鸭 (guǎngshì shāotiányā)

广东菜。将整鸭腹内填入葱、姜、蒜、糖、料酒、花椒、酱油等调料,烤熟。

Broiled stuffed duck. Guangdong dish. Whole duck stuffed with mixture of scallion, ginger, garlic, sugar, cooking wine, Chinese prickly ash, and soy sauce, then broiled.

归丹 (guīdān)

西藏传统点心,即酒羹。将青稞酒与红糖、人参果、碎奶渣、糌粑等搅拌,煮成糊状。

Barley wine chowder. Tibetan traditional snack. Mixture of highland barley wine, brown sugar, silverweed roots, minced dairy dregs, and *zanba* mixed, then boiled into chowder.

龟羊汤 (guīyángtāng)

湖南菜。先将龟肉和羊肉煸炒,然后加入冰糖、党参、当归、枸杞等熬炖。

Stewed turtle and mutton. Hunan dish. Turtle and mutton quick-fried, then stewed with rock sugar and herbs such as pilose asiabell root, Chinese angelica, and Chinese wolfberry.

贵妃鸡 (guìfēijī)

北方菜。先将鸡翅炸至金黄,然后与煸过的猪排入鸡汤焖煮,加盐、姜、葱、红葡萄酒。

Braised chicken wings and pork ribs. Northern dish. Deep-fried chicken wings and stir-fried pork ribs stewed in chicken soup and flavored with salt, ginger, scallion, and red wine.

桂花酒酿元宵 (guìhuā jiǔniàng yuánxiāo)

江浙沪点心。将元宵煮熟,加入酒酿及糖桂花。

Sticky rice dumplings in fermented rice soup. Jiangsu, Zhejiang, and Shanghai snack. Sticky rice dumplings boiled in sweet fermented sticky rice soup and flavored with sugared osmanthus flower.

桂花糯米糖藕 (guìhuā nuòmǐ tángǒu)

江浙沪点心。将糯米塞满藕孔,用切下的藕盖把口封上,放入锅中煮透,然后切片装盘,浇糖浆,撒糖桂花。

Sweet lotus roots with osmanthus flower. Jiangsu, Zhejiang, and Shanghai snack. Lotus roots filled with sticky rice, sealed with lotus root ends, boiled, then sliced. Washed with syrup, then showered with sugared osmanthus flower.

桂花藕粉 (guìhuā ǒufěn)

浙江点心。将藕粉用开水冲成糊状,加入白糖和糖桂花。

Lotus starch gruel with osmanthus flower. Zhejiang snack. Lotus root starch made into gruel with boiling water, then flavored with white sugar and sugared osmanthus flower.

桂花肉 (guìhuāròu)

上海菜。将猪瘦肉片裹淀粉糊油炸,撒上椒盐粉。

Fried lean pork with spicy powder. Shanghai dish. Sliced lean pork coated with starch, deep-fried, then served with salt and prickly ash power.

桂花糖大栗 (guìhuā tángdàlì)

江苏点心。先将栗子肉蒸熟,再加入白糖和猪油烹煮,然后放入糖桂花,勾芡,撒上红绿丝。

Braised chestnuts with sugared osmanthus flower. Jiangsu snack. Shelled chestnuts steamed, braised with sugar and lard, flavored with sugared osmanthus flower, thickened with starch, then sprinkled with shredded candied red and green gourd.

桂花糖芋艿 (guìhuā tángyùnǎi)

浙江点心。先将芋艿用清水氽过,再加食用碱、红糖、白糖煮熟,撒糖桂花。

Sweet taros with osmanthus flower. Zhejiang snack. Scalded taros boiled with baking soda, brown and white sugar, then sprinkled with sugared osmanthus flower.

桂花蹄筋 (guìhuā tíjīn)

湖南菜。先把油炸蹄筋用水发好,焯水,入鲜汤煮熟,然后与蛋浆同炒,撒火腿末、胡椒粉,淋麻油。

Stir-fried pork tendons with eggs. Hunan dish. Fried soaked pork tendons quick-boiled, braised in seasoned soup, stir-fried with egg paste, flaked with ham crumbs and ground black pepper, then washed with sesame oil.

桂花芋泥卷 (guìhuā yùníjuǎn)

广东点心。面皮包入芋泥、猪油、桂花、糖、豆沙等制成的馅,炸至金黄。

Osmanthus flower and taro rolls. Guangdong snack. Flour dough sheets made into rolls filled with mashed taros, lard, dried osmanthus flower, sugar, and bean paste, then deep-fried.

桂林荷叶鸭 (guìlín héyèyā)

广西菜。先将鸭炸至金黄,再将炒过的瘦肉、冬笋、腊肉、香菇等填入鸭腹,用荷叶包裹,蒸熟。

Steamed duck in lotus leaves. Guangxi dish. Deep-fried whole duck stuffed with stir-fried lean pork, winter bamboo shoots, smoked cured pork, and dried mushrooms, wrapped in lotus leaves, then steamed.

桂林田螺 (guìlíntiánluó)

广西菜。将田螺用当地酸辣椒煮过,再加葱、姜、料酒等翻炒。

Stir-fried field snails. Guangxi dish. Field snails boiled with local pickled sour pepper, then stir-fried with scallion, ginger, and cooking wine.

桂平西山茶 (guìpíng xīshānchá)

绿茶。产于广西桂平西山。外形纤细紧结,成龙卷状,茸毫显露。汤色清澈,叶底嫩绿。

Guiping Xishancha, green tea produced in Mount West in Guiping, Guangxi. Its dry leaves are tenuous, tight, curved, and hairy. The tea is crystal with infused leaves tender green.

鳜鱼螃蟹火锅 (guìyú pángxiè huǒguō)

湖北菜。先将螃蟹、酸菜丝、大海米入鸡汤煮开,然后放入海参、鳜鱼片等涮熟。

Crab and fish in fire pot. Hubei dish. Crab, pickled cabbage, and dried prawn meat boiled in chicken soup in a fire pot. Sea cucumbers, Chinese perch fillets, and other ingredients added for quick boiling.

锅巴双脆 (guōbā shuāngcuì)

南方菜。将腌渍过的猪肚片和鸭胗片加调料爆炒,倒在油炸过的糯米锅巴上。

Hog tripe and duck gizzard with rice crust. Southern dish. Pickled and sliced hog tripe and duck gizzard stir-fried with seasonings, then topped over fried sticky rice crust.

锅包肉 (guōbāoròu)

北方菜。将里脊肉腌渍,裹上淀粉糊,炸至金黄,然后加入香菜、红椒丝、糖、料酒、蒜茸等翻炒。

Stir-fried pork fillets. Northern dish. Pickled pork tenderloin coated with starch, deep-fried, then stir-fried with coriander, shredded red pepper, sugar, cooking wine, and minced garlic.

锅爆鲈鱼 (guōbào lúyú)

上海菜。先将鲈鱼煎至金黄,然后入清汤烹煮,撒姜末和葱花。

Braised perch. Shanghai dish. Perch fried until golden, braised in clear soup, then flavored with minced ginger and chopped scallion.

锅魁 (guōkuí)

四川家常点心。将面粉与水、椒盐、糖、葱油调和,烙成饼,配泡椒、大头菜、卤肉。

Salty pancake with appetizer. Home-style snack in Sichuan. Flour mixed with water, spicy salt, sugar, and scallion oil, made into cakes, then baked. Served with pickled pepper, preserved mustard turnips, and marinated pork.

锅烧鸡 (guōshāojī)

四川菜。先将仔鸡煮至将熟,去骨,加入调料蒸熟,再抹上芡粉、盐、黄酒、蛋浆糊,炸透,配椒盐。

Fried chicken with spicy salt. Sichuan dish. Boiled and boned young chicken steamed with seasonings, coated with paste of starch, salt, yellow wine, and egg paste, then deep-fried. Served with spicy salt.

锅烧鳗 (guōshāomán)

浙江菜。用葱段铺底,将鳗鱼段排列其上,加调料煮沸,用小火焖至肉酥汁浓。

Braised eel. Zhejiang dish. Eel sections put on a bed of scallion with other seasonings, then braised until thickened.

锅酥牛肉 (guōsū niúròu)

各地家常菜。将牛肉片裹上加调料的鸡蛋淀粉浆,炸黄,切条,配生菜丝装盘。

Fried beef with lettuce. Home-style dish in many places. Sliced beef coated with seasoned egg and starch paste, deep-fried, cut into strips, then garnished with shredded lettuce.

锅塌豆腐 (guōtā dòufu)

山东菜。先将肉馅塞入豆腐块,蒸熟,裹蛋糊煎至浅黄,然后加少量汤汁焖煮。

Braised pork in bean curd. Shandong dish. Cubed bean curd filled with seasoned pork, steamed, coated with egg paste, fried, then simmered with a little soup.

锅塌鸡片 (guōtā jīpiàn)

山东菜。先将鸡脯肉挂上蛋糊,煎至金黄,再加入少量清汤烹煮,用黄酒、酱油等调味。

Fried and simmered chicken breast. Shandong dish. Chicken breast coated with flour and egg paste, deep-fried, simmered with a little clear soup, then flavored with yellow wine and soy sauce.

锅塌蒲菜 (guōtā púcài)

山东菜。先将蒲菜裹蛋糊,用微火煎至金黄,再放入少量清汤,加盐烹煮。

Fried and simmered cattail roots. Shandong dish. Cattail roots coated with flour and egg paste, fried until golden over low heat, simmered with a little clear soup, then seasoned with salt.

锅鳎鱼盒 (guōtǎ yúhé)

山东菜。先用鳎鱼即比目鱼片夹上肉馅,挂芡粉鸡蛋浆,煎至金黄,再入清汤煨煮。

Simmered fish fillets with pork. Shandong dish. Sliced flatfish sandwiched with ground pork, coated with starch and egg paste, deep-fried, then simmered in clear soup.

锅贴 (guōtiē)

南方点心。面皮包入肉末蔬菜馅,做成饺子形,用少量油和水先煎后蒸。

Fried pork and vegetable buns. Southern snack. Flour dough made into wrappers, filled with ground pork and vegetables, then made into *jiaozi*-shape buns. Fried with a little oil, then steamed with a little water.

锅贴鸡片 (guōtiē jīpiàn)

四川菜。将腌过的鸡肉片和肥猪肉片煎熟,配圆白菜丝装盘。

Fried chicken and fatty pork. Sichuan dish. Pickled chicken and sliced fat pork fried, then served with shredded cabbage.

锅贴石斑夹 (guōtiē shíbānjiá)

广东菜。先将石斑鱼片裹蛋液,放到面包片上,再将粘在一起的鱼片和面包片裹上蛋液,炸至金黄。

Fried grouper with bread. Guangdong dish. Grouper fillets coated with egg paste, each piece stuck to a piece of bread, coated with egg paste, then fried golden.

锅贴鱼 (guōtiēyú)

山东菜。将鱼片挂芡粉鸡蛋浆,撒火腿米,盖上猪肉片,稍焖煎,然后入油锅炸透。

Fried fish fillets and pork. Shandong dish. Fish fillets coated with starch and egg paste, sprinkled with crumbed ham, topped with sliced pork, simmered briefly, then deep-fried.

锅熠银鱼 (guōyì yínyú)

山东菜。将银鱼腌渍,裹上淀粉蛋浆,煎至金黄,淋麻油。

Fried whitebait fish. Shandong dish. Pickled whitebait fish coated with starch and egg white, fried until golden, then washed with sesame oil.

果条筒 (guǒtiáotǒng)

福建食品。米浆蒸熟成薄张,卷成圆柱形,佐大蒜、酱油。

Steamed rice rolls. Fujian food. Rice paste steamed into sheets, then made into rolls. Served with garlic, soy sauce or other sauces.

过桥米线 (guòqiáomǐxiàn)

云南食品。先把热鸡油和调料放入大碗,冲入烧开的上汤,然后放入肉片、鱼片、蔬菜、米线烫熟。

Scalded noodles. Yunnan dish. Heated chicken oil and seasonings put in a big bowl. Boiling broth poured in. Sliced meat, fish, vegetables, and rice noodles added for scalding.

咖喱鸡 (gālíjī)

各地家常菜。先将鸡块爆炒,然后加土豆、清汤、咖喱,小火炖至收汁。

Curry chicken. Home-style dish in many places. Cubed chicken stir-fried, then braised in clear soup with potatoes and curry until reduced.

咖喱牛肉粉丝汤 (gālí niúròu fěnsī tāng)

上海菜。先用蒜茸和姜末爆香牛肉,再把牛肉和粉丝放入牛肉汤煮熟,加黄酒、盐、咖喱粉。

Beef and starch noodles in curry soup. Shanghai dish. Beef quick-fried with mashed garlic and ginger, boiled in beef soup with fine starch noodles, then flavored with yellow wine, salt, and curry powder.

H

哈达饼 (hādábǐng)

内蒙点心。将水油面和油酥面擀成面皮，包上熟面粉、糖、葵花子、芝麻、核桃仁等做成的馅，擀成荷叶状饼，烙熟。

Cake stuffed with nuts and sugar. Inner Mongolian snack. Flour, water, and butter made into dough. Flour and butter made into pastry. Dough and pastry made into wrappers filled with mixture of roasted flour, sugar, sunflower seeds, sesame, and walnuts. Kneaded into cakes in the shape of lotus leaves, then baked.

哈尔滨红肠 (hāěrbīn hóngcháng)

黑龙江菜。先将碎猪肉加调料做成馅，填入肠衣，风干，然后用炭火烤熟。

Red pork sausage. Heilongjiang dish. Seasoned pork in hog casing air-dried, then roasted over charcoal.

哈尔滨啤酒 (hāěrbīn píjiǔ)

中国老牌啤酒，产于黑龙江哈尔滨。以大麦、大米、新疆酒花、特殊处理过的水制成，酒色清澈，泡沫细腻洁白，口味清正爽口。

Harbin Beer. The oldest beer in China produced in Harbin, Heilongjiang. Brewed from barley, rice, hops from Xinjiang, yeast, and prepared water. It is crystal clear with snow white foam, and tastes cool and refreshing.

海底松银肺 (hǎidǐsōng yínfèi)

江苏菜。把猪肺、海蜇头、火腿加调料用砂锅煮熟。

Braised hog lung and jelly fish. Jiangsu dish. Hog lung, jellyfish, and ham braised with seasonings in a casserole.

海蛎包 (hǎilìbāo)

福建福安点心。海蛎即牡蛎。用米浆为皮，海蛎肉、葱等为馅，炸至金黄。

Fried oyster dumplings. Snack in Fu'an, Fujian. Rice paste made into wrappers, packed with oyster and scallion, then deep-fried until golden.

海蛎饼 (hǎilìbǐng)

福建著名点心。先用大米、黄豆磨成浓浆，倒少量浆入球形勺，再放入用海蛎肉、瘦猪肉、姜、葱、盐等做成的馅，面上浇上米浆，然后置入油锅煎至金黄。

Fried oyster cakes. Famous Fujian snack. Rice, soybean, and water ground into paste, put in a scoop, filled with mixture of oyster, lean pork, ginger, scallion, and salt, covered with rice paste, then deep-fried.

海米扒油菜 (hǎimǐ pá yóucài)

山东菜。将海米和汆过的油菜心放入烧沸的鲜清汤煮开,勾芡。

Boiled shrimp meat and rape greens. Shandong dish. Dried shrimp meat and scalded rape greens boiled in seasoned clear soup and thickened with starch.

海米翡翠柿子椒 (hǎimǐ fěicuì shìzijiāo)

东北家常凉菜。先将青椒焯水,然后加海米、酱油、麻油拌匀。

Green pepper with dried shrimp meat. Northeastern home-style cold dish. Green pepper quick-boiled, then flavored with dried shrimp meat, soy sauce, and sesame oil.

海米香干芹菜 (hǎimǐ xiānggān qíncài)

东北家常凉菜。香干即熏豆腐干。先将香干和芹菜段焯熟,然后加海米、酱油、麻油拌匀。

Bean curd with celery and shrimp meat. Northeastern home-style cold dish. Shredded smoked bean curd and sectioned celery quick-boiled and then mixed with dried shrimp meat, soy sauce, and sesame oil.

海米珍珠笋 (hǎimǐ zhēnzhūsǔn)

山东菜。先将珍珠笋焯水,然后放入海米拌炒,烹入高汤。

Sautéed baby corn and shrimp meat. Shandong dish. Baby corn quick-boiled, sautéed with dried shrimp meat, then flavored with broth.

海南胡椒肚 (hǎinán hújiāodǔ)

海南菜。先将猪肚用胡椒粉、白糖、蒜茸、料酒等腌渍,然后填入腌好的猪头肉和猪舌,蒸熟。

Steamed hog tripe and head meat. Hainan dish. Hog tripe pickled with ground black pepper, sugar, minced garlic, and cooking wine. Stuffed with pickled hog head meat and hog tongue, then steamed.

海南鸡饭 (hǎinánjīfàn)

海南食品。先把土鸡用葱、姜、米酒腌过,煮熟,置凉,切块;用鸡汤煮熟米饭,佐辣酱或黑酱油。

Rice with chicken. Specialty in Hainan. Chicken pickled with scallion, ginger, and yellow wine. Boiled, chilled, and chopped. Rice cooked in chicken soup. Served with chili sauce or black soy sauce.

海南椰奶鸡 (hǎinán yēnǎijī)

海南文昌菜。先将土鸡浸入上汤,加盐、姜、葱、黄酒蒸熟,然后浇用牛奶、椰汁、生粉做成的卤汁。

Steamed chicken in coconut milk. Dish originated in Wenchang, Hainan. Local chicken steamed in broth with salt, ginger, scallion, and yellow wine, then washed with gravy of milk, coconut milk, and starch.

海鲜煲 (hǎixiānbāo)

广东菜。将鱿鱼、虾、洋葱、番茄等加入椰汁和调料煲煮,勾芡。

Seafood in coconut milk. Guangdong dish. Squid, shrimp, onion, and tomatoes braised in coconut milk, seasoned, then thickened with starch.

海鲜铁锅烧 (hǎixiān tiěguōshāo)

浙江菜。将鱼肚、海参、鲜虾、冬笋、菜心加调料炒熟,勾芡。

Stir-fried seafood and vegetables. Zhejiang dish. Fish maw, sea cucumbers, fresh shrimp, winter bamboo shoots, and baby cabbage stir-fried with seasonings, then thickened with starch.

含羞丸子 (hánxiū wánzi)

山东菜。将鱼茸和鸡脯茸加调料制成丸子,慢火油炸。

Fried chicken and fish balls. Shandong dish. Meatballs of seasoned ground fish and chicken breast deep-fried over low heat.

杭三鲜 (hángsānxiān)

杭州名菜。名为三鲜,实为多鲜。把鸡、火腿、虾、鱼、猪肉、笋、猪肚、肉皮等加调料炒熟,勾芡。

Three delicacies. Hangzhou famous dish. Its name means 3 delicacies, but it actually contains more. Chicken, ham, shrimp, fish, pork, bamboo shoots, hog tripe, and hog skins stir-fried with seasonings, then thickened with starch.

杭州菊花茶(hángzhōu júhuāchá)

再加工茶,产于浙江杭州,中国著名花茶。以绿茶为茶胚,用菊花窨制而成。香气清芬,汤色澄明。

Hangzhou chrysanthemum tea, produced in Hangzhou, Zhejiang. Famous re-processed flower-scented tea. Made with green tea and chrysanthemum flower. The tea has an elegant aroma, a crystal yellowish green color, and a light sweet aftertaste.

杭州素火腿 (hángzhōu sùhuǒtuǐ)

浙江菜。把豆腐皮切成粒状,加调料拌匀,装入火腿模型中蒸熟。

Soy ham. Zhejiang dish. Chopped bean curd skins seasoned, put into a ham-shape container, then steamed.

蚝菇鸡球 (háogū jīqiú)

广东菜。鸡丁和草菇加入盐、葱、姜、蚝油、黄酒等同炒。

Stir-fried chicken and mushrooms. Guangdong dish. Diced chicken and meadow mushrooms stir-fried, then flavored with ginger, scallion, oyster sauce, and yellow wine.

蚝油鲍脯 (háoyóu bàopú)

山东菜。先将鲍鱼用蚝油、鸡汤等炒熟,放在炒熟的青菜上,然后浇勾芡的卤汁。

Sautéed abalone with greens.

Shandong dish. Abalone sautéed with oyster sauce and chicken soup, put on a bed of prepared green Chinese cabbage, then washed with starched gravy.

蚝油冬菇 (háoyóu dōnggū)

各地家常菜。冬菇片与焯水冬笋

片加蚝油和少许汤烩煮，勾芡，淋猪油。

Mushrooms with oyster sauce. Home-style dish in many places. Sliced winter mushrooms and scalded sliced winter bamboo shoots simmered with oyster sauce and a little soup, thickened with starch, then washed with lard.

蚝油鸡 (háoyóujī)

北京菜。先将仔鸡用沸水稍烫，再煮熟切块，然后浇蚝油。

Chicken with oyster sauce. Beijing dish. Scalded young hen boiled, chopped, then washed with oyster sauce.

蚝油茭白 (háoyóu jiāobái)

浙江菜。先将茭白片过油，再用蚝油煸炒，然后加调料焖煮，勾芡，淋麻油。

Wild rice stems with oyster sauce. Zhejiang dish. Sliced wild rice stems lightly fried, sautéed with oyster sauce, simmered with condiments, thickened with starch, then washed with sesame oil.

蚝油生菜 (háoyóu shēngcài)

广东菜。将生菜汆过，浇上用蚝油、黄酒、糖、味精、酱油、芡粉、清汤等调成的卤汁，淋麻油。

Scalded lettuce with oyster sauce. Guangdong dish. Romaine lettuce scalded, marinated in dressing of oyster sauce, yellow wine, sugar, MSG, soy sauce, starch, and clear soup, then flavored with sesame oil.

蚝汁烧牛肚 (háozhī shāo niúdǔ)

广东菜。将牛肚与蚝油、姜片、葱、红椒、清汤等煮熟，勾芡。

Ox tripe with oyster sauce. Guangdong dish. Ox tripe braised in clear soup with oyster sauce, ginger, scallion, and red pepper, then thickened with starch.

荷包蛋 (hébāodàn)

各地家常菜。分为水煮荷包蛋和煎荷包蛋。将鸡蛋逐个打入锅中，煮或煎至蛋白成固体，蛋黄成半流体。水煮荷包蛋配糖或盐汤，煎荷包蛋佐酱油。

Omelet or poached eggs. Home-style dish around the country. Eggs broken, then boiled or fried until egg white solid and egg yolks soft. Boiled eggs served in sugared or salty soup. Fried eggs served with soy sauce.

合川肉片 (héchuānròupiàn)

四川菜。先将猪瘦肉片腌渍，然后与玉兰片、木耳、泡椒、姜、葱等爆炒。

Stir-fried pork. Sichuan dish. Pickled sliced lean pork stir-fried with sliced bamboo shoots, wood ears, pickled pepper, ginger, and scallion.

河套老窖 (hétàolǎojiào)

白酒，产于内蒙古河套平原，以高粱、小麦为原料，大麦、豌豆、谷壳等制曲酿制。酒精含量为28-52度，属浓香型酒。

Hetaolaojiao, liquor produced in

Hetao plain, Inner Mongolia. White spirit made with sorghum and wheat as the primary materials and yeast of barley, peas, and rice shells. Contains 28-52% alcohol, and has a thick aroma.

河州大饼 (hézhōudàbǐng)

清真食品,源于甘肃河州。面粉加水、盐、香豆粉、花椒粉、胡麻油等做成饼,烤熟。

Baked spicy cakes. Muslim food, originated in Hezhou, Gansu. Flour mixed with water, salt, common fenugreek seeds powder, ground Chinese prickly ash, and flax seed oil, made into cakes, then baked.

核桃仁拌鸡块 (hétaorén bàn jīkuài)

北方传统菜。先将鸡肉煮熟,去骨切条,核桃仁稍烫去皮,然后加酱油、辣椒油、糖、醋等拌匀。

Chicken with walnuts. Traditional dish in northern areas. Boiled and boned chicken slivers mixed with scalded and skinned walnuts, then seasoned with soy sauce, chili oil, sugar, and vinegar.

荷花白嫩鸡 (héhuā báinènjī)

江苏菜。先把油炸过的全鸡蒸酥,再把油炸过的荷花瓣摆放在鸡的周围,浇上卤汁。

Chicken with lotus flower. Jiangsu dish. Whole chicken deep-fried, steamed, decorated with fried lotus flower petals, then washed with special sauce.

荷花吉利鱼片 (héhuā jílì yúpiàn)

安徽、江苏菜。将腌过的鱼片裹上芡粉蛋浆和面包屑,炸熟,用番茄切成的荷花瓣饰盘。

Fried fish fillets with tomatoes. Anhui and Jiangsu dish. Fish fillets coated with starch and egg paste and bread crumbs, fried, then garnished with tomato sections cut like lotus petals.

荷花集锦炖 (héhuā jíjǐndùn)

江苏菜。将鸡脯肉、虾茸蛋卷、火腿、冬笋、鳜鱼肉、鸡肫、猪腰、鱼肚、浆虾仁等先蒸后炖。

Stewed meat varieties. Jiangsu dish. Chicken breast, shrimp and egg rolls, ham, winter bamboo shoots, Chinese perch meat, chicken gizzard, hog kidney, fish maw, and starched shrimp meat steamed, then stewed.

荷花铁雀 (héhuā tiěquè)

江苏菜。先将铁雀用黄酒、盐、姜葱汁腌渍,再裹上糯米粉油炸,然后浇糖醋芝麻油卤汁。

Yellow-breast buntings. Jiangsu dish. Yellow-breast buntings pickled with yellow wine, salt, and ginger-scallion juice, coated with sticky rice flour, deep-fried, then washed with dressing of sugar, vinegar, and sesame oil.

荷香粉蒸肉 (héxiāng fěnzhēngròu)

湖北、湖南菜。先将腌过的猪肉与米粉拌匀,再用烫过的荷叶包成方块,蒸熟。

Steamed pork in lotus leaves. Hubei

and Hunan dish. Pickled pork mixed with rice flour, wrapped into cubes with scalded lotus leaves, then steamed.

荷香笼仔鸭 (héxiānglóng zǐyā)

广东菜。先将鸭块用料酒、麻油、酱油、盐、味精等腌渍,然后用荷叶包好蒸熟。

Steamed duck in lotus leaf.

Guangdong dish. Duck chunks pickled with cooking wine, sesame oil, soy sauce, salt, and MSG, wrapped in lotus leaves, then steamed.

荷叶粉蒸鸡 (héyè fěnzhēngjī)

安徽、江苏菜。先将鸡肉与猪板油用姜、葱、黄酒、酱油腌渍,裹上用炒米粉、麻油、酱油、糖等调成的糊,然后用荷叶包上,蒸熟。

Steamed chicken in lotus leaf. Anhui and Jiangsu dish. Chicken and lard pickled with ginger, scallion, yellow wine, and soy sauce, coated with paste of roasted ground rice, sesame oil, soy sauce, and sugar, wrapped in lotus leaves, then steamed.

黑茶 (hēichá)

后发酵茶。经杀青、揉捻、渥堆、干燥四个工序加工而成,因茶叶为黑褐色而得名。汤色黄中带红、汤味醇和。常被加工成各式压制茶。

Heicha tea. Dark green tea or post-fermented tea. It is processed by gentle-baking, rubbing-twisting, piling-fermenting, and drying. Its tea is yellowish red with a stale and mellow flavor. It is often used to make different types of compressed teas.

黑豆焖猪蹄 (hēidòu mèn zhūtí)

广东菜。将猪蹄与黑豆、调料炖煮。

Stewed pork trotters and black soybean. Guangdong dish. Pork trotters stewed with black soybean and seasonings.

黑椒牛柳 (hēijiāo niúliǔ)

广东菜。先将牛里脊肉腌渍,再与甜豆、芋头、青椒翻炒,撒黑胡椒末。

Stir-fried beef and vegetables.

Guangdong dish. Pickled beef tenderloin stir-fried with snow peas, taros, and green pepper, then sprinkled with ground black pepper.

黑三剁 (hēisānduò)

云南菜。将肉末、碎尖椒、玫瑰大头菜粒翻炒,加鸡精和麻油。

Pork with pepper and mustard turnips. Yunnan dish. Prepared ground pork stir-fried with chopped hot pepper and preserved mustard turnips, then seasoned with chicken essence and sesame oil.

红扒鸡 (hóngpājī)

山东菜。先将鸡腌渍,水焯,再涂上红糖和白酒油炸,然后焖煮,用原汤勾芡。

Braised chicken with sauce.

Shandong dish. Pickled chicken scalded, coated with brown sugar and white spirit, deep-fried, braised, then marinated in starched broth.

红扒鱼唇 (hóngpá yúchún)

北方菜。将鱼唇、猪肉片、鸡块汆过,加调料蒸至鱼唇熟烂,勾芡。

Steamed fish lips with sauce. Northern dish. Fish lips, sliced pork, and chopped chicken quick-boiled, steamed with seasonings, then thickened with starch.

红扒鱼肚 (hóngpá yúdǔ)

北方菜。先将发好的鱼肚用鸡汤煨熟,再油炸,然后入鸡汤加盐、糖、酱油、黄酒、葱、姜烹煮,勾芡。

Braised fish maw with sauce. Northern dish. Prepared fish maw braised in chicken soup, fried, simmered with salt, sugar, soy sauce, yellow wine, scallion, and ginger, then thickened with starch.

红白豆腐汤 (hóngbái dòufu tāng)

四川菜。将胡萝卜、白萝卜、豆腐加油、盐、姜、葱煮汤。

Carrot, turnip, and bean curd soup. Sichuan dish. Carrots, white turnips, and bean curd boiled in clear soup and seasoned with oil, salt, ginger, and scallion.

红斑二吃 (hóngbān èrchī)

广东菜。先将半条红斑鱼滑油,用姜丝、胡萝卜丝、酱油、盐拌炒,再把另外半条鱼加香菇丝、姜、葱、盐、黄酒蒸熟,然后一起装盘。

Steamed and sautéed red spotted grouper. Guangdong dish. Fish halved. One half lightly fried, then sautéed with shredded ginger, carrots, soy sauce, and salt. The other half steamed with shredded dried mushrooms, ginger, scallion, salt, and yellow wine. Served on the same platter.

红茶 (hóngchá)

全发酵茶。鲜茶叶经萎凋、揉捻、发酵、干燥等工序制成。茶叶红中带乌黑,茶汤呈黄红色。

Black tea or fully fermented tea. It is processed by withering, rubbing-twisting, fermenting, and drying. Its dry leaves are blackish red. The tea is reddish yellow.

红豆糕 (hóngdòugāo)

南方点心。将糯米粉、红豆沙、糖、桂花加水调匀,蒸熟,切块。

Sticky rice and red bean cakes. Southern snack. Sticky rice flour mixed with red bean paste, sugar, osmanthus flower, and water, then steamed. Cut into cubes to serve.

红豆黄精乌鸡汤 (hóngdòu huángjīng wūjī tāng)

福建菜。先将乌骨鸡与红豆、黄精、陈皮及调料用旺火煮沸,然后转小火炖烂。

Stewed chicken with herb. Fujian dish. Black-bone chicken boiled with red beans, manyflower solomonseal root, processed tangerine peel, and seasonings, then stewed.

红豆沙包 (hóngdòushābāo)

南方点心。用发酵面皮包入用红豆沙、糖、油等炒制的馅,做成包子,蒸熟。

Buns stuffed with red bean paste.

Southern snack. Fermented flour dough made into buns, stuffed with prepared mashed red beans, sugar, and oil, then steamed.

红豆西米露 (hóngdòu xīmǐ lù)

广东点心。将西米、红豆、冰糖加水熬汤。

Sago and red bean soup. Guangdong snack. Sago boiled into soup with red beans and rock sugar.

红花酸奶 (hónghuā suānnǎi)

西藏点心。酸奶中加入藏红花汁。

Yogurt with saffron. Tibetan snack. Yogurt flavored with saffron tea.

红花羊排 (hónghuā yángpái)

西藏菜。先将羊排骨用酥油煎炸，再加入藏红花水、豆粉等烹煮。

Braised lamb ribs with saffron. Tibetan dish. Chopped lamb ribs fried with Tibetan butter, then braised with Tibetan saffron soup and bean powder.

红椒爆牛肉 (hóngjiāo bào niúròu)

四川菜。先将牛肉腌渍，然后与红辣椒和洋葱及料酒、盐、酱油爆炒。

Stir-fried beef with red pepper. Sichuan dish. Pickled beef stir-fried with red pepper and onion. Flavored with cooking wine, salt, and soy sauce.

红椒肥肠 (hóngjiāo féicháng)

四川菜。先将猪大肠入卤汤煮熟，切段，煸炒至发脆，然后与姜、蒜、青花椒、干红椒、芹菜等爆炒，烹入料酒、酱油、糖、五香粉、辣椒粉。

Stir-fried large intestine with red pepper. Sichuan dish. Hog large intestine boiled in gravy, sectioned, and fried until crispy. Stir-fried with ginger, garlic, green Chinese prickly ash, dried red pepper, and celery, then flavored with cooking wine, soy sauce, sugar, five-spice powder, and chili powder.

红莲炖雪蛤 (hónglián dùn xuěhá)

广东点心。将莲子、泡好的雪蛤、红枣、冰糖等熬炖。

Lotus seed, date, and frog oviduct soup. Guangdong snack. Lotus seeds stewed with Chinese forest frog oviduct, red dates, and rock sugar.

红莲炖银耳 (hónglián dùn yíněr)

广东点心。先将银耳、红枣、莲子、冰糖、姜放入碗里，加水，然后将碗放入蒸锅，隔水炖熟。

Date and white wood ear soup. Guangdong snack. White wood ears, red dates, lotus seeds, rock sugar, and ginger put in a bowl with water, then steamed.

红菱扒冬菇 (hónglíng pá dōnggū)

广东菜。先将冬菇和菱角肉加黄酒、高汤、调料炒熟，勾芡，然后浇热油。

Mushrooms with water caltrop. Guangdong dish. Winter mushrooms and water caltrops sautéed with yellow wine, broth, and condiments, thickened with starch, then washed with heated oil.

红菱子鸡 (hónglíng zǐjī)

浙江菜。先将嫩鸡块和红菱肉稍炸,再与香菇及黄酒、酱油、糖、醋煸炒,勾芡。

Chicken with red water chestnuts. Zhejiang dish. Lightly fried cubed young chicken and shelled red water chestnuts stir-fried with mushrooms, seasoned with yellow wine, soy sauce, sugar, and vinegar, then thickened with starch.

红卤肉皮 (hónglǔ ròupí)

东北菜。先将猪肉皮切条,然后加肉汤、酱油、黄酒、八角,小火焖烂。

Stewed hog skins. Northeastern dish. Slivered hog skins stewed in broth and flavored with soy sauce, yellow wine, and aniseed.

红梅菜胆 (hóngméi càidǎn)

广东菜。先将虾仁用酱油和料酒腌渍,炒熟,然后配鲜汤菜心装盘。

Sautéed shrimp meat with baby greens. Guangdong dish. Shrimp meat pickled with soy sauce and cooking wine, sautéed, then served with baby greens in seasoned soup.

红焖鲈鱼 (hóngmèn lúyú)

广东菜。先将鲈鱼用盐腌过,再用高汤加调料焖熟,撒香菜末。

Braised perch. Guangdong dish. Perch pickled, braised in seasoned broth, then garnished with chopped coriander.

红焖通心河鳗 (hóngmèn tōngxīn hémán)

福建菜。先将鳗鱼炸至赤黄,微焖,再去骨,然后与炒过的香菇、冬笋、五花肉加调料同蒸,浇勾芡的汤。

Fried and steamed eel. Fujian dish. River eel fried red yellow, then simmered briefly. Boned, steamed with quick-fried dried mushrooms, winter bamboo shoots, and streaky pork, then washed with starched soup.

红焖羊肉 (hóngmèn yángròu)

河南菜。先将山羊肉腌渍,然后与几十种植物药焖煮。

Mutton stewed with herbs. Henan dish. Pickled goat meat stewed with dozens of Chinese herbs.

红米糕 (hóngmǐgāo)

台湾点心。先将糯米饭、香菇、虾米、干鱿、卤肉等炒香,然后放上掰开的青蟹,蒸熟。

Sticky rice and seafood cakes. Taiwan snack. Cooked sticky rice stir-fried with dried mushrooms, dried shrimp meat, dried squid, and marinated pork, then steamed with crab chunks.

红米碱水粑 (hóngmǐ jiǎnshuǐbā)

江西点心,又名寒婆粑。先把大米浆加碱水调和,蒸熟成糕,切片,然后与腊肉、蒜等爆炒,或与小白菜、菠菜等入肉汤煮熟。

Steamed rice cakes. Jiangxi snack, also called poor grandma's cakes. Rice paste mixed with soda, steamed into cake, then sliced. It may be stir-fried with smoked cured pork and garlic, or boiled in meat soup with baby cabbage and spinach.

红杞活鱼 (hóngqǐ huóyú)

四川菜。先将鲫鱼用开水烫过，再与枸杞一道放入奶汤，加调料，用文火炖煮。

Crucian and herb soup. Sichuan dish. Scalded crucian carp stewed in milk soup with Chinese wolfberry and flavorings.

红杞蒸鸡 (hóngqǐ zhēng jī)

四川菜。先将整鸡氽过，在鸡腹中装入枸杞子、姜、葱、胡椒粉、盐、清汤，用湿棉纸封口，蒸熟。

Steamed chicken with herb. Sichuan dish. Whole chicken quick-boiled, filled with Chinese wolfberry, ginger, scallion, ground black pepper, salt, and clear soup, sealed with wet cotton paper, then steamed.

红烧鳊鱼 (hóngshāo biānyú)

南方菜。先将鳊鱼煎好，再加入玉兰片、黄酒、酱油、盐、姜、葱烹煮，勾芡。

Bream and bamboo shoots with sauce. Southern dish. Fried bream simmered with soaked bamboo shoots, flavored with yellow wine, soy sauce, salt, ginger, and scallion, then thickened with starch.

红烧大鲍翅 (hóngshāo dàbào chì)

广东菜。先将鲍肉和鱼翅用黄酒和上汤煨熟，然后加胡椒粉、酱油、盐等烹煮，勾芡。

Braised abalone and shark fin. Guangdong dish. Abalone and shark fin braised in broth and yellow wine, simmered with ground black pepper, soy sauce, and salt, then thickened with starch.

红烧大黄鱼 (hóngshāo dàhuángyú)

浙江菜。大黄鱼加葱、姜、蒜、黄酒、酱油、糖、盐、胡椒粉、味精焖煮，浇上煸炒过的青红椒、洋葱、西红柿卤汁。

Braised yellow croaker. Zhejiang dish. Big yellow croaker braised with scallion, ginger, garlic, yellow wine, soy sauce, sugar, salt, ground black pepper, and MSG, then marinated in sauce of sautéed green and red pepper, onion, and tomatoes.

红烧大排 (hóngshāo dàpái)

上海菜。先将猪大排稍煎，加黄酒、冰糖、酱油翻炒，然后焖煮。

Braised pork steaks with soy sauce. Shanghai dish. Pork steaks lightly fried, stir-fried with yellow wine, rock sugar, and soy sauce, then braised.

红烧大乌 (hóngshāo dàwū)

江苏菜。大乌即黑鱼。将乌鱼肉、乌鱼肠加黄酒、姜、葱、酱油、白糖等焖熟。

Braised mullet with soy sauce. Jiangsu dish. Mullet meat and intestine braised and seasoned with yellow wine, ginger, scallion, soy sauce, and sugar.

红烧鲴鱼 (hóngshāo gùyú)

上海菜。先将鲴鱼略煎，再加入黄

酒、酱油、糖等焖煮，然后勾芡，加香醋。

Braised silver carp with soy sauce. Shanghai dish. Silver carp lightly fried, braised with yellow wine, soy sauce, and sugar, thickened with starch, then flavored with vinegar.

红烧龟肉 (hóngshāo guīròu)

湖南菜。先用菜油翻炒龟肉块，然后加入料酒、盐、糖、酱油，用文火炖烂。

Stewed turtle with sauce. Hunan dish. Chopped turtle meat stir-fried with rape seed oil, then stewed with yellow wine, salt, sugar, and soy sauce.

红烧海参 (hóngshāo hǎishēn)

山东菜。将海参、蒜头、芋头入酱汁焖煮，然后与里脊肉片、竹笋、胡萝拌炒。

Braised sea cucumbers with sauce. Shandong dish. Sea cucumbers braised with garlic and taros in soy sauce and fermented flour sauce soup, then stir-fried with sliced pork tenderloin, bamboo shoots, and carrots.

红烧海螺 (hóngshāo hǎiluó)

山东菜。先将海螺肉汆过，再过油，然后与冬笋片、木耳拌炒，烹入酱油和清汤。

Whelks with brown sauce. Shandong dish. Scalded whelk meat sautéed with sliced winter bamboo shoots and wood ears, then flavored with soy sauce and clear soup.

红烧海鲜参 (hóngshāo hǎixiānshēn)

东南方菜。先将海参入姜汤汆熟，然后入冬笋、青红椒、大豆酱、鸡蛋及调料做成的汤焖煮。

Braised sea cucumbers. Southeastern dish. Sea cucumbers scalded in ginger soup, then braised in gravy of winter bamboo shoots, red and green pepper, soybean sauce, eggs, and seasonings.

红烧猴头菇 (hóngshāo hóutóugū)

东北菜。先将猴头菇加玉兰片、葱、姜、酱油、鲜汤等焖煮，再勾芡，淋麻油。

Braised monkey-head mushrooms. Northeastern dish. Monkey-head mushrooms braised with sliced soaked bamboo shoots in seasoned soup, flavored with scallion, ginger, soy sauce, thickened with starch, then washed with sesame oil.

红烧划水 (hóngshāo huáshuǐ)

上海菜。先将草鱼尾即划水用酱油、盐、糖、姜、葱等腌渍，再过油，然后加入黄酒、糖、酱油等焖煮。

Simmered carp tails. Shanghai dish. Grass carp tails pickled with soy sauce, salt, sugar, ginger, and scallion, fried, then simmered with yellow wine, sugar, and soy sauce.

红烧甲鱼 (hóngshāo jiǎyú)

山东菜。先将甲鱼块水焯，在煸炒过的葱、姜、蒜里加入料酒、酱油、清汤，然后将甲鱼、香菇、笋片等入汤焖至汁浓，加入火腿片，淋麻油。

Braised turtle. Shandong dish.

Scalded turtle chunks braised with dried mushrooms and sliced bamboo shoots in gravy of quick-fried scallion, ginger, garlic, cooking wine, soy sauce, and clear soup until reduced. Garnished with sliced ham, then sprinkled with sesame oil.

红烧烤麸 (hóngshāo kǎofū)

上海菜。先将水发烤麸炸至金黄，再加酱油、糖、姜等焖煮，然后与油炸花生、腰果、黄花、竹笋等翻炒。

Sautéed gluten. Shanghai dish. Soaked diced gluten deep-fried, braised with soy sauce, sugar, and ginger, then stir-fried with fried peanuts, cashew nuts, day lily, and bamboo shoots.

红烧鲤鱼 (hóngshāo lǐyú)

河南菜。先将鲤鱼油煎，然后烹入料酒，加入辣椒粉、松蘑丝、盐、酱油、姜，焖透。

Simmered carp with sauce. Henan dish. Carp fried and simmered with cooking wine, chili powder, shredded mushrooms, salt, soy sauce, and ginger.

红烧牛鞭 (hóngshāo niúbiān)

山东菜。先将牛鞭煮透，加入鸡汤、白糖、盐等调料，再用文火煨至汁浓。

Braised beef penis. Shandong dish. Beef penis boiled, braised in chicken soup, then flavored with sugar and salt.

红烧排骨 (hóngshāo páigǔ)

南方菜。排骨加入酱油、糖、盐、葱、姜、花椒、黄酒等烹煮。

Braised pork ribs. Southern dish. Chopped pork ribs braised with soy sauce, sugar, salt, scallion, ginger, Chinese prickly ash, and yellow wine.

红烧圈子 (hóngshāo quānzi)

上海菜。将猪大肠煮熟，切成小圈，加酱油、糖等焖煮，勾芡，淋猪油。

Braised hog large intestine. Shanghai dish. Hog large intestine boiled, cut into rings, braised with soy sauce and sugar, thickened with starch, then washed with lard.

红烧狮子头 (hóngshāo shīzitóu)

南方家常菜。先将大肉圆蒸熟，油煎至金黄，再加入酱油、糖等烹煮，勾芡。

Braised meatballs with sauce. Southern home-style dish. Steamed big meatballs deep-fried, braised with soy sauce and sugar, then marinated in starched sauce.

红烧鲥鱼 (hóngshāo shíyú)

江苏菜。先将鲥鱼略煎，再加入黄酒、酱油、白糖等焖煮，然后勾芡，加香醋。

Braised shad with soy sauce. Jiangsu dish. Reeves shad lightly fried, braised with yellow wine, soy sauce, and sugar, thickened with starch, then flavored with vinegar.

红烧素鸡 (hóngshāo sùjī)

上海菜。将多层豆腐即素鸡煎至两面起硬壳，加入酱油、糖及少量水焖煮。

Braised soy chicken with soy sauce.

Shanghai dish. Multi-layer bean curd or soy chicken deep-fried, then braised with soy sauce, sugar, and a little water.

红烧蹄髈 (hóngshāo típǎng)

各地家常菜。先将蹄髈加调料汆过,抹酱油炸至泛红,然后放入料汤焖煮至汤汁近干,冷却切片,配轻炒过的青菜装盘,浇勾芡的原汤。

Braised hog knuckle. Home-style dish in many places. Hog knuckle quick-boiled in seasoned soup, coated with soy sauce, fried until red, then braised in seasoned soup until reduced. Chilled and sliced. Garnished with lightly sautéed baby greens, then washed with starched broth.

红烧武昌鱼 (hóngshāo wǔchāngyú)

湖北菜。先将鱼腌渍,再将葱、蒜、姜等煸炒,然后同鱼焖煮,加酱油、醋、黄酒、糖、鸡精。

Braised Wuchang carp with sauce. Hubei dish. Local carp pickled, braised with quick-fried scallion, garlic, and ginger, then flavored with soy sauce, vinegar, yellow wine, sugar, and chicken essence.

红烧羊排 (hóngshāo yángpái)

四川菜。先将羊排和萝卜汆过,再加入葱、姜、辣椒酱、红枣、酱油、白糖、胡椒粉等炖煮,然后勾芡,撒香菜。

Stewed spicy lamb ribs. Sichuan dish. Quick-boiled lamb ribs and turnips stewed with scallion, ginger, chili sauce, red dates, soy sauce, sugar, and ground black pepper, thickened with starch, then sprinkled with coriander.

红烧油面筋 (hóngshāo yóumiànjīn)

上海菜。先将油面筋泡填满肉末、葱、盐、芡粉等馅料,再加酱油与大白菜焖煮。

Simmered stuffed gluten with soy sauce. Shanghai dish. Fried gluten puffs stuffed with mixture of minced pork, chopped scallion, salt, and starch, simmered with Chinese cabbage, then flavored with soy sauce.

红汤羊肉 (hóngtāng yángròu)

四川菜。先将豆瓣酱爆香,再加入羊肉和调料炖煮。

Mutton with fermented broad bean sauce. Sichuan dish. Mutton stewed with quick-fried fermented broad bean sauce and other seasonings.

红煨羊蹄花 (hóngwēi yángtíhuā)

湖南邵阳菜。将羊蹄加入秘制调料煨烂。

Braised lamb hooves. Specialty of Shaoyang, Hunan. Lamb hooves braised with special seasonings.

红血肠 (hóngxuècháng)

北京菜。先将猪血与香菜及调料拌和,灌入猪小肠,煮熟,然后切片,入汤烹煮。

Blood sausage. Beijing dish. Hog small intestine stuffed with mixture of hog blood, coriander, and other seasonings, boiled, sliced, then simmered in soup.

红油抄手 (hóngyóu chāoshǒu)

四川菜。将肉馅馄饨煮熟,在汤里加红辣椒油。

Meat wontons in chili oil soup. Sichuan dish. Wontons filled with seasoned ground meat, boiled, then served in chili oil soup.

红油豆瓣鱼 (hóngyóu dòubànyú)

四川菜。先将鲫鱼腌渍,炸透,然后加入红辣椒油、豆瓣酱等焖煮。

Simmered crucian with chili oil. Sichuan dish. Pickled crucian deep-fried, then simmered with chili oil and fermented broad bean sauce.

红油肚片 (hóngyóu dǔpiàn)

四川凉菜。将牛肚煮熟,切片,加入红辣椒油、酱油、芝麻酱等拌匀。

Ox tripe with chili oil. Sichuan cold dish. Ox tripe boiled, sliced, then mixed with chili oil, soy sauce, and sesame sauce.

红糟鱼丝 (hóngzāo yúsī)

山东菜。先将黑鱼丝入油锅划散,再用红糟、黄酒、清汤兑成汁,放入鱼丝翻炒。

Sautéed snakehead fish. Shandong dish. Lightly fried shredded snakehead fish sautéed with sauce of sweet fermented sticky rice, yellow wine, and clear soup.

红枣菊花粥 (hóngzǎo júhuā zhōu)

南方点心。将白米、红枣、当归、菊花用清水煮成粥,加红糖。

Sticky rice and herb porridge. Southern snack. Sticky rice boiled into porridge with red dates, Chinese angelica, and chrysanthemum flower, then sweetened with brown sugar.

红枣软糕 (hóngzǎo ruǎngāo)

四川菜。先将糯米粉、红枣泥、糖、豆沙等调匀,蒸熟,置冷,然后切成菱形块。

Sticky rice and red date cakes. Sichuan dish. Sticky rice flour mixed with red date paste, sugar, and bean paste, steamed, cooled, then cut into diamond cubes.

红枣烧兔腿 (hóngzǎo shāo tùtuǐ)

东北菜。将兔腿肉块炸至金黄,加入红枣和煸过的干辣椒,用小火焖煮。

Braised rabbit legs with dates. Northeastern dish. Fried chopped rabbit legs braised with dates and quick-fried chili pepper.

洪都素烩 (hóngdūsùhuì)

江西菜。将红枣、冬笋、萝卜、菜心、香菇、白莲、荸荠、银耳等烹煮,装盘时红枣居中,其他配料环盘。

Braised assorted vegetables. Jiangxi dish. Red dates, bamboo shoots, turnips, baby greens, dried mushrooms, lotus seeds, water chestnuts, and white wood ears braised together. Served on a platter with red dates at the center.

洪山菜苔 (hóngshāncàitái)

湖北菜。将葱、蒜、姜煸香,放入菜苔及盐爆炒。

Sautéed rape blossoms. Hubei dish.

Scallion, garlic, and ginger quick-fried, then stir-fried with rape blossoms and salt.

鸿茅药酒 (hóngmáoyàojiǔ)

药酒。产于山西榆次。用鸿茅白酒加入中草药浸泡制成，酒精含量为38-40度，既可外用，也可内服。鸿茅白酒产于内蒙古凉城县。

Hongmao folk-medicinal liquor, produced in Yuci, Shanxi. It is made with herb infusion, and contains 38-40% alcohol. It may be used as a health drink or topical medicine. Hongmao liquor is produced in Liangcheng, Inner Mongolia.

猴蘑牛头方 (hóumó niútóufāng)

四川菜。先将牛头皮煮至胶质脱落，再与鸡鸭骨、排骨、火腿骨、干贝等用小火煨炖，用盐、冰糖、料酒、姜、葱调味，配汆过的猴蘑和菜心装盘。

Stewed ox head skins with mushrooms. Sichuan dish. Sliced boiled ox head skins stewed with chicken and duck bones, pork spareribs, ham bones, and dried scallops, then seasoned with salt, rock sugar, cooking wine, ginger, and scallion. Served with quick-boiled monkey-head mushrooms and baby greens.

猴头炖乌鸡 (hóutóu dùn wūjī)

四川菜。将猴头菇与乌鸡及盐、料酒、姜、葱等炖煮。

Stewed mushrooms and black-bone chicken. Sichuan dish. Monkey-head mushrooms stewed with black-bone chicken, then seasoned with salt, cooking wine, ginger, and scallion.

猴头四宝 (hóutóu sìbǎo)

浙江菜。先将虾仁、猴头菇片、火腿片、鸡脯肉片挂浆，入油锅划散，然后与青菜及调料翻炒，勾芡。

Mushrooms with four treasures. Zhejiang dish. Shrimp meat, sliced monkey-head mushrooms, ham, and chicken breast coated with starch paste, lightly fried, sautéed with green vegetable and condiments, then thickened with starch.

胡羊肉 (húyángròu)

西北菜。先将羊肉入五香汤煮至近熟，切薄片，然后加原汤、花椒粉、盐等蒸烂，勾芡，淋麻油，撒香菜。

Steamed mutton. Northwestern dish. Mutton boiled in five-spice soup, sliced, steamed with ground Chinese prickly ash and salt in broth, thickened with starch, then sprinkled with sesame oil and coriander.

湖彩蛋 (húcǎidàn)

江苏菜。先将鲜鸭蛋裹上用沸水、纯碱、生石灰、红茶末、盐、植物灰等调成的腌泥，瓮封40-50天，制成硬心皮蛋，然后去壳，配卤汁。

Preserved duck eggs with sauce. Jiangsu dish. Duck eggs coated with paste of soda, quick lime, minced black tea, salt, hay ash, and boiling water. Sealed in clay jar for 40-50 days. Shelled and served with sauce.

湖南黑茶 (húnánhēichá)

黑茶。产于湖南安化、益阳等地。香味醇厚，带松烟香，汤色橙黄，叶底黄褐。

Hunan heicha, darh green tea produced in Anhua and Yiyang, Hunan. Its dry leaves have a pine smoke aroma. The tea is orange red. The infused leaves are yellowish brown.

湖南腌三样 (húnán yānsānyàng)

湖南菜。先将茄子、豆角、辣椒焯水，盐腌，晒干，然后用植物油煸炒。

Sautéed three dried vegetables. Hunan dish. Eggplants, young legumes, and hot pepper scalded, pickled with salt, sun-dried, then sautéed with vegetable oil.

葫芦大吉翅子 (húlu dàjí chìzi)

山东名菜。先将鸡脯泥、肥肉泥与盐、蛋清、高汤等调匀，做成葫芦状，顶端置海米，蒸熟，然后配蒸熟的鱼翅装盘。

Shark fin with chicken and pork. Famous Shandong dish. Ground chicken breast and minced fat pork mixed with salt, egg white, and broth, made into big meatballs in the shape of bottle gourd, then garnished with dried shrimp on the top. Steamed. Served with cooked shark fin.

蝴蝶海参 (húdié hǎishēn)

山东菜。先将水发刺参刻成蝴蝶状片，略炒，然后加浓汤，与虾仁、熟鸡肉片、蛋白片、火腿片等烩煮。

Sea cucumbers with other meat varieties. Shandong dish. Sea cucumbers carved into butterfly-shape pieces, quick-fried, boiled in gravy with shrimp meat, prepared sliced chicken, cooked egg white, and sliced ham, then simmered.

蝴蝶烩鳝 (húdié huì shàn)

江苏菜。先将猪后腿蝴蝶骨焖煮至汤稠，蒸烂；再将鳝鱼段过油，与葱、姜、酱油、糖、香菇、竹笋等焖酥；然后将鳝鱼段与蝴蝶骨同烩，淋麻油。

Simmered hog bones and eel. Jiangsu dish. Hog hind-leg bones stewed until thickened, then steamed. Eel sections lightly fried, then braised with scallion, ginger, soy sauce, sugar, dried mushrooms, and bamboo shoots. Prepared bones and eel simmered together, then sprinkled with sesame oil.

虎皮蛋 (hǔpídàn)

各地家常菜。先将鸡蛋煮熟，去壳，再炸至蛋白起皱，然后入卤汤焖煮。

Marinated eggs. Home-style dish in many places. Eggs boiled, shelled, fried until egg white wrinkled, then marinated in gravy.

虎皮豆腐 (hǔpí dòufu)

东北家常菜。先将豆腐块炸至发黄，然后加调料及少许汤烧至汁浓，勾芡。

Braised bean curd. Northeastern home-style dish. Bean curd deep-fried, simmered with seasonings and a little broth, then thickened with starch.

虎皮鸽蛋 (hǔpí gēdàn)

东北家常菜。先将土豆泥与熟冬菇丁和笋丁拌成馅,做成鸽蛋形,然后裹上芡浆和面包渣,炸熟。

Fried potato balls. Northeastern home-style dish. Mashed potatoes mixed with prepared diced winter mushrooms and bamboo shoots, made into small balls like pigeon eggs, coated with starch paste and bread crumbs, then deep-fried.

琥珀蛋 (hǔpòdàn)

湖南菜。将咸鸭蛋、松花蛋、小葱合蒸而成。

Steamed egg varieties. Hunan dish. Salty duck eggs and preserved duck eggs steamed with scallion.

琥珀莲子 (hǔpò liánzǐ)

江苏点心。先将莲子嵌入桂圆肉,加冰糖煮熟,然后撒糖桂花。

Lotus seeds and longan soup. Jiangsu snack. Lotus seeds inlaid in longans, boiled with rock sugar, then sprinkled with sugared osmanthus flower.

琥珀桃仁 (hǔpò táorén)

广东点心。先将核桃仁炸至金黄,然后裹上糖衣。

Sugarcoated walnuts. Guangdong dessert. Walnut meat fried until golden, then coated with heated melting sugar.

护国菜 (hùguócài)

广东菜。将蘑菇与红薯叶翻炒。

Sautéed sweet potato leaves and mushrooms. Guangdong dish. Sweet potato leaves sautéed with mushrooms.

花茶 (huāchá)

再加工茶。用绿茶、红茶或乌龙茶做茶胚,用茉莉、菊花、玫瑰等能散发香味的鲜花通过特殊工艺窨制而成。通常以花命名,如茉莉花茶。

Flower-scented tea or re-processed tea. It is made from green tea, black tea, or oolong tea with aromatic flowers such as jasmine, chrysanthemum, and rose. It is usually named after the flower, such as jasmine tea.

花雕鸡 (huādiāojī)

广东菜。将鸡用花雕酒、蚝油、蜂蜜、葱等腌渍,放在煎过的猪肥肉片上焖熟。

Chicken with yellow wine.

Guangdong dish. Chicken pickled with *huadiao* yellow wine, oyster sauce, honey, and scallion, put on a bed of fried sliced fat pork, then braised.

花雕酒 (huādiāojiǔ)

黄酒,产于浙江绍兴。将加饭酒装入雕花坛子,封存多年而成,酒精含量 15 度左右。

Huadiao yellow wine from Shaoxing, Zhejiang. It is made with Jiafan yellow wine sealed in engraved crocks for many years. Contains about 15% alcohol.

花椒鸡 (huājiāojī)

四川菜。将嫩鸡块与香菇、花椒、白

果、盐、酱油、料酒、姜、葱等焖烧。

Chicken with Chinese prickly ash. Sichuan dish. Chopped young chicken simmered with dried mushrooms, Chinese prickly ash, ginkgoes, salt, soy sauce, cooking wine, ginger, and scallion.

花椒魔芋 (huājiāo móyù)

四川菜。将魔芋糕条与花椒、辣椒、火腿、姜、葱、酱油等爆炒。

Stir-fried spicy konjac jelly. Sichuan dish. Konjac jelly strips stir-fried with Chinese prickly ash, chili pepper, ham, ginger, scallion, and soy sauce.

花椒牛肉 (huājiāo niúròu)

四川菜。先将牛肉腌渍,油炸,然后加入煸过的干辣椒和花椒、盐、糖、酱油、胡椒粉、牛肉汤等焖烧。

Beef with Chinese prickly ash. Sichuan dish. Pickled beef fried, then stewed in beef broth with prepared chili pepper, Chinese prickly ash, salt, sugar, soy sauce, and ground black pepper.

花椒肉 (huājiāoròu)

四川菜。将五花肉块腌渍,炸透,然后加入花椒、姜、葱、红辣椒、香菇、酱油、料酒等焖烧。

Pork with Chinese prickly ash. Sichuan dish. Pickled streaky pork deep-fried, then braised with Chinese prickly ash, ginger, scallion, red pepper, dried mushrooms, soy sauce, and cooking wine.

花椒乳鸽 (huājiāo rǔgē)

四川菜。先将乳鸽腌渍,油炸,然后加入花椒、白果、姜、葱、香菇等焖烧。

Pigeon with Chinese prickly ash. Sichuan dish. Pickled young pigeon fried, then braised with Chinese prickly ash, ginkgoes, ginger, scallion, and dried mushrooms.

花揽鳜鱼 (huālǎn guìyú)

山东菜。先将鱼身两边劈出刀口,一边的刀口里嵌入火腿片,另一边填入鸡茸;再将鱼身抹遍酱汁,用猪网油包好,裹上面糊,用炭火慢烤,佐姜醋。

Grilled Chinese perch with ham and chicken. Shandong dish. Chinese perch cut on sides, inlaid with sliced ham on one side and minced chicken on the other, brushed with sauce, wrapped in web lard, coated with flour paste, then grilled over charcoal. Served with ginger and vinegar.

花生炖猪脚 (huāshēng dùn zhūjiǎo)

广东菜。将猪脚、花生加入盐、姜、葱、黄酒等炖煮。

Stewed hog trotters with peanuts. Guangdong dish. Hog trotters stewed with peanuts and seasoned with salt, ginger, scallion, and yellow wine.

花生糕 (huāshēnggāo)

河南点心。将炒过的花生仁与热糖浆拌和,结成大块,置凉,切块。

Peanut cookie. Henan snack. Roasted peanuts mixed with heated syrup, made into a lump, cooled, then cut into cubes.

滑炒里脊丝 (huáchǎo lǐjisī)

东北菜。先将猪里脊丝上浆,入油锅划散,然后与冬笋丝、青蒜及调料翻炒,淋麻油。

Sautéed shredded pork tenderloin. Northeastern dish. Shredded pork tenderloin coated with starch paste, lightly fried, sautéed with shredded winter bamboo shoots, garlic sprouts, and condiments, then sprinkled with sesame oil.

滑熘里脊 (huáliū lǐji)

上海菜。先把猪里脊肉片挂蛋清芡粉浆,滑熟,然后放入黄瓜片及调料翻炒,勾芡。

Sautéed pork tenderloin. Shanghai dish. Sliced pork tenderloin coated with starch and egg paste, lightly fried over low heat, sautéed with sliced cucumbers and seasonings, then thickened with starch.

滑熘鱼片 (huáliū yúpiàn)

上海菜。先把鱼片挂蛋清芡粉浆,滑熟,然后放入黄瓜片及调料翻炒,勾芡。

Sautéed fish fillets. Shanghai dish. Fish fillets coated with starch and egg paste, lightly fried over low heat, sautéed with sliced cucumbers and seasonings, then thickened with starch.

滑熘鸡片 (huáliū jīpiàn)

东北菜。先将鸡脯肉片上浆,入油锅划散,然后与黄瓜片及调料翻炒,加调料,勾芡。

Sautéed chicken. Northeastern dish. Sliced chicken breast coated with starch paste, lightly fried, sautéed with sliced cucumbers and seasonings, then thickened with starch.

话梅嫩姜泡杂菜 (huàméi nènjiāng pào zácài)

北京凉菜。将熟毛豆、炸白果、鲜莴笋、嫩姜、葱头等放入用盐、糖、醋、话梅等制成的汁,加入各式蔬菜泡制。

Pickled vegetables with processed plums. Beijing dish. Boiled green soybean, fried ginkgoes, stem lettuce, baby ginger, and green onion heads pickled with assorted vegetables in brine of salt, sugar, vinegar, and processed plums.

淮杞莲藕牛腩煲 (huái qǐ liánǒu niúnǎn bāo)

广东菜。将牛腩、淮山、枸杞、莲藕等加调料炖煮。

Stewed beef flank with herbs. Guangdong dish. Beef flank stewed with Chinese yam, Chinese wolfberry, lotus roots, and seasonings.

淮鱼干丝 (huáiyú gānsī)

江苏菜。将豆腐干丝、鳝丝、鸡皮加调料炒熟。

Stir-fried bean curd and eel. Jiangsu dish. Shredded dried bean curd, eel, and chicken skins stir-fried with seasonings.

槐茂酱菜 (huáimàojiàngcài)

河北保定菜。各种蔬菜配花生仁、杏仁、核桃仁、姜等腌制的酱菜。

Pickled vegetables. Specialty in Baoding, Hebei. Various vegetables, peanuts, almonds, walnuts, and ginger pickled with seasonings.

还丝汤 (huánsītāng)

江苏菜。先将鳝鱼煎脆,再用猪肉汤烧煮还软。

Fried eel in soup. Jiangsu dish. Eel deep-fried crisp, then boiled in pork soup.

换心乌贼 (huànxīn wūzéi)

山东菜。先将乌贼肚内填入虾仁、猪肉、盐、酱油、葱、姜等制成的馅,用鱼茸封口,蒸熟,然后用原汤勾芡,淋麻油。

Steamed stuffed cuttlefish. Shandong dish. Cuttlefish stuffed with mixture of shrimp meat, pork, salt, soy sauce, scallion, and ginger, sealed with fish paste, steamed, then washed with starched broth and sesame oil.

黄粑 (huángbā)

云贵川传统点心。将糯米饭、米粉或玉米粉用生黄豆浆调和,做成筒状或块状糕,蒸熟并焖至深黄色。

Steamed rice, corn, and soybean cakes. Traditional snack in Yunnan, Guizhou, and Sichuan. Steamed sticky rice, rice flour, or corn flour mixed with soybean paste, made into big rolls or cubes, then steamed into deep-yellow cakes.

黄茶 (huángchá)

轻发酵茶。经过杀青、闷堆渥黄、干燥三个过程制成,形成黄叶黄汤的独特品质和风格。

Yellow tea or light-fermented tea. It is processed by baking, fermenting, and drying. Tea leaves and tea have a unique yellow color.

黄豆炖猪蹄 (huángdòu dùn zhūtí)

北方家常菜。先将猪蹄入卤水煮至上色入味,然后剁块,与黄豆及盐、料酒炖烂。

Stewed hog trotters and soybean. Home-style dish in northern areas. Hog trotters marinated in gravy until brown, then chopped. Stewed with soybean, salt, and cooking wine.

黄菇鱼火锅 (huánggūyú huǒguō)

湖北菜。以黄菇鱼为主料,放入高汤火锅,加多种配料烫熟。

Yellow catfish in fire pot. Hubei dish. Yellow catfish boiled in fire pot soup with seasonings. A variety of ingredients added.

黄瓜拌干豆腐丝 (huángguā bàn gāndòufusī)

东北凉菜。将黄瓜丝和豆腐干丝加酱油、盐、麻油拌匀。

Cucumbers with dried bean curd. Northeastern cold dish. Shredded cucumbers and dried bean curd mixed, then seasoned with soy sauce, salt, and sesame oil.

黄鹤楼酒 (huánghèlóujiǔ)

白酒,产于湖北武汉。以高粱为主

料，用大麦、豌豆制曲酿造，酒精含量为39-62度，属清香型酒。

Huangheloujiu, liquor produced in Wuhan, Hubei. White spirit made with sorghum and yeast of barley and peas. Contains 39-62% alcohol, and has a delicate aroma.

黄花菜炒肉丝 (huánghuācài chǎo ròusī)

四川菜。将黄花菜与肉丝、辣椒、榨菜等爆炒。

Sautéed day lily and pork. Sichuan dish. Day lily stir-fried with shredded pork, chili pepper, and preserved mustard tubers.

黄金桂 (huángjīnguì)

乌龙茶，又名黄旦。产于福建安溪虎邱镇。外形条索紧细，色泽润亮。汤色金黄透明，叶底柔软明亮，边缘朱红。

Golden oolong tea, also called *huangdan*. It is produced in Huqiu Town of Anxi, Fujian. Its dry leaves are slim, tight, sleek, and lustrous black. The tea is crystal golden. The infused leaves are soft with red fringes.

黄酒 (huángjiǔ)

世界三大古酒之一，源于浙江绍兴。以糯米、大米、黍米等为主料酿造，酒精含量在20度以下，因其色黄得名。

Rice wine or yellow wine originated in Shaoxing, Zhejiang. It is one of the 3 oldest alcoholic drinks in the world. It is made from grains such as glutinous rice, plain rice, and corn. Its alcohol content is usually less than 20%. It has a yellowish color, and thus is named as such.

黄焖着甲 (huángmèn zhuójiǎ)

江苏菜。甲鱼肉、明骨、山药、冬菇、冬笋等入猪肉汤炖煮。

Stewed turtle and shark bones. Jiangsu dish. Turtle meat and prepared shark soft bones stewed with Chinese yam, dried mushrooms, and winter bamboo shoots in pork soup.

黄焖子铜鹅 (huángmèn zǐtóngé)

湖南邵阳菜。将腌渍过的子鹅肉加入嫩姜、蒜、红椒、酱油翻炒。

Stir-fried goose. Specialty of Shaoyang, Hunan. Pickled young goose meat stir-fried with baby ginger, garlic, red pepper, and soy sauce.

黄米切糕 (huángmǐ qiēgāo)

黑龙江点心。将红黏米浆和芸豆泥间层铺放，蒸制成糕，切片，佐白糖。

Steamed rice and bean cakes. Heilongjiang snack. Layered red rice and kidney bean paste steamed into a lump cake, then cut into pieces. Served with sugar.

黄桥烧饼 (huángqiáoshāobǐng)

江苏点心，源于江苏泰兴黄桥镇。先将面粉、水及猪油和成水面团，用面粉与猪油和成油面团，再用水面团包裹油面团，擀成面皮，包入用猪油丁、火腿、葱、盐、麻油拌成的馅，做成饼，涂上蛋液，撒芝麻，烤熟。

Flour and lard cakes. Jiangsu snack

originated in Huangqiao Town of Taixing, Jiangsu. Flour, water, and lard made into water dough. Flour and lard made into oil dough. Oil dough rolled in water dough, then kneaded into wrappers. Filled with mixture of diced lard, ham, leek, salt, and sesame oil, then made into cakes. Brushed with egg paste, topped with sesame, then baked.

黄山毛峰 (huángshānmáofēng)

绿茶。中国名茶,产于安徽歙县黄山。外形肥壮匀齐,银毫显露。汤色清澈明亮,叶底匀嫩成朵。

Huangshanmaofeng, famous green tea produced in Mount Huangshan in Shexian, Anhui. Its leaves are stout, even and hairy. The tea is crystal clear with infused leaves like flowers.

黄石港饼 (huángshígǎngbǐng)

湖北黄石点心。用精面粉做成饼皮,用冰糖、桔饼、糖桂花、麻油等做成馅,包馅成饼,撒上芝麻,烤熟。

Baked stuffed cakes. Snack in Huangshi, Hubei. Fine flour dough made into cake wrappers, filled with mixture of rock sugar, candied tangerines, sugared osmanthus flower, and sesame oil, topped with sesame, then baked.

黄松糕 (huángsōnggāo)

江浙沪点心。用糯米粉、粳米粉、糖、核桃仁、松子、红枣加水调和,蒸熟。

Yellow rice cakes. Jiangsu, Zhejiang, and Shanghai snack. Sticky rice and round grain rice flour, sugar, walnuts, pine nuts, red dates, and water mixed, then steamed into cakes.

徽州桃脂烧肉 (huīzhōu táozhī shāo ròu)

安徽菜。先将五花肉丁略煎,加调料用小火焖煮,然后投入桃脂,略焖。

Braised pork with peach glue. Anhui dish. Lightly fried diced streaky pork braised with seasonings, then simmered briefly with peach glue.

回锅肉 (huíguōròu)

云贵川湘鄂家常菜。先将带皮半肥瘦猪肉煮熟,切片,然后加入辣椒、葱、蒜、盐、酱油、料酒、豆瓣酱等爆炒。

Stir-fried spicy pork. Home-style dish in Yunnan, Guizhou, Sichuan, Hunan, and Hubei. Boiled and sliced fatty pork with skins stir-fried with hot pepper, scallion, garlic, salt, soy sauce, cooking wine, and fermented broad bean sauce.

茴香豆 (huíxiāngdòu)

浙江点心。先把干蚕豆用水泡胀,再加茴香、桂皮、盐等煮熟,然后晾干。

Fennel-scented broad beans.

Zhejiang snack. Dried broad beans soaked, boiled with fennel seeds, cinnamon, and salt, then air-dried.

茴香炖鲤鱼 (huíxiāng dùn lǐyú)

东北菜。先将鲤鱼炸至金黄,然后加清汤、辣椒、茴香及调料焖烧。

Simmered carp with fennel. Northeastern dish. Deep-fried carp simmered in clear soup with hot pepper, fennel, and other condiments.

会友发包子 (huìyǒu fābāozi)

吉林点心。精粉面发酵,做成面皮,包入用猪肉、蔬菜、调料、鸡汤冻等做成的馅,蒸熟。

Steamed pork and vegetable bun. Jilin snack. Fine flour dough made into buns, stuffed with mixture of pork, vegetables, seasonings, and chicken soup jelly, then steamed.

烩芙蓉三丁 (huì fúróng sāndīng)

东北菜。先将鸡蛋清蒸成蛋羹,然后将海参丁、鸡肉丁、兰片丁炒熟,放在蛋羹上。

Egg white with diced meat.

Northeastern dish. Steamed egg white covered with stir-fried diced sea cucumbers, chicken, and bamboo shoots.

烩明珠 (huìmíngzhū)

内蒙菜。将汆熟的羊眼切片,加入调料及高汤烹煮,勾芡。

Braised lamb eyes. Inner Mongolian dish. Sliced scalded lamb eyes braised in broth with seasonings, then thickened with starch.

烩三鲜 (huìsānxiān)

东北菜。将海参、鸡脯肉片、大虾、冬笋片等入汤烩熟,加盐、味精、酱油,勾芡。

Braised three meat varieties.

Northeastern dish. Sea cucumbers, sliced chicken breast, and prawns braised with sliced winter bamboo shoots. Seasoned with salt, MSG, and soy sauce, then thickened with starch.

烩酸菜豆汤 (huì suāncài dòutāng)

西南家常菜。先将干四季豆用清水煮至酥烂,再与切碎的酸菜烩煮,加豆汁原汤、猪油或植物油、盐等调料。

Boiled kidney beans with pickled mustard. Home-style dish in southwestern areas. Boiled dried kidney beans simmered in bean soup with chopped sour pickled vegetable, and seasoned with lard or vegetable oil, and salt.

惠明茶 (huìmíngchá)

绿茶。产于浙江景宁敕木山的惠明寺村一带。外形条索紧缩壮实,色泽翠绿,全芽披毫,汤色清澈,叶底黄绿明亮。

Huimingcha, green tea produced in the area around Huiming Temple Village by Chimu Hill in Jingning, Jiangsu. Its dry leaves are tight, stout, jade-green, and hairy. The tea is crystal clear with infused leaves bright yellowish green.

荤罗汉 (hūnluóhàn)

福建菜。先将焯过的鱼翅、鲍鱼、海参、鸡、猪蹄、干贝、冬菇加调料焖熟,然后加入轻炒过的菜心烧开。

Braised sea food and meat varieties. Fujian dish. Scalded shark fin, abalone, chicken, hog trotters, dried scallops, and dried mushrooms braised with seasonings, then boiled with quick-fried baby greens.

馄饨 (húntún)

江浙沪皖家常点心。面粉加水做成方形面皮,包入咸鲜味菜肉馅,清水煮熟,放入加蛋丝、紫菜、榨菜丝、葱花的清汤。

Wontons. Home-style snack in Jiangsu, Zhejiang, Shanghai, and Anhui. Flour mixed with water, made into square thin wrappers, stuffed with fresh salted vegetables and meat, then boiled. Served in soup of shredded fried eggs, laver, shredded tuber mustard, and chopped scallion.

火爆双脆 (huǒbào shuāngcuì)

四川菜。先将猪肚片和鸡胗片腌渍,然后加入红辣椒、葱、蒜等爆炒。

Stir-fried hog tripe and chicken gizzard. Sichuan dish. Sliced and pickled hog tripe and chicken gizzard stir-fried with red pepper, scallion, and garlic.

霍山黄芽 (huòshānhuángyá)

黄茶。产于安徽霍山。外形条直微展,嫩绿披毫。汤色黄绿清澈,叶底嫩黄匀实。

Huoshanhuangya, yellow tea produced in Mount Huoshan, Anhui. Its dry leaves are straight, flat, tender green, and hairy. The tea is yellowish green and crystal with infused leaves tender yellow and even.

J

鸡火百合 (jī huǒ bǎihé)

山东凉菜。将煸香的葱撒在熟百合、鸡丝、火腿丝上,加调料拌匀。

Lily bulbs with chicken and ham. Shandong cold dish. Prepared lily bulbs and shredded chicken and ham mixed with seasonings, then garnished with quick-fried scallion.

鸡火煮干丝 (jī huǒ zhǔ gānsī)

上海菜。先用鸡汤和骨头汤把豆腐干丝和干虾仁煮熟,然后放入熟鸡丝、熟火腿丝、汆过的豆苗。

Bean curd and chicken in soup. Shanghai dish. Shredded dried bean curd and dried shrimp meat cooked in chicken and hog bone soup. Garnished with shredded prepared chicken, ham, and scalded pea sprouts.

鸡皮虾丸汤 (jīpí xiāwán tāng)

曹雪芹《红楼梦》菜谱。先将虾肉、鲜贝、肥肉捣茸,加入盐、黄酒、葱油、姜汁、清汤、蛋清搅匀,挤成丸子,然后入清汤煮熟,放入鸡皮、菠菜。

Chicken skins and shrimp ball soup. Menu from *A Dream of Red Mansions* by Cao Xueqin. Minced shrimp meat, fresh shellfish meat, and fat pork mixed with salt, yellow wine, scallion oil, ginger juice, clear soup, and egg white, made into balls, then boiled in clear soup. Chicken skins and spinach added.

鸡茸蛋 (jīróngdàn)

江苏菜。先将鸡茸和肥肉丁加调料搅成糊,做成蛋形,炸熟,然后加火腿、竹笋、香菇等煸炒。

Stir-fried chicken and pork balls. Jiangsu dish. Minced chicken breast and ground fat pork mixed with seasonings, made into egg-shape balls, deep-fried, then stir-fried with ham, bamboo shoots, and dried mushrooms.

鸡茸金丝笋 (jīróng jīnsīsǔn)

福建菜。先把金丝冬笋用鸡汤煨熟,再与鸡茸及调料爆炒,撒火腿米。

Stir-fried chicken and bamboo shoots. Fujian dish. Winter bamboo shoots braised in chicken soup, stir-fried with ground chicken and condiments, then sprinkled with ham crumbs.

鸡血豆腐 (jīxuè dòufu)

四川菜。先将鸡血蒸熟,然后与嫩豆腐煮汤,加盐、葱花、猪油。

Chicken blood and bean curd soup.

Sichuan dish. Steamed chicken blood boiled with tender bean curd. Seasoned with salt, scallion, and lard.

鸡油菜心 (jīyóu càixīn)

江苏菜。先把青菜心放入鸡清汤煮熟，再把熟火腿片盖在菜心上，浇鸡油。

Shanghai greens in chicken soup. Jiangsu dish. Baby Shanghai greens boiled in clear chicken soup, covered with sliced ham, then washed with chicken oil.

鸡汁燕窝 (jīzhī yànwō)

浙江菜。先将燕窝用温水发好，再焖酥，然后与火腿、蔬菜、鸡汤煮熟。

Bird's nests in chicken soup. Zhejiang dish. Edible bird's nests saturated in warm water, braised, then boiled in chicken soup with ham and vegetable.

鸡粥菜心 (jīzhōu càixīn)

江苏菜。将鸡茸、肥膘泥、淀粉、蛋清、盐等调和，倒入煮沸的鸡清汤，加入碎菜心，翻搅煮熟。

Chicken and vegetable chowder. Jiangsu dish. Mixture of ground chicken breast, fat pork, egg white, starch, and salt poured in boiling clear chicken soup, then boiled with chopped baby cabbage. Stirred till ready.

鸡粥花菜 (jīzhōu huācài)

湖北菜。先将鸡茸加调料拌成糊状，推入鸡汤煮熟，然后放入轻炒过的花菜，淋麻油，撒火腿末。

Chicken and broccoli. Hubei dish. Ground chicken made into paste with condiments, then boiled in chicken soup. Quick-fried broccoli added. Sprinkled with sesame oil and ham crumbs.

鸡煮干丝 (jī zhǔ gānsī)

河南家常菜。火腿丝、鸡丝与豆腐干丝同煮，加调料，撒香菜和胡椒粉。

Boiled chicken and bean curd. Home-style dish in Henan. Shredded chicken and ham boiled with shredded dried bean curd, seasoned, then sprinkled with coriander and ground black pepper.

鸡仔饼 (jīzǎibǐng)

广东点心。用白面粉与麦芽糖做成饼皮，用肥肉粒、核桃仁、芝麻、糖、盐等做成馅，用模具定型，烤熟。

Stuffed salty and sweet cakes. Guangdong snack. Flour and maltose made into small cakes stuffed with minced fat pork, walnuts, sesame, sugar, and salt, shaped in mould, then baked.

及第粥 (jídìzhōu)

广东点心。将米、猪骨、瘦猪肉、腐竹熬粥。

Rice porridge with meat varieties. Guangdong snack. Rice boiled into porridge with hog bones, lean pork, and dried bean curd skins.

吉利虾 (jílìxiā)

福建菜。把对虾做成虾球，裹上鸡蛋面粉浆和面包屑，炸至金黄，用辣椒、香菇、冬笋、洋葱、胡萝卜、葱、猪骨汤、淀粉、猪油等做成卤汁，浇在虾球上。

Prawn balls in gravy. Fujian dish. Prawn balls coated with flour egg paste and bread crumbs, deep-fried, then washed with dressing of hot pepper, dried mushrooms, bamboo shoots, onion, carrots, scallion, hog bone soup, starch, and lard.

吉列海鲜卷 (jílìe hǎixiānjuǎn)

广东菜。先将腌渍过的虾仁和带子、胡萝卜轻炒，再与果肉、沙拉酱等拌匀，然后用威化纸包裹成卷，裹上干淀粉、蛋浆、面包屑，炸熟。

Fried sea food and fruit rolls. Guangdong dish. Pickled shrimp meat and scallops lightly fried with carrots, then mixed with fruit meat and mayonnaise. Rolled in rice paper, coated with starch, egg paste, and bread crumbs, then fried.

即墨老酒 (jímòlǎojiǔ)

黄酒，中国古典名酒之一，产于山东即墨。以黄米、麦曲、崂山矿泉水为原料，经自然发酵后压榨所得的原汁，酒精含量为11-16度。

Jimolaojiu, a kind of traditional yellow wine produced in Jimo, Shandong, one of the renowned classical alcoholic drinks in China. Made with yellow rice, wheat yeast, and Laoshan mineral spring water through natural fermentation and extraction. Contains 11-16% alcohol.

济南肴鸡 (jǐnányáojī)

山东菜。将小公鸡腹内塞入葱、姜、花椒，放入有丁香、桂皮、白芷等香料的卤锅，小火炖煮。

Stewed cockerels. Shandong dish. Cockerels stuffed with green onion, garlic, and Chinese prickly ash, then marinated in gravy of clove, cinnamon, and angelica dahurica.

继光饼 (jìguāngbǐng)

福建福安点心。将面粉发酵，揉进盐、芝麻、葱，做成中间带孔的饼，烤熟。

Baked flour cakes. Snack in Fu'an, Fujian. Fermented flour dough flavored with salt, sesame, and scallion, made into cakes with holes in the middle, then baked.

加饭酒 (jiāfànjiǔ)

黄酒，产于浙江绍兴。酒液深黄带红，透明晶莹，酒精含量15度左右。

Jiafanjiu, yellow wine from Shaoxing, Zhejiang. Its alcohol content is about 15%. It has a crystal reddish yellow color.

家常豆腐 (jiācháng dòufu)

各地家常菜。将炒熟的肉片、冬笋、木耳等与炸过的豆腐片烩煮，用盐、糖、酱油、辣椒等调味，勾芡。Sautéed pork and bean curd. Home-style dish in many places. Deep-fried bean curd sautéed with prepared sliced pork, winter bamboo shoots, and wood ears. Flavored with salt, sugar, soy sauce, and hot pepper, then thickened with starch.

袈裟鱼 (jiāshāyú)

北方菜。先将鱼片裹上甜豆沙，然后挂糊炸至金黄。佐番茄汁。

Fried fish with tomato sauce. Northern dish. Fish fillets coated with sweetened bean paste, covered with starch, then deep-fried. Served with tomato sauce.

嘉禾脆皮鸡 (jiāhé cuìpíjī)

福建菜。将肉丁、猪肚、香菇、冬笋、火腿、虾米、糯米饭拌匀，填入鸡腹，蒸熟，抹糖和醋，炸至棕红。

Fried stuffed chicken. Fujian dish. Whole chicken stuffed with diced pork, hog tripe, dried mushrooms, winter bamboo shoots, ham, dried shrimp meat, and cooked sticky rice, then steamed. Coated with sugar and vinegar sauce, then deep-fried red brown.

尖椒肉丝 (jiānjiāo ròusī)

北京菜。将腌过的肉丝与葱、姜、尖椒、酱油、味精、. 料酒翻炒。

Stir-fried pork with pepper. Beijing dish. Pickled shredded lean pork stir-fried with hot pepper, ginger, and scallion, then flavored with soy sauce, MSG, and cooking wine.

煎馄饨 (jiānhúntun)

江浙沪点心。先将菜肉馄饨煮熟，过冷水，稍晾，然后用少量油煎脆，佐酱油和醋。

Fried wontons. Jiangsu, Zhejiang, and Shanghai snack. Wontons filled with minced vegetable, pork, and seasonings, boiled, cooled in cold water, then drained. Fried with a little oil until crisp. Served with soy sauce and vinegar.

煎牛柳 (jiānniúliǔ)

东北菜。先将牛里脊肉切成柳叶片，稍腌，再用文火煎熟，然后与洋葱、胡椒粉、麻油等翻炒。

Fried beef. Northeastern dish. Beef cut into narrow strips, lightly pickled, lightly fried, then stir-fried with onion, ground black pepper, and sesame oil.

碱酥饼 (jiǎnsūbǐng)

湖北点心。用面粉、米糖浆、猪油做成饼皮，面粉与猪油做成酥面，用糖、熟面、陈皮、桂圆、小苏打等做成馅。用饼皮包酥面与馅做成饼，撒芝麻，烤熟。

Stuffed soda cakes. Hubei snack. Flour, rice syrup, and lard made into wrappers. Flour and lard made into pastry. Sugar, half-roasted flour, processed tangerine peels, longan, and soda mixed into stuffing. Wrappers made into cakes filled with pastry over stuffing, then baked.

剑南春 (jiànnánchūn)

白酒，产于四川绵竹。以高粱、大米、糯米、玉米、小麦为原料酿制而成。酒精含量在 28-52 度之间，属浓香型酒。

Jiannanchun, liquor produced in Mianzhu, Sichuan. White spirit distilled from fermented sorghum, rice, glutinous rice, corn, and wheat. Contains 28-52% alcohol, and has a thick aroma.

江南百花鸡 (jiāngnán bǎihuājī)

广州名菜。先在整张鸡皮的内皮上拍满淀粉，抹上虾茸蟹肉糊，蒸熟，然后切条，淋麻油，配鲜花瓣装盘。

Chicken and shrimp with flower. Specialty in Guangzhou, Guangdong. Whole chicken skin starched inside, spread with shrimp and crab meat paste, then steamed. Cut into strips, washed with sesame oil, then served with flower petals.

江山绿牡丹 (jiāngshān lǜmǔdān)

绿茶。中国名茶，产于浙江江山市仙霞岭裴家地、龙井一带。外形紧结挺直，色泽绿翠显毫。汤色碧绿，叶底嫩绿。

Jiangshan lumudan, famous green tea produced in the areas of Peijiadi and Longjing of Mount Xianxialing in Jiangshan, Zhejiang. Its dry leaves are tight, straight, emerald green, and hairy. The tea is crystal green with infused leaves tender green.

江糟羊 (jiāngzāoyáng)

福建菜。先将姜末、红糟炒香，再加冬笋、汆过的羊肉稍炒，然后放入猪骨汤焖煮，淋麻油。

Braised mutton with bamboo shoots. Fujian dish. Winter bamboo shoots and scalded mutton quick-fried with ginger and fermented sticky rice juice, simmered in hog bone soup, then washed with sesame oil.

姜葱爆羊肉 (jiāngcōng bào yángròu)

四川菜。先将羊肉片腌渍，然后加入花椒、姜、葱、辣椒等爆炒。

Stir-fried mutton with ginger and scallion. Sichuan dish. Pickled sliced mutton stir-fried with Chinese prickly ash, ginger, scallion, and chili pepper.

姜豆腐 (jiāngdòufu)

山东菜。豆腐与葱、姜、蒜、虾、海米、豆豉等翻炒，烹入鸡汤，加入南荠末，勾芡。

Sautéed bean curd with assorted flavorings. Shandong dish. Bean curd stir-fried with green onion, garlic, shrimp, dried shrimp meat, and fermented soybean. Flavored with mashed water chestnuts and chicken soup, then thickened with starch.

将军过桥 (jiāngjūn guòqiáo)

江苏菜。将军指黑鱼。先将黑鱼片划油，加调料炒熟，再将鱼头和鱼骨煮至汤色变白，然后加入菜心和香菇煮开。

Sautéed snakehead fish fillets and fish soup. Jiangsu dish. Lightly fried snakehead fish fillets sautéed with flavorings. Fish head and bones stewed until soup milky, then boiled with baby greens and mushrooms.

豇豆烧肉 (jiāngdòu shāo ròu)

浙江菜。先将五花肉稍煸，加调料煮熟，再放入干豇豆、蒜瓣、酱油、糖等炖烂。

Stewed streaky pork with cowpeas. Zhejiang dish. Quick-fried streaky pork boiled with seasonings, then stewed with dried cowpeas, garlic cloves, soy sauce, and sugar.

酱板鸡 (jiàngbǎnjī)

湖南菜。先将整鸡腌渍,压扁,烘烤近熟,然后入卤水煮熟。

Marinated chicken. Hunan dish. Pickled and flattened chicken half-roasted, then marinated in gravy.

酱爆肉丁 (jiàngbào ròudīng)

山东菜。先将猪肉丁过油,然后与玉兰片、蒲菜、青菜心及调料翻炒,加甜面酱,勾芡。

Pork and vegetables with sauce. Shandong dish. Lightly fried diced pork stir-fried with sliced bamboo shoots, cattail roots, baby greens, and seasonings. Flavored with sweet fermented flour sauce, then thickened with starch.

酱爆肉丝 (jiàngbào ròusī)

北方家常菜。将葱段爆香,与腌过的瘦肉丝翻炒,加甜面酱。

Stir-fried pork with sauce. Northern home-style dish. Pickled shredded lean pork stir-fried with sectioned leek and flavored with sweet fermented flour sauce.

酱爆鱿鱼卷 (jiàngbào yóuyújuǎn)

上海菜。将炸过的鱿鱼用肉末和甜面酱翻炒,放入酱油、糖、味精。

Stir-fried squid. Shanghai dish. Fried squid stir-fried with minced meat and sweet fermented flour sauce, then flavored with soy sauce, sugar, and MSG.

酱排骨 (jiàngpáigǔ)

江苏菜。先将排骨腌渍,焯水,然后加黄酒、葱、姜、茴香、桂皮、酱油、糖等焖煮。

Braised spareribs with sauce. Jiangsu dish. Pork spareribs pickled, quick-boiled, then braised with yellow wine, scallion, ginger, fennel, cinnamon, soy sauce, and sugar.

酱味烤海鱼 (jiàngwèi kǎohǎiyú)

广东菜。将海鱼涂上用豆瓣酱、糖、沙拉油、酱油、蒜茸、姜汁等调制的酱料,烤熟。

Grilled sea fish with sauce. Guangdong dish. Sea fish coated with mixture of fermented broad-bean sauce, sugar, salad oil, soy sauce, minced garlic, and ginger juice, then grilled.

酱鸭 (jiàngyā)

上海菜。将鸭与酱油、糖、黄酒、葱、姜煮熟,浸泡入味,冷却后切块。

Marinated duck. Shanghai dish. Duck boiled, then marinated in gravy of soy sauce, sugar, yellow wine, scallion, and ginger. Cooled and chopped before serving.

酱汁烤鳗 (jiàngzhī kǎo mán)

江苏菜。先将鳗鱼用黄酒、酱油、糖等腌渍,然后烤熟。

Broiled eel. Jiangsu dish. Eel pickled with yellow wine, soy sauce, and sugar, then broiled.

茭白炒肉 (jiāobái chǎo ròu)

苏、浙、沪、皖菜。先将腌渍过的猪瘦肉片滑熟,然后放入茭白片翻炒,用盐、鸡精调味。

Sautéed wild rice stems and pork. Jiangsu, Zhejiang, Shanghai, and Anhui dish. Pickled sliced lean pork lightly fried, then sautéed with sliced wild rice stems. Seasoned with salt and chicken essence.

椒盐八宝鸡 (jiāoyán bābǎojī)

四川菜。先将整只母鸡去骨,腌渍,填入新鲜豌豆、糯米、虾米、火腿、香菇等八种配料,然后蒸熟,炸透,佐椒盐。

Fried stuffed chicken. Sichuan dish. Boned and pickled whole hen stuffed with fresh peas, sticky rice, dried shrimp meat, ham, and dried mushrooms, steamed, then deep-fried. Served with spicy salt.

椒盐里脊 (jiāoyán lǐji)

湖南菜。先将猪里脊肉油炸,然后加葱、花椒、盐翻炒。

Fried pork tenderloin. Hunan dish. Deep-fried pork tenderloin stir-fried with scallion, Chinese prickly ash, and salt.

焦溜菊花里脊 (jiāoliū júhuālǐji)

广东菜。将猪里脊肉切成菊花状片,腌渍,油炸,浇上芡汁。

Fried pork tenderloin with sauce. Guangdong dish. Pork tenderloin carved into pieces in the shape of chrysanthemum flower, pickled, fried, then washed with starched sauce.

蕉城鱼丸 (jiāochéngyúwán)

福建点心。鱼茸与地瓜粉做成皮,瘦肉末为馅,包馅成丸,煮熟。

Fish and pork balls. Fujian snack. Mashed fish and sweet potato powder mixed, made into balls filled with ground lean pork, then boiled.

饺子 (jiǎozi)

北方家常点心。面粉加水做成圆形面皮,包入各种菜馅或肉馅,清水煮熟,佐蒜茸酱油、香菜麻油等食用。

Jiaozi. Home-style snack in northern areas. Flour mixed with water, made into round thin wrappers, stuffed with a variety of fillings made with vegetables or meat, then boiled. Served with dressing of minced garlic and soy sauce or coriander and sesame oil.

叫化鸡 (jiàohuājī)

江苏菜。又称黄泥煨鸡。先将鸡腹内填入用虾仁、猪肉、鸡肫、火腿等炒制的馅,再用泥把鸡裹好,烤熟。

Beggar's chicken. Jiangsu dish, also called yellow mud chicken. Chicken stuffed with shrimp meat, pork, chicken gizzard, and ham. Coated with yellow mud, then baked.

芥菜缨炖豆腐 (jiècàiyīng dùn dòufu)

东北菜。将腌制的芥菜叶切碎,加豆腐及调料炖熟。

Leaf mustard with bean curd. Northeastern dish. Chopped pickled leaf mustard braised with bean curd and flavorings.

金蝉吐丝 (jīnchán tǔsī)

家常菜。蝉即蚕蛹，丝即土豆丝。将蚕蛹和土豆丝加调料炒熟。

Sautéed silkworms and potatoes. Home-style dish in many places. Silkworms sautéed with shredded potatoes and seasoned.

金钩油菜 (jīngōu yóucài)

东北家常菜。先将油菜焯水，再与海米及调料翻炒，然后与切成条的煎蛋皮拌匀。

Rape greens with dried shrimp meat. Home-style dish in northeastern areas. Scalded rape greens sautéed with dried shrimp meat and flavorings, then mixed with fried eggs strips.

金钱饼 (jīnqiánbǐng)

江苏菜。先将面粉和豆粉做成的饼剖成两个圆片，夹上用猪肉和虾仁做成的馅，边上用淀粉发蛋糊封住，炸至金黄，然后加葱花、椒盐、麻油快炒。

Fried sandwich cakes. Jiangsu dish. Wheat and soybean cake sandwiches filled with pork and shrimp meat, sealed with whipped starch and egg paste, then deep-fried. Quick-fried with chopped scallion, spicy salt, and sesame oil.

金钱鹅掌 (jīnqián ézhǎng)

四川菜。将鹅掌腌渍，挂鸡蛋芡粉糊，炸至金黄。

Fried goose feet. Sichuan dish. Pickled goose feet coated with starch and egg paste, then deep-fried.

金钱肉 (jīnqiánròu)

湖南菜。先将猪肥、瘦肉切成大小相同的圆片，叠起来下锅油炸，然后撒上花椒粉、麻油。热炒和冷盘均可。

Fried fat and lean pork. Hunan dish. Fat and lean pork cut into round pieces, fried, then flavored with ground Chinese prickly ash and sesame oil. May be served cold or stir-fried.

金腿脊梅炖腰酥 (jīntuǐ jí méi dùn yāosū)

江苏菜。将猪里脊肉、猪腰、火腿等放入汤盅，加入调料和水，蒸至酥烂。

Steamed pork varieties. Jiangsu dish. Pork tenderloin, hog kidney, and ham steamed with condiments.

金线吊葫芦 (jīnxián diào húlu)

江西点心。将精制面丝和馄饨煮熟，放入鸡汤、葱等调料。

Boiled noodles and wontons. Jiangxi snack. Fine noodles and wontons boiled, then combined with chicken soup and scallion.

金银菜煲猪肺 (jīnyíncài bāo zhūfèi)

广东菜。将猪肺、猪骨、干菜、豆芽、杏仁、罗汉果、蜜枣加盐、姜、葱、料酒熬煮。

Hog lung and bones with vegetable varieties. Guangdong soup. Hog lung and bones boiled with dried vegetables, bean sprouts, apricots, momordica fruit, and honey dates, then flavored with salt, ginger, scallion, and cooking wine.

锦绣玉鸳鸯 (jǐnxiù yùyuānyāng)

广东菜。将莴笋丝、青椒丝、红椒丝、黄椒丝、鸡丝、烤鸭丝、海蜇丝加盐、料酒、花生酱翻炒。

Chicken, duck, and jellyfish with vegetables. Guangdong dish. Shredded chicken, broiled duck, and jellyfish stir-fried with stem lettuce and shredded green, red and yellow pepper, then flavored with salt, cooking wine, and peanut sauce.

京葱焖牛方 (jīngcōng mèn niúfāng)

北京菜。先将京葱段炸黄，将牛肉块煎至金黄，然后将牛肉、京葱、笋片加黄酒及调料焖烂。

Beef with leek. Beijing dish. Cubed beef deep-fried, then braised with fried leek stems, sliced bamboo shoots, and yellow wine.

京葱烧虾球 (jīngcōng shāo xiāqiú)

南方菜。先将新鲜大虾仁用蛋浆、芡粉、盐等腌渍，然后与葱白翻炒。

Stir-fried prawn meat with leek. Southern dish. Shelled prawns pickled with egg paste, starch, and salt, then stir-fried with leek stems.

京葱羊糕 (jīngcōng yánggāo)

北京凉菜。先将羊腿肉与牛筋同煮，将煮熟的羊肉撕碎，再将牛筋煮化入汤，把汤浇在羊肉上，冷却成冻，然后将肉冻切块，配葱丝装盘。

Beef and mutton jelly with leek. Beijing cold dish. Boiled gigot torn into pieces, immersed in thick broth with melted beef tendon, then chilled into jelly. Cubed and served with shredded leek.

京酱肉丝 (jīngjiàng ròusī)

北京菜。先将猪里脊肉丝用蛋清与生粉腌过，入油划散，然后加甜面酱等炒熟，倒在葱丝上。

Stir-fried pork with leek. Beijing dish. Shredded pork tenderloin pickled with egg white and starch, lightly fried, stir-fried with sweet fermented flour sauce, then served over shredded leek.

京味驴肉 (jīngwèi lǘròu)

北京凉菜。先将驴肉入料汤卤熟，切片，然后浇上用盐、味精、蒜等调成的汁。

Donkey meat with sauce. Beijing cold dish. Sliced marinated donkey meat served with dressing of salt, MSG, and garlic.

荆沙鱼糕 (jīngshāyúgāo)

湖北名菜。先将青鱼肉、猪肥膘捣成茸，与绿豆粉、盐、胡椒粉等搅拌成糊状，急火蒸熟，然后抹上蛋黄浆稍蒸，冷却后切片。

Fish cakes. Hubei dish. Black carp meat and fat pork mashed, made into paste with mung bean powder, salt, and ground black pepper, steamed over high heat, coated with egg yolks, then re-steamed briefly. Cooled and sliced.

精武鸭颈 (jīngwǔyājǐng)

湖北菜。将鸭颈腌渍、汆烫，放入辣味卤汁煮熟、浸泡入味。

Marinated duck necks. Hubei dish. Scalded pickled duck necks boiled, then marinated in spicy gravy.

井冈山烟笋烧肉 (jǐnggāngshān yānsǔn shāo ròu)

江西菜。先将煮过的井冈山竹笋用炭火熏烤成褐黄色,然后同五花肉焖煮,加盐、料酒、辣椒、酱油。

Braised pork and smoked bamboo shoots. Jiangxi dish. Bamboo shoots from the Jinggangshan Mountains boiled, smoked over charcoal until dark brown, braised with streaky pork, then flavored with salt, cooking wine, chili pepper, and soy sauce.

敬亭绿雪 (jìngtínglǜxuě)

绿茶。产于安徽宣城敬亭山。外形色泽翠绿,两叶抱一芽,全身披白毫。汤色清碧,叶底匀嫩。

Jingtingluxue, green tea produced in Mount Jingting in Xuancheng, Anhui. Its dry leaves are jade green and hairy, each piece having 2 leaves holding a leaf bud. The tea is crystal green with infused leaves even and tender.

揪片子 (jiūpiànzi)

新疆食品。将面团揪成面片,与炒好的羊肉、姜、土豆、萝卜、西红柿等煮熟。

Noodles with mutton and vegetables. Specialty in Xinjiang. Dough hand-made into small thin pieces, then boiled with stir-fried mutton, ginger, potatoes, radishes, and tomatoes.

九江封缸酒 (jiǔjiāngfēnggāngjiǔ)

黄酒,产于江西九江。用优质糯米和当地矿泉水酿制而成,色泽晶莹淡黄,酒精含量 11-13 度。

Jiujiang yellow wine from Jiujiang, Jiangxi. Made from glutinous rice and local mineral water. Has a crystal light yellow color. Its alcohol content is 11-13%.

九江煎堆 (jiǔjiāngjiānduī)

江西点心。先将爆玉米、碎桔饼、花生仁、糖浆等调和,压成煎堆坯,然后裹上糯米浆,撒上芝麻,炸至金黄。

Fried corn and peanut cakes. Jiangxi snack. Pop corn, chopped candied tangerines, peanuts, and syrup mixed, pressed into lump cakes, coated with sticky rice paste, showered with sesame, then fried until golden.

九转大肠 (jiǔzhuǎn dàcháng)

山东菜。先将大肠煮熟,切段,用糖和油煎炒,再将葱、姜、蒜爆香,放入大肠翻炒,烹入醋、盐、酱油、料酒、清汤焖煮,撒胡椒粉、浇花椒油。

Savoury hog large intestine. Shandong dish. Boiled and sectioned hog large intestine fried with sugar and lard, stir-fried with quick-fried scallion, ginger, and garlic, braised in clear soup with vinegar, salt, soy sauce, and cooking wine, then sprinkled with ground black pepper and Chinese prickly ash oil.

久玛 (jiǔmǎ)

西藏菜,即血肠。将新鲜牛血、羊血、糌粑、盐、野葱等混合,灌入牛肠或羊肠,煮熟。

Boiled Tibetan blood sausage. Tibetan dish. Ox or goat intestine filled with mixture of fresh ox and goat blood, *zanba*, salt, and wild onion, then boiled.

韭菜炒蛋 (jiǔcài chǎo dàn)

各地家常菜。将韭菜和蛋浆加盐调匀,煎熟。

Fried Chinese chive and eggs. Home-style dish in many places. Chinese chive and egg paste mixed, seasoned with salt, then fried.

韭菜干丝 (jiǔcài gānsī)

东北家常菜。先将豆腐干用素油加高汤和调料慢炒,然后加入韭菜快炒。

Sautéed bean curd with chive. Home-style dish in northeastern areas. Dried bean curd sautéed with vegetable oil, flavorings, and broth, then quick-fried with Chinese chive.

韭黄炒肉丝 (jiǔhuáng chǎo ròusī)

四川菜。将韭黄与蛋浆调匀,与肉丝和红椒丝及调料煎炒。

Sautéed chive and pork. Sichuan dish. Yellow Chinese chive and egg paste mixed, then stir-fried with shredded pork, red pepper, and condiments.

酒鬼酒 (jiǔguǐjiǔ)

白酒,产于湖南吉首。以糯高粱为原料,用传统药曲酿造,酒精含量50度左右,属兼香型酒。

Jiugui liquor, produced in Jishou, Hunan. White spirit made with sticky sorghum and traditional herb yeast. Contains about 50% alcohol, and has a mixed aroma.

酒酿清蒸鸭子 (jiǔniàng qīngzhēng yāzi)

曹雪芹《红楼梦》菜谱。先将鸭肉块用酒酿汁和盐拌腌,再加入葱、姜、清汤,蒸烂。

Steamed duck with fermented sticky rice juice. Menu from *A Dream of Red Mansions* by Cao Xueqin. Chopped duck pickled with fermented rice juice and salt, then steamed in clear soup with scallion and ginger.

酒酿圆子 (jiǔniàng yuánzi)

南方点心。将小糯米粉圆子用酒酿煮熟,加糖桂花。

Sticky rice balls in fermented rice soup. Southern snack. Mini sticky rice flour balls boiled in sweet fermented sticky rice soup and flavored with sugared osmanthus flower.

菊花荸荠 (júhuā bíqí)

湖南菜。将荸荠切成菊花形汆熟,再加入番茄汁、白糖、高汤煸炒。

Sautéed water chestnuts. Hunan dish. Water chestnuts cut like chrysanthemum flower, scalded, then sautéed with tomato sauce, sugar, and broth.

菊花火锅 (júhuā huǒguō)

山东菜。将白菊花、腌韭菜花、鱼

肉、鸡脯肉、里脊肉、鸡蛋、水饺等入火锅汤涮熟。

Chrysanthemum fire pot. Shandong dish. White chrysanthemum petals, preserved Chinese chive flower, fish fillets, chicken breast, pork tenderloin, eggs, dumplings, and other ingredients quick-boiled in fire pot soup.

菊花青鱼 (júhuā qīngyú)

湖南菜。先将带皮青鱼刻成菊花状，油煎，然后浇上特制卤汁。

Blue carp with sauce. Hunan dish. Blue carp carved like a chrysanthemum flower, deep-fried, then washed with special sauce.

菊花五蛇羹 (júuā wǔshégēng)

广东名菜。将五种新鲜蛇肉与鸡丝、菇丝加调料烩煮，放入菊花瓣出锅。

Snake, chicken, and mushroom soup. Famous Guangdong dish. Five kinds of fresh snake meat braised with shredded chicken, mushrooms, and seasonings. Served with chrysanthemum flower petals.

菊花鸭肫 (júhuā yāzhūn)

东北家常菜。先将鸭肫入清汤加葱、姜、黄酒等煮熟，然后切片，摆成菊花状。

Boiled duck gizzard. Northeastern home-style dish. Duck gizzard boiled in clear soup, flavored with scallion, ginger, and yellow wine, then sliced. Arranged like a chrysanthemum flower on a platter.

菊叶玉板 (júyè yùbǎn)

江苏菜。将菊叶加盐和白糖轻炒，四周缀上煮熟的鲜笋片与火腿片。

Chrysanthemum leaves with bamboo shoots. Jiangsu dish. Chrysanthemum leaves sautéed gently with salt and sugar, then decorated with prepared bamboo shoots and sliced ham.

橘瓣鱼圆 (júbàn yúyuán)

湖北菜。先将鱼茸、荸荠末、蛋清调成糊，做成橘瓣形丸子，氽过，然后与香菇入鸡汤煮熟。

Fish ball and mushroom soup. Mashed fish meat, diced water chestnuts, and egg white blended, made into pieces like orange sections, then scalded. Boiled with dried mush-rooms in chicken soup.

橘皮鸡块 (júpí jīkuài)

东北家常菜。先将橘皮、葱、姜煸炒，然后放入鸡块，加清汤、盐、黄酒，用慢火焖煮。

Orange-taste chicken. Northeastern home-style dish. Cubed chicken braised in clear soup with quick-fried orange peel, scallion, and ginger, then flavored with salt and yellow wine.

君山银针 (jūnshānyínzhēn)

黄茶。中国名茶，产于湖南岳阳洞庭湖中的君山。外形芽壮，芽身金黄，披银毫。汤色澄黄明净，叶底嫩亮。

Junshanyinzhen, yellow tea produced on Junshan Hill in Dongting Lake in Yueyang, Hunan. Its dry leaves are stout, golden, and silvery hairy. The tea is crystal yellow with infused leaves tender and lustrous.

开花献佛 (kāihuā xiànfó)

山东菜。先将菜花和台蘑用花椒、麻油等炒熟，然后将台蘑置于菜花块中央，用西红柿瓣盘边。

Sautéed cauliflower and mushrooms with tomatoes. Shandong dish. Cauliflower sautéed with mushrooms from Mount Wutaishan, Chinese prickly ash, and sesame oil. Mushrooms arranged at the center of cauliflower cuts, then surrounded by sectioned tomatoes.

开胃藕片 (kāiwèi ǒupiàn)

湖南凉菜。先将莲藕用开水烫过，然后加入葱、姜、蒜、酱油、盐、味精拌匀。

Lotus roots with dressing. Cold dish in Hunan. Lotus roots quick-boiled and flavored with scallion, ginger, garlic, soy sauce, salt, and MSG.

开阳拌三丝 (kāiyáng bànsānsī)

山东菜。将腌过的高丽菜丝与五香豆腐干丝、虾米、香菜、麻油、蒜茸等拌匀。

Cabbage with special dressing. Shandong cold dish. Pickled shredded cabbage mixed with shredded dried bean curd, dried shrimp meat, coriander, sesame oil, and minced garlic.

康巴汉子 (kāngbāhànzi)

西藏食品，即青稞卷子。用青稞面薄饼卷入黄瓜片、香菜、烤牛肉等，佐酱汁。

Barley rolls with sauce. Tibetan food. Highland barley pancakes made into rolls filled with sliced cucumbers, coriander, and broiled beef. Served with sauce.

烤包子 (kǎobāozi)

新疆点心。用面粉与水和成面皮，包入用羊肉、羊尾油、洋葱、盐、孜然粉等做成的馅，烤熟。

Baked mutton buns. Xinjiang snack. Flour dough made into buns, stuffed with mixture of mutton, sheep-tail fat, onion, salt, and ground cumin, then baked.

烤扁担肉 (kǎobiǎndanròu)

四川菜。先将肉块扎出多个小孔，塞进调料腌渍，然后用文火慢烤，淋麻油。

Grilled pork. Sichuan dish. Pricked pickled pork grilled over low heat, then sprinkled with sesame oil.

烤广式三层肉 (kǎo guǎngshì sāncéngròu)

广东菜。将五花腊肉切片，抹上用

果酱和蚝油制成的调料，和青蒜烤熟。

Grilled smoked streaky pork.

Guangdong dish. Sliced smoked cured streaky pork coated with paste of fruit jam and oyster sauce, then grilled with garlic sprouts.

烤红薯 (kǎohóngshǔ)

各地家常点心。将红薯洗净，烤熟。

Baked sweet potatoes. Home-style snack in many places. Sweet potatoes cleaned, then baked.

烤加吉鱼 (kǎojiājíyú)

山东菜。先将加吉鱼稍炸，然后浇上用清汤、肉末、盐、黄酒等做成的卤汁，烤熟。

Broiled red sea bream. Shandong dish. Lightly fried red sea bream coated with sauce of clear soup, minced pork, salt, and yellow wine, then broiled.

烤鹿肉 (kǎolùròu)

曹雪芹《红楼梦》菜谱。将鹿腿肉用黄酒、盐、丁香、酱油、大料、花椒、葱、姜、茴香等腌渍，淋花生油和鸡汤，烤熟，缀生菜装盘，用红油蒜泥汁或酸辣汁佐食。

Broiled leg of venison. Menu from *A Dream of Red Mansions* by Cao Xueqin. Leg meat of venison pickled with yellow wine, salt, clove, soy sauce, aniseed, Chinese prickly ash, scallion, ginger, and fennel. Washed with peanut oil and chicken soup, then broiled. Garnished with lettuce, then served with dressing of mashed garlic and chili oil or spicy sour sauce.

烤馕 (kǎonáng)

新疆维吾尔族日常食品，即烤面饼。面粉加入水、油、鸡蛋、肉末、芝麻、糖或盐做成各式饼，或称之为油馕、肉馕、芝麻馕、窝窝馕等，在馕坑中烤熟。

Baked flour cakes. Uygur daily food in Xinjiang. Flour mixed with water, oil, eggs, ground meat, sesame, sugar, or salt, made into a variety of cakes called oil *nang*, meat *nang*, sesame *nang*, or dimpled *nang*, then baked in a special oven.

烤全羊 (kǎoquányáng)

内蒙、新疆传统菜。将羊去毛洗净，在腹内塞入葱、姜、辣椒、盐等，烤熟。

Broiled whole lamb. Traditional dish in Inner Mongolia and Xinjiang. Whole lamb dressed out, stuffed with onion, ginger, chili, salt, and other spices, then broiled.

烤肉串 (kǎoròuchuàn)

内蒙传统点心。将羊肉和羊油切片，串在铁丝上烤熟，裹上盐、辣椒、孜然粉等食用。

Grilled mutton and suet with spices. Inner Mongolian traditional snack. Sliced mutton and suet strung on iron bars, grilled, then sprinkled with salt, pepper, and ground cumin.

烤乳猪 (kǎorǔzhū)

广东菜。将乳猪涂上用五香粉、八角、豆酱、甜酱、红腐乳、芝麻酱等制成的酱料,烤熟。

Broiled baby hog. Guangdong dish. Baby hog coated with mixture of five-spice powder, aniseed, fermented soybean sauce, sweet fermented flour sauce, red fermented bean curd, and sesame sauce, then broiled.

烤土豆 (kǎotǔdòu)

各地家常点心。将土豆洗净,烤熟。佐椒盐。

Baked potatoes. Home-style snack in many places. Potatoes cleaned, then baked. Served with spicy salt.

烤网油签 (kǎowǎngyóuqiān)

山东菜。先将瘦肉丝加调料拌匀,再用猪网油包裹肉丝,用蛋清芡粉糊封口,然后烤至金黄。

Grilled lean pork in lard. Shandong dish. Shredded lean pork pickled with seasonings, wrapped in web lard, sealed with starch and egg white paste, then grilled until golden.

焅鹌鹑 (kàoānchún)

各地家常菜。先将鹌鹑炸成红色,然后加入鲜汤,与煸过的桂皮、冬笋片等用小火慢焅。

Fried and braised quails. Home-style dish in many places. Quails deep-fried, then braised in seasoned soup with quick-fried cinnamon and sliced winter bamboo shoots.

蚵仔煎 (kēzǎijiān)

台湾著名点心。将新鲜蚵仔肉即牡蛎肉与番薯粉和水调和,用蛋皮包裹,煎至金黄。

Oyster in omelet. Famous Taiwan snack. Fresh oyster meat mixed with sweet potato flour and water, wrapped in omelet, then fried golden.

客家封鸡 (kèjiāfēngjī)

广东菜。先将鸡加入香菇、蒜、酱油、味精等蒸熟,然后放在焯过的空心菜上,浇上汤,配蒸熟的香菇和蒜。

Steamed chicken on water spinach. Guangdong dish. Chicken steamed with dried mushrooms, garlic, soy sauce, and MSG. Put on a bed of scalded water spinach, washed with broth, then served with steamed mushrooms and garlic.

空心煎堆 (kōngxīnjiānduī)

广东点心。先将糯米粉加水揉成团,放入糖水中煮熟,然后包入用花生仁、橘饼、白芝麻做成的馅,炸至金黄。

Fried rice buns with sweet fillings. Guangdong snack. Sticky rice flour and water made into balls, boiled in sweetened water, stuffed with mixture of peanuts, candied tangerines, and white sesame, then deep-fried.

空心琉璃丸子 (kōngxīn liúlíwánzi)

山东菜。先在面糊中掺入鸡蛋黄搅匀,做成丸子,再将丸子炸至空心并呈金色,然后将白糖熬起泡,倒入炸好的丸子翻匀。

Sugarcoated flour and yolk balls. Shandong dish. Flour and egg yolks

mixed, made into small balls, fried until hollow and golden, then coated with heated melting sugar.

孔府家酒 (kǒngfǔjiājiǔ)

白酒,产于山东曲阜。以高粱为原料,用小麦制曲酿造,酒精含量为40度左右,属浓香型酒。

Kongfu jiajiu liquor produced in Qufu, Shandong. White spirit made with sorghum and wheat yeast. Contains about 40% alcohol, and has a thick aroma.

孔府一品锅 (kǒngfǔ yīpǐnguō)

山东曲阜孔府菜。将龙须粉、山药、鸡、鸭、海参、鱼肚、鱿鱼卷等逐层放入锅内,加调料蒸熟。

Assorted meat with yam and vermicelli. Specialty of the Kong Family in Qufu, Shandong. Vermicelli, Chinese yam, chicken, duck, sea cucumbers, fish maw, and squid layered in a container, then steamed with seasonings.

口蘑鸡块 (kǒumó jīkuài)

四川菜。先将鸡块焯水,然后与口蘑、火腿、鲜笋及调料炖煮。

Stewed chicken and mushrooms. Sichuan dish. Quick-boiled chicken stewed with mushrooms, ham, and fresh bamboo shoots.

口水鸡 (kǒushuǐjī)

四川菜。先将鸡加调料煮熟,迅速冷却,斩块,然后浇红油、麻油、芝麻酱。

Chicken with spicy sauce. Sichuan dish. Chicken boiled with seasonings, quickly chilled, chopped, then washed with chili oil, sesame oil, and sesame sauce.

口子坊 (kǒuzifáng)

白酒,产于安徽淮北,以高粱为原料,用小麦、大麦、豌豆等制曲酿造。酒精含量为36-52度,属兼香型酒。

Kouzifang liquor, produced in Huaibei, Anhui. White spirit made with sorghum and yeast of wheat, barley, and peas. Contains 36-52% alcohol, and has a mixed aroma.

口子窖 (kǒuzijiào)

白酒,产于安徽淮北,以高粱为原料,用小麦、大麦、豌豆等制曲酿造。酒精含量为38-50度,属浓香型酒。

Kouzijiao liquor produced in Huaibei, Anhui. White spirit made with sorghum and yeast of wheat, barley, and peas. Contains 38-50% alcohol, and has a thick aroma.

扣肉 (kòuròu)

南方菜。将五花肉厚片炸至深红,肉皮朝下码在碗内,放入梅菜、酱油、糖,蒸烂。

Steamed pork with preserved mustard. Southern dish. Thick pieces of streaky pork deep-fried until deep red, put in a bowl with skins down, covered with preserved mustard, soy sauce, and sugar, then steamed.

扣三丝 (kòusānsī)

上海菜。鸡丝、火腿丝、笋丝入鸡汤先煮后蒸,配汆过的绿豆苗装盘。

Shredded chicken, ham, and bamboo shoots. Shanghai dish. Shredded chicken, ham, and bamboo shoots boiled, then steamed in chicken soup. Served with scalded pea sprouts.

苦丁茶 (kǔdīngchá)

南方传统清凉饮料，产于华南和西南地区，常用作保健茶。将苦丁茶叶经过萎凋、杀青、揉捻、干燥四个过程制成。外形条索粗壮，无茸毫，汤色黄绿，叶底摊张，呈深绿，味先苦后甘，耐冲泡。

Kuding tea. Traditional summer beverage in southern areas. Produced in southern and southwestern areas, also used as a tonic drink. Made with leaves of broadleaf holly trees through withering, gentle-baking, rubbing-twisting, and drying. Its dry leaves are stout, tight, and hairless. The tea is crystal dark green with a bitter taste then a sweet aftertaste. The infused leaves are flat. Good for repeated brewing.

苦瓜炒腊肉 (kǔguā chǎo làròu)

湖南家常菜。苦瓜片和腊肉片加姜、蒜、红辣椒、盐翻炒，勾芡。

Bitter gourd and smoked pork. Hunan home-style dish. Sliced smoked cured pork and bitter gourd stir-fried and flavored with ginger, garlic, red pepper, salt, and starched broth.

库车汤面 (kùchētāngmiàn)

新疆食品。将羊尾油丁汤浇在细拉面上，放入熟羊肉片和蛋皮丝。

Noodles with mutton. Xinjiang food. Fine hand-made noodles washed with soup of diced lamb tail oil, then topped with sliced cooked mutton and shredded fried egg sheets.

侉炖目鱼 (kuǎdùn mùyú)

山东菜。先将肉片、鹿角菜、冬笋等煸炒，然后加入鸡汤、米醋，放入炸熟的目鱼炆煮。

Braised flatfish with vegetables. Shandong dish. Sliced pork, harthorn vegetables, and winter bamboo shoots quick-fried, then braised with fried flatfish, chicken soup, and vinegar.

L

拉条子 (lātiáozi)

新疆食品。先将面粉加水做成粗拉面,煮熟,然后放入过油肉、西红柿炒鸡蛋、辣椒炒羊肉、或芹菜炒肉片等。

Thick noodles with dressing.

Xinjiang food. Flour and water hand-made into thick noodles, boiled, then served with fried and braised pork, stir-fried eggs with tomatoes, stir-fried mutton with chili pepper, or sautéed sliced pork with celery.

腊味拼盘 (làwèi pīnpán)

西藏菜。把腌过的猪舌、猪心、猪肠、猪肉煮熟,切片,配黄瓜、鲜橙、胡萝卜装盘。

Assorted hog offal. Tibetan dish. Pickled hog tongue, heart, and intestine boiled with pork, sliced, then served with cucumbers, oranges, and carrots.

腊羊肉 (làyángròu)

陕西菜。先将肥羊肉腌渍,然后放入老卤汤焖煮,佐蒜泥酱油或麻油香醋。

Marinated mutton. Shaanxi dish. Pickled fatty mutton braised in aged gravy, then served with dressing of garlic and soy sauce or sesame oil and vinegar.

辣白菜 (làbáicài)

朝鲜族凉菜。先将大白菜用盐和其他调料腌渍,再将大量辣椒酱从里到外抹到腌白菜上,然后装坛密封15天左右。

Pickled hot Chinese cabbage.

Korean ethnic minority cold dish. Salted Chinese cabbage drained, pickled thoroughly with lots of chili sauce, then sealed in a jug for about 15 days before serving.

辣拌野桔梗 (làbàn yějiégěng)

东北凉菜。先将桔梗泡开撕成条,然后加盐、味精、辣椒油等拌匀。

Pickled balloon flower roots.

Northeastern cold dish. Balloon flower roots soaked, torn into pieces, then seasoned with salt, MSG, and chili oil.

辣酒羊肉炉 (làjiǔ yángròulú)

广东菜。将羊腩与草菇、姜、蒜、红椒、香叶、香菇加盐、料酒焖熟。

Stewed mutton flank. Guangdong dish. Mutton flank stewed with mushrooms, ginger, garlic, red pepper, laurel leaves, and dried mushrooms. Seasoned with salt and cooking wine.

辣味过江鸡 (làwèi guòjiāngjī)

四川菜。先把鸡汆过，再用黄酒、盐、味精、胡椒等腌渍，然后蒸熟，切块，浇原汁，佐红油辣椒、油酥豆瓣。

Steamed spicy chicken. Sichuan dish. Scalded chicken pickled with yellow wine, salt, MSG, and pepper, steamed, chopped, then washed with broth. May be served with fried chili pepper and fried board beans.

辣汁酱香肘 (làzhī jiàngxiāngzhǒu)

湖南菜。将猪前肘加入南乳汁、干椒、红曲米、鲜汤等熬炖，佐辣椒酱。

Stewed hog joint with chili sauce. Hunan dish. Hog front joint stewed with southern-style fermented bean curd juice, chili pepper, red yeast rice, and seasoned soup, then served with chili sauce.

辣子炒熏肉 (làzi chǎo xūnròu)

江西菜。先将干辣椒用油煸香，然后放入薰猪肉片，加葱、姜、料酒、酱油翻炒。

Stir-fried smoked pork with pepper. Jiangxi dish. Sliced smoked pork stir-fried with chili pepper and flavored with scallion, ginger, cooking wine, and soy sauce.

辣子鸡 (làzijī)

贵州菜。先将鸡块腌渍，再爆炒，然后加入生辣椒酱、姜、葱、盐、酱油、料酒等焖烧。

Braised chili chicken. Guizhou dish. Pickled and stir-fried chopped chicken braised with raw chili paste, ginger, scallion, salt, soy sauce, and cooking wine.

辣子鸡丁 (làzi jīdīng)

山东菜。先将鸡丁挂浆，入油锅拨散，再加辣椒、酱油、黄酒、葱头、蒜等煸炒。

Stir-fried hot chicken. Shandong dish. Diced chicken coated with starch and egg paste, lightly fried, then stir-fried with hot pepper, soy sauce, yellow wine, green onion heads, and garlic.

兰度鸽脯 (lándù gēpú)

广东菜。将鸽脯与甘笋炒熟，用上汤、料酒、蒜、姜、蚝油、麻油、胡椒粉等调味，配汆过的芥蓝装盘。

Stir-fried pigeon and Chinese broccoli. Guangdong dish. Pigeon breast stir-fried with carrots, then washed with sauce of broth, cooking wine, garlic, ginger, oyster sauce, sesame oil, and ground black pepper. Served with scalded Chinese broccoli.

兰花春笋 (lánhuā chūnsǔn)

浙江菜。先将虾泥和鱼茸搅匀，做成莲子状丸子，再把丸子镶嵌在兰花笋尖上，撒上熟火腿末，蒸熟，配焯熟的菜心装盘。

Bamboo shoots with shrimp and fish balls. Zhejiang dish. Minced shrimp meat and fish meat mixed, made into small balls like lotus seeds, put on orchid-scent bamboo shoots, sprinkled with ham crumbs, then steamed. Served with scalded baby greens.

兰花根 (lánhuāgēn)

江西菜。将兰花根和糯米粉、白糖、清油、水调和，擀成皮，切丝，油煎。

Fried orchid roots. Jiangxi dish. Orchid roots mixed with sticky rice flour, white sugar, vegetable oil, and water, made into sheets, shredded, then fried.

兰陵美酒 (lánlíngměijiǔ)

白酒，产于山东苍山县兰陵镇。以高粱、小麦、大米为原料酿造，酒精含量为 39-42 度，属浓香型酒。

Lanling meijiu, liquor produced in Lanling Town of Cangshan County, Shandong. White spirit made with sorghum, wheat, and rice. Contains 39-42% alcohol, and has a thick aroma.

榄仁鸡丁 (lǎnrén jīdīng)

广东菜。将鸡肉丁与榄仁、姜、蒜、料酒爆炒。

Stir-fried chicken and olive.

Guangdong dish. Diced chicken stir-fried with olives, ginger, garlic, and cooking wine.

郎酒 (lángjiǔ)

白酒，产于四川古蔺县二郎滩镇。以高粱为主料，用小麦制曲酿造，酒精含量为 39-53 度，属酱香型酒。

Langjiu, liquor produced in Erlangtan Town of Gulin County, Sichuan. White spirit made with sorghum and wheat yeast. Contains 39-53% alcohol, and has a soy aroma.

醪糟 (láozāo)

各地传统点心。又称甜酒酿。将糯米蒸熟，放入酒曲，密封酿成。加水煮熟食用。亦常用作调料。

Sweet fermented sticky rice soup. Traditional snack in many places. Steamed sticky rice mixed with Chinese yeast, then sealed until fermented. Boiled before serving. Often used as a condiment.

醪糟鸡蛋 (láozāo jīdàn)

各地家常点心。先将清水烧沸后放入醪糟，然后打入鸡蛋煮熟。可加入白糖、枸杞、桂圆等。

Poached eggs in fermented rice soup. Home-style snack around the country. Eggs poached in fermented sticky rice soup. May be enriched with sugar, Chinese wolfberry, and longan meat.

老醋花生 (lǎocù huāshēng)

山西家常菜。以生菜丝、香干丁做底，上摆炸花生，浇香醋、白糖、麻油、酱油、葱等调成的卤汁。

Peanuts with vinegar. Home-style dish in Shanxi. Fried peanuts on a bed of shredded lettuce and diced dried bean curd washed with dressing of vinegar, sugar, sesame oil, soy sauce, and scallion.

老婆饼 (lǎopobǐng)

广东点心。用猪油和面粉制成饼皮，包入用糖冬瓜丁、芝麻仁等做成的馅，烤熟。

Wife's cakes. Guangdong snack.

Flour and lard made into cake wrappers, filled with diced candied wax gourd and sesame, then baked.

老谦记牛肉豆丝 (lǎoqiānjì niúròu dòusī)

湖北点心。大米和绿豆做成的粉丝与牛肉、水发香菇及调料用麻油烹炒。

Sautéed beef and starch noodles. Hubei snack. Rice and mung bean vermicelli sautéed with beef, soaked mushrooms, seasonings, and sesame oil.

老青茶 (lǎoqīngchá)

黑茶。主要产于湖北咸宁。分面茶与里茶两种。面茶较精细,里茶较粗放。主要用来压制茶砖。

Laoqingcha, dark green tea mainly produced in Xianning, Hubei. It includes fine wrapping tea and coarse filling tea. Usually used to make brick tea.

老通城豆皮 (lǎotōngchéng dòupí)

湖北武汉名点。用绿豆、糯米磨浆制成薄皮,用鲜肉、鲜菇、鲜笋、糯米饭等制馅,做成厚馅夹饼,煎至皮脆色黄,切成方块。

Fried stuffed mung bean cakes. Famous snack in Wuhan, Hubei. Mung bean and sticky rice paste made into a sandwich filled with a thick layer of fresh pork, mushrooms, bamboo shoots, and cooked sticky rice. Fried until crisp and golden, then cut into cubes.

老竹大方 (lǎozhúdàfāng)

绿茶。又称铁色大方或竹叶大方,产于安徽歙县老竹岭大方山,外形扁平匀齐,色泽深绿褐润,满披金毫。汤色清澈微黄,叶底嫩匀肥壮。

Laozhudafang, green tea produced in the area of Dafang Hill of Mount Laozhu in Shexian, Anhui. Its dry leaves are flat, straight, dark green, and golden haired. The tea is crystal clear and light yellow with infused leaves even, tender, and stout.

擂茶 (léichá)

南方饮料。先将大米、黄豆、花生、蔬菜、茶叶、生姜、芝麻等捣碎,然后加入茶叶、盐或糖煮沸。

Pestled tea. Southern drink. Rice, soybean, peanuts, vegetables, tea leaves, ginger, and sesame pestled. Boiled with tea leaves, salt, or sugar.

擂茶糕 (léichágāo)

南方点心。将米浆、红枣、枸杞、绿豆粉、茶叶粉、菊花、人参等调和,做成饼,烤熟,佐擂茶。

Special cake for pestled tea. Southern snack. Cakes made of rice paste, red dates, Chinese wolfberry, mung bean flour, tea powder, chrysanthemum flower, and ginseng. Baked. Served with pestled tea.

黎家竹筒饭 (líjiā zhútǒngfàn)

海南黎族传统食品。将山兰米和猪肉加调料拌匀,装入竹筒,密封蒸熟。

Sticky rice and pork in bamboo tube. Traditional food of Li ethnic minority in Hainan. Local sticky rice and pork seasoned, mixed, sealed in a bamboo tube, then steamed.

藜蒿炒腊肉 (líhāo chǎo làròu)

江西菜。将腊肉与藜蒿、盐、料酒、味精等拌炒。

Preserved smoked pork with celery wormwood. Jiangxi dish. Smoked cured pork stir-fried with celery wormwood and seasoned with salt, cooking wine, and MSG.

荔荷炖鸭 (lì hé dùn yā)

南方菜。将肥鸭、猪瘦肉、鲜荔枝、鲜荷花加调料炖煮。

Duck with litchi and lotus flower. Southern dish. Fat duck stewed with lean pork, fresh litchi, fresh lotus flower, and seasonings.

荔浦扣肉 (lìpǔkòuròu)

广东菜。先将煮熟的五花肉稍炸，加蒜茸、豆腐乳等调味，再与炸过的芋头蒸熟。

Steamed streaky pork and taros. Guangdong dish. Boiled streaky pork lightly fried, flavored with minced garlic and fermented bean curd, then steamed with fried taros.

荔枝虾球 (lìzhī xiāqiú)

广东菜。先将虾仁腌渍，滑熟，再与荔枝、红椒、甘笋、蒜茸等轻炒，烹上汤，加调料，勾芡。

Sautéed litchi and prawns. Guangdong dish. Pickled shelled prawns lightly fried, sautéed with litchi, red pepper, carrots, and mashed garlic. Flavored with broth and seasonings, then thickened with starch.

栗子白菜 (lìzi báicài)

山东菜。将白菜稍炸，加栗子、高汤微火煨煮。

Simmered Chinese cabbage with chestnuts. Shandong dish. Lightly fried Chinese cabbage simmered in broth with shelled chestnuts.

栗子鸡 (lìzijī)

各地家常菜。先将栗子和姜爆透，然后与鸡块和豆豉爆炒，加少量水焖煮，勾芡，放香菜、香葱。

Chicken with chestnuts. Home-style dish in many places. Chopped chicken stir-fried with fermented soybean, fried chestnuts, and ginger, simmered with a little water, then flavored with starched soup, coriander, and scallion.

连锅羊肉 (liánguōyángròu)

湖南菜。羊腿肉加入菠菜、豆腐及调料炖煮。

Stewed gigot. Hunan dish. Lamb leg meat stewed with spinach, bean curd, and seasonings.

莲荷童鸡 (liánhé tóngjī)

江苏菜。先把油炸过的仔鸡同莲子入鸡汤焖炖，加入番茄酱、黄酒、八角、酱油、糖，然后用荷叶包住鸡与莲子，蒸透。

Chicken in lotus leaves. Jiangsu dish. Fried whole young chicken stewed with lotus seeds in chicken soup, flavored with tomato sauce, yellow wine, aniseed, soy sauce, and sugar, wrapped in lotus leaves with lotus seeds, then steamed.

莲藕糯米粥 (liánǒu nuòmǐzhōu)

南方水乡家常点心。将藕和糯米熬煮成粥,加入白糖和糖桂花。

Sticky rice and lotus root porridge. Home-style food in southern region of rivers and lakes. Lotus roots and sticky rice cooked into porridge, then flavored with sugar and sugared osmanthus flower.

莲藕养生汤 (liánǒu yǎngshēngtāng)

台湾菜。先将莲藕和海带用清水炖熟,然后放入白萝卜、枸杞、素鸡等焖煮。

Lotus root and kelp soup. Taiwan dish. Lotus roots and kelp stewed in water, then braised with white turnips, Chinese wolfberry, and multi-layer bean curd or soy chicken.

莲蓬豆腐 (liánpéng dòufu)

江苏、浙江菜。将豆腐泥与鸡茸、黄酒、蛋清、淀粉等调匀,装入碗中,面上嵌 7 粒豌豆,蒸熟。因其形如莲蓬而得名。

Steamed bean curd cup. Jiangsu and Zhejiang dish. Mashed bean curd mixed with ground chicken, yellow wine, egg white, and starch, put in a bowl, topped with seven peas, then steamed. It looks like a lotus pod and thus is named as such.

莲蓉包 (liánróngbāo)

广东、福建点心。用发酵面制皮,莲蓉做馅,包成圆形蒸熟。

Steamed stuffed sweet buns.

Guangdong and Fujian snack. Round buns made with flour dough, stuffed with sweetened mashed lotus seeds, then steamed.

莲枣薏仁粥 (liánzǎo yìrén zhōu)

南方点心。将薏仁、莲子、红枣、大米与冰糖熬煮。

Porridge of seeds of Job's tears and others. Southern snack. Seeds of Job's tears, lotus seeds, red dates, and rice cooked into porridge, then sweetened with rock sugar.

莲子鸡丁 (liánzǐ jīdīng)

福建菜。先把鸡丁用蛋清、淀粉拌匀,煸至半熟,然后加入蒸熟的莲子、香菇、火腿丁及调料翻炒。

Sautéed chicken with lotus seeds. Fujian dish. Diced chicken mixed with egg white and starch, half-fried, then sautéed with steamed lotus seeds, dried mushrooms, diced ham, and seasonings.

莲子扣肉 (liánzǐ kòuròu)

湖南菜。将带皮五花肉、莲子、干豆角加入盐、红油、蚝油等佐料,蒸烂。

Steamed pork with lotus seeds. Hunan dish. Streaky pork steamed with lotus seeds and dried legumes, then flavored with salt, chili oil, and oyster sauce.

莲子雪耳 (liánzǐ xuěěr)

各地家常点心。先将莲子、银耳、芡实、山药炖烂,然后倒入蛋浆搅匀,加糖。

Lotus seed and white wood ear soup. Home-style snack in many places. Husked lotus seeds stewed with white fungus, water chestnuts, and Chinese yam, mixed with egg paste, then flavored with sugar.

莲子薏仁汤 (liánzǐ yìrén tāng)

东南方点心。莲子、薏米、桂圆肉等炖汤,加蜂蜜。

Lotus seeds, seeds of Job's tear, and longan soup. Southeastern snack. Seeds of Job's tears, lotus seeds, and longan pulp stewed and flavored with honey.

凉拌怪味兔肉丝 (liángbàn guàiwèi tùròusī)

四川凉菜。将兔肉用清水煮熟,切丝,然后与葱白丝、芝麻酱、酱油、辣椒油、花椒粉等拌匀。

Rabbit with spicy sauce. Sichuan cold dish. Shredded boiled rabbit meat mixed with shredded scallion stems, sesame sauce, soy sauce, chili oil, and ground Chinese prickly ash.

凉拌米豆腐 (liángbàn mǐdòufu)

湖北点心。先把米豆腐切成小块,将大头菜、咸菜、酥黄豆、酥花生、葱花等撒在米豆腐上,然后浇上特制的卤汁。

Rice curd with dressing. Hubei snack. Rice curd cut into small cubes, topped with preserved mustard turnips, salty cabbage, fried soybean, fried peanuts, and chopped scallion, then washed with special dressing.

凉拌皮蛋 (liángbàn pídàn)

各地家常菜。先将皮蛋切片,浇糖、醋、酱油,撒葱花。

Preserved eggs with dressing. Home-style dish in many places. Preserved duck eggs sectioned, washed with sugar, vinegar, and soy sauce, then sprinkled with chopped scallion.

凉拌三丝 (liángbàn sānsī)

东北凉菜。将大头菜丝、胡萝卜丝、粉丝加醋、糖、蒜等拌匀。

Pickled vegetables with vermicelli. Northeastern cold dish. Shredded mustard turnips and carrots mixed with vermicelli, then seasoned with vinegar, sugar, and garlic.

凉拌四宝 (liángbàn sìbǎo)

浙江凉菜。先将虾仁,鸡肫片、鸡脯片、猪肚片、鸭掌、鞭笋、香菇等用开水煮沸,然后加调料凉拌,盖上火腿片。

Four cold meat varieties. Zhejiang cold dish. Shelled shrimp, sliced chicken gizzard, chicken breast, hog tripe, duck feet, summer bamboo shoots, and dried mushrooms boiled, cooled, then mixed with seasonings. Covered with sliced ham to serve.

凉拌莴笋 (liángbàn wōsǔn)

各地家常菜。先将莴笋片用盐腌渍,再控干水分,然后加入酱油、醋、糖、辣油、麻油等拌匀。

Lettuce stems with dressing. Home-style cold dish in many places. Sliced lettuce stems pickled with salt, then drained. Seasoned with soy sauce, vinegar, sugar, chili oil, and sesame oil.

凉拌五毒 (liángbàn wǔdú)

山东凉菜。先将葱、姜、蒜、韭菜、辣椒丝用水浸泡，然后加入调料拌匀，略腌。

Five spicy vegetables with sauce. Shandong cold dish. Scallion, ginger, garlic, Chinese chive, and shredded hot pepper steeped in water, then lightly pickled with seasonings.

凉粉 (liángfěn)

云贵川传统点心。先将绿豆浆加石灰水煮熟，置凉成冻，然后切成条或丁，放入辣椒油、油炸黄豆、酱油、醋等。

Mung bean jelly with sauce. Traditional snack in Yunnan, Guizhou, and Sichuan. Mung bean milk mixed with lime water, boiled, then chilled into jelly. Cut into strips or dices. Flavored with chili oil, fried soybean, soy sauce, and vinegar.

梁王鱼 (liángwángyú)

江苏菜。先将鳡鱼头油炸，再加板油丁、酱油、白糖、黄酒等烹煮，用火腿片、海参片、冬菇片饰盘。

Braised carp head. Jiangsu dish. Yellowcheek carp head deep-fried, braised with diced leaf lard, soy sauce, sugar, and yellow wine, then garnished with sliced ham, sea cucumbers, and winter mushrooms.

两吃大虾 (liǎngchī dàxiā)

山东菜。将对虾切成两段，虾头段用黄酒、高汤爆炒，虾尾段裹上蛋液和面包屑，炸至金黄。

Two-taste prawns. Shandong dish. Prawns cut into 2 sections. Head sections stir-fried with yellow wine and broth. Tail sections coated with egg paste and bread crumbs, then deeply fried.

蓼花 (liǎohuā)

福建点心，也称麻筒。将糯米、芋头、蔗糖、苏打、麦芽糖等调和，制成条状，油炸，然后撒满芝麻。

Fried sticky rice and taro bars. Fujian snack. Sticky rice, taros, cane sugar, soda, and maltose made into bars, deep-fried, then coated with sesame seeds.

淋糖千层酥 (líntáng qiāncéngsū)

广东点心。用酥皮包上莲蓉馅，裹蛋液，入炉烤熟，出炉后浇上糖浆。

Multi-layer cookie. Guangdong snack. Flour and oil dough made into cake wrappings, stuffed with mashed lotus seeds, coated with egg paste, baked, then washed with syrup.

灵芝糯团 (língzhī nuòtuán)

江西点心。糯米粉、灵芝粉、水和成面团，包白糖豆沙馅，做成糕团，蒸熟，淋少许麻油。

Sticky rice, herb, and bean paste buns. Jiangxi snack. Sticky rice flour, ganoderma powder, and water mixed, made into buns filled with sugar and bean paste, steamed, then sprinkled with a little sesame oil.

灵芝酥 (língzhīsū)

四川点心。将面团擀成饼皮,包入甜、咸豆沙馅,炸至金黄。

Fried bean-paste cakes. Sichuan snack. Flour dough made into cakes filled with sweet or salty bean paste, then deep-fried.

凌川白酒 (língchuānbáijiǔ)

白酒,产于辽宁锦州。以高粱、麸曲为原料酿造,酒精含量为 53-55 度,属酱香型酒。

Lingchuanbaijiu, liquor produced in Jinzhou, Liaoning. White spirit made with sorghum and bran yeast. Contains 53-55% alcohol, and has a soy aroma.

溜藕片 (liūǒupiàn)

浙江菜。将藕片炒熟,烹入酱油、白糖、香醋,勾芡。

Sautéed lotus roots. Zhejiang dish. Sliced lotus roots sautéed, flavored with soy sauce, sugar, and vinegar, then thickened with starch.

溜鱼片 (liūyúpiàn)

山东菜。先将鱼片入油锅滑透,再与南芥、绿菜心、高汤、黄酒等翻炒,勾芡。

Sautéed fish fillets with vegetables. Shandong dish. Lightly fried fish fillets sautéed with Chinese broccoli, baby Chinese cabbage, broth, and yellow wine, then thickened with starch.

溜猪肝 (liūzhūgān)

湖南菜。先将猪肝腌渍,熘炒,然后与黄瓜、木耳、青椒、姜、蒜等翻炒。

Stir-fried hog liver. Hunan dish. Pickled hog liver lightly fried, then stir-fried with cucumbers, wood ears, green pepper, ginger, and garlic.

浏阳豆豉蒸排骨 (liúyángdòuchǐ zhēng páigǔ)

湖南浏阳菜。将排骨拌上浏阳豆豉,平放在豆腐上,蒸熟。

Pork ribs with fermented soybean. Specialty in Liuyang, Hunan. Pork ribs sections mixed with Liuyang fermented soybean, put on cubed bean curd, then steamed.

浏阳河酒 (liúyánghéjiǔ)

白酒,产于湖南浏阳。以高粱、大米、糯米、小麦、玉米为主要原料酿造,酒精含量为 28-52 度,属浓香型酒。

Liuyanghejiu, liquor produced in Liuyang, Hunan. White spirit made with sorghum, rice, glutinous rice, wheat, and corn. Contains 28-52% alcohol, and has a thick aroma.

流黄青蟹 (liúhuáng qīngxiè)

浙江菜。先将蟹块煸至发红,加入黄酒、盐、白汤等烹煮,勾芡,然后放入鸡蛋浆和熟花生油。

Stir-fried crab. Zhejiang dish. Crab chunks stir-fried, boiled in clear soup with yellow wine and salt, thickened with starch, then enriched with egg paste and prepared peanut oil.

琉璃蛋球 (liúlí dànqiú)

河南菜。将面粉调蛋浆做成丸子,油煎至金黄,然后倒入熬化的糖汁翻匀。

Sugarcoated flour balls. Henan dish. Flour and egg balls deep-fried until golden, then quick-fried with heated melting sugar.

榴莲富贵虾 (liúlián fùguìxiā)

广东菜。将虾肉和榴莲肉加调料拌匀,装在虾壳里,烤至金黄。

Baked prawns with durian.

Guangdong dish. Prawn meat and durian meat mixed with seasonings, put into prawn shells, then baked until golden.

六安瓜片 (liùānguāpiàn)

绿茶。产于安徽省六安、金寨两县。外形呈片状,叶缘向背面翻卷。汤色清澈晶亮,叶底嫩绿。

Liuanguapian green tea produced in Liuan and Jinzhai, Anhui. Its dry leaves are flat with edges slightly back curved. The tea is crystal clear with infused leaves tender green.

六堡散茶 (liùbǎosǎnchá)

黑茶。产于广西梧州地区。外形条索长整,色泽黑褐光润。汤色红浓,带有松烟味和槟榔味。叶底呈棕褐色。

Liubaosancha dark green tea, produced in Wuzhou area, Guangxi. Its dry leaves are long, tidy, and black bloom. The tea is rich red with a mixed aroma of pine and areca. The infused leaves are brownish auburn.

龙虎斗 (lónghǔdòu)

广东菜。将备好的蛇肉、猫肉、鸡丝、鱼肚、冬菇、木耳与调料同煲,配薄脆、柠檬叶、白菊花上桌。

Stewed snake, cat, and chicken. Guangdong dish. Prepared snake meat, cat meat, shredded chicken, fish maw, winter mushrooms, and wood ears stewed with seasonings. Served with fried crackers, lemon leaves, and white chrysanthemum petals.

龙井虾仁 (lóngjǐng xiārén)

浙江菜。先将新鲜河虾仁挂芡,再用猪油、盐、黄酒轻炒至半熟,然后与龙井新茶叶速炒,勾芡。

Stir-fried shrimp meat with green tea. Zhejiang dish. Fresh river shrimp meat pickled with starch, lightly fried with lard, quick-fried with fresh *Longjingcha* tea leaves, then thickened with starch.

龙身凤尾虾 (lóngshēn fèngwěixiā)

安徽菜。先将大虾去壳留尾,从背部切开,嵌入火腿条,裹芡粉,过油,然后与香菇、冬笋、葱翻炒。

Stir-fried prawns with ham. Anhui dish. Shelled prawns slit down back, inlaid with ham strips, then coated with starch. Fried, then sautéed with mushrooms, winter bamboo shoots, and scallion.

龙虾二吃 (lóngxiā èr chī)

海南菜。将龙虾身蒸熟,切段装盘,佐姜、醋、酱油、麻油调成的卤汁。将龙虾头与大米熬粥,佐腐乳或榨菜。

One lobster for two dishes. Hainan dish. Lobster body steamed, chopped, then served with dressing of ginger, vinegar, soy sauce, and sesame oil. Lobster head and rice cooked into porridge, then served with fermented bean curd or preserved mustard tubers.

隆果卡查 (lóngguǒkǎchá)

西藏菜,即咖喱羊头。将羊头煮熟,去骨,加入咖喱、茴香、味精、盐等拌匀。

Lamb head with curry. Tibetan dish. Lamb head boiled, boned, then flavored with curry, fennel, MSG, and salt.

露笋炒蛇片 (lùsǔn chǎo shépiàn)

广东菜。将蛇片、露笋、胡萝卜、姜、葱等加盐、料酒、酱油翻炒。

Sautéed snake and asparagus. Guangdong dish. Sliced snake sautéed with asparagus, carrots, ginger, and scallion, then seasoned with salt, cooking wine, and soy sauce.

庐山石鸡 (lúshānshíjī)

江西菜。先将庐山石鸡即棘胸蛙过油,然后入高汤加黄酒、酱油、盐焖熟,淋麻油。

Simmered frogs. Jiangxi dish. Lightly fried Lushan frogs simmered in broth with yellow wine, soy sauce, and salt, then sprinkled with sesame oil.

庐山云雾 (lúshānyúnwù)

绿茶。中国名茶,产于江西庐山。外形条索紧凑,嫩绿多毫。汤色清澈,叶底嫩绿匀齐。

Lushanyunwu, famous green tea produced in Mount Lushan, Jiangxi. Its dry leaves are straight, tight, green, and hairy. The tea is crystal clear with infused leaves even and tender green.

泸州老窖特曲 (lúzhōu lǎojiào tèqū)

白酒,中国传统四大名酒之一,产于四川泸州。用当地糯高粱和小麦为原料,在有特殊酒泥的百年老窖里发酵酿制,酒精含量在38-68度之间,属浓香型酒。

Luzhou laojiao special liquor, one of the top 4 traditional alcoholic drinks in China. White spirit produced in Luzhou, Sichuan. Made from local sticky sorghum and wheat. Fermented in special cellars that retain centuries-old ferment mud. Contains 38-68% alcohol, and has a thick aroma.

卤鸡 (lǔjī)

各地传统菜。先将鸡焯水,然后放入用盐、糖、姜、葱、茴香、桂皮、酱油、黄酒等制成的卤汁煮熟。

Marinated chicken. Traditional dish in many places. Scalded chicken boiled and then marinated in gravy of salt, sugar, ginger, scallion, fennel, cinnamon, soy sauce, and yellow wine.

卤兰花干素鸡 (lǔ lánhuāgān sùjī)

山东菜。将切成兰花状的豆腐干和素鸡即千层豆腐放入卤汤中煮透。

Marinated bean curd and soy chicken. Shandong dish. Dried bean curd carved into orchid-shape pieces, then marinated in gravy with multi-layer bean curd or soy chicken.

卤肉 (lǔròu)

各地传统菜。先将猪头肉、蹄髈等焯水，然后放入用盐、糖、姜、葱、茴香、桂皮、酱油、黄酒等制成的卤汁煮熟。

Marinated pork. Traditional dish in many places. Scalded hog head and knuckle boiled and marinated in gravy of salt, sugar, ginger, scallion, fennel, cinnamon, soy sauce, and yellow wine.

卤五香牛肉 (lǔ wǔxiāngniúròu)

各地传统菜。将牛肉放入用盐、糖、姜、葱、茴香、桂皮、酱油、黄酒等制成的卤汁煮熟。

Marinated beef. Traditional dish in many places. Beef boiled and marinated in gravy of salt, sugar, ginger, scallion, fennel, cinnamon, soy sauce, and yellow wine.

绿茶 (lǜchá)

不发酵茶。先用高温破坏鲜茶叶中的酶，制止发酵，再经过炒、烘、晒或蒸等深加工，使成品茶叶保持鲜茶叶的绿色。

Unfermented tea or green tea. Made with fresh tea leaves. Heated to destroy enzymes and prevent fermentation. Baked, sun-dried, or steamed to retain a green color.

绿豆粥 (lǜdòuzhōu)

各地传统点心。将绿豆与大米用清水煮成粥，加糖。

Mung bean and rice porridge. Traditional snack in many places. Mung beans and rice boiled into porridge with water, then sweetened with sugar.

绿叶豆腐羹 (lǜyè dòufu gēng)

江苏菜。将豆腐丁、焯过的芹菜叶和鸡蛋浆放入猪骨汤煮熟，加葱姜汁、黄酒，勾芡。

Bean curd and celery soup. Jiangsu dish. Diced bean curd, quick-boiled celery leaves, and egg paste boiled in hog bone soup, flavored with scallion-ginger juice and yellow wine, then thickened with starch.

罗汉鸭 (luóhànyā)

福建菜。将冬菇、菜心、栗子、白果肉、冬笋、青笋、腐竹加调料煸炒，填入蒸熟去骨的鸭肚，蒸熟，浇芡汁。

Duck stuffed with vegetables. Fujian dish. Steamed and boned duck stuffed with quick-fried winter mushrooms, baby cabbage, chestnuts, ginkgoes, winter bamboo shoots, stem lettuce, bean curd skins, and seasonings. Steamed, then washed with starched soup.

萝卜糕 (luóbogāo)

南方点心。将氽过的萝卜丝与面粉加水调和，蒸熟成糕，浇上用番茄酱、葱、姜、白糖、白醋、芡粉制成的汁。

Turnip cakes. Southern snack. Scalded shredded turnips mixed with flour and water, steamed into cakes, then washed with dressing of tomato sauce, scallion, ginger, sugar, white vinegar, and starch.

萝卜烧排骨 (luóbo shāo páigǔ)

各地家常菜。将过油萝卜和排骨入鲜汤加调料炖熟,勾芡。

Stewed turnips and spareribs. Home-style dish in many places. Turnips and pork spareribs lightly fried, stewed in seasoned soup, then thickened with starch.

萝卜丝饼 (luóbosībǐng)

南方传统点心。用面皮包入白萝卜丝、肉末、虾仁、香菇、盐等做成的馅,做成饼,煎至金黄。

Fried turnip cakes. Traditional snack in southern areas. Flour wrappers packed with mixture of shredded turnips, ground pork, shrimp meat, dried mushrooms, and salt. Made into cakes, then deep-fried until golden.

洛阳水席 (luòyángshuǐxí)

河南洛阳素席。用萝卜、白菜、绿豆粉丝等制作。每席 24 种素菜,因以汤菜为主得名。

Water banquet or vegetable banquet. Vegetarian banquet in Luoyang, Henan. A banquet of 24 dishes made with vegetables such as turnip, cabbage, and mung bean noodles. Most dishes are soups.

洛阳燕菜 (luòyángyàncài)

河南洛阳菜。将萝卜丝拌芡粉蒸熟,浇鲜汤。其形似燕窝。

Shredded turnips in soup. Specialty of Luoyang, Henan. Shredded turnips mixed with starch, steamed, then washed with seasoned soup. It looks like a bird's nest.

落叶琵琶虾 (luòyè pípáxiā)

山东菜。先将河虾腌渍,压成琵琶状氽过,再加入清汤、黄酒煮熟,撒上焯水的豌豆苗。

Sautéed river shrimp. Shandong dish. River shrimp pickled, pressed flat like lutes, scalded, sautéed with yellow wine and clear soup, then decorated with scalded pea sprouts.

麻城肉糕 (máchéngròugāo)

湖北菜。先将鱼茸和肉末加调料制成糊状,摊在油豆腐皮上蒸熟,抹上鸡蛋黄,撒上红辣椒末,再略蒸,然后切片,浇上特制的肉汤。

Fish and pork cakes. Hubei dish. Mashed fish and ground pork mixed into paste with seasonings, laid on dried bean curd skins, then steamed into cakes. Cakes painted with egg yolks, topped with ground chili pepper, re-steamed briefly, sliced, then washed with gravy.

麻烘糕 (máhōnggāo)

湖北点心。先把米粉、绵白糖、麻油、熟黑芝麻、糖桂花等拌匀,做成糕块,慢火烘烤,然后切成相连的薄片。

Sweet sticky rice cookie. Hubei snack. Sticky rice flour, white soft sugar, sesame oil, fried black sesame, and sweet osmanthus flower mixed, made into lumps, baked, then cut into connected thin pieces.

麻辣蚕豆 (málà cándòu)

四川点心。先将蚕豆煮熟,再用盐腌渍,然后加入花椒粉、辣椒粉、麻油、葱花等拌匀。

Spicy broad beans. Sichuan snack. Boiled broad beans pickled with salt, then mixed with ground Chinese prickly ash, chili powder, sesame oil, and chopped scallion.

麻辣鸡 (málàjī)

四川菜。先将整鸡焯水,然后放入用茴香、桂皮、酱油、糖、料酒、花椒、姜、葱等制成的料汤卤熟,切片,浇花椒油和辣油。

Marinated spicy chicken. Sichuan dish. Quick-boiled whole chicken marinated in gravy of fennel, cinnamon, soy sauce, sugar, cooking wine, Chinese prickly ash, ginger, and scallion. Chopped, then sprinkled with Chinese prickly ash oil and chili oil.

麻辣牛肉 (málà niúròu)

四川菜。将牛肉氽过,放入用茴香、桂皮、酱油、糖、料酒、花椒、姜、葱等制成的卤汤煮熟,切片,浇花椒油和辣油。

Marinated spicy beef. Sichuan dish. Scalded beef marinated in gravy of fennel, cinnamon, soy sauce, sugar, cooking wine, Chinese prickly ash, ginger, and scallion. Chopped, then sprinkled with Chinese prickly ash oil and chili oil.

麻辣烫 (málàtàng)

重庆点心。先将猪肉、牛肉、氽过的牛血、鱼、鲜藕、鹌鹑蛋、青瓜等串在竹签上，入火锅烫熟，然后蘸麻油，裹红辣椒粉和花椒粉。

Quick-boiled spicy meat and vegetables. Chongqing snack. Pork, beef, scalded ox blood, fish, lotus roots, quail eggs, and cucumbers strung on bamboo sticks, quick-boiled in fire pot soup, sprinkled with sesame oil, then coated with chili powder and ground Chinese prickly ash.

麻辣兔丁 (málà tùdīng)

四川成都菜。先将兔肉丁用酱油腌渍，再与花椒、红椒、香菇等爆炒。

Stir-fried spicy rabbit. Specialty in Chengdu, Sichuan. Diced rabbit pickled with soy sauce, then stir-fried with Chinese prickly ash, red pepper, and dried mushrooms.

麻婆豆腐 (mápódòufu)

四川菜。将豆腐丁、肉末、辣酱、花椒粉、豆瓣酱等爆炒。

Stir-fried spicy bean curd. Sichuan dish. Diced bean curd stir-fried with ground meat, chili sauce, ground Chinese prickly ash, and fermented broad bean sauce.

麻森 (másēn)

西藏点心。先将糌粑、碎奶渣、酥油和红糖拌和，装入木盒压成块，然后切片。

Barley and dairy cookie. Tibetan snack. *Zanba* mixed with minced dairy dregs, Tibetan butter, and brown sugar, packed in a wooden box, pressed into a hard lump, then sliced.

麻食子 (máshízi)

西北回族食品。先将面粉和水做成麻雀蛋大小的面疙瘩，用羊肉汤煮熟，然后同芹菜、胡萝卜、羊肉丁等烩煮，加辣椒油。

Dumplings and vegetables in mutton soup. Hui ethnic minority food in northwestern areas. Flour and water mixed, made into small balls like sparrow eggs, boiled in mutton soup, simmered with celery, carrots, and diced mutton, then flavored with chili oil.

麻酥里脊 (másū lǐji)

北京菜。将猪里脊肉条挂糊，炸至金黄，佐椒盐。

Fried pork with spicy salt. Beijing dish. Strips of pork tenderloin coated with starch paste, deep-fried, then served with spicy salt.

麻油鸡 (máyóujī)

山东菜。将鸡块加酱油、麻油、黄酒、糖等翻炒至汁干。

Chicken with sesame oil. Shandong dish. Chopped chicken stir-fried with soy sauce, sesame oil, yellow wine, and sugar.

麻圆 (máyuán)

湖南点心。将米粉、淀粉、苏打、砂糖、水调和，做成小团，裹上芝麻，煎至色黄壳脆。

Sticky rice balls with sesame.

Hunan snack. Rice flour, starch, soda, sugar, and water mixed, hand-kneaded into small balls, coated with sesame, then deep-fried until golden and crisp.

马奶酒 (mǎnǎijiǔ)

蒙古族传统饮料。将鲜马奶发酵，待马奶变清淡透明即成。烈性马奶酒经多次蒸酿而成，酒精含量可达50度。

Horse milk wine. Traditional drink of Mongolian ethnic minority. Fresh horse milk fermented until it becomes clear and transparent. Strong horse milk wine may contain 50% alcohol through repeated fermentation and distillation.

马蹄糕 (mǎtígāo)

广东点心。将马蹄制成浆，撒上葡萄干、枸杞、瓜子等蒸熟。

Water chestnut cakes. Guangdong snack. Water chestnuts made into paste, sprinkled with raisins, Chinese wolfberry, and sunflower seeds, then steamed.

马蹄酥 (mǎtísū)

各地点心。先用面粉与水和成皮面，用面粉、猪油、糖等和成酥面，再将酥面与皮面糅合，擀成皮、做成卷，弯成马蹄形，然后炸熟。

Fried flour cookie. Snack in many places. Flour and water made into dough. Flour, lard, and sugar made into pastry. Dough and pastry kneaded into sheets, rolled into horseshoe-shape pieces, then deep-fried.

玛瑙鸡片 (mǎnǎo jīpiàn)

浙江菜。先将鸡脯、香菇、荸荠、绿叶菜加调料烩煮，然后浇在炸成棕黄色的豆腐皮上。

Chicken and vegetable on bean curd skins. Zhejiang dish. Chicken breast, mushrooms, water chestnuts, and green vegetable simmered with flavorings, then put on a bed of deep-fried bean curd skins.

玛瑙银杏 (mǎnǎo yínxìng)

山东点心。先将银杏炸至金黄，然后与炒化的白糖迅速拌匀，撒上芝麻、青红丝。

Sugarcoated ginkgoes. Shandong snack. Ginkgoes deep-fried until golden, then quickly mixed with heated melting sugar. Sprinkled with sesame and shreds of candied green and red gourd.

蚂蚁上树 (mǎyǐ shàngshù)

四川菜。先将肉末腌渍，然后与粉丝、辣椒丝等加盐、酱油、葱、姜爆炒。

Ground pork and bean noodles. Sichuan dish. Pickled ground pork stir-fried with bean noodles and shredded hot pepper, then seasoned with salt, soy sauce, scallion, and ginger.

馒头 (mántou)

北方主食之一。将发酵面团做成半圆形或长方形馒头，蒸熟。

Steamed buns. A staple food in northern areas. Fermented flour dough made into half-round or oblong buns. Steamed.

鳗鱼扎蹄 (mányú zhātí)

山东菜。将豆腐皮加入调料煮熟，搅碎为馅，用豆腐皮和紫菜卷馅，扎紧，蒸熟。

Steamed bean curd skins and laver. Shandong dish. Bean curd skins boiled with seasonings, smashed, packed with bean curd skins and laver, bound tightly, then steamed.

芒果鸡 (mángguǒjī)

广东菜。将鸡丁与干辣椒、姜、蒜、青椒、盐等爆炒，加入芒果丁，撒葱花。

Stir-fried chicken with mangoes. Guangdong dish. Diced chicken stir-fried with chili pepper, ginger, garlic, green pepper, salt, and mangoes, then garnished with chopped scallion.

猫耳朵 (māoěrduo)

山西食品。先将菜炒好，然后将形状如猫耳的薄面片倒在菜上焖熟。

Flour dough pieces over dishes. Shanxi food. Flour dough made into thin pieces like cat ears, then simmered over prepared dishes.

毛氏红烧肉 (máoshì hóngshāoròu)

湖南菜。将半肥瘦猪肉加入酱油、糖、辣椒炖熟。

Stewed spicy streaky pork. Hunan dish. Streaky pork stewed with soy sauce, sugar, and chili pepper.

毛蟹 (máoxiè)

乌龙茶。产于福建安溪，属安溪色种。外形紧结，多白毫，色泽绿黄。汤色青黄，叶底叶张圆，呈黄绿色。

Maoxie oolong, one of the oolong varieties from Anxi, Fujian. Its dry leaves are yellowish green, tight, and hairy. The tea is crystal yellowish green with infused leaves in good shape.

毛血旺 (máoxuèwàng)

重庆菜。先将猪血或鸡鸭血入骨头汤烫熟，然后与煸炒过的干辣椒、花椒、姜、大葱、猪肉、火腿肠、木耳、莴苣、黄豆芽等共煮，加盐、酱油、辣椒油、花椒油。

Blood, meat, and vegetables in spicy soup. Chongqing dish. Hog, chicken, or duck blood boiled in bone soup, simmered with quick-fried chili pepper, Chinese prickly ash, ginger, leek, pork, sausage, wood ears, stem lettuce, and soybean sprouts, then seasoned with salt, soy sauce, chili oil, and Chinese prickly ash oil.

茅台酒 (máotáijiǔ)

白酒，世界名酒之一，也是中国传统四大名酒之一，产于贵州仁怀茅台镇，以高粱为原料，经反复发酵、原汁蒸馏勾兑而成。酒精含量在 33-53 度之间，属酱香型酒。

Maotaijiu, one of the world's top, also one of China's top 4, traditional high alcoholic drinks. White spirit produced in Maotai Town of Renhuai, Guizhou. It is made from sorghum after repeated fermentation and distillation. Con-tains 33-53%

alcohol, and has a soy aroma.

玫瑰大虾 (méigui dàxiā)

福建菜。先将生菜叶涂上蛋黄,卷入紫菜和明虾,然后用豆腐皮包裹成长条,小火炸至金黄,切段。

Fried lettuce, laver, and prawn rolls. Fujian dish. Lettuce leaves brushed with egg yolks, made into rolls filled with laver and shelled prawns, wrapped into long strips with bean curd skins, deep-fried, then sectioned.

玫瑰锅炸 (méigui guōzhá)

山东点心。先将鸡蛋面粉糊烘成糕,炸至金黄,然后浇上加白糖和玫瑰酱的芡汁。

Fried cake with sweet rose sauce. Shandong snack. Egg paste and flour mixture baked into cake, deep-fried, then washed with sauce of sugar, rose jam, and starch.

玫瑰蔗香鸡 (méiguī zhèxiāngjī)

广东菜。先将鸡用玫瑰酒腌制,蒸熟,再用玫瑰花瓣和糖焖熏,然后浇上玫瑰酒汁。

Rose-smoked chicken. Guangdong dish. Chicken pickled with rose wine and steamed. Smoked with rose petals and sugar in a covered wok, then washed with rose wine dressing.

梅菜鸡 (méicàijī)

广东菜。先将整鸡腹内填入梅菜、冬笋、冬菇、瘦肉、火腿及调料,然后用蛋清淀粉浆涂遍鸡身,油炸,淋芡汁。

Fried chicken with preserved mustard. Guangdong dish. Chicken stuffed with preserved mustard, winter bamboo shoots, winter mushrooms, lean pork, ham, and seasonings, coated with starched egg white, deep-fried, then washed with starched soup.

梅菜扣肉 (méicài kòuròu)

南方菜。先将猪肉油炸,再与梅干菜入鸡汤蒸烂。

Streaky pork with dried vegetables. Southern dish. Streaky pork deep-fried, then steamed with dried preserved vegetable in chicken soup.

梅占 (méizhàn)

乌龙茶。产于福建安溪,属安溪色种。外形壮实,色泽褐绿稍带暗红。汤色透明呈黄,味厚浓。叶底叶张粗大,长而渐尖。

Meizhan oolong, one of the oolong varieties from Anxi, Fujian. Its dry leaves are stout, solid, and dark green over gentle dark red. The tea is crystal yellow with infused leaves broad and strong with long points.

美宫山药 (měigōng shānyào)

福建菜。先将山药片和京糕片摆成房脊形,青梅摆成绿瓦状,稍蒸,然后浇上用白糖和桂花糖酱做的芡汁。

Chinese yam and rice cakes. Fujian dish. It is arranged like a palace with Chinese yam and rice cakes as roof and green plums as tiles. Steamed, then washed with starched soup flavored with sugar and sweet osmanthus sauce.

美味茄鲞 (měiwèi qiéxiǎng)

曹雪芹《红楼梦》菜谱。先将茄子丁用鸡油炸至金黄,再加入糟酒、酱油、盐、糖、醋、清汤煨透,勾芡,然后浇上用鲜蘑、香菇、豆腐干、红绿椒、干果仁、糟酒、糖、清汤、水淀粉做的卤汁。

Simmered savoury eggplants. Menu from *A Dream of Red Mansions* by Cao Xueqin. Diced eggplants deep-fried in chicken oil. Simmered in clear soup seasoned with yellow wine, soy sauce, salt, sugar, and vinegar. Thickened with starch, then marinated in gravy made with fresh and dried mushrooms, dried bean curd, green and red pepper, nuts, yellow wine, sugar, clear soup, and wet starch.

焖辣子鸡腿 (mèn làzi jītuǐ)

南方菜。先将鸡腿加姜、干辣椒、酱油煸炒,再入清汤焖煮,然后和青椒翻炒。

Spicy drumsticks. Southern dish. Drumsticks lightly fried with ginger, chili pepper, and soy sauce, braised in clear soup, then sautéed with green pepper.

焖炉烤鸭 (mènlú kǎoyā)

山东菜。先将鸭腌渍,然后放入秫秸炉内,用炉壁和炉灰的高温将鸭焖熟。

Baked duck. Shandong dish. Pickled duck baked by the heat of the walls and ashes in the hearth of a sorgam stove.

焖酿鳝卷 (mènniàng shànjuǎn)

广东菜。先将瘦肉、虾茸、香菇等加调料拌成馅,用鳝鱼片裹馅成卷,炸至金黄,然后与香菇入上汤烹煮,加糖、酱油、味精,勾芡。

Simmered eel rolls. Guangdong dish. Sliced field eel made into rolls filled with mixture of seasoned lean pork, shrimp meat, and dried mushrooms. Deep-fried, simmered in broth with dried mushrooms, sugar, soy sauce, and MSG, then thickened with starch.

蒙顶黄芽 (méngdǐnghuángyá)

黄茶。产于四川蒙山。外形扁直,色泽微黄,芽毫显露。汤色黄亮,叶底嫩黄匀齐。

Mengding huangya, yellow tea produced in Mount Mengshan, Sichuan. Its dry leaves are flat, straight, light yellow, and hairy. The tea is crystal yellow with infused leaves yellowish, tender, and even.

蒙古八珍 (měnggǔbāzhēn)

蒙古族盛宴,包括八种美食:醍醐即乳酪上凝聚的油珠、麆沆即獐的幼羔、野驼蹄、鹿唇、驼乳、麋、烤天鹅肉、马奶。配西域葡萄酒。

Mongolian eight delicacies. Banquet of Mongolian ethnic minority. Contains eight delicacies, including cream, broiled baby river deer, wild camel hooves, deer lips, camel milk, elk, broiled swan, and horse milk.

Served with wine from western regions of China.

蒙古包子 (měnggǔbāozi)

内蒙点心。用面粉和水做成包子皮,用羊肉和大葱为馅,做成包子,蒸熟。

Steamed mutton and leek buns. Inner Mongolian snack. Flour and water mixed into dough, made into buns stuffed with mutton and leek, then steamed.

蒙古果子 (měnggǔguǒzi)

内蒙点心。面粉、水、糖和成面团,擀成面张,切成条,用羊油炸至金黄。

Fried Mongolian flour strips. Inner Mongolian snack. Flour, water, and sugar mixed into dough, made into thick sheets, then cut into strips. Deep-fried with suet oil.

蒙古奶酪 (měnggǔnǎilào)

蒙古族食品。用牦牛奶、普通牛奶、山羊奶或绵羊奶制成。先在煮开的奶里加入牛乳酒酵使奶发酵,然后取出奶里的豆腐状物质,滤水,压制成块,晾干。干奶酪可以用来做奶茶。

Mongolian cheese. Delicacy of Mongolian ethnic minority. Mongolian cheese made with yak, cow, goat, or sheep milk. Kefir grain is added to boiled milk. Curd is removed, drained, pressed into lumps, then dried. Dried cheese may be used to make milk tea.

蒙古馅饼 (měnggǔxiànbǐng)

内蒙点心。将面粉与水和成面团,做成饼,包入牛、羊肉馅,烙熟。

Pancake stuffed with meat. Inner Mongolian snack. Flour and water made into cakes stuffed with beef or mutton, then pan fried.

咪达 (mīdá)

西藏点心。将米与酥油、肉丁、红枣、杏干、葡萄干等煮成粥。

Rice and butter porridge. Tibetan snack. Rice cooked into porridge with a variety of ingredients such as Tibetan butter, diced meat, red date, dried apricot, or raisin.

米包子 (mǐbāozi)

福建点心,也叫韭菜包。用精米粉做成面皮,用韭菜、瘦肉、笋、虾米、盐等做成馅,包成形似水饺的包子,蒸熟。

Steamed rice and chive buns. Fujian snack. Fine rice flour made into *jiaozi*-shape buns, stuffed with mixture of Chinese chive, lean pork, bamboo shoots, shrimp meat, and salt, then steamed.

米肠子 (mǐchángzi)

新疆维吾尔族菜。羊肠内灌入加调料的羊肝、羊心、羊油、大米等馅料,煮熟。

Boiled lamb offal and rice sausage. Uygur dish in Xinjiang. Lamb intestine stuffed with seasoned lamb liver, heart, suet, and rice, then boiled.

米豆腐 (mǐdòufu)

南方点心。先把大米加水磨成浆,煮熟,再加入石灰水调匀,冷却成糕状。凉拌、炒、煎、炸均可。

Rice curd. Southern snack. Soaked rice ground into paste, boiled, mixed with lime water, then cooled into curd. Served cold or hot.

米饭 (mǐfàn)

南方各地主食。大米加水煮熟或蒸熟。

Boiled or steamed rice. Staple food in southern areas. Rice boiled or steamed.

米酒汤圆 (mǐjiǔ tāngyuán)

各地传统点心。各种馅料的汤圆放入甜米酒煮熟。

Tangyuan in sweet fermented sticky rice soup. Traditional snack in many places. Sticky rice flour made into *tangyuan* with a variety of fillings, then boiled in sweet fermented sticky rice soup.

蜜枣桂圆粥 (mìzǎo guìyuán zhōu)

南方点心。将糯米、桂圆、红枣、姜熬煮成粥,加蜂蜜。

Date and longan porridge. Southern snack. Sticky rice boiled into porridge with longan pulp, red dates, and ginger. Sweetened with honey.

蜜汁叉烧 (mìzhī chāshāo)

广东菜。猪瘦肉加糖、酱油、蚝油等调料烤熟。

Broiled sweet and salty pork. Guangdong dish. Lean pork seasoned with sugar, soy sauce, and oyster sauce, then broiled.

棉花糕 (miánhuāgāo)

广东点心。将米浆和蛋清、糖、苏打粉调匀,倒入模具,蒸熟。

Steamed rice cakes. Guangdong snack. Rice paste mixed with egg white, sugar, and soda powder, put in a mould, then steamed.

沔阳三蒸 (miǎnyángsānzhēng)

湖北沔阳菜。即蒸青鱼、蒸猪肉、蒸蔬菜。将鱼、肉、蔬菜分别加入调料,蒸熟,同装一盘。

Steamed fish, pork, and vegetables. Specialty in Mianyang, Hubei. Fish, pork, and vegetables seasoned and steamed separately. Served on the same platter.

面条 (miàntiáo)

北方主食之一。面粉加水合成面团,擀成面皮,切成细条,煮熟,加各种调料和小菜。

Wheat noodles. A staple food in northern areas. Flour mixed with water, kneaded into sheets, then shredded. Boiled. Served with a variety of flavorings and side dishes.

面线糊 (miànxiànhú)

福建点心。先将虾、蛏、淡菜加调料熬汤,再放入面线煮成糊。

Sea food and rice noodles. Fujian snack. Shrimp, razor clam, and dried clam meat cooked into soup with flavorings, combined with rice noodles, then boiled into paste.

苗家三色饭 (miáojiā sānsèfàn)

海南苗族食品。先把山兰米用黑色的桑叶汁、红色的红藤叶汁和黄色的姜汁分别浸泡,然后蒸熟,加入椰浆。

Three-color rice with coconut milk. Specialty of Miao ethnic minority in Hainan. Local sticky rice soaked separately in black mulberry leaf juice, red sargent glory vine leaf juice, and yellow ginger juice, then steamed. Washed with coconut milk.

苗族竹板鱼 (miáozú zhúbǎnyú)

广西苗族菜。先将鲮鱼用盐腌渍，然后用粽叶或藕叶包好，夹在两块竹板间蒸熟。

Steamed carp in bamboo clips. Specialty of Miao ethnic minority in Guangxi. Mud carp pickled with salt, wrapped in reed leaves or lotus leaves, clamped with 2 bamboo slips, then steamed.

妙计锦囊 (miàojì-jǐnnáng)

山东菜。用豆腐皮裹卷鸭血、粟米、西芹，蒸熟。

Duck blood and vegetable rolls. Shandong dish. Bean curd skins made into rolls filled with prepared duck blood, corn, and celery, then steamed.

闽北水仙 (mǐnběishuǐxiān)

乌龙茶。产于福建建瓯、建阳两县。外形壮实匀整，尖端扭结，色泽黄绿油润。汤色红艳明亮，叶底柔软，红边明显。

Minbeishuixian, oolong tea produced in Jianzhen and Jianyang, Fujian. Its dry leaves are stout, even, twisted at the tips, and yellowish green. The tea is crystal red, and the infused leaves are soft with red fringes.

闽红功夫茶 (mǐnhóng gōngfuchá)

红茶。政和功夫茶、坦洋功夫茶、白琳功夫茶的统称。均系福建特产。

Minhong gongfucha, a general name for *zhenghe gongfu*, *tanyang gongfu*, and *bailin gongfu* black teas produced in Fujian.

明炉烧鸭 (mínglú shāoyā)

广东菜。先把姜、葱、花椒、虾酱、味精、韭黄等拌成馅，填入鸭腹，然后入炉烤熟至皮红。

Broiled duck. Guangdong dish. Duck stuffed with mixture of ginger, scallion, Chinese prickly ash, shrimp sauce, MSG, and yellow Chinese chive, then broiled in an oven until skins red.

茉莉花茶 (mòlìhuāchá)

再加工茶。主要产区在江苏。以绿茶为茶胚，用茉莉花窨制而成。香气清芬，汤色澄明。

Jasmine tea. Re-processed tea mainly produced in Jiangsu. Made with green tea by adding jasmine flowers through special techniques. Its tea has an elegant aroma and a crystal yellowish green color.

木瓜凉粉 (mùguā liángfěn)

江西点心。先将木瓜粉加水和芡粉调成浆煮熟，冷却成糕状，然后切片，加入薄荷水、白糖。

Papaya curd with peppermint tea. Jiangxi snack. Papaya powder made into paste with water and starch,

boiled, then chilled into curd. Cut into small pieces, then immersed in sugared peppermint tea.

木鱼渡僧 (mùyú dùsēng)

河南菜。将夹馅茄片裹面糊，炸至金黄，佐椒盐食用。

Fried eggplant cakes. Henan dish. Eggplant segments layered with stuffing, coated with flour paste, deep-fried, then served with spicy salt.

目鱼炒韭菜 (mùyú chǎo jiǔcài)

湖南菜。将目鱼氽烫后切成丝，加入韭菜、盐煸炒，勾芡。

Flatfish with Chinese chive. Hunan dish. Scalded flatfish stir-fried with Chinese chive and salt. Thickened with starch.

苜蓿肉 (mùxùròu)

北京菜，又称木须肉。先将里脊肉片腌渍，再用油滑散，然后与黄花、黄瓜、黑木耳、鸡蛋等翻炒。

Sautéed pork with eggs. Beijing dish. Sliced pickled pork tenderloin lightly fried, then sautéed with day lily, cucumbers, wood ears, and eggs.

纳仁 (nàrén)

新疆食品,即手抓羊肉面。先将面条或面片用加调料的羊肉汤煮熟,再拌入熟羊肉块,撒上辣椒粉、碎洋葱等,用手抓食。

Noodles with mutton. Xinjiang food. Noodles or sliced flour dough boiled in seasoned mutton broth, mixed with cooked mutton chops, then flavored with chili powder and chopped onion. Normally eaten by hand.

奶茶 (nǎichá)

蒙古族、哈萨克族点心。用鲜奶加入砖茶煮制,可加盐、黄油、炒米或奶制品等。

Milk tea. Snack of Mongolian and Kazak ethnic minorities. Fresh milk boiled with brick tea and flavored with salt, butter, roasted rice, or dairy products.

奶豆腐 (nǎidòufu)

蒙古族点心。用牦牛奶、普通牛奶、山羊奶或绵羊奶制成。先将奶里凝结成豆腐状的物质取出,滤水,放在模子里轻压成型,然后晾干。

Milk curd. Snack of Mongolian ethnic minority. Milk curd made with yak, cow, goat, or sheep milk. Milk is allowed to curdle, then curd is removed, drained, lightly pressed, and dried.

奶酪蒸虾仁 (nǎilào zhēng xiārén)

广东菜。将鲜牛奶、鸡蛋、鸡粉调和,撒上干虾仁、火腿茸,蒸熟。

Steamed shrimp meat in milk. Guangdong dish. Fresh milk mixed with egg paste and chicken powder, topped with dried shrimp meat and ham crumbs, then steamed.

奶皮子 (nǎipízi)

蒙古族、柯尔克孜族食品。分生奶皮和熟奶皮,即新鲜牛、羊奶或煮过的牛、羊奶面上的脂皮,从奶里挑出,晾干。

Milk fat. Delicacy of Mongolian and Kirgiz ethnic minorities. Layers of fat on surfaces of fresh or boiled milk removed from milk, then air-dried.

奶油煎克戟 (nǎiyóu jiān kèjǐ)

广东点心,即奶油煎饼。将面粉、鲜奶、鸡蛋、水果、牛油、水等调成糊,倒入平底锅,用慢火煎至微黄。

Milk and flour pancakes. Guangdong snack. Flour, milk, eggs, fruit, butter, and water mixed into paste and fried into light yellow pancakes over low heat.

奶油松瓤卷酥 (nǎiyóu sōngráng juǎnsū)

曹雪芹《红楼梦》菜谱。将面粉、发酵粉、水、油分别和成水油面团和油酥面团。用水油面团包油酥面团，擀成面皮，铺上用鸡蛋、猪油、芝麻、盐、奶油、松仁、白糖调成的馅，卷成如意卷形，切块，烤熟。

Baked cream and nuts rolls. Menu from *A Dream of Red Mansions* by Cao Xueqin. Flour, baking powder, water, and oil mixed into water dough and oil dough separately. Oil dough packed in water dough, then made into wrapping sheets. Each sheet spread with mixture of eggs, lard, sesame, salt, cream, pine nuts, and sugar, then rolled from two ends to the middle in the shape of a curved cloud symbolizing good luck. Rolls sectioned, then baked.

奶渣蜕 (nǎizhātuì)

西藏点心。将干奶渣与酥油和红糖调和，压制成形。

Dairy dreg cookies. Tibetan snack. Dried dairy dregs, Tibetan butter, and brown sugar mixed, then pressed in mould to take shape.

南瓜饼 (nánguābǐng)

南方点心。将南瓜蒸熟捣成泥，加入蜂蜜和面粉，做成小圆饼，煎至金黄。

Fried pumpkin cakes. Guangdong snack. Mashed steamed pumpkin mixed with honey and flour, made into small cakes, then deep-fried.

南翔小笼包 (nánxiáng xiǎolóngbāo)

上海点心。又名南翔小笼馒头，源于上海南翔镇，形状小巧，皮薄呈半透明状。用猪肉末配竹笋丁、虾仁丁、蟹肉或蟹黄及调料为馅，用小竹笼蒸熟，佐姜醋或蛋丝汤。

Steamed stuffed mini buns. Shanghai snack. Also called Nanxiang steamed mini bread, originated in Nanxiang Town, Shanghai. Small and translucent, stuffed with seasoned minced pork, diced bamboo shoots, shrimp meat, and crab meat or roe. Steamed in small bamboo steamers, then served with ginger vinegar or egg soup.

馕坑烤羊肉 (nángkēng kǎoyángròu)

新疆维吾尔族食品。将羊肉块裹上面粉鸡蛋糊，入馕坑焖烤。

Baked mutton. Delicacy of the Uygur ethnic minority in Xinjiang. Mutton chunks coated with flour and egg paste, then baked in a special covered oven.

年糕 (niángāo)

江浙沪春节传统点心。粳米饭捣烂压成条形，晾干，切成片，与其他配料煎炒。

Spring Festival *niangao* or rice cakes. Traditional Spring Festival snack in Jiangsu, Zhejiang, and Shanghai. Cooked round grain rice pestled, made into flat oblong cakes, then air dried. Sliced and cooked with other ingredients.

酿冬瓜圈 (niàngdōngguāquān)

四川菜。先将冬瓜切成圆环,煮熟,再将虾茸酿入冬瓜圈,用温油浸熟,然后浇上用蟹肉、蛋清、芡粉及调料做成的卤汁,撒葱花。

Wax gourd with shrimp. Sichuan dish. Wax gourd cut into rings, boiled, filled with minced shrimp meat, then fried over low heat. Washed with sauce of crab meat, egg white, and starch, then sprinkled with chopped scallion.

酿金钱发菜 (niàng jīnqián fàcài)

西北菜。先将发菜和蛋清拌匀,摊在蛋皮上,中间放蛋糕条,卷成柱形蒸熟,然后浇鸡清汤,加调料再稍蒸。

Black moss with egg cakes.

Northwestern dish. Black moss and egg white paste spread evenly on prepared egg sheets, made into rolls filled with yellow cake strips, then steamed. The rolls are washed with clear chicken soup, seasoned, then re-steamed briefly.

宁城老窖 (níngchénglǎojiào)

白酒,产于内蒙古宁城。以当地红高粱为主料,用玉米、小麦、大米、蚕豆、荞麦等制曲酿造。酒精含量为38-52度,属浓香型酒。

Ningchenglaojiao, liquor produced in Ningcheng, Inner Mongolia. White spirit made with local red sorghum and yeast of corn, wheat, rice, broad beans, and buckwheat. Contains 38-52% alcohol, and has a thick aroma.

宁红功夫茶 (nínghóng gōngfuchá)

红茶。产于江西修水县,该地古称宁州。外形紧细多毫,乌黑油润,汤色红艳明亮,叶底柔嫩。

Ninghong gongfucha, black tea produced in Xiushui, the modern name for ancient Ningzhou, Jiangxi. Its dry leaves are slim, tight, hairy, and glittery black. The tea is bright red with infused leaves soft and tender.

宁式鳝丝 (níngshì shànsī)

浙江宁波菜。先将鳝丝用芡粉腌渍,然后过油,加姜、竹笋、韭菜、葱丝及调料煸炒,稍焖,浇滚麻油。

Stir-fried eel. Specialty in Ningbo, Zhejiang. Lightly fried starch-pickled shredded eel stir-fried with ginger, bamboo shoots, Chinese chive, scallion, and seasonings. Simmered briefly, then washed with heated sesame oil.

柠檬鸡 (níngméngjī)

广东菜。将鸡脯肉条用姜、葱、料酒、胡椒粉腌渍,裹上鸡蛋淀粉糊,炸熟,佐柠檬、盐、糖、芡粉做成的卤汁。

Fried chicken breast with lemon sauce. Guangdong dish. Strips of chicken breast pickled with ginger, scallion, cooking wine, and ground black pepper, coated with starched egg white, then fried. Served with sauce of lemon juice, salt, sugar, and starch.

牛百叶烧卖 (niúbǎiyè shāomài)

广东点心。先将氽过的牛百叶用盐、糖、鸡精等腌渍,再加入胡萝卜丝、胡椒粉、麻油、豆豉油等拌匀,冷冻数小时,然后用面皮包成鸡蛋大小,蒸熟。

Steamed ox tripe dumplings.

Guangdong snack. Scalded ox tripe pickled with salt, sugar, and chicken essence, mixed with shredded carrots, ground black pepper, sesame oil, and fermented soybean oil, then chilled for several hours. Wrapped into dumplings with flour sheets, then steamed.

牛肉炒粉 (níuròu chǎo fěn)

江西菜。先将牛肉丝煸炒,再与米粉拌炒,用盐、辣椒、葱、姜调味。

Stir-fried rice noodles and beef.

Jiangxi dish. Quick-fried shredded beef stir-fried with rice noodles, then flavored with salt, hot pepper, scallion, and ginger.

牛蛙煲 (niúwābāo)

湖南菜。先将牛蛙与青、红椒翻炒,然后加入汤及调料煲煮。

Stewed bullfrog. Hunan dish. Bullfrogs stir-fried with red and green pepper, then stewed in soup with condiments.

牛肉泡馍 (niúròu pàomó)

陕西点心。将牛肉和牛骨加调料炖煮,配烙饼或馍。

Steamed bread in beef soup. Shaanxi snack. Beef and ox bones stewed with seasonings. Served with pancakes or steamed bread.

牛中三杰 (niú zhōng sānjié)

湖南套菜。由发丝牛百叶、红烧牛蹄筋、烩牛脑髓三道菜组成。

Three ox dishes. Hunan dish. A set of 3 ox dishes. Shredded ox tripe, beef tendons with brown sauce, and braised ox brains.

糯米扣排骨 (nuòmǐ kòu páigǔ)

湖南菜。将排骨腌渍后裹上糯米蒸熟,撒葱花。

Steamed pork ribs in sticky rice. Hunan dish. Pickled pork ribs coated with sticky rice, steamed, then sprinkled with chopped scallion.

O

藕香芹味 (ǒuxiāng qínwèi)

浙江菜。将粉丝、银耳、西芹、莲藕、胡萝卜等分别用开水烫过，摆盘，淋上特制的卤汁。

Vegetables and vermicelli with sauce. Zhejiang dish. Vermicelli, white wood ears, American celery, lotus roots, and carrots scalded separately, put on the same platter, then washed with special sauce.

P

扒白菇 (pábáigū)

山东菜。将新鲜白菇加高汤煮入味，勾芡。

Boiled white mushrooms. Shandong dish. Fresh white mushrooms boiled in broth and thickened with starch.

扒鲍鱼冬瓜球 (pá bàoyú dōngguāqiú)

山东菜。先将鲍鱼、冬瓜球、冬菇、火腿、青豆等放入清汤加调料烧沸，然后勾芡，微火扒透。

Braised abalone with gourd balls. Shandong dish. Abalone boiled in clear soup with wax gourd balls, winter mushrooms, sliced ham, string beans, and flavorings. Thickened with starch, then braised.

扒鲍鱼芦笋 (pá bàoyú lúsǔn)

山东菜。将鲍鱼肉、芦笋、火腿、冬菇等蒸熟，浇上勾芡的鲜汤。

Steamed abalone with asparagus. Shandong dish. Abalone, asparagus, ham, and winter mushrooms steamed, then marinated in seasoned starched soup.

扒大肠油菜 (pá dàcháng yóucài)

浙江菜。将大肠和油菜用黄酒、盐、糖等烹煮并勾芡。

Braised large intestine with kale. Zhejiang dish. Hog large intestine braised with kale, flavored with yellow wine, salt, and sugar, then thickened with starch.

扒方肉 (páfāngròu)

江苏菜。五花肉加酱油、糖、盐、姜、黄酒等炖酥。

Stewed cubed streaky pork. Jiangsu dish. Streaky pork stewed with soy sauce, sugar, salt, ginger, and yellow wine.

扒广肚 (páguǎngdǔ)

河南菜。将涨发的广肚与菜心、香菇、冬笋、火腿等汤焯，然后加调料扒煮。

Fish maw with ham and vegetables. Henan dish. Soaked sea fish maw quick-boiled in soup with baby cabbage, mushrooms, winter bamboo shoots, and ham, then simmered with seasonings.

扒海参猴头蘑 (pá hǎishēn hóutóumó)

东北菜。先将猴头蘑、海参发好切条，再将猴头蘑煮熟，海参焯好，然后放入鲜汤煨烂，勾芡，淋麻油。

Stewed sea cucumbers with monkey-head mushrooms. Northeastern dish. Strips of boiled monkey-head mushrooms and sea cucumbers stewed in seasoned soup, thickened with starch, then washed with sesame oil.

扒海羊 (páhǎiyáng)

山东菜。将煮熟的海参和羊肉放入鸡汤烹煮,加调料,勾芡,淋鸡油。

Braised sea cucumbers and mutton. Shandong dish. Boiled sea cucumbers and mutton braised in chicken soup, seasoned, thickened with starch, then washed with chicken oil.

扒蚝油生菜 (pá háoyóu shēngcài)

广东菜。将焯过的生菜用色拉油、蚝油、酱油煸炒,浇芡汁。

Sautéed romaine lettuce with oyster sauce. Guangdong dish. Scalded romaine lettuce sautéed with salad oil, oyster sauce, and soy sauce, then washed with starched soup.

扒猴头 (páhóutóu)

河南菜。将猴头菇氽透,加入火腿、冬笋、菜心、鸡腿及调料炖煮,勾芡。

Braised monkey-head mushrooms. Henan dish. Quick-boiled monkey-head mushrooms braised with ham, winter bamboo shoots, baby cabbage, drumsticks, and flavorings, then thickened with starch.

扒鸡茸发菜 (pá jīróng fàcài)

内蒙菜。先将当地发菜与鸡肉末、蛋清、盐等调成糊,氽熟,然后浇上芡汁。

Chicken with black moss. Inner Mongolian dish. Local black moss, ground chicken, egg white, and salt mixed into paste, quick-boiled, then marinated in starched soup.

扒鸡腿海参 (pá jītuǐ hǎishēn)

东北菜。先将鸡腿肉加调料蒸熟,海参焯水,然后与冬笋、冬菇、火腿等入鸡汤煨煮,勾芡,淋油。

Braised chicken and sea cucumbers. Northeastern dish. Drumsticks steamed with seasonings, braised in chicken soup with quick-boiled sea cucumbers, winter bamboo shoots, winter mushrooms, and ham, thickened with starch, then washed with oil.

扒栗子白菜 (pá lìzi báicài)

山东菜。先将栗子和白菜焯水,再加高汤、白糖焖煮,勾芡。

Braised chestnuts and Chinese cabbage. Shandong dish. Scalded chestnuts and Chinese cabbage braised in sugared broth and thickened with starch.

扒龙须鲍鱼 (pá lóngxū bàoyú)

山东菜。先将鲍鱼和龙须菜用鸡汤氽熟,然后勾芡。

Boiled abalone and asparagus. Shandong dish. Abalone and asparagus boiled in chicken soup and thickened with starch.

扒酿海参 (pániàng hǎishēn)

甘肃酒泉菜。先将鸡肉茸及调料塞入海参,稍蒸,然后用微火烤熟,勾芡,淋油。

Chicken in sea cucumbers. Specialty in Jiuquan, Gansu. Sea cucumbers filled with ground chicken and seasonings, half-steamed, roasted over low heat, thickened with starch, then washed with oil.

扒酿猴头 (pániàng hóutóu)

山东菜。先将猴头蘑、鸡骨架、猪肉加黄酒、盐、姜、葱等蒸烂，然后将煮沸的汤汁勾芡，浇在蘑菇上。

Steamed mushrooms in broth. Shandong dish. Monkey-head mushrooms steamed with chicken bones, pork, yellow wine, salt, ginger, and green onion, then marinated in starched boiled broth.

扒牛肉条 (pániúròutiáo)

各地家常菜。先将牛肉氽过，加姜、葱、清水煮熟，切成长条，再加调料入原汤蒸酥，然后勾芡、淋麻油。

Steamed boiled beef. Home-style dish in many places. Scalded beef boiled with ginger and scallion in water, slivered, then steamed in broth with seasonings. Thickened with starch, then washed with sesame oil.

扒三白 (pásānbái)

山东菜。先将鸡脯肉片腌渍，入油锅滑散，然后和鲍鱼片、芦笋片入鲜汤煮开，用小火扒透，勾芡。

Braised chicken breast, abalone, and asparagus. Shandong dish. Pickled sliced chicken breast lightly fried, boiled in seasoned soup with sliced abalone and asparagus, braised, then thickened with starch.

扒三样 (pásānyàng)

浙江菜。将蒸鸡腿肉、蒸鲜酱肉和海参加调料焖熟，勾芡。

Braised chicken, pork, and sea cucumbers. Zhejiang dish. Steamed drumsticks and pickled pork braised with sea cucumbers and seasonings, then thickened with starch.

扒山虎 (páshānhǔ)

内蒙菜。先将羊蹄煮熟，再加入醋、高汤、盐、酱油等翻炒。

Sautéed lamb hooves. Inner Mongolian dish. Boiled lamb hooves sautéed with vinegar, broth, salt, and soy sauce.

扒烧牛筋 (páshāo niújīn)

东北菜。先将牛蹄筋蒸烂，再与香菇、冬笋、火腿入鸡汤煨煮，勾芡，淋麻油。

Braised beef tendons. Northeastern dish. Steamed beef tendons braised in chicken soup with dried mushrooms, winter bamboo shoots, and ham, thickened with starch, then washed with sesame oil.

扒双素 (páshuāngsù)

四川菜。先将油菜和芦笋用加调料旺火翻炒，然后加入清汤烧开，用微火炖烂。

Simmered rape greens and asparagus. Sichuan dish. Rape greens and asparagus sautéed with flavorings over high heat, boiled in clear soup, then simmered.

扒驼掌 (pátuózhǎng)

内蒙名菜。先将骆驼前掌焯水，加调料炖或蒸，然后去骨，切片，浇上特制的卤汁，配汆过的蘑菇、熟鸡脯肉、青菜装盘。

Steamed camel feet with sauce. Famous Inner Mongolian dish. Scalded camel front feet steamed or stewed with seasonings, boned, and sliced. Washed with special sauce. Served with quick-boiled mushrooms, cooked chicken breast, and green vegetable.

扒鱼卷 (páyújuǎn)

东北菜。先用虾仁、肥膘肉、鸡蛋清和调料做成虾泥，再用鱼肉片裹虾泥成卷，放入汤中煨熟，勾芡，淋猪油。

Braised fish rolls. Northeastern dish. Fish rolls filled with mixture of ground shrimp, fat pork, and egg white, braised in broth, thickened with starch, then drenched with lard.

扒原壳鲍鱼 (pá yuánkébàoyú)

东南方菜。将鲍鱼去壳焯水，加火腿、笋片、菜心、冬菇及调料蒸熟，然后放回原壳中，勾芡，淋油。

Steamed abalone. Southeastern dish. Scalded shelled abalone steamed with ham, sliced bamboo shoots, baby cabbage, winter mushrooms, and flavorings. Put back in shells, then drenched with starched soup and sesame oil.

扒肘子 (pázhǒuzi)

东北菜。先将猪肘煮熟，切成连皮的大片，然后放入汤中加调料扒至收汁，勾芡，淋猪油。

Braised hog joint. Northeastern dish. Boiled hog joint cut into big connected pieces, braised in broth with seasonings, thickened with starch, then washed with lard.

扒猪头 (pázhūtóu)

江苏菜。猪头加黄酒、冰糖、香醋、酱油、桂皮、八角等焖熟。

Stewed hog head. Jiangsu dish. Hog head stewed with yellow wine, rock sugar, vinegar, soy sauce, cinnamon, and aniseed.

帕查麻枯 (pàchámákū)

西藏点心。先将面粉做成面疙瘩煮熟，再与酥油、红糖和碎奶渣拌匀。

Dumplings with butter and dairy dregs. Tibetan snack. Flour and water made into small pieces, boiled, then mixed with Tibetan butter, brown sugar, and minced dairy dregs.

帕尔木丁 (pàěrmùdīng)

新疆维吾尔食品，即羊油馅饼。用面粉、发酵粉、鸡蛋、盐、水等做成饼皮，包入用肥羊肉、洋葱、盐、胡椒粉、孜然粉等做成的馅，做成马鞍形，蘸醋后放入馕坑中烤熟，抹上熟羊油。

Fat mutton cakes. Uygur delicacy in Xinjiang. Flour, baking powder, eggs, salt, and water made into cake wrappers filled with mixture of fat mutton, chopped onion, salt, ground black pepper, and ground cumin, arranged like horse saddles, quickly dipped in vinegar, then baked in a special oven. Brushed with prepared suet oil before serving.

排骨海带汤 (páigǔ hǎidài tāng)

南方家常菜。先将氽过的排骨加调料焖烧,然后放入海带炖煮。

Pork spareribs and kelp soup. Home-style dish in southern areas. Scalded pork spareribs braised with flavorings, then stewed with kelp.

排骨年糕 (páigǔ niángāo)

上海点心。猪排裹上面粉、芡粉、五香粉、蛋浆调成的糊油煎,然后同炸过的年糕加酱油、糖翻炒。

Fried pork steak and rice cakes. Shanghai snack. Pork steaks coated with paste of flour, starch, five-spice powder, and egg paste, deep-fried, then stir-fried with deep-fried *niangao*, soy sauce, and sugar.

泡椒蚝油炖鸡 (pàojiāo háoyóu dùn jī)

四川菜。先将母鸡过油,然后加入泡椒、蚝油等炖熟。

Hen with pickled pepper and oyster sauce. Sichuan dish. Lightly fried hen stewed with pickled pepper, oyster sauce, and other condiments.

泡椒肉丝 (pàojiāo ròusī)

四川菜。先将肉丝滑散,再与姜、蒜、泡椒、笋片、木耳翻炒,勾芡。

Stir-fried pork with pickled pepper. Sichuan dish. Lightly fried shredded pork stir-fried with ginger, garlic, pickled pepper, sliced bamboo shoots, and wood ears, then thickened with starch.

沛公狗肉 (pèigōnggǒuròu)

江苏菜,源于汉高祖刘邦(256-195 BC)的故乡沛县。先将狗肉煮熟,再放入甲鱼块和甲鱼蛋,加调料炖至酥烂。

Braised dog and turtle. Jiangsu dish, originated in Peixian, the hometown of Liu Bang (256-195 BC), the first emperor of Han Dynasty. Dog meat boiled, seasoned, then braised with chopped turtle and turtle eggs.

彭城鱼丸 (péngchéngyúwán)

江苏菜。先将鲤鱼茸与猪肥膘茸加调料做成丸子,入鸡汤煮熟,然后加入菜心、火腿、冬菇。

Fish and fat pork balls. Jiangsu dish. Minced carp and fat pork mixed with seasonings, made into balls, then boiled in chicken soup with baby greens, ham, and winter mushrooms.

皮蛋瘦肉粥 (pídàn shòuròu zhōu)

广东点心。将米、瘦肉、皮蛋、香菜与高汤熬煮。

Pork and preserved egg porridge. Guangdong snack. Rice cooked into porridge with lean pork, preserved eggs, coriander, and broth.

啤酒鸭 (píjiǔyā)

西南地区菜肴。将鸭块煸炒后加入啤酒、姜、八角、桂皮、丁香、蒜、高汤,用小火焖熟。

Braised duck with beer. Southwestern dish. Quick-fried chopped duck braised in broth with beer, ginger, aniseed, cinnamon, clove, and garlic.

琵琶豆腐 (pípádòufu)

山东菜。将豆腐泥、肉末、蛋浆等调成糊,倒入汤匙,撒火腿末和碎冬菇,蒸熟。

Steamed bean curd with pork and eggs. Shandong dish. Mashed bean curd, ground pork, egg paste, and starch mixed, put in spoons, topped with ham crumbs and chopped winter mushrooms, then steamed.

葡挞 (pútà)

澳门点心。用鸡蛋、水、面粉调成糊,倒入面饼皮做成的小碟,烤熟。

Eggs and flour cakes. Macao snack. Eggs, water, and flour mixed, put in small dishes made with flour dough, then baked.

葡萄干炒饭 (pútáogān chǎofàn)

新疆食品。将米饭与熟羊肉、葡萄干、胡萝卜等用植物油拌炒。

Fried rice with raisins and mutton. Xinjiang food. Cooked rice stir-fried in vegetable oil with prepared mutton, raisins, and carrots.

蒲棒里脊 (púbàng lǐji)

山东菜。将猪瘦肉末加调料做成肉棒,中间插入竹签,裹上蛋浆和面包屑,炸至金黄。

Fried lean pork. Shandong dish. Ground lean pork seasoned, made into meat lollies on bamboo sticks, coated with egg paste and bread crumbs, then deep-fried.

蒲酥全鱼 (púsū quányú)

山东菜。先将鱼切成鱼头、鱼身、鱼尾三段,鱼身剔骨切片,再将全部鱼料挂鸡蛋面粉浆,小火炸熟,然后浇上用火腿、菜心、冬菇、料酒、酱油、芡粉等调制的卤汁。

Fried fish with special sauce. Shandong dish. Fish cut into 3 sections of head, body, and tail. Body boned and sliced. Fish head, body, and tail coated with starch and egg paste, fried over low heat, then marinated in dressing of ham, Chinese baby cabbage, winter mushrooms, cooking wine, soy sauce, and starch.

普洱茶 (pǔ'ěrchá)

黑茶,亦称滇青茶。中国名茶,因原加工运销集散地在云南普洱县而得名。外形条索粗壮,色泽乌润。汤色棕红,汤味醇厚回甘,具有独特的陈香。以普洱茶为原料加工的压制茶有普洱沱茶、七子饼茶、普洱砖茶等。

Puercha, also called Yunnan dark green tea, produced in Yunnan. Its dry leaves are stout with a black bloom. The tea is brownish red with a unique stale flavor and a rich sweet aftertaste. Used to make several compressed teas such as *puer tuocha*, *qizibingcha*, and *puer zhuancha*.

普洱紧茶(pǔ'ěr jǐnchá)

云南普洱茶的一种类别。其制作方法是将散茶经过蒸或炒,模压成型。

普洱紧茶按照发酵方法，分为“生普洱”即自然发酵陈化茶，和“熟普洱”即高温、高湿速发酵茶。

Pue. compressed tea. One category of *puer* tea produced in Yunnan. It is made with *puer* loose tea by steaming or baking, then compressing into different shapes. There are 2 kinds of compressed *puer* tea: raw compressed *puer* tea fermented naturally after compression; ripe compressed *puer* tea quickly fermented at high temperature and dense humidity.

普洱散茶 (pǔ'ěr sǎnchá)

云南普洱茶的一种类别。传统品类为毛尖、粗叶，今已发展为普洱绿茶、普洱青茶、普洱红茶、普洱黑茶、普洱黄茶、普洱白茶 6 个品种。

Puer loose tea. One category of *puer* tea produced in Yunnan. Traditional kinds are *puer maojian* with fine leaves and *puer cuye* with coarse leaves. There are six new kinds: *puer* green tea; *puer* dark green tea; *puer* red tea; *puer* black tea; *puer* yellow tea; and *puer* white tea.

七彩吊片丝 (qīcǎi diàopiànsī)

香港菜。将鱿鱼、青椒、红辣椒、黄花菜用盐、蚝油、料酒翻炒。

Stir-fried squid with vegetables. Hong Kong dish. Squid stir-fried with green and red pepper and day lily buds, then flavored with salt, oyster sauce, and cooking wine.

七彩烘蛋 (qīcǎi hōngdàn)

西北菜。将切碎并用盐水汆过的西芹、黄甜椒、洋葱、黑木耳、青豆、火腿与蛋浆调和，倒入平底锅慢火烘熟。

Fried eggs with vegetables. Northwestern dish. Chopped American celery, sweet yellow pepper, onion, black wood ears, green beans, and ham scalded in salty water, mixed with egg paste, then fried over low heat.

青岛啤酒 (qīngdǎopíjiǔ)

中国著名啤酒，产于山东青岛。用大麦、大米、崂山泉水及特殊酵母酿造，有多种口味和香型，酒精含量在0.8-7.5度之间。

Qingdao beer, famous beverage produced in Qingdao, Shandong. Brewed from barley, rice, Laoshan Spring water, hops, and special yeast. There are several types with different tastes and aromas. Contains 0.8-7.5% alcohol.

七星拼盘 (qīxīng pīnpán)

山东菜。先将炸熟的椰子蛋黄卷切片，将蒸熟的猪舌卷、竹叶鱼卷切片，干贝蒸熟，虾仁炒熟，黄瓜卷切片，海蜇切丝，然后将这七样分别装盘，置于旋转餐桌上，配不同的卤汁。

Seven dishes on a turntable. Shandong dish. Salty egg yolk and coconut rolls deep-fried, then sliced. Hog tongue rolls and bamboo fish rolls steamed, then sliced. Dried scallops steamed. Shrimp meat sautéed. Sliced cucumbers rolled. Jellyfish shredded. The 7 dishes put on different platters, then served with different dressing on a turntable.

七星丸 (qīxīngwán)

福建菜。把猪肉末和虾仁末塞入海鳗鱼丸，将鱼丸煮熟，放入鸡汤烧开。

Stuffed eel balls. Fujian dish. Pike eel meat made into balls filled with ground pork and shrimp meat, scalded, then boiled in chicken soup.

七星紫蟹 (qīxīng zǐxiè)

广东菜。先将七个紫蟹蒸熟，再将蛋清加盐、味精、清汤调和并蒸熟，然后将蟹放在蛋羹上。

Steamed seven crab over egg custard. Guangdong dish. Seven purple crab steamed, then served on steamed custard made with clear soup, egg white, salt, and MSG.

祁门红茶 (qíménhóngchá)

红茶。产于安徽祁门地区，外形条索紧秀，色泽乌黑泛灰光。汤色红艳，叶底微软红亮。

Qimen, or *Keemum*, black tea produced in Qimen, Anhui. Its dry leaves are tight, straight, and off-black. The tea is red with infused leaves soft and red.

祁阳笔鱼 (qíyángbǐyú)

湖南祁阳菜。先将本地特产笔鱼与红椒、姜、葱白、黄酒、酱油等煸炒，然后添入鲜汤焖烧。

Simmered pen fish. Dish in Qiyang, Hunan. Local pen fish or bronze gudgeon stir-fried with red pepper, ginger, scallion stems, yellow wine, and soy sauce, then simmered in seasoned soup.

奇兰 (qílán)

乌龙茶。产于福建安溪，属安溪色种。外形细瘦，梗窄，色泽黄绿鲜润。汤色青黄，叶底叶张圆，有光泽。

Qilan oolong tea, one of the oolong varieties from Anxi, Fujian. Its dry leaves are slender, yellowish green, sleek, and lustrous on slim stalks. The tea is crystal yellowish green with infused leaves in good shape.

麒麟石斑 (qílínshíbān)

广东菜。先将石斑鱼用盐、黄酒、姜、葱腌渍，再配火腿、香菇、竹笋蒸熟，浇芡汁。

Steamed grouper with mushrooms and ham. Guangdong dish. Grouper pickled with salt, yellow wine, ginger, and scallion, steamed with ham, dried mushrooms, and bamboo shoots, then washed with starched soup.

麒麟脱胎 (qílín tuōtāi)

福建菜。先将乳狗放入猪肚腌制，再放入陶钵加调料炖熟。

Stewed baby dog in hog tripe. Fujian dish. Pickled baby dog put in hog tripe, then stewed with seasonings in an earthen container.

气糕 (qìgāo)

福建点心。将粳米粉加水、糖调成糕浆，蒸熟，面上撒红糖或白糖。

Steamed rice cakes. Fujian snack. Round grain rice flour, water, and sugar made into paste, steamed into cake, then showered with brown or white sugar.

气锅滑嫩丸子 (qìguō huánèn wánzi)

江苏菜。将炸好的猪肉圆与火腿片、冬菇、笋尖等放入鸡清汤蒸熟。

Meatballs in chicken soup. Jiangsu dish. Deep-fried pork meatballs, sliced ham, dried mushrooms, and tender bamboo shoot tips steamed in chicken soup.

汽锅鸡 (qìguōjī)

南方菜。将肥母鸡块放入陶瓷汽锅,加入冬笋丁、火腿丁、清汤,然后将锅盖严,置于蒸锅内,将鸡蒸熟。

Steamed hen. Southern dish. Cubed hen, diced winter bamboo shoots, and diced ham steamed in clear soup in a covered pottery container.

荠菜干贝羹 (jìcài gānbèi gēng)

浙江菜。先将干贝蒸透,碾碎,入高汤煮沸,勾芡,然后将煸炒过的碎荠菜放入贝羹,淋猪油。

Scallop and shepherd's purse soup. Zhejiang dish. Minced steamed dried scallops boiled in broth, thickened with starch, then garnished with quick-fried chopped shepherd's purse and lard.

荠菜肉丝春卷 (jìcài ròusī chūnjuǎn)

上海点心。将薄面皮包加调料的荠菜猪肉馅成卷状,炸至金黄,佐醋。

Pork and shepherd's purse rolls. Shanghai snack. Flour sheets made into rolls filled with seasoned ground pork and chopped shepherd's purse, then deep-fried crisp. Served with vinegar.

荠菜肉丝豆腐羹 (jìcài ròusī dòufu gēng)

上海菜。将豆腐丁、肉丝、荠菜放入肉汤煮熟,加黄酒和盐,勾芡。

Shepherd's purse and bean curd soup. Shanghai dish. Diced bean curd, shredded meat, and shepherd's purse boiled in pork soup, flavored with yellow wine and salt, then thickened with starch.

千层泡菜烩鸡 (qiāncéngpàocài huì jī)

河南菜。先将腌好的鸡块过油,炒至近熟,然后加泡菜翻炒。

Sautéed chicken and pickled cabbage. Henan dish. Quick-fried pickled cubed chicken stir-fried, then sautéed with pickled Chinese cabbage.

千层猪耳 (qiāncéng zhūěr)

广东凉菜。将猪耳加姜、葱、盐、料酒等煮熟,切成薄片,佐麻油、蒜茸、米醋、酱油。

Boiled hog ears with sauce.
Guangdong cold dish. Hog ears boiled with ginger, scallion, salt, and cooking wine. Sliced, then served with dressing of sesame oil, minced garlic, vinegar, and soy sauce.

千山酒 (qiānshānjiǔ)

白酒,产地辽宁辽阳。以辽东高粱为主要原料酿造,酒精含量为38-53度,分酱香、浓香、兼香三种香型。

Qianshanjiu, liquor produced in Liaoyang, Liaoning. White spirit with 38-53% alcohol made with sorghum grown in Eastern Liaoning. There are 3 types according to aroma: Soy, thick, and mixed.

千页糕 (qiānyègāo)

福建点心。将面团擀成薄片,对折,中间放进熟肉、白糖等馅料,蒸熟。

Steamed pork and sugar cakes.

Fujian snack. Flour dough made into thin sheets, folded around cooked pork and sugar, then steamed.

千张肉 (qiānzhāngròu)

湖北菜。先将五花肉煮熟,用金酱涂猪皮,煎至发红,切成薄片,皮朝下码入碗,放入红腐乳汁、豆豉、姜、葱,蒸熟。

Steamed sliced pork. Hubei dish. Boiled streaky pork coated with fermented flour sauce on skins, fried until red, then sliced. Arranged in a bowl skin down, then steamed with red fermented bean curd juice, fermented soybean, ginger, and scallion.

钱江鲈鱼 (qiánjiānglúyú)

浙江菜。先将钱塘江鲈鱼用沸水烫过,然后在鱼身上撒肥肉丁,放火腿、香菇、葱结、姜片,蒸熟。佐姜末与醋。

Steamed perch. Zhejiang dish. Scalded perch from Qiantang River steamed with diced fat pork, ham, dried mushrooms, scallion knots, and sliced ginger. Served with minced ginger and vinegar.

芡实薏仁粥 (qiànshí yìrén zhōu)

香港点心。先将芡实和薏米煮烂,再加入瘦肉和槟榔,用盐调味。

Gorgon euryale and seeds of Job's tears porridge. Hong Kong snack. Gorgon euryale seeds and seeds of Job's tears boiled into porridge with lean pork and betel nuts, then seasoned with salt.

炝蛏鼻 (qiàngchēngbí)

江苏菜。先将蛏鼻用白酒、盐、姜末等腌渍,再炝炒,烹入酱油、香醋,撒上白胡椒粉和香菜叶。

Stir-fried razor clam. Jiangsu dish. Pipe noses of razor clam pickled with Chinese liquor, salt, and minced ginger, stir-fried, flavored with soy sauce and vinegar, then sprinkled with ground white pepper and coriander leaves.

炝鸡丝兰片 (qiàng jīsī lánpiàn)

东北菜。先将鸡脯肉丝上浆,入油锅划散,然后与焯过的玉兰片丝、海米及调料拌匀。

Chicken breast with bamboo shoots. Northeastern dish. Shredded chicken breast coated with starch paste, lightly fried, then mixed with scalded shredded soaked bamboo shoots, dried shrimp meat, and seasonings.

炝三丝 (qiàngsānsī)

东北菜。将土豆丝、海带丝、干豆腐丝分别焯好,加盐、姜、蒜、花椒油炒匀。

Potatoes with kelps and bean curd. Northeastern dish. Shredded potatoes, kelp, and dried bean curd scalded separately, then sautéed with salt, ginger, garlic, and Chinese prickly ash oil.

炝糟五花肉 (qiàngzāo wǔhuāròu)

南方家常菜。将五花肉煮至半熟,再加葱、蒜茸、辣椒、红糟等炝炒。

Stir-fried streaky pork. Home-style

dish in southern areas. Streaky pork half-boiled, then stir-fried with scallion, garlic, hot pepper, and sweet fermented sticky rice juice.

荞面条 (qiáomiàntiáo)

内蒙食品。把荞麦面条用羊肉汤煮熟,加香菜、葱、酱油、醋等调料。

Buckwheat noodles in mutton soup. Inner Mongolian food. Buckwheat noodles boiled in mutton soup and flavored with coriander, leek, soy sauce, and vinegar.

巧炒豆渣 (qiǎochǎo dòuzhā)

山东家常菜。先将姜、葱炒香,放入大豆渣炒匀,然后加盐、胡椒、鸡精、酱油、青豆。

Stir-fried bean pomace. Home-style dish in Shandong. Bean pomace stir-fried with quick-fried ginger and scallion, then flavored with salt, pepper, chicken essence, soy sauce, and green beans.

巧手长寿汤 (qiǎoshǒu chángshòutāng)

湖南菜。先将面粉炒熟,再加入南瓜、胡萝卜、玉米、青豆煮成汤。

Corn and vegetable soup. Hunan dish. Roasted flour boiled into soup with pumpkin, carrots, corn, and green beans.

茄汤焗香鸡 (qiétāng júxiāngjī)

浙江菜。先将鸡块炸至金黄,与爆香的葱花、蒜茸、番茄酱拌匀,然后用箔纸包好,焗透。

Broiled fried chicken with tomato sauce. Zhejiang dish. Chicken fried until golden, mixed with quick-fried scallion, garlic, and tomato sauce, wrapped in foil, then broiled.

茄汁石斑鱼 (qiézhī shíbānyú)

浙江菜。将石斑鱼肉裹上生粉,炸至金黄,淋上用番茄汁调制的酱。

Fried grouper with tomato sauce. Zhejiang dish. Grouper coated with starch, deep-fried, then marinated in tomato sauce.

茄汁虾球 (qiézhī xiāqiú)

广东菜。先将腌好的虾仁与葱段同炒,再加入用茄汁、麻油、胡椒粉等调成的汁炒匀。

Sautéed shrimp meat. Guangdong dish. Pickled shrimp meat sautéed with scallion, then flavored with gravy of tomato sauce, sesame oil, and ground black pepper.

茄子煮花甲 (qiézi zhǔ huājiǎ)

山东菜。先将茄子煮熟,然后加入煸炒过的蛤、盐、味精、白糖、红辣椒,淋麻油。

Boiled eggplants and shellfish. Shandong dish. Boiled eggplants combined with quick-fried shellfish, salt, MSG, sugar, red pepper, and sesame oil.

芹菜香干肉丝 (qíncài xiānggān ròusī)

江浙皖沪家常菜。先将肉丝煸过,再与姜丝、泡辣椒、香干、芹菜同炒,用酱油、味精调味。

Sautéed celery and bean curd. Home-style dish in Jiangsu, Zhejiang, Anhui, and Shanghai. Quick-fried shredded pork sautéed with ginger, pickled pepper, dried bean curd, and celery, then flavored with soy sauce and MSG.

芹香牛肉 (qínxiāng niúròu)

广东菜。将腌渍过的牛肉与芹菜、红辣椒、豆瓣酱、麻油、胡椒粉翻炒。

Beef stir-fried with celery. Guangdong dish. Pickled beef stir-fried with celery, red pepper, fermented broad bean sauce, sesame oil, and ground black pepper.

青椒肉丝 (qīngjiāo ròusī)

各地家常菜。先将肉丝用酱油芡汁腌渍,然后与青椒、姜、葱爆炒。

Stir-fried pork and green pepper. Home-style dish in many places. Shredded lean pork pickled with soy sauce and starch paste, then stir-fried with green pepper, ginger, and scallion.

青椒瘦肉土豆丝 (qīngjiāo shòuròu tǔdòusī)

各地家常菜。先将腌渍过的肉丝爆炒,然后放入土豆丝和青椒丝炒匀,加醋和鸡精。

Sautéed green pepper, pork, and potatoes. Home-style dish in many places. Quick-fried shredded pork sautéed with shredded potatoes and green pepper, then flavored with vinegar and chicken essence.

青椒兔柳 (qīngjiāo tùliǔ)

东北菜。先将兔肉切成柳叶片,挂蛋清芡粉浆,入油锅划散,再与青椒片及调料翻炒,勾芡。

Rabbit meat with green pepper. Northeastern dish. Rabbit meat cut into narrow strips, coated with starched egg white, fried over low heat, stir-fried with green pepper and seasonings, then thickened with starch.

青椒蒸香菌 (qīngjiāo zhēng xiāngjūn)

南方家常菜。将青椒和香菇加鸡油及调料拌匀,蒸熟。

Steamed green pepper and mushrooms. Home-style dish in southern areas. Green pepper and dried mushrooms mixed with chicken oil and flavorings, then steamed.

青椒盅 (qīngjiāozhōng)

四川菜。先将青椒汆过,再填入蔬菜肉末馅,撒上盐和胡椒粉,浇猪油,烤熟。

Roasted stuffed green pepper. Sichuan dish. Scalded green pepper stuffed with vegetables and minced meat, sprinkled with salt and ground black pepper, washed with lard, then roasted.

青稞酒 (qīngkējiǔ)

水酒,产于西藏和青海。用青稞酿制,酒精含量一般在 20 度以下。

Qingke wine produced in Qinghai and Tibet. Light alcohol drink made with highland barley. Its alcohol content is usually under 20%.

青稞糌粑 (qīngkē zānbā)

藏族传统主食。将青稞炒熟成花,与糌粑粉拌匀。

Roasted barley with ground barley. Traditional food of Tibetan ethnic minority. Baked Tibetan pop barley mixed with baked ground Tibetan barley.

青蒜炒腊肉 (qīngsuàn chǎo làròu)

云贵川湘家常菜。将蒜薹、腊肉、豆豉、辣椒等爆炒。

Stir-fried smoked pork and garlic stems. Home-style dish in Yunnan, Guizhou, Sichuan, and Hunan. Garlic stems stir-fried with smoked cured pork, fermented soybean, and chili pepper.

青团 (qīngtuán)

江浙沪点心。用青艾汁调糯米粉做成小圆团,包入用甜豆沙、甜芝麻粉等做成的馅,蒸熟。

Stuffed green sticky rice balls. Jiangsu, Zhejiang, and Shanghai snack. Sticky rice flour and green wormwood soup made into round buns stuffed with sweet bean paste or sweet sesame powder, then steamed.

青鱼下巴甩水 (qīngyú xiàba shuǎi-shuǐ)

上海菜。先将青鱼的嘴和尾油煎,然后放入肉汤,加姜末、酱油、糖焖煮。

Braised black carp chins and tails. Shanghai dish. Black carp chins and tails deep-fried, then braised in pork soup with minced ginger, soy sauce, and sugar.

清煲蛇汤 (qīngbāo shétāng)

澳门菜。将蛇肉、甘蔗段、龙眼、姜等加盐、料酒炖煮。

Snake, cane, and longan soup. Macao dish. Snake meat braised with sugar cane sections, longan pulp, and ginger, then seasoned with salt and cooking wine.

清炒蛏子 (qīngchǎo chēngzi)

上海菜。将葱段与蛏子翻炒,烹入黄酒和盐,勾芡。

Stir-fried razor clam. Shanghai dish. Razor clam stir-fried with scallion sections, flavored with yellow wine and salt, then thickened with starch.

清炒海蚌 (qīngchǎo hǎibàng)

福建菜。先将腌渍过的蚌肉爆至近熟,再将青菜心和香菇煸炒,然后放进蚌肉翻匀,勾芡。

Sautéed clam with baby greens and dried mushrooms. Fujian dish. Half fried pickled clam meat sautéed with quick-fried baby greens and dried mushrooms, then thickened with starch.

清炒金针菇 (qīngchǎo jīnzhēngū)

南方菜。将金针菇用素油清炒,用盐、姜、葱、酱油调味。

Sautéed golden mushrooms. Southern dish. Golden mushrooms sautéed with vegetable oil, salt, ginger, scallion, and soy sauce.

清炒芦笋 (qīngchǎo lúsǔn)

南方菜。先煸炒葱花,再放入芦笋及调料翻炒,勾芡。

Sautéed asparagus. Southern dish. Asparagus sautéed with lightly fried scallion chops and other condiments, then thickened with starch.

清炒驼峰丝 (qīngchǎo tuófēngsī)

宁夏菜。先将驼峰丝汆透,然后与芫荽、辣椒及调料翻炒。

Stir-fried camel hump. Ningxia dish. Boiled shredded camel hump stir-fried with coriander, hot pepper, and other flavorings.

清炒羊肝 (qīngchǎo yánggān)

东北菜。先将羊肝切成薄片焯水,然后加青椒、姜、葱、黄酒、盐翻炒。

Stir-fried lamb liver. Northeastern dish. Scalded sliced lamb liver stir-fried with green pepper, ginger, scallion, yellow wine, and salt.

清炒鱿鱼卷 (qīngchǎo yóuyújuǎn)

各地家常菜。先将鱿鱼片焯水至起卷,然后加调料翻炒。

Stir-fried squid. Home-style dish in many places. Sliced squid scalded until curly, then stir-fried with seasonings.

清汆赤鳞鱼 (qīngcuān chìlínyú)

山东菜。先将泰山赤鳞鱼汆熟,再将清汤加花椒、黄酒等煮沸,浇在鱼上。

Red squama fish in clear soup. Shandong dish. Red squama fish from Mount Taishan quick-boiled, then marinated in clear soup flavored with Chinese prickly ash and yellow wine.

清炖鸡块 (qīngdùn jīkuài)

各地家常菜。鸡块加葱、姜、料酒、盐、味精,小火炖煮。

Stewed chicken. Home-style dish in many places. Cubed chicken stewed with scallion, ginger, cooking wine, salt, and MSG.

清炖石斛螺 (qīngdùn shíhú luó)

广东菜。将石螺肉与猪里脊肉、石斛加盐煲炖。

Stewed gray snails with herb. Guangdong dish. Gray snail meat, pork tenderloin, and noble dendrobium stem stewed, then seasoned with salt.

清炖羊肉 (qīngdùn yángròu)

新疆菜。羊肉和萝卜用清汤炖熟,加盐、料酒、洋葱、姜调味,配香菜、醋、胡椒粉、麻油等调成的卤汁。

Stewed mutton and turnips. Xinjiang dish. Mutton and turnips stewed in clear soup, flavored with salt, cooking wine, onion, and ginger, then served with sauce of coriander, vinegar, ground black pepper, and sesame oil.

清烩海参 (qīnghuì hǎishēn)

广东菜。先将汆过的海参切片,再与嫩豌豆角、胡萝卜、笋片等入高汤加调料烩煮,勾芡。

Braised sea cucumbers with vegetables. Guangdong dish. Scalded sliced sea cucumbers braised with snow peas, carrots, and sliced bamboo shoots in broth, seasoned, then thickened with starch.

清烩鲈鱼片 (qīnghuì lúyúpiàn)

江苏菜。先将韭黄、荸荠、木耳等入

鲈鱼骨鲜汤烧沸，然后倒入鱼片汆熟，放胡椒粉，撒香菜叶。

Perch fillets in soup. Jiangsu dish. Perch fillets scalded in seasoned soup of yellow Chinese chive, water chestnuts, wood ears, and perch bones. Flavored with ground black pepper and coriander.

清水鱼 (qīngshuǐyú)

各地家常菜。将鱼入清水煮熟，加盐、醋、料酒，撒葱丝、香菜，浇特制的卤汁。

Boiled fish with sauce. Home-style dish in many places. Fish boiled with salt, vinegar, and cooking wine, covered with shredded scallion and coriander, then washed with special sauce.

清汤鲍鱼 (qīngtāng bàoyú)

山东菜。将鲍鱼片与笋片、菜心等入清汤煮沸。

Abalone in clear soup. Shandong dish. Sliced abalone boiled in clear soup with sliced bamboo shoots and baby Chinese cabbage.

清汤柴把鸭 (qīngtāng cháibǎyā)

湖南菜。将鲜鸭肉与熟火腿入鸡汤蒸炖，加入香菇、玉兰片、青笋、熟猪油。

Stewed duck and ham. Hunan dish. Duck and ham stewed in chicken soup with dried mushrooms, sliced soaked bamboo shoots, stem lettuce, and lard.

清汤柳叶燕菜 (qīngtāng liǔyè yàncài)

山东菜。用燕窝、火腿、香菜、鸽蛋入清汤煮熟。

Pigeon egg and bird's nest soup. Shandong dish. Edible bird's nests boiled in clear soup with ham, coriander, and pigeon eggs.

清汤龙须菜 (qīngtāng lóngxūcài)

山东菜。将清汤加盐和黄酒，放入龙须菜煮沸。

Asparagus soup. Shandong dish. Asparagus boiled in clear soup seasoned with salt and yellow wine.

清汤泡糕 (qīngtāng pàogāo)

江西菜。先用面皮包入咸味肉馅，入清汤煮熟，再放入甜桂花糕。

Salty dumplings with sweet cakes. Jiangxi dish. Flour sheets made into dumplings filled with salty ground pork, then boiled in clear soup. Served with sweet osmanthus flower cakes.

清汤全家福 (qīngtāng quánjiāfú)

山东菜。将海参、鱼肚、蹄筋、鱼翅、鸡片、鱼片、虾片汆熟，加清汤、料酒、盐、味精。

Meat varieties in soup. Shandong dish. Sea cucumbers quick-boiled with fish maw, pork tendons, shark fin, sliced chicken, fish fillets, and shrimp chips, then seasoned with broth, cooking wine, salt, and MSG.

清汤肉丸 (qīngtāng ròuwán)

湖南菜。将肉泥与蛋浆、盐、芡粉做成丸子，加入葱、蒜、姜、辣椒炖煮。

Boiled pork meatballs. Hunan dish. Ground pork, egg paste, salt, and starch made into balls, then boiled with scallion, garlic, ginger, and hot pepper.

清汤鸭条 (qīngtāng yātiáo)

山东菜。先将熟鸭肉条和鸭骨蒸透,再将冬菇、玉兰片、菜心等入鲜汤煮沸,浇在鸭条上。

Duck meat in clear soup. Shandong dish. Prepared duck meat strips steamed with duck bones, then covered with winter mushrooms, soaked bamboo shoots, and baby Chinese cabbage boiled in seasoned soup.

清汤鱼丸 (qīngtāng yúwán)

南方菜。将白鲢鱼肉及调料做成丸,煮沸,放入熟笋片、香菇、豆苗。

Silver carp meat balls. Southern dish. Silver carp meat and seasonings made into balls, then boiled with prepared bamboo shoots, dried mushrooms, and bean shoots.

清鲜爽嫩烟筒鸭 (qīngxiān shuǎngnèn yāntǒngyā)

广东菜。将鸡肉茸、虾肉茸、香菇、发菜及调料拌匀,塞入去骨的鸭颈,蒸熟。

Chicken and shrimp in duck neck. Guangdong dish. Duck necks boned, filled with mixture of minced chicken, mashed shrimp meat, dried mushrooms, black moss, and seasonings, then steamed.

清炸黄河刀鱼 (qīngzhá huánghédāoyú)

山东菜。将黄河刀鱼腌渍,裹上面粉,炸至金黄,佐椒盐。

Fried saury with spicy salt. Shandong dish. Yellow River saury pickled, coated with flour, then deep-fried. Served with spicy salt.

清蒸八宝甲鱼 (qīngzhēng bābǎo jiǎyú)

山东菜。将海参、猪肉、蛋糕、冬笋等八种原料填入甲鱼腹,蒸烂。

Steamed turtle with eight treasures. Shandong dish. Soft-shell turtle stuffed with eight ingredients, such as sea cucumbers, pork, eggs cake, and winter bamboo shoot, then steamed.

清蒸大闸蟹 (qīngzhēng dàzháxiè)

上海菜。将湖蟹蒸熟,佐姜醋。

Steamed lake crab with vinegar. Shanghai dish. Lake crab steamed. Served with ginger vinegar.

清蒸刀鱼 (qīngzhēng dāoyú)

江苏菜。在刀鱼上放竹笋、火腿、冬菇及调料,浇上烧滚的鸡汤,蒸熟。

Steamed saury. Jiangsu dish. Saury covered with bamboo shoots, ham, winter mushrooms, and seasonings, washed with boiling chicken soup, then steamed.

清蒸黄河鲤 (qīngzhēng huánghélǐ)

河南菜。在黄河鲤鱼腹中填入用猪肉、酱油、麻油、姜、香菇调制的馅,蒸熟,淋卤汁。

Steamed stuffed carp. Henan dish. Yellow River carp stuffed with mixture of pork, soy sauce, sesame oil, ginger, and dried mushrooms, steamed, then drenched with special sauce.

清蒸鸡 (qīngzhēngjī)

东北菜。先将鸡加清汤及调料蒸烂,然后将兰片和油菜入鸡汤烧开,浇在鸡上。

Steamed chicken. Northeastern dish. Chicken steamed in seasoned clear soup, then covered with sliced bamboo shoots and rape greens boiled in chicken soup.

清蒸九孔鲍 (qīngzhēng jiǔkǒngbāo)

海南菜。将九孔鲍鱼放回原壳,撒上碎香菜和姜末,浇少量色拉油,急火蒸熟,撒盐。

Steamed nine-eye abalone. Hainan dish. Nine-eye abalone put in shells, topped with minced coriander and ginger, washed with a little sald oil, steamed over high heat, then sprinkled with salt.

清蒸鲈鱼 (qīngzhēng lúyú)

沿海一带家常菜。先将鲈鱼稍腌,再把葱丝、姜丝、冬菇丝等放在鱼身上,急火蒸熟。

Steamed perch. Home-style dish in coastal areas. Lightly pickled perch covered with shredded scallion, ginger, and winter mushrooms, then steamed over high heat.

清蒸石斑鱼 (qīngzhēng shíbānyú)

南方家常菜。先将石斑鱼用盐腌渍,然后加姜、葱、酱油、味精等蒸熟,淋麻油。

Steamed grouper. Home-style dish in southern areas. Grouper pickled with salt, steamed with ginger, scallion, salt, soy sauce, and MSG, then flavored with sesame oil.

清蒸鲥鱼 (qīngzhēng shíyú)

江苏名菜。先将鲥鱼氽过,再将火腿片、香菇片、笋片、猪网油铺放在鱼身上,加入鸡清汤和调料,蒸熟,然后撒上胡椒粉和香菜。

Steamed reeves shad. Famous Jiangsu dish. Scalded reeves shad covered with sliced ham, mushrooms, bamboo shoots, and webbed lard, steamed in clear chicken soup, then flavored with ground black pepper and coriander.

清蒸武昌鱼 (qīngzhēng wǔchāngyú)

湖北菜。先将鱼用盐腌渍,再放入黄酒、葱结、姜、冬笋,蒸熟,然后浇上加调料的鸡汤。

Steamed Wuchang carp. Hubei dish. Local carp pickled, steamed with yellow wine, knotted scallion, ginger, and winter bamboo shoots, then washed with seasoned chicken soup.

清蒸西瓜鸡 (qīngzhēng xīguājī)

河南菜。将西瓜掏空,装进用盐、料酒、味精腌过的鸡块,蒸熟。

Chicken steamed in watermelon. Henan dish. Cubed chicken pickled with salt, cooking wine, and MSG, then steamed in a hollowed watermelon.

清蒸鸭饺 (qīngzhēng yājiǎo)

江苏菜。先将鸭肉茸加调料制成馅,再用蛋皮包上鸭肉馅成饺子状,蒸熟。

Steamed duck *jiaozi*. Jiangsu dish. Minced duck meat mixed with seasonings, wrapped into *jiaozi* with egg sheets, then steamed.

鳅鱼附豆腐汤 (qiūyú fù dòufu tāng)

湖南菜。先将鳅鱼酒炝后煸炒,再加入鸡汤、豆腐片炖煮。

River eel with bean curd. Hunan dish. River eel drowned in Chinese liquor, stir-fried, then braised with bean curd in chicken soup.

曲瑞 (qǔruì)

西藏传统食品,即麦片粥。先将羊肉、面粉、干辣椒、干奶酪用文火煮烂,然后加入煮熟的燕麦和青稞,搅匀,稍煮。

Oatmeal porridge. Traditional Tibetan food. Cooked oat and barley added to stewed mutton, flour, chili pepper, and dried cheese. Boiled briefly and mixed well before serving.

全虾三做 (quánxiā sān zuò)

山东菜。先将虾头炆熟,虾尾煎熟,虾腰炸至金黄,再与红辣椒、发菜、冬菇等摆成芙蓉虾扇,蒸熟。

Three-taste prawns. Shandong dish. Prawns cut into 3 sections with heads braised, tails lightly fried, and bodies deep-fried. Arranged like a fan on a platter with red pepper, black moss, and winter mushrooms, then steamed.

全兴大曲 (quánxīngdàqū)

白酒,产于四川成都。以高粱为主料,用小麦制曲酿造,酒精含量为38-60度,属浓香型酒。

Quanxingdagu, liquor produced in Chengdu, Sichuan. White spirit made with sorghum and wheat yeast. Contains 38-60% alcohol, and has a thick aroma.

群龙戏珠 (qúnlóng xìzhū)

海南菜。先将虾身剥壳留尾,再用生鱼片包裹,同虾头、虾胶珠一并炸熟,然后摆成群龙戏珠图案。

Fried prawns with fish fillets. Hainan dish. Prawns shelled save tails, wrapped in fish fillets, then deep-fried with prawn heads and shrimp gelatin balls. Arranged like a flock of dragons playing with balls on a platter.

R

热干面 (règānmiàn)

湖北点心。先将煮熟的面条拌麻油摊晾,然后放在沸水里烫热,加调料。当地的热干面有多种做法。

Sesame oil noodles. Hubei snack. Noodles boiled, processed with sesame oil, then cooled. Scalded with boiling water, then served with flavorings. There are other ways to make sesame oil noodles in Hubei.

热泡泥鳅 (rèpào níqiū)

南方家常菜。先将泥鳅腌渍过油,然后与芹菜、姜、葱、调味酱翻炒。

Stir-fried loach. Home-style dish in southern areas. Quick-fried pickled loach stir-fried with celery, ginger, scallion, and special sauce.

人参果奶渣糕 (rénshēnguǒ nǎizhā gāo)

西藏点心。将人参果、酥油、细奶渣、红糖、红枣、葡萄干等调和,煮熟,冷却成糕。

Silverweed root and dairy dreg cakes. Tibetan snack. Silverweed roots Tibetan butter, minced dairy dregs, brown sugar, red dates, and raisins mixed, boiled, then cooled into cakes.

仁丹糕 (réndāngāo)

福建点心。将糯米粉、糖、盐、山楂汁拌和,蒸熟,撒白糖。

Sticky rice and haw cakes. Fujian snack. Sticky rice flour, sugar, salt, and haw soup mixed, steamed into cakes, then showered with white sugar.

日月套三环 (rìyuè tào sānhuán)

河南开封菜。将海参、鱿鱼、玉兰片、鱼肚、蹄筋、香菇及调料装入鸡腹,再把鸡装入鸭腹,然后把鸭装入冬瓜盅,蒸熟。

Steamed chicken and duck in gourd. Dish in Kaifeng, Henan. Duck stuffed with a whole chicken that is stuffed with sea cucumbers, squid, soaked bamboo shoots, fish maw, pork tendons, dried mushrooms, and seasonings. Put in a wax gourd, then steamed.

茸汤广肚 (róngtāng guǎngdǔ)

福建菜。广肚即干鱼肚。先将水发广肚入清水文火炖煮,然后入烧沸的上汤浸泡,用黄酒和味精调味。

Marinated fish maw. Fujian dish. Soaked fish maw stewed in water, then marinated in boiling broth seasoned with yellow wine and MSG.

肉末炝泡菜 (ròumò qiàng pàocài)

四川菜。先将肉末与干辣椒煸炒，然后加入泡菜翻炒，加盐、酱油。

Stir-fried ground pork and pickled vegetables. Sichuan dish. Ground pork quick-fried with chili pepper, then sautéed with pickled vegetables, salt, and soy sauce.

肉片鲜蘑 (ròupiàn xiānmó)

江浙沪菜。先把鲜蘑菇片炒熟，再把猪肉片加调料炒熟并勾芡，然后把肉片浇在蘑菇片上。

Stir-fried pork and mushrooms. Jiangsu, Zhejiang, and Shanghai dish. Sliced pork stir-fried with seasonings, thickened with starch, then poured over sautéed sliced fresh mushrooms.

肉蓉鹿茸鸡汤 (ròuróng lùróng jītāng)

湘式药膳。将肉苁蓉、熟地、菟丝子、山萸肉、远志煎成的药汤冲入用鸡肉和鹿茸炖成的汤中。

Chicken and herb soup. Folk-medicinal dish in Hunan. Chicken and young stag antler soup combined with herb tea made with saline cistanche, prepared rehmannia root, dodder seeds, dogwood fruit pulp, and milkwort root.

肉丝兰片 (ròusī lánpiàn)

各地家常菜。将猪肉丝和兰片加盐、黄酒、味精等炒熟。

Sautéed pork and bamboo shoots. Home-style dish in many places. Sliced pork sautéed with sliced bamboo shoots, then flavored with salt, yellow wine, and MSG.

肉松 (ròusōng)

福建、台湾菜。先将红糟、黄酒、糖、盐略炒，然后放入焯过的瘦肉炒透，加入清汤，慢烧至肉发松，汁收干。

Pork floss. Fujian and Taiwan dish. Scalded lean pork sautéed with quick-fried red fermented sticky rice juice, yellow wine, sugar, and salt. Simmered in clear soup. Reduced until dry and the meat becomes floss.

如意金砖 (rúyìjīnzhuān)

福建菜。先将炒熟的菜胆放在炸好的豆腐块之间，再加入黄耳、榆耳煨煮，然后调味、勾芡。

Braised bean curd with mushrooms. Fujian dish. Prepared mustard tender put between deep-fried cubed bean curd, simmered with yellow mushrooms and elm mushrooms, then flavored with seasonings and starched soup.

如意双味卷 (rúyì shuāngwèijuǎn)

广东点心。将薄面皮摊开，两端各放一条枣泥、芋泥，分别向内卷起，做成如意云形卷，炸至金黄。

Taro and date rolls. Guangdong snack. Flour sheets made into rolls filled with mashed taros from one end and mashed dates from the other end, made in the shape of a curved cloud symbolizing good luck, then deep-fried.

如意鸭卷鲜 (rúyì yājuǎnxiān)

上海菜。先将蛋皮抹上虾茸，从两

头向中间对卷成如意云状，再蒸熟，切成厚片，与鸭块装碗，然后将蛋卷鸭块连碗入笼蒸透，淋鸡汤。

Good-luck duck rolls. Shanghai dish. Egg sheets spread with minced shrimp meat, rolled from two ends to the middle in the shape of a curved cloud symbolizing good luck, then steamed. Sectioned into big pieces, put in a bowl with chopped duck, re-steamed, then washed with chicken soup.

汝南风干兔肉（rǔnán fēnggān tùròu）

河南汝南菜。先将野兔肉过油，再入卤汤煨煮，然后自然风干。

Air-dried hare. Specialty of Runan, Henan. Hare meat fried, marinated in gravy, then air-dried.

汝南鸡汁豆腐干（rǔnán jīzhīdòufugān）

河南汝南菜。先将豆腐干放入鸡汤、酱油、麻油制成的卤汤中煮熟，然后晒干。

Marinated dried bean curd. Specialty of Runan, Henan. Dried bean curd marinated in gravy of chicken broth, soy sauce, and sesame oil, then air-dried.

乳牛肉（rǔniúròu）

云南菜。将乳牛肉块用清水炖至皮烂，切片，用辣椒粉、盐、酱油、薄荷、花椒等调成的卤汁佐食。

Stewed calf with spicy sauce. Yunnan dish. Calf chunks stewed in water till skins soft, then sliced. Served with dressing of chili powder, salt, soy sauce, peppermint, and Chinese prickly ash.

乳香吊烧鸡（rǔxiāng diàoshāojī）

广东菜。先将鸡用沸水略烫，再涂上烧烤酱，炸至金黄。

Deep-fried chicken. Guangdong dish. Quick-boiled chicken coated with barbecue sauce and deep-fried.

乳猪大拼盘（rǔzhū dàpīnpán）

广东名菜。将熟乳猪、油鸡、烧鸭、叉烧、海蜇皮等切块，铺在甜豆上，用香菜和萝卜花装饰。

Assorted meat plate. Famous Guangdong dish. Broiled baby hog, fried chicken, broiled duck, barbecued pork, and jellyfish chopped, then put on a bed of snow peas. Garnished with coriander and carrot cuts.

软熘珠廉鱼（ruǎnliū zhūliányú）

福建菜。将腌过的黄鱼片裹黄鱼茸成卷状，裹上蛋清淀粉糊，蒸熟。

Steamed yellow croaker. Fujian dish. Pickled yellow croaker fillets made into rolls filled with mashed yellow croaker, coated with starched egg white, then steamed.

软熘鲤鱼（ruǎnliū lǐyú）

北京菜。先将鲤鱼稍炸，然后加汤、黄酒、醋、糖等焖至汁浓。

Simmered carp. Beijing dish. Carp lightly fried, then simmered in soup with yellow wine, vinegar, and sugar until reduced.

软炸鸡（ruǎnzhájī）

河北家常菜。将鸡块炸至深黄色，

佐椒盐。

Fried chicken with spicy salt. Home-style dish in Hebei. Cubed chicken deep-fried, then served with spicy salt.

软炸里脊 (ruǎnzhá lǐji)

山东菜。将腌渍好的里脊肉片裹上蛋糊炸熟，淋麻油。

Fried pork fillets. Shandong dish. Pickled pork tenderloin slices coated with egg and starch paste, fried, then sprinkled with sesame oil.

润饼 (rùnbǐng)

福建、台湾点心，亦称春饼。面粉或米粉加水调成浆，烙成薄饼，卷入炒好的春笋、胡萝卜、高丽菜、韭黄、绿豆芽、香菇、海苔、猪肉、虾仁、蛋皮等馅料。

Vegetable rolls. Fujian and Taiwan snack. Rice or wheat flour and water mixed into paste, then fried into thin pancakes over low heat. Rolled in cooked spring bamboo shoots, carrots, baby Chinese cabbage, yellow Chinese chive, mung bean sprouts, dried mushrooms, laver, pork, shrimp meat, and eggs skins.

S

萨其马 (sàqímǎ)

南方点心。先将鸡蛋面糊油炸成条状，再与特制的糖浆混合，模压，切块。

Sweet flour strip cookie. Southern snack. Egg and flour paste deep-fried into crisp strips, mixed with special syrup, put in a mould box, pressed, then cut into cubes.

萨吾尔干锅焖羊腿 (sàwúěr gānguō mènyángtuǐ)

新疆菜。用羔羊前腿加土豆、胡萝卜等焖制。

Braised lamb foreleg with vegetables. Xinjiang dish. Lamb forelegs braised with potatoes and carrots.

赛螃蟹 (sàipángxiè)

北方菜。先将葱、姜煸香，再倒入鸭蛋黄和鱼肉文火翻炒，淋麻油。

Sautéed duck egg yolks with fish. Northern dish. Fish meat and duck egg yolks sautéed with quick-fried scallion and ginger over low heat, then sprinkled with sesame oil.

三不沾 (sānbùzhān)

河南安阳菜。将鸡蛋黄加芡粉、白糖、桂花糖水、盐等调匀，烘炒至深黄色呈糕状。

Sweet egg yolk pudding. Specialty in Anyang, Henan. Egg yolks mixed with starch, white sugar, sweet osmanthus flower soup, and salt, then fried yellow-brown.

三彩菠菜 (sāncǎi bōcài)

浙江菜。将烫过的菠菜、粉丝、蛋皮丝加调料拌匀，撒上虾米。

Three-color spinach. Zhejiang dish. Quick-boiled spinach mixed with starch noodles and shredded egg sheets, seasoned, then garnished with prepared dried shrimp meat.

三彩大虾 (sāncǎi dàxiā)

山东菜。先将比目鱼茸和猪肉末调成馅，抹在虾肉片上，再放上香菜、火腿、冬菇，制成三色大虾，然后炸至金黄。

Three-color prawns. Shandong dish. Sliced prawn meat layered with mixture of minced flatfish and ground pork, garnished with coriander, ham, and winter mushrooms, then deep-fried.

三丁大包 (sāndīng dàbāo)

江苏点心。将发酵面团和肉丁、鸡丁、笋丁拌成的馅做成包子，蒸熟。

Stuffed big buns. Jiangsu snack. Fermented flour dough made into big buns, stuffed with diced pork, chicken, and bamboo shoots, then steamed.

三合汤 (sānhétāng)

湖北郧阳点心。先将熟牛肉片和红薯粉烫熟,然后加入调料、熟水饺、鲜汤。

Beef, noodles, and *jiaozi* soup. Specialty of Yunyang, Hubei. Scalded cooked sliced beef, yam noodles, seasonings, and cooked *jiaozi* put in seasoned boiling soup.

三花酒 (sānhuājiǔ)

白酒,产于广西桂林。以大米为原料,用香草药曲酿造,酒精含量在45-58度之间,属米香型酒。

Sanhuajiu, liquor produced in Guilin, Guangxi. White spirit made with rice and fragrant herb yeast. Contains 45-58% alcohol, and has a rice aroma.

三皮丝 (sānpísī)

广东凉菜。将熟乌鸡皮、熟猪皮、海蜇皮用葱、花椒、酱油、芝麻酱等拌匀,配熟火腿丝装盘。

Skin varieties with sauce.

Guangdong cold dish. Cooked chicken skins, hog skins, and jellyfish mixed with scallion, Chinese prickly ash, soy sauce, and sesame sauce. Served with prepared shredded ham.

三仁烧鸡翅 (sānrén shāo jīchì)

山东菜。先将鸡翅用高汤、料酒等烧熟,再放入炸杏仁、炸核桃仁和炸花生仁,然后勾芡。

Chicken wings with nuts. Shandong dish. Chicken wings braised in broth and cooking wine. Fried almonds, walnuts, and peanuts added. Thickened with starch.

三色百叶 (sānsè bǎiyè)

湖南菜。先将焯过的牛百叶用鸡汤煨透,然后与香菇、红椒及调料爆炒。

Three-color ox tripe. Hunan dish. Scalded ox tripe simmered in chicken soup, then stir-fried with dried mushrooms, red pepper, and seasonings.

三色明虾片 (sānsè míngxiāpiàn)

山东菜。先将明虾切片,轻炒,烹入料酒和上汤,再分成3份,将其中两份分别加番茄酱、咖喱拌炒,然后三份同装一盘。

Three-color prawn meat. Shandong dish. Sliced prawn meat lightly fried and flavored with cooking wine and seasoned soup. Divided into 3 parts. One part sautéed with tomato sauce. One sautéed with curry. Served on the same platter.

三色沙虫 (sānsè shāchóng)

广东菜。先将沙虫油炸,再同青、红椒丝及调料翻炒,勾芡。

Stir-fried sea worm with pepper. Guangdong dish. Fried sea worms stir-fried with shredded green and red pepper and seasonings, then thickened with starch.

三丝拌蛏 (sānsī bàn chēng)

浙江菜。先将鲜蛏肉汆熟,再将韭菜和香菇焯水,然后将蛏肉和韭菜、香菇、火腿丝加调料拌匀。

Razor clam with chive and mushrooms. Zhejiang dish. Fresh razor clam meat, Chinese chive, and mushrooms quick-boiled, then mixed with shredded ham and seasonings.

三丝玻璃 (sānsī bōlí)

台湾菜。先将火腿、瘦肉、冬菇切丝,加盐和生粉拌匀,再用冬瓜片裹丝成卷,蒸熟,浇芡汁。

Wax gourd rolls with meat and mushrooms. Taiwan dish. Sliced wax gourd made into rolls stuffed with mixture of shredded ham, lean pork, winter mushrooms, salt, and starch. Steamed, then washed with starched soup.

三丝敲鱼 (sānsī qiāo yú)

浙江菜。先将脘鱼切片、敲打、焯水,然后加菜心及调料煮熟,放入香菇丝、熟鸡脯丝、熟火腿丝。

Boiled carp fillets. Zhejiang dish. Carp fillets pounded gently, scalded, then boiled with baby greens and flavorings. Garnished with shredded mushrooms, prepared chicken breast, and ham.

三丝燕菜 (sānsī yàncài)

江苏菜。先将燕窝蒸熟,放在蒸熟的鸡腿丝、火腿丝、冬菇丝上面,然后浇入鸡清汤。

Steamed bird's nests with meat. Jiangsu dish. Edible bird's nests steamed, put on a bed of shredded steamed drumsticks, ham, and dried mushrooms, then washed with clear chicken soup.

三套鸡 (sāntàojī)

山东菜。先将鸽子装入童子鸡腹,再将童子鸡装入母鸡腹,加佐料煮沸,改小火焖烂。

Braised hen, young chicken, and pigeon. Shandong dish. Hen stuffed with a young chicken over a pigeon. Boiled with seasonings, then braised.

三套鸭 (sāntàoyā)

江苏扬州菜。先将鸽子塞入野鸭腹内,再将野鸭塞入家鸭腹中,然后用冬菇、火腿片、冬笋片等填满鸭腹,稍烫,加调料焖熟。

Braised duck with pigeon. Specialty in Yangzhou, Jiangsu. Duck stuffed with a wild duck over a pigeon. Ducks filled fully with winter mushrooms, ham, and winter bamboo shoots. Quick-boiled, then braised with seasonings.

三味鸡丝 (sānwèi jīsī)

南方家常菜。将熟鸡肉切丝,佐怪味汁、红油、姜汁。

Three-taste chicken. Home-style dish in southern areas. Shredded cooked chicken served with 3 kinds of dressing, savoury dressing, chili oil, and ginger juice.

三下锅 (sānxiàguō)

山东菜。先将甜面酱、八角等调料煸炒,再放进猪肉、鸡块、清汤、黄酒、盐等炖煮,然后加入胡萝卜焖熟。

Braised pork, chicken, and carrot. Shandong dish. Pork and chicken stewed in clear soup with quick-fried sweet fermented flour sauce, aniseed, yellow wine, and salt, then braised with carrots.

三鲜豆腐 (sānxiān dòufu)

山东菜。将豆腐、大虾、海参、鲜蘑入高汤煨煮,用黄酒、姜、葱、盐调味。

Braised bean curd with three fresh things. Shandong dish. Bean curd braised in broth with prawns, sea cucumbers, and fresh mushrooms. Seasoned with yellow wine, ginger, green onion, and salt.

馓子 (sǎnzi)

各地清真点心。先将面粉、清油、盐、花椒水调和,搓成细长条,再将条做成并列环状,油煎至黄脆。

Fried crisp cookie. Muslim snack in many places. Flour, vegetable oil, salt, and Chinese prickly ash soup mixed, kneaded into strings, arranged into connected rings, then deep-fried until golden and crisp.

桑康帕里 (sāngkāngpàlǐ)

西藏点心,即油炸面皮。将面粉与酥油调和,擀成面皮,切块,用酥油或菜油炸脆。

Fried flour and butter pieces. Tibetan snack. Flour and Tibetan butter made into sheets, sliced, then deep-fried with Tibetan butter or rape seed oil until crisp.

臊子烘蛋 (sàozi hōngdàn)

广东菜。将肉末、木耳、榨菜、盐、玉米粉与蛋浆调匀,烘至金黄,浇上用肉末、木耳、榨菜、葱、高汤等炒成的哨子。

Fried eggs with dressing. Guangdong dish. Minced pork, wood ears, preserved tuber mustard, salt, and corn powder mixed with egg paste, baked until golden, then washed with dressing of sautéed minced pork, wood ears, preserved tuber mustard, scallion, and broth.

杀猪菜 (shāzhūcài)

东北菜。将五花肉、血肠、酸菜、粉丝、冻豆腐加调料炖煮。

Stewed pork medley. Northeastern dish. Streaky pork stewed with blood sausage, sour pickled Chinese cabbage, starch noodles, frozen bean curd, and seasonings.

沙茶牛肉 (shāchá niúròu)

广东菜。将牛肉片用上汤烫熟,佐沙茶酱、熟猪油、麻油、辣椒油等调成的卤汁。

Quick-boiled beef with barbecue sauce. Guangdong dish. Sliced beef quick-boiled in broth. Served with dressing of barbecue sauce, lard, sesame oil, and chili oil.

沙道观鸡 (shādàoguànjī)

湖北菜。先将鸡块稍煎,再加入干红辣椒焖熟,然后放入土钵加调料炖煮。

Braised spicy chicken. Hubei dish. Chicken lightly fried, braised with chili pepper, then stewed with seasonings in an earthen pot.

沙河王 (shāhéwáng)

白酒,源于安徽界首,以高粱为原料,用小麦、大麦、豌豆、绿豆制曲酿造,酒精含量为 28-50 度,属浓香型酒。

Shahewang liquor produced in Jieshou, Anhui. White spirit made with sorghum and yeast of wheat, barley, peas, and mung beans. Contains 28-50% alcohol, and has a thick aroma.

沙律片皮鸡 (shālǜ piànpíjī)

广东菜。将整鸡涂上烧烤酱,炸至金黄,斩成块,配生菜、芒果肉等装盘。

Chicken with salad. Guangdong dish. Whole chicken coated with barbecue sauce, deep-fried, chopped, then served with salad of lettuce and mangoes.

砂锅豆腐 (shāguō dòufu)

湖南菜。将豆腐块、粉丝、腐竹、胡萝卜块放入砂锅,用鸡汤炖煮。

Bean curd in clay pot. Hunan dish. Bean curd braised in a clay pot with chicken soup and starch noodles, dried bean curd skins, and carrots.

砂锅三味 (shāguō sānwèi)

山东菜。先将去壳熟鸡蛋炸至金黄,再与猪肘和仔鸡加调料煨炖。

Stewed eggs, hog knuckle, and chicken. Shandong dish. Shelled boiled eggs fried until golden, then stewed with hog knuckle, young chicken, and flavorings.

砂锅娃娃菜 (shāguō wáwacài)

湖南菜。将炒过的五花肉与娃娃菜及调料用砂锅焖煮。

Streaky pork with baby turnip leaves. Hunan dish. Stir-fried streaky pork simmered with baby turnip leaves and flavorings in a clay pot.

砂锅鱼头 (shāguō yútóu)

广东菜。先将鱼头用黄酒和酱油略腌,然后与猪肉、冬菇、豆腐、粉皮等加辣豆瓣酱用砂锅煲煮。

Stewed fish head. Guangdong dish. Fish head pickled with yellow wine and soy sauce, then stewed in an earthen casserole with pork, winter mushrooms, bean curd, starch noodles, and hot fermented broad bean sauce.

纱纸钵仔鸽 (shāzhǐbō zǐgē)

广东菜。先将洋葱丝铺在钵底,再将腌渍过的乳鸽放在洋葱上,用芹菜盖上,用绵纸密封蒸熟。

Steamed young pigeon. Guangdong dish. Pickled young pigeon put on a bed of shredded onion in an earthen container, topped with celery, covered with *goupi* tissue paper, then steamed.

山斑鱼豆腐汤 (shānbānyú dòufu tāng)

广东菜。将山斑鱼煎至金黄,加入豆腐和汤烹煮,用盐和胡椒粉调味。

Fish and bean curd soup. Guangdong dish. Deep-fried snakehead fish braised in soup with bean curd. Seasoned with salt and ground black pepper.

山东酥肉 (shāndōngsūròu)

山东菜。先将猪肉片挂浆,过油,然后入鸡汤加调料炖烂。

Stewed pork in chicken soup. Shandong dish. Sliced pork coated with starch and egg paste, quick-fried, then stewed in chicken soup with seasonings.

山东丸子 (shāndōngwánzi)

山东菜。将肉末、鹿角菜、香菜、海米、盐、酱油、蛋浆等调和,做成丸子,蒸熟。

Pork and vegetable balls. Shandong dish. Ground pork, hart-horn vegetable, coriander, dried shrimp meat, salt, soy sauce, and egg paste mixed, made into balls, then steamed.

珊瑚金钩 (shānhú jīngōu)

山东菜。将黄豆芽汆熟,浇上用辣椒、木耳、酱油、黄酒、醋等做成的卤汁。

Soybean sprouts with sauce. Shandong dish. Soybean sprouts quick-boiled, then washed with dressing of red pepper, wood ears, soy sauce, yellow wine, and vinegar.

珊瑚鱼球 (shānhú yúqiú)

福建菜。先将鳜鱼片、青红椒片、香菇片加芡粉及盐拌和,做成鱼球,然后加高汤和调料翻炒,勾芡。

Chinese perch and pepper balls. Fujian dish. Chinese perch fillets mixed with sliced green and red pepper, dried mushrooms, starch, and salt, then made into balls. Stir-fried with broth and flavorings, then thickened with starch.

山鸡炖白蘑 (shānjī dùn báimó)

东北菜。先将鸡块炒变色,然后加白蘑、粉条、盐、黄酒、猪油等炖熟。

Stewed chicken and white mushrooms. Northeastern dish. Cubed chicken quick-fried, stewed with white mushrooms and starch noodles, then seasoned with salt, yellow wine, and lard.

善酿酒 (shànniàngjiǔ)

黄酒,产于浙江绍兴。用存储一至三年的元红酒兑水酿成,酒精含量13-14度。

Shanniang, yellow wine from Shaoxing, Zhejiang. Made with 1-3 year old Yuanhong yellow wine and distilled water. Its alcohol content is 13-14%.

山西过油肉 (shānxī guòyóuròu)

山西菜。先将炸过的肉片加调料煸香,然后与笋片、木耳、菠菜翻炒,勾芡,淋油。

Stir-fried pork with vegetables. Shanxi dish. Deep-fried sliced pork stir-fried with sliced bamboo shoots, wood ears, spinach, and quick-fried seasonings, then washed with starched soup and lard.

山药红枣紫米露 (shānyào hóngzǎo zǐmǐ lù)

广东点心。将山药、红枣、紫米、何首乌、花生熬煮,加红糖。

Chinese yam, date, and purple rice soup. Guangdong snack. Chinese yam, red dates, purple rice, fleeceflower root, and peanuts cooked into chowder, then sweetened with brown sugar.

山药烩鱼头 (shānyào huì yútóu)

西北菜。先将香菇煸过,再放入鱼头、山药、调料,加清水焖煮收汁,勾芡。

Chinese yam with fish head. Northwestern dish. Fish head braised in water with quick-fried dried mushrooms, Chinese yam, and seasonings until reduced, then thickened with starch.

山药丸子 (shānyàowánzi)

台湾菜。先将煮熟的山药捣成泥,与糯米粉揉成面皮,包入用虾皮、干贝、虾仁、香菇及调料制成的馅,做成丸子,蒸熟。

Steamed Chinese yam balls. Taiwan dish. Mashed boiled Chinese yam and sticky rice flour kneaded into balls stuffed with mixture of cooked dried small shrimp, dried scallops, shrimp meat, dried mushrooms, and seasonings, then steamed.

山楂白菜 (shānzhā báicài)

北方凉菜。将白菜丝和山楂糕丝加糖拌匀。

Cabbage with haw. Northern cold dish. Shredded Chinese cabbage and haw cakes mixed with sugar.

山楂核桃茶 (shānzhā hétao chá)

广东饮料。将核桃仁捣烂,与山楂和冰糖煲煮。

Walnut and haw tea. Guangdong drink. Minced walnuts boiled with haws and rock sugar.

山竹牛肉卷 (shānzhú niúròu juǎn)

广东菜。先将鸡蛋摊成蛋皮,然后用蛋皮裹上牛肉、蘑菇、木耳、竹笋丝及调料,蒸熟。

Egg rolls filled with beef. Guangdong dish. Egg sheets made into rolls stuffed with beef, mushrooms, wood ears, shredded bamboo shoots, and seasonings, then steamed.

商芝肉 (shāngzhīròu)

陕西菜。商芝又名柴萁叶。先将商芝加调料煮软,然后盖在肉片上,加鸡汤蒸熟。

Pork steamed with potherb. Shaanxi dish. Potherb is locally called *shangzhi*. Sliced pork covered with boiled seasoned *shangzhi*, then steamed in chicken soup.

上素福袋 (shàngsù fúdài)

西北菜。福袋即豆皮。用豆皮包裹炒过的蔬菜馅,用烫软的芹菜封口成袋,放入高汤炖煮。

Boiled bean curd bag. Northwestern dish. Cooked vegetables filled in bags of dried bean curd skins, sealed with scalded celery leaves, then braised in broth.

上素仙鹤 (shàngsù xiānhè)

浙江菜。将蘑菇、银耳、玉米笋、冬笋、马蹄等填入乳鸽腹,用微火煨熟,配汆过的西蓝花装盘。

Simmered pigeon with vegetables. Zhejiang dish. Young pigeon stuffed with mushrooms, white fungus, baby corn, bamboo shoots, and water chestnuts, then simmered. Served with scalded broccoli.

烧犴鼻 (shāoànbí)

内蒙菜。先将驼鹿鼻焯过,腌渍,再油煎至半熟,然后炖煮。

Stewed moose nose. Inner Mongolian dish. Moose nose scalded, pickled, half-fried, then stewed.

烧凤眼肝 (shāofèngyǎngān)

广东菜。先将猪肝挖孔,用白酒、盐、酱油等浸渍,再将焯熟的猪肥肉条沾盐塞入孔内,烤熟。

Broiled hog liver with fat pork. Guangdong dish. Hog liver holed, pickled with liquor, salt, and soy sauce, filled with scalded shredded salty fat pork, then grilled.

烧鲑鱼头 (shāoguīyútóu)

江浙皖家常菜。将萝卜、香菇、牛蒡置锅底,上放煎过的鲑鱼头,加水和调料炖煮,撒青蒜。

Braised salmon head with vegetables. Home-style dish in Jiangsu, Zhejiang, and Anhui. Fried salmon heads braised with turnips, dried mushrooms, burdock, water, and seasonings, then garnished with garlic sprouts.

烧禾虫 (shāohéchóng)

广东菜。先将禾虫用花生油浸泡,再用盐、蒜、陈皮、黄酒、麻油腌渍,然后蒸熟,浇麻油,撒胡椒粉。

Steamed clam worms. Guangdong dish. Clam worms steeped in peanut oil, pickled with salt, garlic, processed tangerine peels, yellow wine, and sesame oil, steamed, then flavored with sesame oil and ground black pepper.

烧罗汉面筋 (shāo luóhàn miànjīn)

山东菜。将煮软的油炸面筋加入姜、葱、料酒、酱油、盐等煨煮,放入冬菇片和玉兰片,略蒸,浇芡汁、花椒油。

Braised gluten. Shandong dish. Deep-fried gluten softened in boiling water, simmered with ginger, scallion, cooking wine, soy sauce, and salt, then steamed briefly with sliced winter mushrooms and soaked bamboo shoots. Drenched with starched soup and Chinese prickly ash oil before serving.

烧千里风 (shāo qiānlǐfēng)

内蒙菜。将熟羊耳、玉兰片、香菜梗加调料拌炒。

Sautéed lamb ears with bamboo shoots. Inner Mongolian dish. Boiled lamb ears sautéed with soaked bamboo shoots, coriander stalks, and seasonings.

烧生巧 (shāoshēngqiǎo)

江苏菜。将炸好的鳝段加酱油、糖、黄酒焖煮,撒胡椒粉。

Simmered field eel. Jiangsu dish. Fried sectioned field eel simmered and seasoned with soy sauce, sugar, yellow wine, and ground black pepper.

烧五丝 (shāowǔsī)

山东烟台菜。先爆炒葱、姜,烹入黄酒,然后放入腌渍过的海参丝、鱼肚丝、玉兰片丝、鸡丝、猪肉丝翻炒，勾芡。

Stir-fried meat varieties. Specialty in Yantai, Shandong. Pickled shredded sea cucumbers, fish maw, soaked bamboo shoots, chicken, and pork stir-fried with quick-fried scallion, ginger, and yellow wine, then thickened with starch.

烧胸口 (shāoxiōngkǒu)

东北清真菜。将牛、羊胸脯肉用鸡汤烹煮，加入料酒、酱油、盐、白糖，浇芡汁，淋麻油，撒香菜。

Beef and mutton in chicken soup. Muslim dish in northeastern areas. Beef loin and mutton loin boiled in chicken soup, seasoned with cooking wine, soy sauce, salt, and sugar, washed with starched soup and sesame oil, then garnished with coriander leaves.

艄公号子鱼 (shāogōnghàoziyú)

四川菜。先将鱼腌渍，然后撒上红椒丝、姜丝、葱丝等蒸熟，淋麻油。

Steamed spicy fish. Sichuan dish. Pickled fish covered with shredded red pepper, ginger, and scallion. Steamed, then sprinkled with sesame oil.

绍式虾球 (shàoshìxiāqiú)

浙江绍兴菜。先将虾仁裹盐味芡粉浆涨发，再同芡粉蛋浆调匀，徐徐倒入油锅，用中火煎炸，然后用筷子拨松，配香菜装盘，佐葱白、甜面酱食用。

Fried shrimp balls with sauce. Specialty in Shaoxing, Zhejiang. Shelled shrimp pickled with salt and wet starch, coated with starch and egg paste, gently poured into frying pan, then fried into balls over medium heat. Loosened with chopsticks, garnished with coriander, then served with scallion stems and sweet fermented flour sauce.

绍兴黄酒 (shàoxīnghuángjiǔ)

浙江绍兴产各种黄酒的总称。主要品种有元红酒、加饭酒、善酿酒、香雪酒等，酒精含量在14-18度之间。

Shaoxinghuangjiu, rice wine or yellow wine. A general name for all kinds of yellow wines produced in Shaoxing, Zhejiang. Major types are Yuanhong, Jiafan, Shanniang, and Xiangxue. The alcohol contents of these kinds vary from 14-18%.

赊店老酒 (shēdiànlǎojiǔ)

白酒，产于河南南阳赊店。用当地高粱、小麦为原料酿造，酒精含量为32-63度，属浓香型酒。

Shedianlaojiu, liquor produced in Shedian of Nanyang, Henan. White spirit made with local sorghum and wheat. Contains 32-63% alcohol, and has a thick aroma.

蛇羹 (shégēng)

广东菜。将备好的蛇肉丝、鸡肉丝、果子狸肉丝、鳘鱼肚丝、木耳等放入高汤烩煮，用黄酒、盐、味精等调味。

Snake soup. Guangdong dish.

Shredded snake, chicken, civet cat, and cod croaker maw boiled in broth with wood ears. Seasoned with yellow wine, salt, and ginger.

蛇龟鸡益寿汤 (shé guī jī yìshòutāng)

广东菜。先将龟肉、老母鸡、蛇肉与葱、姜、料酒翻炒,再放入汤锅内熬煮,用盐调味。

Stewed snake, turtle, and hen. Guangdong dish. Turtle, mature hen, and snake meat stir-fried with scallion, ginger, and cooking wine, then stewed. Seasoned with salt.

参果炖瘦肉 (shēnguǒ dùn shòuròu)

广东菜。将猪瘦肉、太子参、无花果等用文火炖煮,加盐、黄酒。

Pork and herb soup. Guangdong dish. Lean pork stewed with falsestarwort root and dried figs. Seasoned with salt and yellow wine.

深沪水丸 (shēnhùshuǐwán)

福建、广东点心。用鳗鱼茸或嘉腊鱼茸加精粉、盐等制成丸子,用清汤煮熟。

Fish ball soup. Fujian and Guangdong snack. Eel or red seabream meat mashed, mixed with fine flour and salt, made into balls, then boiled in clear soup.

神奇香巴拉 (shénqí xiāngbālā)

西藏菜。将煮熟的人参果、牛肉、火腿、羊肝、香菇、心里美等摆成峡谷、动物、瀑布、雪山状装盘。

Assorted meat and vegetables. Tibetan dish. Cooked silverweed roots beef, ham, lamb liver, dried mushrooms, and pink turnips arranged on a platter like valleys, animals, waterfalls, and snow mountains.

参茸熊掌汤 (shēnróng xióngzhǎng tāng)

湖南菜。将熊掌、瘦猪肉、鸡肉、人参、鹿茸熬炖,用蜂蜜、姜、葱等调味。

Bear paw and other treasures in soup. Hunan dish. Bear paw stewed with lean pork, chicken, ginseng, and young stag antler, then flavored with honey, ginger, and scallion.

神仙蛋 (shénxiāndàn)

江苏菜。将煮熟的鸡蛋磕开一小孔,取出蛋黄,填入猪肉末,用淀粉糊封口,先炸后蒸。

Fried stuffed eggs. Jiangsu dish. Eggs boiled, yolks removed from small openings, stuffed with minced pork, then sealed with starch paste. Deep-fried, then steamed.

神仙鸭子 (shénxiānyāzi)

山东菜。先将鸭腌渍,然后加入上汤、料酒、盐、葱、姜、花椒、小茴香等蒸熟,用原汤勾芡。

Steamed duck. Shandong dish. Pickled duck steamed in soup with cooking wine, salt, scallion, ginger, Chinese prickly ash, and fennel, then marinated in starched broth.

生煸草头 (shēngbiān cǎotóu)

上海菜。草头即苜蓿。将草头加盐急火煸炒。

Sautéed clover. Shanghai dish. Clover quickly sautéed with salt.

生炒海蚌 (shēngchǎo hǎibàng)

福建菜。先将蚌肉汆过,与煸过的

冬笋和香菇及调料翻炒，勾芡。

Clam with bamboo shoots and mushrooms. Fujian dish. Scalded clam meat stir-fried with quick-fried bamboo shoots, dried mushrooms, and flavorings, then thickened with starch.

生炒鳝鱼丝 (shēngchǎo shànyúsī)

广东菜。新鲜鳝鱼丝用猪油爆炒，加入香菜、酱油、黄酒、盐。

Stir-fried eel. Guangdong dish. Shredded eel stir-fried in lard and flavored with coriander, soy sauce, yellow wine, and salt.

生炒四丝 (shēngchǎo sìsī)

台湾菜。先将豆腐干、粉皮、榨菜、木耳等切丝，再与韭黄和辣椒丝拌炒，烹入高汤、盐，勾芡。

Sautéed shredded vegetables. Taiwan dish. Shredded dried bean curd, rice jelly sheets, preserved tuber mustard, and wood ears sautéed with yellow Chinese chive and shredded pepper. Flavored with broth and salt, then thickened with starch.

生吃黑鱼 (shēngchī hēiyú)

山东菜。将新鲜黑鱼去皮剔骨，切成薄片，佐辣椒、红油、芥末油、酱油等调成的卤汁。

Raw snakehead fillets with spicy sauce. Shandong dish. Snakehead fish skinned, boned, sliced, then served with sauce of red pepper, chili oil, mustard oil, and soy sauce.

生汆丸子 (shēngcuān wánzi)

各地家常菜。将肉末加盐、芡粉、蛋浆、葱花、姜末等做成肉丸，入清汤煮熟。

Meatball soup. Home-style dish in many places. Ground meat mixed with salt, starch, egg paste, chopped scallion, and minced ginger, made into meatballs, then boiled.

生麸肉圆 (shēngfū ròuyuán)

江苏菜。将水面筋球填入加调料的肉馅，用小火煮熟。

Boiled stuffed gluten balls. Jiangsu dish. Wet gluten balls stuffed with seasoned meat, then boiled over low heat.

生蚵腐皮卷 (shēngkē fǔpí juǎn)

广东菜。先把生蚵即牡蛎、瘦肉、甘薯及调料制成馅，然后用豆腐皮裹馅成卷，用面糊封口，炸至金黄。

Fried oyster in bean curd skins. Guangdong dish. Bean curd skins made into rolls filled with mixture of oyster meat, lean pork, sweet potatoes, and seasonings, sealed with flour paste, then deep-fried.

生煎鲑鱼 (shēngjiān guīyú)

江苏菜。先将鲑鱼用黄酒、酱油、盐等腌渍，然后裹上鸡蛋面粉糊，油煎。

Fried ban fish. Jiangsu dish. Ban fish pickled with yellow wine, soy sauce, and salt, coated with flour and egg paste, then deep-fried.

生煎鳜鱼 (shēngjiān guìyú)

湖南菜。先将鳜鱼腌渍，然后加生

姜、葱、青红椒、黄酒、酱油等爆炒。

Stir-fried Chinese perch. Hunan dish. Pickled Chinese perch stir-fried with ginger, scallion, green and red pepper, yellow wine, and soy sauce.

生煎馒头 (shēngjiān mántou)

上海点心。用发酵面团包加调料的猪肉馅做成小馒头,用少量油和少量水边煎边蒸。

Fried and steamed stuffed buns. Shanghai snack. Fermented flour dough made into small buns stuffed with seasoned minced pork. Fried and steamed with a little oil and a little water at the same time.

生扣鸳鸯鸡 (shēngkòu yuānyāngjī)

广东菜。先将鸡块和火腿用盐、猪油、麻油等腌渍,然后蒸熟。

Steamed chicken and ham.

Guangdong dish. Chicken and ham pickled with salt, lard, and sesame oil, then steamed.

生梨肉片 (shēnglí ròupiàn)

华南菜。先将猪肉片裹上蛋清淀粉浆,滑熟,再加汤汁焖烂,然后加入生梨片和西红柿片煮沸。

Simmered pork with pear. Southern dish. Sliced pork pickled with starched egg white, lightly fried, simmered, then boiled with pear and tomato sections.

生焖狗肉 (shēngmèn gǒuròu)

广东菜。先将狗肉、豆腐乳、青蒜、芝麻酱、姜、陈皮、八角等爆炒,然后入砂锅加调料焖煮。

Braised dog meat. Guangdong dish. Dog meat stir-fried with fermented bean curd, garlic sprouts, sesame sauce, ginger, processed tangerine peels, and aniseed, then braised with seasonings in a casserole.

生焖鲈鱼 (shēngmèn lúyú)

广东菜。先将肉丝、菇丝、姜、葱煸炒,再与鲈鱼肉入汤焖煮,用盐、白糖调味。

Braised perch. Guangdong dish. Perch braised in soup with quick-fried shredded pork, mushrooms, ginger, and scallion, then flavored with salt and sugar.

生鱼片 (shēngyúpiàn)

山东菜。将罗非鱼肉用白醋浸泡洗净,切成薄片,佐酱油、醋、麻油、芥末等调成的卤汁。

Raw tilapia fillets with sauce. Shandong dish. Tilapia meat steeped in white vinegar, cleaned, sliced, then served with dressing of soy sauce, vinegar, sesame oil, and mustard.

生鱼片湿炒河粉 (shēngyúpiàn shīchǎo héfěn)

广东点心。河粉与生鱼片炒熟,用特制芡汁、胡椒粉调味。

Stir-fried fish and rice noodles. Guangdong snack. Rice noodles and sliced fish stir-fried, then flavored with special starched sauce and ground black pepper.

生醉蟹油拌佛手海蜇皮（shēngzuì xièyóu bàn fóshǒu hǎizhépí）

江苏菜。将氽过的海蜇皮切成手掌状，与醉蟹肉拌和，用醉雄蟹油调味。

Jellyfish with crab oil. Jiangsu dish. Scalded jellyfish carved like palms, mixed with liquor-saturated crab meat, then flavored with liquor-saturated male crab oil.

诗礼银杏（shīlǐyínxìng）

山东点心。先将孔府白果焯水，煮酥，然后加白糖、蜂蜜、桂花酱焖至汁浓。

Braised sweet ginkgoes. Shandong snack. Ginkgoes from Confucius' hometown quick-boiled, then braised with white sugar, honey, and osmanthus flower sauce until reduced.

狮子头（shīzitóu）

江浙沪皖菜。猪肉末加芡粉、面粉、盐等做成大肉圆，油炸，略焖煮，配氽过的菜心装盘，浇酱油、糖、味精、芡粉做成的浓汁。

Braised big meatballs. Dish in Jiangsu, Zhejiang, Shanghai, and Anhui. Ground pork, starch, flour, and salt made into big meatballs, deep-fried, then simmered briefly. Garnished with scalded baby greens, then washed with thick dressing of soy sauce, sugar, MSG, and starch.

十全大补汤（shíquán dàbǔtāng）

各地传统药膳。将猪肉、墨鱼、猪肚、猪排骨与党参、黄芪、肉桂、当归、甘草等十种中药煨炖。

Stewed pork and inkfish with herbs. Traditional folk-medicinal soup in many places. Pork, inkfish, hog tripe, and pork spareribs stewed with ten herbs such as pilose asiabell root, milkvetch root, cinnamon, Chinese angelica, and licorice root.

十香醉排骨（shíxiāng zuìpáigǔ）

福建菜。带骨里脊肉条和马蹄片挂浆，炸至金黄，然后倒入葱、蒜、酱油、白糖、醋、番茄酱、芝麻酱、上汤、桔汁等制成的卤汁拌匀。

Ten-taste pork loin. Fujian dish. Sectioned pork tenderloin with bones and sliced water chestnuts coated with starch paste, then fried golden. Mixed with sauce of scallion, garlic, soy sauce, sugar, vinegar, tomato sauce, sesame sauce, broth, and tangerine juice.

什锦炒米线（shíjǐn chǎo mǐxiàn）

华南食品。将米线、猪肉丝与青红椒、葱、豆芽加盐、酱油翻炒。

Stir-fried rice noodles and vegetables. Southern food. Rice noodles and shredded pork stir-fried with green and red pepper, scallion, and bean sprouts. Seasoned with salt and soy sauce.

什锦炒面（shíjǐn chǎomiàn）

江苏食品。先将面条煮熟，过凉水，用沙拉油拌匀，再入锅炕至发脆，然后与煸炒过的胡萝卜、香菇、竹笋、西芹等翻炒。

Stir-fried noodles and assorted vegetables. Jiangsu food. Noodles boiled, quickly cooled in cold water, mixed with salad oil, fried until crisp over low heat, then stir-fried with prepared carrots, dried mushrooms, bamboo shoots, and American celery.

什锦冬瓜盅 (shíjǐn dōngguāzhōng)

广东、广西菜。先将猪肉片裹上芡粉,放入冬瓜盅,然后加入火腿、冬菇、薏米、干贝、白果、鸡汤等蒸熟,用盐和胡椒粉调味。

Assorted meat steamed in wax gourd. Guangdong and Guangxi dish. Sliced pork coated with starch, then steamed in wax gourd with ham, winter mushrooms, seeds of Job's tears, dried scallops, gingkoes, and chicken soup. Seasoned with salt and ground black pepper.

什锦豆腐 (shíjǐn dòufu)

南方素菜。先将豆腐煎至金黄,再同汆过的胡萝卜丁、玉米、豌豆、木耳、香菇等拌炒,用盐、糖、酱油、葱等调味,勾芡。

Sautéed bean curd and vegetables. Southern vegetarian dish. Bean curd fried golden, sautéed with scalded carrots, corn, peas, wood ears, and mushrooms, flavored with salt, sugar, soy sauce, and scallion, then thickened with starch.

什锦锅巴 (shíjǐn guōbā)

浙江菜。先将锅巴炸至金黄,再将烩好的叉烧、里脊肉、豌豆夹、小黄瓜等趁热浇在锅巴上。

Fried rice crusts with sauce. Zhejiang dish. Sticky rice crusts deep-fried, then washed with boiling mixture of barbecued pork, pork tenderloin, pea pods, and cucumbers.

什锦果盘 (shíjǐn guǒpán)

各地水果点心。将各色新鲜水果洗净,控干水,切块摆盘。

Assorted fruits plate. Fruit snack around the country. Assorted fruits cleaned, drained, and then arranged in a large plate.

什锦海参 (shíjǐn hǎishēn)

广东菜。将海参、鸡肫、鸟蛋、黄瓜、竹笋、胡萝卜等炖煮,用盐、料酒、葱、姜调味。

Stewed sea cucumbers. Guangdong dish. Sea cucumbers stewed with chicken gizzard, bird eggs, cucumbers, bamboo shoots, and carrots, then seasoned with salt, cooking wine, ginger, and scallion.

什锦烩面 (shíjǐn huìmiàn)

四川食品。先将薄面片煮熟,然后与火腿、鲜笋、黄花、香菇、鸡肉等入鸡汤烩煮,加盐、酱油、姜末、辣椒油、胡椒粉。

Boiled flour pieces with various ingredients. Sichuan food. Boiled thin flour pieces simmered in chicken soup with ham, fresh bamboo shoots, day lily, dried mushrooms, and chicken, then flavored with salt, soy sauce, minced ginger, chili oil, and ground black pepper.

什锦火锅 (shíjǐn huǒguō)

各地家常菜。将白菜、豆腐、粉丝、

猪肉、牛肉、羊肉、海参、蹄筋、冬菇、干贝等放入火锅汤内，边烫煮边食用。
Assorted treasures in fire pot. Home-style dish in many places. Chinese cabbage, bean curd, starch noodles, pork, beef, mutton, sea cucumbers, pork tendons, winter mushrooms, and soaked scallops quick-boiled in fire pot soup.

什锦凉菜 (shíjǐn liángcài)
各地家常凉菜。先将胡萝卜丝和黄瓜丝盐腌，然后与烫过的芹菜、绿豆芽、木耳、豆腐丝、蛋皮丝、鸡肉丝拌匀，加酱油、醋、麻油。
Assorted vegetables with dressing. Home-style cold dish in many places. Salted shredded carrots and cucumbers mixed with scalded celery, mung bean sprouts, wood ears, shredded bean curd, egg sheets, and chicken, then seasoned with soy sauce, vinegar, sesame oil, and other condiments.

什锦卤菜 (shíjǐn lǔcài)
台湾菜。先将猪肉、白菜、香菇、胡萝卜、虾皮等拌炒，然后放入蛋丝，添入高汤烹煮。
Stir-fried pork with vegetables. Taiwan dish. Pork, Chinese cabbage, dried mushrooms, carrots, and dried small shrimp stir-fried, flavored with shredded egg sheets, then boiled in broth.

什锦肉丁 (shíjǐn ròudīng)
各地家常菜。先将里脊肉丁滑熟，然后与辣椒、青豆、香菇加调料翻炒，勾芡。
Pork tenderloin with vegetables. Home-style dish in many places. Diced pork tenderloin lightly fried over low heat, stir-fried with pepper, green beans, dried mushrooms, and flavorings, then thickened with starch.

什锦水果糕 (shíjǐn shuǐguǒgāo)
广东点心。先将莲子、白果、红枣煮烂，然后与香蕉、菠萝、桔子、鲜桃入糖桂花水煮沸，勾芡，置凉。
Assorted fruit cakes. Guangdong snack. Lotus seeds, ginkgoes, and red dates stewed, boiled with bananas, pineapple, tangerines, and peaches in sugared osmanthus flower soup, thickened with starch, then chilled.

什锦素菜煲 (shíjǐn sùcàibāo)
广东家常菜。先将冬笋、蘑菇、香菇、西蓝花、胡萝卜、鲜芦笋切条汆过并煸炒，然后加入高汤稍煮。
Simmered assorted vegetables. Home-style dish in Guangdong. Scalded shredded winter bamboo shoots, fresh and dried mushrooms, broccoli, carrots, and asparagus quick-fried, then simmered in broth briefly.

石耳炖鸡 (shíěr dùn jī)
安徽菜。先将鸡肉用文火炖至半熟，再加入石耳焖煮，用盐、黄酒、姜、葱等调味。
Braised chicken with stone fungi.

Anhui dish. Half-cooked chicken braised with stone fungi and flavored with salt, yellow wine, ginger, and scallion.

石家酱方 (shíjiā jiàngfāng)

山东菜。将五花肉加大料、桂皮、黄酒、酱油等炖熟，勾芡。

Stewed streaky pork with sauce. Shandong dish. Streaky pork stewed with aniseed, cinnamon, yellow wine, and soy sauce, then thickened with starch.

石榴鸡 (shíliujī)

福建菜。将鸡肉末、猪肉末、虾、冬笋、冬菇及调料做成馅，用鸡皮包成石榴形，用芹菜扎口，蒸熟。

Steamed meat and vegetables. Fujian dish. Ground chicken and pork mixed with shrimp, winter bamboo shoots, winter mushrooms, and seasonings, wrapped like pomegranates with chicken skins, fastened with celery, then steamed.

石门肥肠煲 (shímén féichángbāo)

湖南石门菜。先将肥肠段加红椒、大蒜煸炒，然后加调料用砂锅炖煮。

Stewed hog large intestine. Specialty in Shimen, Hunan. Hog large intestine stir-fried with red pepper and garlic, then stewed with seasonings in a clay pot.

石首鸡茸鱼肚 (shíshǒu jīróng yúdǔ)

湖北石首菜。先将鱼肚和香菇用鸡汤芡汁腌渍，然后放入鸡茸蛋泡汤煮沸，淋麻油，撒火腿末、葱花、胡椒粉。

Fish maw and chicken soup. Specialty in Shishou, Hubei. Local longsnout catfish maw and dried mushrooms pickled with starched chicken soup, boiled in ground chicken and whipped egg soup, then sprinkled with sesame oil, ham crumbs, chopped scallion, and ground black pepper.

手掰肝 (shǒubāigān)

东北菜。将猪肝煮熟后掰成块，佐酱油、蒜、味精等调成的卤汁。

Hog liver with sauce. Northeastern dish. Hog liver boiled, hand-torn into pieces, then served with dressing of soy sauce, garlic, and MSG.

手撕鸡 (shǒusījī)

四川菜。先将腌渍过的柴鸡稍煎，再加调料烹煮，然后去皮去骨，把鸡肉撕成条装盘，淋鸡汤。

Hand-stripped braised chicken. Sichuan dish. Lightly fried pickled local young chicken braised with flavorings, skinned, boned, hand-stripped, then washed with chicken soup.

手撕牦牛肉干 (shǒusī máoniúròugān)

西藏菜。牦牛肉和松茸烤熟，用手撕食。

Broiled yak meat with mushrooms. Tibetan dish. Yak meat and pine mushrooms broiled, then served as finger food.

手撕泥鳅 (shǒusī níqiū)

东北菜。先将泥鳅炸至皮脆，再加泡椒、酱油等烹煮至收汁，然后用手撕成条。

Fried and simmered loach. Northeastern dish. Loach deep-fried until crisp, simmered with pickled pepper and soy sauce, then hand-torn into strips.

手抓牛掌 (shǒuzhuā niúzhǎng)

西藏菜。先将牛掌切开，煮熟，去蹄壳，然后入卤汤煮熟，手拿进食。

Marinated ox hooves. Tibetan dish. Ox hooves cut into halves, boiled, stripped off hard nails, then marinated in gravy. Served as finger food.

手抓羊肉 (shǒuzhuā yángròu)

西藏、内蒙传统菜。将带骨嫩羊肉用清水煮熟，用手撕食。

Boiled lamb. Traditional dish in Tibet and Inner Mongolia. Lamb chunks with bones boiled, then served as finger food.

首乌炖蛋 (shǒuwū dùn dàn)

浙江菜。煎首乌水与鸡蛋浆和鸡肉末调匀，蒸熟。

Egg and herb custard. Zhejiang dish. Tuber fleeceflower root tea mixed with egg paste and ground chicken, then steamed.

首乌麻地瘦肉汤 (shǒuwū má dì shòuròu tāng)

湖南药膳，又称长寿汤。将瘦肉配何首乌、黑芝麻、生地炖汤。

Longevity soup. Hunan folk-medicinal dish. Lean pork stewed with tuber fleeceflower root, black sesame, and rehmannia root.

寿星水鱼 (shòuxīng shuǐyú)

江苏菜。将甲鱼、桂圆肉、红枣、人参等用鸡清汤炖熟。

Longevity turtle soup. Jiangsu dish. Turtle stewed in chicken soup with longan meat, red dates, and ginseng.

寿字鸭羹 (shòuzì yāgēng)

山东菜。将蛋清摊在盘中，用火腿在上面摆出“寿”字，蒸熟后放入鸡、鸭、猪肘煮成的汤中。

Longevity duck soup. Shandong dish. Ham shreds arranged into a Chinese character of longevity “寿” on egg white on a platter, then steamed. Boiled in soup of chicken, duck, and pork joint.

鼠曲果 (shǔqūguǒ)

南方点心。将蒸熟的鼠曲草又称毛香与糯米粉混合，做成面皮，包入芋头、豆沙、干香菇、肉丝、笋丝、腌菜等馅，蒸熟。

Sticky rice and herb buns. Southern snack. Steamed cudweed herb and sticky rice flour mixed, made into buns stuffed with taros, mung bean paste, dried mushrooms, shredded pork, bamboo shoots, and pickled vegetables, then steamed.

涮九品 (shuànjiǔpǐn)

福建菜。把牛里脊、牛舌、牛百叶、牛心、牛肝、牛腰等切成薄片，用姜汁、芝麻酱、花椒、沙茶酱、醋等制蘸卤，用黄酒和数味药草制汤，边涮边吃。

Ox varieties in fire pot. Fujian dish. Beef tenderloin and ox offal, such as tongue, tripe, heart, liver, and kidney, cut into fine slices and quick-boiled in soup of yellow wine and herbs in a fire pot. Served with dressing of ginger juice, sesame sauce, Chinese prickly ash, barbecue sauce, and vinegar.

双冬辣鸡球 (shuāngdōng làjīqiú)

山东菜。先将去骨鸡腿肉炸熟，再与煎过的冬菇和冬笋入鸡汤烹煮，然后与辣椒、葱、姜、蒜翻炒，用酱油、盐、味精调味。

Braised drumsticks, bamboo shoots, and mushrooms. Shandong dish. Deep-fried boned drumsticks braised in chicken soup with fried winter mushrooms and winter bamboo shoots, then stir-fried with hot pepper, scallion, ginger, and garlic. Seasoned with soy sauce, salt, and MSG.

双沟大曲 (shuānggōudàqū)

白酒，产于江苏泗洪县双沟镇，中国十大名酒之一。以高粱为主料，用小麦、大麦、豌豆制曲酿造，酒精含量为 42-53 度，属浓香型酒。

Shuanggoudagu, liquor produced in Shuanggou Town of Sihong County, Jiangsu. White spirit made with sorghum and yeast of wheat, barley, and peas. Contains 42-53% alcohol, and has a thick aroma.

双菇扒菜胆 (shuānggū pá càidǎn)

东北家常菜。先将油菜烫熟，然后浇上用香菇、草菇、胡椒粉、酱油、麻油等制成的卤汁。

Rape greens with mushrooms. Home-style dish in northeastern areas. Rape greens quick-boiled, then marinated in gravy of black and meadow mushrooms, ground black pepper, soy sauce, and sesame oil.

双椒炸藕圆 (shuāngjiāo zhá ǒuyuán)

湖北菜。先把藕丁蒸熟，再与肉末、芡粉、盐等拌匀，做成丸子，炸至金黄，然后加入红辣椒、青椒、酱油等烩煮。

Spicy lotus roots and pork balls. Hubei dish. Steamed diced lotus roots mixed with ground pork, starch, and salt, made into balls, then deep-fried. Braised with chili and green peppers, soy sauce, and other seasonings.

双烤肉 (shuāngkǎoròu)

山东菜。先将带骨带皮硬肋猪肉煮熟，再抹上蛋清芡粉浆，烤至棕红色，然后去皮，切块，配春饼或荷叶饼装盘，用葱白、椒盐、萝卜条、甜面酱等佐食。

Broiled pork with sauce. Shandong dish. Pork with ribs and skins boiled, coated with starched egg white, broiled red brown, skinned, then cut into pieces. Served with thin small pancakes, leek stems, spicy salt, slivered turnips, and sweet fermented flour sauce.

双轮池 (shuānglúnchí)

白酒,产于安徽涡阳高炉。以高粱为原料,用小麦、大麦、豌豆制曲酿造,酒精含量53度,属浓香型酒。

Shuanglunchi liquor produced in Gaolu Town of Woyang, Anhui. White spirit made with sorghum and yeast of wheat, barley, and peas. Contains 53% alcohol, and has a thick aroma.

双麻萝卜饼 (shuāngmá luóbobǐng)

湖南点心。面团包入用白萝卜丝、熟火腿、熟鸡肉、香葱制成的馅,做成饼,油炸。

Turnip, ham, and chicken cakes. Hunan snack. Flour dough made into cakes stuffed with shredded white turnips, prepared ham, cooked chicken, and scallion, then fried.

双酿团 (shuāngniángtuán)

上海点心。在一个糯米团里装进两个馅,一个是糖芝麻粉,另一个是甜豆沙。

Buns with double-fillings. Shanghai snack. Sticky rice buns stuffed with two kinds of fillings in each. One filling is sugared ground sesame, the other is sweetened bean paste.

双皮刀鱼 (shuāngpí dāoyú)

江苏菜。先将刀鱼茸、猪肥膘茸、蛋清拌和,摊在刀鱼皮上,再将鱼皮合上成鱼原样,盖上火腿、春笋、冬菇,放入葱、姜、黄酒、盐,蒸熟。

Steamed saury with fat pork. Jiangsu dish. Minced saury meat and fat pork mixed with egg white, then spread on saury skins. The skins are folded back to fish shape, covered with ham, bamboo shoots, and winter mushrooms, then steamed with scallion, ginger, yellow wine, and salt.

双味蝤蠓 (shuāngwèi yóuměng)

浙江菜。蝤蠓即青蟹。先将猪肥膘煮熟,切片,裹上鱼茸虾胶,文火煎熟,然后与蒸熟的蝤蠓一同装盘,配姜醋。

Steamed crabs with fried fat pork. Zhejiang dish. Boiled sliced fat pork coated with fish and shrimp paste, fried over low heat, put in the same platter with steamed green crabs, then served with ginger vinegar.

双鱼茄子 (shuāngyú qiézi)

南方菜。将圆茄子剖成两半,腌入味,炸透,浇上特制卤汁。

Fried eggplants with sauce. Southern dish. Round eggplants halved, pickled with seasonings, deep-fried, then washed with special sauce.

爽脆水晶爪 (shuǎngcuì shuǐjīngzhuǎ)

四川凉菜。将鸡爪煮熟,放入泡菜坛,与芹菜、姜、辣椒等泡渍至脆。

Pickled chicken feet and vegetables. Sichuan cold dish. Boiled chicken feet steeped in pickling brine with celery, ginger, pepper, and other vegetables until crisp.

爽口老坛子 (shuǎngkǒu lǎotánzi)

四川菜。先将凤爪、鸭舌、猪耳、鹅肫等汆熟,置凉,然后放进用野山椒、花椒、盐等制成的泡菜水里泡渍。

Marinated offal. Sichuan dish. Chicken feet, duck tongue, hog ears, and goose gizzard quick-boiled, cooled, then pickled in brine flavored with wild chili and wild Chinese prickly ash.

水八块 (shuǐbākuài)

四川菜。先将小公鸡煮熟,砍成八块,然后加调料拌匀,撒上花椒粉。

Boiled chicken with spice. Sichuan dish. Young rooster boiled, chopped into eight chunks, mixed with flavorings, then dusted with ground Chinese prickly ash.

水炒苜蓿虾仁 (shuǐchǎo mùxù xiārén)

山东菜。亦称木须虾仁。先将虾仁切片,水焯,再与蛋浆调匀,然后加上汤、葱姜末、盐、料酒等轻炒。

Shrimp meat with eggs. Shandong dish. Scalded sliced shelled prawns mixed with egg paste, then quick-fried with broth, minced ginger, chopped scallion, salt, and yellow wine.

水豆花 (shuǐdòuhuā)

西南家常菜。将水发黄豆磨成浆,煮开后点酸卤成嫩水豆腐,佐各式辣酱食用。

Jellied bean curd. Home-style dish in southwestern areas. Soaked soybean ground into paste, then boiled. Made into jellied bean curd with sour bittern. Served with a variety of spicy sauces.

水瓜煮泥鳅 (shuǐguā zhǔ níqiū)

广东菜。将汆过的泥鳅与丝瓜、蘑菇入清汤煮熟,用盐、胡椒粉、生姜、花生油、黄酒调味。

Loach and luffa soup. Guangdong dish. Scalded loach boiled with luffa and mushrooms in clear soup, then flavored with ground black pepper, ginger, peanut oil, and yellow wine.

水果鸡 (shuǐguǒjī)

湖南菜。先将菠萝片、葡萄干、青红椒、洋葱煸炒,然后放入鸡肉,加鲜汤慢火炆熟。

Stewed chicken with fruit. Hunan dish. Chicken stewed in seasoned soup with quick-fried pineapple, raisins, green and red pepper, and onion.

水浒肉 (shuǐhǔròu)

四川菜。先将腌渍过的里脊肉片炒熟,倒在轻炒过的豌豆苗上,然后撒上煸炒过的干辣椒和花椒。

Stir-fried pork with pea sprouts. Sichuan dish. Pickled sliced pork tenderloin stir-fried, put on a bed of sautéed pea sprouts, then sprinkled with quick-fried chili pepper and Chinese prickly ash.

水晶明虾球 (shuǐjīng míngxiāqiú)

广东菜。先将明虾仁腌渍,稍炸,再用开水烫过,加调料稍炒并勾芡。

Stir-fried prawn meat. Guangdong dish. Shelled pickled prawns lightly fried, scalded, quick-fried with seasonings, then thickened with starch.

水晶排骨 (shuǐjīng páigǔ)

湖南菜。将猪肋骨加入香葱、生姜、盐等翻炒,勾浓芡。

Crystal pork ribs. Hunan dish. Chopped pork ribs stir-fried with scallion, ginger, and salt, then marinated in thick starched soup.

水晶皮冻 (shuǐjīng pídòng)

北京凉菜。先将肉皮切小块煮烂,连汤置凉成冻,然后切成长条,浇用香菜、酱油等调成的卤汁。

Hog skin jelly. Beijing cold dish. Diced hog skins boiled thoroughly, chilled into jelly, cut into strips, then washed with dressing of coriander and soy sauce.

水晶肉 (shuǐjīngròu)

北京凉菜。先将猪肉片和猪皮条加水及调料蒸烂,将肉皮捞出做他用;然后将肉连同汤冷却成冻,切块,佐酱油、麻油、辣酱。

Crystal pork. Beijing cold dish. Sliced pork steamed with slivered pork skins and flavorings in water. Pork and broth chilled into jelly without the skins. Jelly cut into pieces, then served with soy sauce, sesame oil, and chili sauce.

水晶蹄髈 (shuǐjīng típǎng)

山东菜。先把猪蹄用特制卤料腌过,再入老卤汤煮熟。

Marinated hog trotters. Shandong dish. Hog trotters pickled with special seasonings and boiled in aged gravy.

水晶虾仁 (shuǐjīng xiārén)

江浙沪传统菜。先将河虾仁与蛋清及调料拌和,冰镇,再将虾仁入油锅划散,加高汤和调料翻炒,勾芡。

Sautéed shrimp meat. Traditional dish in Jiangsu, Zhejiang, and Shanghai. River shrimp meat mixed with egg white and seasonings, then chilled. Lightly fried, sautéed with broth and seasonings, then thickened with starch.

水晶鲜虾饺 (shuǐjīng xiānxiājiǎo)

广东点心。用精面粉和太白粉制皮,包入用虾仁、笋丝、肥肉及调料做成的馅,蒸熟。

Crystal shrimp *jiaozi*. Guangdong snack. *Jiaozi* made with fine flour and potato starch, stuffed with shrimp meat, shredded bamboo shoots, fat pork, and seasonings, then steamed.

水晶蟹粉卷 (shuǐjīng xièfěnjuǎn)

广东凉菜。将蟹肉、胡萝卜、青瓜等做成馅,用腌白萝卜片裹馅成卷,放入加糖的青柠檬汁浸泡。

Pickled turnip and crab meat rolls. Guangdong cold dish. Pickled sliced turnips made into rolls filled with mixture of crab meat, carrots, and cucumbers, then steeped in sugared green lemon juice.

水晶鸭 (shuǐjīngyā)

福建菜。先把煮熟的鸭切成条,皮朝下码放在大碗里,再把煮熟的猪皮切块，与鸡茸、姜、葱、盐、黄酒等熬成汤，然后将汤倒入鸭条，冷冻。

Duck jelly. Fujian dish. Chopped boiled duck put in a big bowl with skins down, then marinated in soup of boiled pork skins, ground chicken, ginger, scallion, salt, and yellow wine. Chilled into jelly before serving.

水晶肴肉 (shuǐjīng yáoròu)

江苏镇江菜。先将猪蹄膀腌渍，然后加盐、矾及调料小火煨炖。

Braised hog knuckle. Specialty in Zhenjiang, Jiangsu. Hog knuckle pickled, then stewed in soup of salt, alum, and other condiments.

水泡鱼圆 (shuǐpào yúyuán)

江西菜。先将鱼茸加调料做成鱼丸，煮熟，然后烩煮。

Simmered fish balls. Jiangxi dish. Minced fish meat and seasonings made into balls, boiled, then simmered.

水煮回头鱼 (shuǐzhǔ huítóuyú)

湖南菜。洞庭湖回头鱼入鲜汤烹煮，加莴笋、青椒、豉油、剁椒、紫苏叶。

Boiled spicy fish. Hunan dish. *Huitou* fish from Dongting Lake boiled in seasoned soup with stem lettuce, green pepper, fermented soybean oil, pickled red pepper, and purple basil leaves.

水煮肉片 (shuǐzhǔ ròupiàn)

四川菜。将肉片入辣椒油汤煮熟，加入青菜、香菜、蒜苗。

Pork boiled in chili soup. Sichuan dish. Sliced pork boiled in soup flavored with chili oil. Green vegetable, coriander, and garlic sprouts added.

丝瓜扒竹荪 (sīguā pá zhúsūn)

广东菜。先将丝瓜、竹荪、香菇、白果、胡萝卜等轻炒，然后加入水、盐、味精、酱油、麻油，用小火扒至瓜软，勾薄芡。

Simmered luffa and long net stinkhorn. Guangdong dish. Luffa, long net stinkhorn, dried mushrooms, ginkgoes, and carrots quick-fried, simmered with water, salt, MSG, soy sauce, and sesame oil, then lightly starched.

四宝豆腐 (sìbǎo dòufu)

南方菜。先在豆腐上撒盐和糖，再堆上虾仁、青豆、冬菇、火腿，蒸熟，淋滚猪油。

Bean curd with four treasures. Southern dish. Salted and sugared bean curd covered with shrimp meat, green beans, winter mushrooms, and ham, steamed, then washed with boiling lard.

四宝凤凰卷 (sìbǎo fènghuángjuǎn)

台湾菜。先将鱼茸、冬菇、虾米、蛋浆等拌成馅，再用蛋皮裹馅成卷，蒸熟，然后浇上用姜酒、栗粉、糖、麻油等调制的卤汁。

Stuffed egg rolls with sauce. Taiwan dish. Egg rolls filled with mixture of minced fish, winter mushrooms, shrimp meat, and egg paste, steamed, then washed with dressing of ginger wine, chestnut powder, sugar, and sesame oil.

四神桂圆汤 (sìshén guìyuántāng)

广东菜。将甲鱼肉、枸杞子、熟地黄、桂圆慢火炖煮。

Turtle stewed with herbs.

Guangdong soup. Turtle meat stewed with Chinese wolfberry, prepared rehmannia root, and longan pulp.

四特酒(sìtèjiǔ)

白酒,产于江西清江。以大米为原料,用大麦制曲酿造,酒精含量为38-54度,属特香型酒。

Sitejiu, liquor produced in Qingjiang, Jiangxi. White spirit made with rice and barley yeast. Contains 38-54% alcohol, and has a heavy aroma.

四味素烩 (sìwèi sùhuì)

福建菜。将青菜心、白菜心、莴笋、白萝卜入清汤加调料煮熟,勾芡。

Braised four vegetables. Fujian dish. Baby greens, baby Chinese cabbage, stem lettuce, and white turnips braised with seasonings, then thickened with starch.

四喜饺 (sìxǐjiǎo)

湖南点心。用精面粉做成饺皮,包入用鲜肉、火腿、青菜、蛋皮、香菇及调料做成的四种馅,清水煮熟。

Jiaozi with four kinds of fillings. Hunan snack. Fine flour dough made into *jiaozi* stuffed with 4 kinds of fillings made with pork, ham, vegetables, egg sheets, and dried mushrooms, then boiled.

四喜丸子 (sìxǐ wánzi)

山东菜。先将猪肉、南荠、玉兰片等切碎,加水、盐、芡粉等做成丸子,再裹蛋糊油炸,然后加酱油、葱、姜、糖、上汤等焖煮。

Pork and vegetable balls. Shandong dish. Mixture of minced pork, water chestnuts, soaked bamboo shoots, water, salt, and starch made into balls, coated with egg paste, deep-fried, then simmered in broth with soy sauce, scallion, ginger, and sugar.

四鲜白菜墩 (sìxiān báicàidūn)

上海菜。将蒸过的白菜心装碗,上面堆放鸡肉、火腿、烤鸭、笋片、撒虾米,加入鸡清汤、黄酒、猪油、盐,用猛火速蒸。

Steamed Chinese cabbage with four treasures. Shanghai dish. Chinese cabbage steamed, covered with chicken, ham, broiled duck, and sliced bamboo shoots, sprinkled with dried shrimp meat, seasoned with clear chicken soup, yellow wine, lard, and salt, then quickly steamed.

四样荤素 (sìyàng hūnsù)

浙江菜。将葱白、青菜、笋片、豆腐皮、肉片同锅煮熟,勾芡,淋猪油。

Four vegetables with pork. Zhejiang dish. Leek stems, green vegetable, sliced bamboo shoots, dried bean curd skins, and sliced pork boiled, thickened with starch, then washed with lard.

松鹤延年 (sōnghè yánnián)

北方菜。用熟香菇和生黄瓜拼成松树样,用鸡脯、熟蛋白拼成鹤样。

Mushrooms and cucumbers with chicken breast. Northern dish. Prepared mushrooms and fresh cucumbers arranged like a pine tree. Boiled chicken breast and egg white arranged like a crane.

松仁猴菇 (sōngrén hóugū)

浙江菜。将猴头菇丁和松仁加黄酒、盐、味精、鲜汤翻炒。

Sautéed mushrooms and pine nuts. Zhejiang dish. Diced monkey-head mushrooms sautéed with pine nuts and seasoned with yellow wine, salt, MSG, and broth.

松仁青稞 (sōngrén qīngkē)

西藏点心。将青稞、青椒、松子、面粉加水调和,用色拉油炸。

Fried barley and pine nut cakes. Tibetan snack. Tibetan barley, green pepper, pine nuts, and flour mixed, then deep-fried in salad oil.

松仁玉米 (sōngrén yùmǐ)

各地家常菜。将半熟的玉米粒和焙过的松仁加调料翻炒,添少许清汤稍煮。

Sautéed pine nuts and corn. Home-style dish in many places. Half-cooked corn and roasted pine nuts sautéed with seasonings, then quick-boiled in a little clear soup.

松鼠鳜鱼 (sōngshǔ guìyú)

江苏苏州名菜。先将鳜鱼剔骨、刻花,挂蛋黄面粉浆,炸至金黄,然后浇糖醋卤汁。

Chinese perch in sweet and sour sauce. Famous dish in Suzhou, Jiangsu. Chinese perch boned, carved, coated with egg yolk and flour paste, then fried golden. Marinated in sweet and sour sauce.

松鼠黄鱼 (sōngshǔ huángyú)

北方菜。黄鱼身刻菱状花纹,裹浓芡炸透,浇卤汁。

Fried croaker with sauce. Northern dish. Yellow croaker carved into diamond patterns, coated heavily with starch, deep-fried, then marinated in special sauce.

松子鸡 (sōngzǐjī)

江苏菜。先将鸡脯和鸡腿肉去骨斩块,扑淀粉,铺上五花肉末,嵌入松仁,文火炸透,然后加入鸡清汤及调料焖煮。

Chicken with pine nuts. Jiangsu dish. Boned chicken breast and drumsticks cut into thick pieces, starched, spread with ground streaky pork, inlaid with pine nuts, then deep-fried over low heat. Braised in clear chicken soup with seasonings.

松子马哈鱼 (sōngzǐ mǎhāyú)

东北菜。先将马哈鱼块腌渍,挂生粉油炸,然后浇白糖番茄汁,撒油炸松子。

Fried salmon with pine nuts. Northeastern dish. Pickled salmon chunks coated with starch, deep-fried, washed with sugared tomato sauce, then showered with fried pine nuts.

松子肉 (sōngzǐròu)

浙江菜。先将腌渍过的五花肉块抹上蛋糊和虾肉茸,撒上松仁,煎至金黄,然后入清汤加调料焖烂,配轻炒的豌豆苗装盘。

Braised streaky pork with pine nuts. Zhejiang dish. Pickled cubed streaky pork coated with egg paste and mixture of minced shrimp and pork, garnished with pine nuts, then fried golden. Braised in seasoned clear soup, then served with lightly sautéed pea sprouts.

松子虾仁 (sōngzǐ xiārén)

福建菜。先将虾仁过油,西芹氽过,然后将虾仁和西芹翻炒,勾芡,撒上微炸的松子。

Sautéed shrimp meat with pine nuts. Fujian dish. Lightly fried shrimp meat sautéed with scalded American celery, thickened with starch, then sprinkled with prepared pine nuts.

松子香蘑 (sōngzǐ xiāngmó)

北方菜。先将松子炸香,然后加香菇、鸡汤、料酒、白糖、味精煨熟,勾芡,淋猪油。

Pine nuts with mushrooms. Northern dish. Quick-fried pine nuts simmered in chicken soup with dried mushrooms, cooking wine, sugar, and MSG, thickened with starch, then flavored with lard.

松子熏肉 (sōngzǐ xūnròu)

江苏菜。将猪去骨肋条肉用杉木屑、糖、茶叶或桂皮等为发烟材料熏熟。

Smoked pork. Jiangsu dish. Boned streaky pork grilled over smoke of China fir chips, sugar, tea leaves, or cinnamon bark.

松子鸭颈 (sōngzǐ yājǐng)

江苏菜。先将烤鸭片拍上干淀粉,上置加调料的松子虾茸馅,裹成鸭颈状卷,然后炸至皮脆,配香菜装盘。

Fried duck and pine nut rolls. Jiangsu dish. Sliced broiled duck dusted with dry starch, spread with seasoned mixture of pine nuts and minced shrimp meat, made into rolls like duck necks, then fried crisp. Served with coriander.

松子鱼米 (sōngzǐ yúmǐ)

江苏菜。将鱼肉切成豌豆状粒,与炸过的松子同炒,加调料,勾芡。

Sautéed fish and pine nuts. Jiangsu dish. Diced fish sautéed with fried pine nuts, seasoned, then thickened with starch.

宋河粮液 (sònghéliángyè)

白酒,产于河南鹿邑。以当地高粱为原料,用小麦、大麦、豌豆制曲酿造,酒精含量为 38-54 度,属浓香型酒。

Songheliangye, liquor produced in Luyi, Henan. White spirit made with sorghum and yeast of wheat, barley, and peas. Contains 38-54% alcohol, and has a thick aroma.

宋嫂鱼羹 (sòngsǎoyúgēng)

浙江杭州名菜。鳜鱼蒸熟,取肉拨碎,添入笋丝和香菇丝煮沸,勾芡。

Chinese perch fish soup. Famous dish in Hangzhou, Zhejiang. Chinese perch steamed and boned. Fish meat stripped into small pieces, boiled with shredded bamboo shoots and dried mushrooms, then thickened with starch.

苏州茉莉花茶 (sūzhōu mòlìhuāchá)

再加工茶,产于江苏苏州,中国著名花茶。以绿茶为茶胚,用茉莉花窨制而成。香气清芬,汤色澄明。

Suzhou jasmine tea produced in Suzhou, Jiangsu. Famous flower-scented tea in China. It is made with green tea and jasmine flower. Its tea has an elegant aroma and a crystal yellowish green color.

酥肥鸭块 (sūféiyākuài)

江苏菜。先将腌渍过的鸭蒸烂,去骨切块,然后油炸,配炸土豆丝装盘,撒五香粉。

Steamed and fried fatty duck. Jiangsu dish. Steamed pickled fatty duck boned, chopped, then deep-fried. Garnished with deep-fried shredded potatoes, then dusted with five-spice powder.

酥海带 (sūhǎidài)

广东菜。将海带用小火焖烂,加葱、麻油、醋、糖、酱油。

Braised kelp. Guangdong dish. Kelp braised with scallion, sesame oil, vinegar, sugar, and soy sauce.

酥糊四季豆 (sūhú sìjìdòu)

浙江菜。四季豆挂面糊炸透。

Fried jack beans. Zhejiang dish. Jack beans coated with flour paste and deep-fried.

酥鲫鱼 (sūjìyú)

浙江家常菜。先将咸菜和胡萝卜置碗底,再放上鲫鱼和葱、姜、蒜,然后用海带盖住,焖熟。

Crucian with vegetables and kelp. Home-style dish in Zhejiang. Crucian put on a bed of preserved vegetable and carrots, surrounded with scallion, ginger, garlic, and other condiments, covered with kelp, then braised.

酥橘圆 (sūjúyuán)

江苏菜。将肥猪肉、蛋黄、橘饼末、面粉、水混合做成丸子,油炸,挂糖霜。

Fatty pork and tangerine balls. Jiangsu dish. Fatty pork, egg yolks, minced candied tangerines, flour, and water made into balls, deep-fried, then coated with sugar powder.

酥蜜鲜果夹 (sūmì xiānguǒjiá)

浙江点心。先将苹果切成厚夹片,裹上干面粉,嵌入百果仁糖桂花馅,然后裹上蛋糊,炸至金黄。

Fried apples with nuts. Zhejiang snack. Apples cut into thick clamps, dusted with dry flour, filled with mixture of nuts and sweet osmanthus flower, coated with egg paste, then fried golden.

酥嫩一品煲 (sūnèn yīpǐnbāo)

广东菜。将鲍鱼、鱼肚、虾仁、带子、乌参过油，入上汤煲煮，用盐、黄酒、葱、姜调味。

Stewed assorted seafood. Guangdong dish. Lightly fried abalone, fish maw, shelled shrimp, scallops, and sea cucumbers stewed in broth, then seasoned with salt, yellow wine, scallion, and ginger.

酥皮蛋挞 (sūpí dàntà)

广东点心。用鸡蛋、炼奶、白糖等调成浆，放在用面粉、油、水烤制的小碟里，烤熟。

Egg and milk cakes. Guangdong snack. Egg paste, condensed milk, and sugar put in small dishes made with flour, oil, and water, then baked.

酥皮莲蓉包 (sūpí liánróngbāo)

广东点心。用发酵面和油酥面制皮，莲蓉制馅，包成圆形，蒸熟。

Steamed sweet buns. Guangdong snack. Round buns made of fermented flour dough and oil dough, stuffed with sweetened mashed lotus seeds, then steamed.

酥皮香椿 (sūpí xiāngchūn)

河南、山东家常菜，亦称香椿鱼。香椿挂面糊炸至金黄，佐椒盐。

Fried Chinese toon leaves. Home-style dish in Henan and Shandong. Tender Chinese toon leaves coated with flour paste, then fried until golden and crisp. Served with spicy salt.

酥皮月饼 (sūpí yuèbǐng)

广东点心。用面粉、油、蛋浆制成饼皮，包入莲蓉、豆沙等制成的馅，做成饼，烤至金黄皮酥。

Crisp moon cakes. Guangdong snack. Flour, oil, and egg paste made into cake wrappers, filled with mixture of sweetened lotus seeds and bean paste, then baked golden and crisp.

酥香豆腐 (sūxiāng dòufu)

湖南菜。将水豆腐浇上用酥肉和香辣花生酱制成的调料，蒸熟。

Steamed tender bean curd. Hunan dish. Tender bean curd washed with sauce of fried pork and spicy peanut sauce, then steamed.

酥香凤翅 (sūxiāng fèngchì)

浙江菜。先将鸡翅放入加盐、胡椒粉、黄酒的汤中煮至近熟，然后去骨，裹上淀粉蛋浆和面包屑油炸，佐辣酱。

Crisp and savoury chicken wings. Zhejiang dish. Chicken wings boiled in soup seasoned with salt, ground black pepper, and yellow wine. Boned, coated with starch and egg paste and bread crumbs, then deep-fried. Served with chili sauce.

酥小鲫鱼 (sūxiǎojìyú)

浙江菜。将小鲫鱼用葱白、姜片、桂皮、花椒、酱油、陈醋、麻油、料酒等腌渍，文火煨酥。

Braised small crucian carp. Zhejiang dish. Small crucian carp pickled with scallion stems, ginger, cinnamon,

Chinese prickly ash, soy sauce, aged vinegar, sesame oil, and cooking wine, then braised.

酥油茶 (sūyóuchá)

藏族传统饮料。先将煮好的茶水倒入特制的桶,加入盐和酥油搅拌,然后熬煮。

Tibetan butter tea. Traditional drink of Tibetan ethnic minority. Tea poured into a special bucket, mixed with salt and Tibetan butter, stirred, then boiled.

酥油炒青稞 (sūyóu chǎo qīngkē)

藏族食品。先将青稞炒熟,磨成粉,然后与酥油焙炒。

Ground barley with Tibetan butter. Specialty of Tibetan ethnic minority. Tibetan barley baked, ground, then stir-fried with Tibetan butter.

酥鱼 (sūyú)

南方菜。先将鲢鱼块炸至酥脆,然后放入清汤,加酱油、料酒、五香粉、糖、醋、姜汁,用小火焖煮收汁。

Fried and braised silver carp. Southern dish. Silver carp chunks deep-fried crisp, then braised in clear soup with soy sauce, cooking wine, five-spice powder, sugar, vinegar, and ginger juice until reduced.

酥炸蚕豆 (sūzhá cándòu)

南方家常菜。将蚕豆炸至酥脆,撒椒盐。

Fried broad beans. Home-style dish in southern areas. Broad beans deep-fried until crisp, then sprinkled with spicy salt.

酥炸金花 (sūzhá jīnhuā)

台湾菜。将虾米、果皮、猪肉、鱼肉等拌匀,塞入掏空的豆腐泡,炸至金黄。

Fried stuffed bean curd buns. Taiwan dish. Fried cubed bean curd hollowed, stuffed with mixture of dried shrimp meat, fruit peels, pork, and fish meat, then fried golden.

素菜包子 (sùcài bāozi)

上海点心。用精白面粉做面皮,用青菜、面筋、冬菇、冬笋、五香豆腐干及盐、麻油为馅,做成包子,蒸熟。

Vegetables buns. Shanghai snack. Fine flour dough made into buns stuffed with mixture of fresh vegetable, gluten, dried mushrooms, winter bamboo shoots, dried spicy bean curd, salt, and sesame oil, then steamed.

素菜干锅 (sùcài gānguō)

湖南菜。用凉薯、土豆、黄瓜、莲藕、胡萝卜、茶树菌及辣味调料炒制。

Vegetables in chili pot. Hunan dish. Assorted vegetables, such as bean yam, potato, cucumber, lotus root, carrot, and tea tree mushroom, braised with hot spices in a wok.

素菜汤 (sùcàitāng)

各地家常菜。用清水将多种蔬菜煮熟,加盐、素油、酱油、葱、姜等。

Vegetable soup. Home-style dish in many places. Assorted vegetables boiled and flavored with salt, vegetable oil, soy sauce, scallion, and ginger.

素烩茄子块 (sùhuì qiézikuài)

河南家常菜。将茄块、大蒜、青椒、芹菜、洋葱等煸炒,加高汤及调料烩煮,收汁后放入西红柿。

Simmered eggplants. Home-style dish in Henan. Sectioned eggplants stir-fried with garlic, green pepper, celery, onion, and other ingredients, simmered in broth with flavorings until reduced, then garnished with tomatoes.

素烩五圆 (sùhuì wǔyuán)

山东菜。将胡萝卜、白萝卜、莴笋、口蘑、草菇切成圆片,煮熟,勾芡。

Boiled vegetables and mushrooms. Shandong dish. Carrots, turnips, stem lettuce, Mongolian mushrooms, and meadow mushrooms sliced into round pieces, boiled, then thickened with starch.

素鸡 (sùjī)

各地家常菜。将豆腐丝或片煮软,压成整块,即成素鸡。将素鸡切成厚片,加不同调料烹煮成不同风格的菜。

Soy chicken. Home-style dish in many places. Shredded or sliced bean curd boiled soft, pressed into long lumps called soy chicken. Cut into thick pieces, then cooked with different flavorings as desired.

素熘鹅皮 (sùliū épí)

江苏菜。将油面筋泡用刀切开,里朝外翻面,用糖醋汁和麻油炸熟。

Sweet and sour gluten. Jiangsu dish. Fried gluten puffs cut open and turned inside out, then deep-fried in sesame oil and sweet-and-sour sauce.

素牛肉 (sùniúròu)

东北家常菜。先将豆腐块炸至金黄,然后加酱油、姜、葱、五香粉等焖煮。

Soy beef. Northeastern home-style dish. Deep-fried bean curd braised with soy sauce, ginger, scallion, and five-spice powder.

粟米粥 (sùmǐzhōu)

各地点心。先将粟米用小火煮烂,然后加入红枣、桔饼、糖熬煮。

Sweet corn porridge. Home-style snack in many places. Ground corn boiled into porridge with red dates, candied tangerines, and sugar.

酸八宝菜 (suānbābǎocài)

东北家常菜。将黄瓜、莴笋、青椒、藕、土豆等八种蔬菜焯过,控干水,加醋和其他调料拌匀。

Eight vegetables with vinegar sauce. Northeastern home-style dish. Eight vegetables, such as cucumber, lettuce, green pepper, lotus root, and potato, scalded, drained, then seasoned with vinegar and other flavorings.

酸白菜 (suānbáicài)

北方家常菜。先将白菜氽过,撒盐,然后放入容器,加入凉开水,密封三周。

Sour pickled cabbage. Northern

home-style dish. Scalded Chinese cabbage pickled with salt, put in a container with cooled boiled water, then sealed for 3 weeks.

酸菜炒豆腐丝 (suāncài chǎo dòufusī)

东北家常菜。将酸菜、笋、豆腐干分别切丝，加调料炒熟。

Pickled cabbage with bean curd. Home-style dish in northeastern areas. Shredded sour pickled Chinese cabbage, bamboo shoots, and dried bean curd sautéed and seasoned.

酸菜炖鸭 (suāncài dùn yā)

东北菜。先将鸭块过油，然后加入清汤、酸菜等炖煮。

Pickled cabbage with duck. Northeastern dish. Duck chunks quick-fried, then stewed with sour pickled Chinese cabbage in clear soup.

酸菜粉丝冻豆腐 (suāncài fěnsī dòngdòufu)

东北菜。将松蘑、粉丝、酸菜、焯过的冻豆腐、香菇等逐层放入火锅，添入蘑菇汤和调料烧开，撒香菜。

Bean curd with pickled cabbage and vermicelli. Northeastern dish. Mushrooms, vermicelli, sour pickled Chinese cabbage, quick-boiled frozen bean curd, and dried mushrooms layered in a fire pot, boiled in seasoned mushroom soup, then garnished with coriander.

酸菜工梅鱼 (suāncài gōng méiyú)

福建菜。先将梅鱼焯过，再与酸菜末一道煮熟，然后把鱼汤加香菇、葱段稍煮，浇在鱼上。

Mackerel with sour pickled mustard. Fujian dish. Spanish mackerel scalded, boiled with chopped sour pickled mustard, then marinated in fish soup flavored with dried mushrooms and scallion.

酸菜鱼 (suāncàiyú)

四川菜。先将煸炒过的泡菜与鱼头和鱼骨熬煮，然后放入带皮鱼片煮熟，用盐、料酒、花椒、泡椒、姜、蒜等调味。

Fish with sour pickled vegetables. Sichuan dish. Quick-fried pickled vegetables, fish heads, and fish bones stewed, then boiled with fish fillets with fish skins. Flavored with salt, cooking wine, Chinese prickly ash, pickled red pepper, ginger, and garlic.

酸菜猪肉炖粉条 (suāncài zhūròu dùn fěntiáo)

东北菜。将五花猪肉片加酸菜丝、粉条及调料炖煮。

Pickled cabbage with pork and starch noodles. Northeastern dish. Streaky pork stewed with sour pickled Chinese cabbage, starch noodles, and seasonings.

酸豆角 (suāndòujiǎo)

东北家常菜。将酸豆角和猪肉分别切丁，加辣椒、酱油、蒜等翻炒。

Stir-fried pickled string beans. Home-style dish in northeastern areas. Chopped pickled string beans stir-fried with diced pork, then flavored with hot pepper, soy sauce, and garlic.

酸豆角田螺肉 (suāndòujiǎo tiánluóròu)

湖南菜。腌渍过的田螺肉加入酸豆角、香菜、葱爆炒。

Field snails with sour cowpeas. Hunan dish. Pickled field snail meat stir-fried with sour pickled young cowpeas, coriander, and scallion.

酸辣粉 (suānlàfěn)

四川点心。先将米粉烫熟,然后放入用酸菜末、红油、辣椒粉等煮成的汤。

Sour and spicy rice noodles. Sichuan snack. Boiled rice noodles served in soup of sour pickled vegetables, chili oil, and chili powder.

酸辣坛子肉 (suānlà tánziròu)

四川菜。先将猪肉、鸡肉、猪骨焯水,放入陶罐,加盐、糟汁、酱油、冰糖、姜、葱、酸菜、辣椒,放入炸过的鸡蛋和肉丸煨煮。

Braised sour and spicy pork. Sichuan dish. Quick-boiled pork, chicken, and pork bones braised with fried eggs and meatballs in an earthen pot. Seasoned with salt, sweet fermented sticky rice juice, soy sauce, rock sugar, ginger, scallion, sour pickled vegetables, and chili pepper.

酸辣汤 (suānlàtāng)

四川家常菜。将肉丝、火腿丝、酸菜、辣椒、豆腐、香菇等煮汤,用酱油、醋、糖、味精调味。

Sour and spicy soup. Home-style dish in Sichuan. Shredded pork and ham boiled with sour pickled vegetables, chili pepper, bean curd, and dried mushrooms, then seasoned with soy sauce, vinegar, sugar, and MSG.

酸梅汤 (suānméitāng)

江浙沪传统消暑饮品。先将乌梅与水、桂花、甘草、山楂等熬煮,再加入冰糖和柠檬汁搅匀,然后加开水稀释,置凉后饮用。

Smoked plum tea. Traditional summer beverage in Jiangsu, Zhejiang, and Shanghai. Smoked plums boiled in water with osmanthus flower, licorice root, and haw, mixed with rock sugar and lemon juice, then made thin with boiled water. Cooled before serving.

酸奶子 (suānnǎizi)

蒙古族传统饮料。将鲜奶烧开,放阴凉处发酵。

Fermented milk. Traditional drink of Mongolian ethnic minority. Fresh milk boiled, then fermented in a cool place.

蒜苗回锅肉 (suànmiáo huíguōròu)

各地家常菜。将汆过的猪肉片配豆瓣酱、甜酱等拌炒,再加入蒜苗快炒出锅。

Stir-fried pork and garlic sprouts. Home-style dish in many places. Scalded sliced pork stir-fried with fermented broad bean sauce or sweet flour sauce and other condiments, then quickly fried with garlic sprouts.

蒜泥白肉 (suànní báiròu)

四川菜。先将带皮五花肉氽过,入汤煮至皮软,然后切薄片,浇酱油、辣椒油、蒜泥。

Boiled streaky pork with spicy sauce. Sichuan dish. Scalded streaky pork boiled in soup until skin soft, then sliced. Served with soy sauce, chili oil, and mashed garlic.

蒜泥黄瓜 (suànní huángguā)

北方家常凉菜。先将黄瓜盐渍,再加入蒜泥、酱油、辣椒油、味精、麻油。

Cucumbers with garlic dressing. Home-style cold dish in northern areas. Cucumbers pickled with salt and flavored with mashed garlic, soy sauce, chili oil, MSG, and sesame oil.

蒜泥梅肉 (suànní méiròu)

四川菜。先将猪脖颈肉即梅肉煮熟,切成薄片,然后将蒜泥放在肉片上,浇上特制的卤汁和红油。

Boiled pork with sauce. Sichuan Dish. Boiled hog neck meat sliced, spread with mashed garlic, then washed with special sauce and chili oil.

蒜茸煎马鲛鱼 (suànróng jiān mǎjiāoyú)

海南、广东菜。马鲛鱼用盐和胡椒粉略腌,煎至金黄,佐蒜茸、蚝油食用。

Fried Spanish mackerel with garlic. Hainan and Guangdong dish. Spanish mackerel pickled with salt and ground black pepper, then deep-fried. Served with mashed garlic and oyster sauce.

蒜茸蜗鱼嘴 (suànróng wōyúzuǐ)

广东菜。先将鱼嘴腌渍,过油,然后加蒜、姜、尖椒、辣酱、黄酒等翻炒,浇芡汁。

Stir-fried spicy fish mouths. Guangdong dish. Pickled fish mouths lightly fried, stir-fried with garlic, ginger, hot pepper, chili sauce, and yellow wine, then marinated in starched soup.

蒜茸蒸生蚝 (suànróng zhēng shēngháo)

海南、广东菜。鲜蚝加蒜茸、粉丝、葱花蒸熟。

Steamed oyster with garlic. Hainan and Guangdong dish. Oyster steamed with mashed garlic, starch noodles, and chopped scallion.

蒜薹腊肉 (suàntái làròu)

南方家常菜。腊肉与蒜薹合炒,加入少许酱油、料酒。

Stir-fried smoked pork and garlic stems. Southern home-style dish. Smoked cured pork stir-fried with garlic stems, then flavored with a little soy sauce and cooking wine.

蒜味皇帝菜 (suànwèi huángdìcài)

西北菜。先将皇帝菜氽过,然后加蒜泥和醋拌匀。

Quick-boiled daisy with garlic dressing. Northwestern dish. Crown daisy quick-boiled, then flavored with mashed garlic and vinegar.

蒜香鲶鱼 (suànxiāng niányú)

广东菜。先将姜、蒜、葱、豆瓣酱煸过,加入高汤、盐、糖、酱油,然后放入鲶鱼焖煮,撒葱、辣油、醋,勾芡。

Braised catfish. Guangdong dish. Catfish braised in broth with quick-fried garlic, ginger, and fermented broad bean sauce, flavored with salt, sugar, soy sauce, scallion, chili oil, and vinegar, then washed with starched soup.

蒜子扣瑶柱 (suànzǐ kòu yáozhù)

广东菜。将瑶柱即干贝加入煸炒过的蒜末、姜片等蒸炖,然后浇上用高汤、蚝油等调成的卤汁。

Scallops with garlic in gravy. Guangdong dish. Dried scallops steamed with quick-fried mashed garlic and sliced ginger, then marinated in gravy of broth and oyster sauce.

蒜子焖甘鱼 (suànzǐ mèn gānyú)

广东菜。先将蒜瓣和姜煸炒,加清汤,然后放入炸好的甘鱼,慢火煲至汤成奶白色,加盐。

Braised fish with garlic. Guangdong dish. Yellow tail fish deep-fried, simmered in clear soup with quick-fried garlic pulps and ginger, then seasoned with salt.

蒜子牛蹄黄 (suànzǐ niútíhuáng)

山东菜。先将牛蹄煮烂,加蒜、盐、胡椒粉、红糖,勾芡。

Stewed ox hooves with garlic. Shandong dish. Ox hooves stewed with garlic, salt, ground black pepper, and brown sugar, then thickened with starch.

碎烧鲤鱼 (suìshāo lǐyú)

东北菜。先将鲤鱼块炸至金黄,然后加汤、黄酒、盐、葱、姜等焖煮。

Fried and braised carp. Northeastern dish. Deep-fried carp chunks braised in soup with yellow wine, salt, scallion, and ginger.

荪角四宝汤 (sūn jiǎo sìbǎo tāng)

浙江菜。先将竹荪、刺参、鲍贝、鲜鱿、火腿等切片,焯水,然后入高汤煮沸,用盐、味精、紫角叶调味。

Sea food and long net stinkhorn soup. Zhejiang dish. Scalded sliced long net stinkhorn, sea cucumbers, scallops, fresh squid, and ham boiled in broth, then flavored with salt, MSG, and climbing spinach.

笋炒虾片 (sǔn chǎo xiāpiàn)

四川菜。先将虾片炸好,然后加入笋片、红辣椒等速炒。

Quick-fried shrimp chips and bamboo shoots. Sichuan dish. Fried shrimp chips quickly fried with sliced bamboo shoots and red pepper.

笋丝百叶 (sǔnsī bǎiyè)

湖南菜。将牛百叶丝与冬笋、红椒、韭黄及调料翻炒。

Ox tripe with vegetables. Hunan dish. Shredded ox tripe stir-fried with bamboo shoots, red pepper, yellow Chinese chive, and seasonings.

梭梭柴烤肉 (suōsuōchái kǎo ròu)

新疆菜。用戈壁滩梭梭柴烤制的羊肉串。

Grilled mutton. Xinjiang dish. Mutton chops strung on iron bars and grilled over fire of saxoul firewood from Gobi desert.

索康必喜 (suǒkāngbìxǐ)

西藏点心。面粉与酥油调和,做成面皮,包入碎肉,用酥油炸熟。

Butter and meat cakes. Tibetan snack. Flour and Tibetan butter mixed, made into cakes filled with ground meat, then deep-fried with Tibetan butter.

T

塌菇菜烧豆腐 (tāgūcài shāo dòufu)

江浙沪家常菜。先将塌菇菜轻炒，再放入煮过的冻豆腐、笋、火腿，加调料，用小火焖煮。

Simmered winter greens with bean curd. Home-style dish in Jiangsu, Zhejiang, and Shanghai. Chinese winter greens quickly fried, then simmered with frozen bean curd, bamboo shoots, ham, and seasonings.

苔菜明虾 (táicài míngxiā)

广东菜。将明虾去壳留头尾，裹上干面粉、蛋液、苔菜屑炸熟，配椒盐。

Fried prawns with spicy salt. Guangdong dish. Prawn bodies shelled save heads and tails, coated with flour, egg paste, and laver crumbs, then fried. Served with spicy salt.

太白鸡 (tàibáijī)

四川菜。先将鸡腿肉炸熟，然后入鸡汤焖煮，放入泡椒、干辣椒、姜、葱、盐、料酒、酱油、冰糖。

Braised tasty drumsticks. Sichuan dish. Fried drumsticks braised in chicken soup seasoned with pickled pepper, chili pepper, ginger, scallion, salt, cooking wine, soy sauce, and rock sugar.

太和蘸鸡 (tàihézhàn jī)

湖北菜。将童子鸡煮至半熟，抹上料酒和糖浆，炸至金黄，配武当山太和茶和特制卤汁。

Fried chicken with tea and sauce. Hubei Dish. Young rooster half-boiled, coated with cooking wine and syrup, then deep-fried. Served with special sauce and Taihe tea from Mount Wudangshan.

太湖银鱼 (tàihúyínyú)

江苏菜。先将煸炒过的太湖银鱼与蛋液调和，煎至金黄，再与笋丝、木耳及调料稍焖，放入韭菜和猪油。

Fried and simmered icefish. Jiangsu dish. Icefish from Lake Taihu quick-fried, mixed with egg paste, fried golden, then simmered with shredded bamboo shoots, wood ears, Chinese chive, and lard.

太极碧螺春 (tàijí bìluóchūn)

江苏菜。先将鸡茸、鱼茸、干贝茸加调料煮成肉羹，再用菜泥和碧螺春茶汤在肉羹里勾勒出太极图。

Meat chowder with vegetable and tea. Jiangsu dish. Ground chicken, fish, scallops, and seasonings boiled into chowder. Drawn on surface a *taiji* diagram with mashed vegetable and Biluochun tea soup.

太极双球 (tàijí shuāngqiú)

台湾菜。先将花枝浆和虾浆捏成圆形,用高汤煮熟,然后摆成太极图状。

Boiled cuttlefish and shrimp meatballs. Taiwan dish. Minced cuttlefish and shrimp meat made into balls, boiled in broth, then arranged into a *taiji* diagram on a platter.

太极芋头 (tàijí yùtou)

福建菜。先将芋泥与糖、熟猪油、清水调匀,蒸熟,然后浇熟猪油,用瓜子仁和樱桃在芋泥上缀出太极图案。

Steamed taro cakes. Fujian dish. Mashed taros mixed with sugar, lard, and water, steamed into cake, then washed with prepared lard. Decorated with a *taiji* diagram made of sunflower seeds and dried cherries.

太极组合盘 (tàijí zǔhépán)

山东菜。将干贝松、菜松、圆形白蛋糕片、香菇在圆盘中堆放成太极形,将肉干、黄瓜皮、鱿鱼丝、香菇、鸭掌、菠菜分别装入六个扇形组合盘,摆在太极盘周围。

Meat and vegetable varieties. Shandong dish. Scallop floss, vegetables floss, and round white cakes arranged into a *taiji* diagram. Surrounded by six fan-shape dishes containing dried meat, cucumber peels, shredded squid, dried mushrooms, duck feet, and spinach.

太平猴魁 (tàipínghóukuí)

绿茶。产于安徽黄山市新明乡猴坑一带。外形两叶抱一芽,平扁挺直,全身披白毫。汤色明澈,叶底成朵。

Taiping Houkui Tea. Green tea produced in Houkeng area of Xinming Town in Huangshan city, Anhui. Its dry leaves are flat, straight, and hairy, each piece having 2 leaves holding a leaf bud. The tea is crystal clear with infused leaves opened like flowers.

太平石笑 (tàipíngshíxiào)

福建菜。将石斑鱼片涂上虾胶,每片裹一个去壳鸽蛋,与鱼头、鱼尾及调料一道蒸熟。

Steamed grouper with pigeon eggs. Fujian dish. Banded grouper fillets spread with shrimp glue. Each fillet rolled with a shelled pigeon egg, then steamed with fish head, tail, and seasonings.

太平燕 (tàipíngyàn)

福州菜。将猪肉茸与甘薯粉混合,制成薄肉面皮,用鱼茸、五花肉、虾干、芹菜及调料等制成馅,做成燕状饺子,煮熟,汤里加入紫菜、虾油、葱花,配煮熟去壳鸭蛋上席。

Starch and pork dumplings with duck eggs. Fujian dish. Ground pork and sweet potato starch made into thin sheets, cut into squared wrappings, made into swallow-shape *jiaozi* stuffed with mixture of mashed fish, streaky pork, dried shrimp meat, celery, and seasonings. Boiled in clear soup, then flavored with laver, shrimp oil, and chopped scallion. Served with boiled shelled duck eggs.

太爷鸡 (tàiyéjī)

广东菜。先将母鸡用卤水煮熟,再放入铁锅用茶叶和糖焖熏,然后浇上特制的卤汁。

Sugar-tea smoked chicken. Guangdong dish. Hen marinated in gravy, put in a covered wok, smoked with sugar and tea leaves, then washed with special sauce.

泰安三美豆腐 (tàiān sānměi dòufu)

山东泰安家常菜。先将豆腐蒸熟,再同白菜一道入鲜汤煮熟。

Bean curd and cabbage soup. Home-style dish in Taian, Shandong. Steamed bean curd boiled in seasoned soup with Chinese cabbage.

汤圆 (tāngyuán)

南方春节点心,亦称元宵。将糯米粉加水揉成面团,包入甜豆沙、糖芝麻粉、或猪油拌五仁末等馅,做成汤圆,入沸水煮熟。

Tangyuan, or boiled sweet stuffed sticky rice dumplings. Southern snack especially for the Spring Festival. Sticky rice flour kneaded into round dumplings filled with mashed sweet beans, sugared ground sesame, or larded sugared ground five nuts, then boiled.

摊黄菜 (tānhuángcài)

山东菜。把汆过的海参和轻炒过的虾仁同鸡蛋浆调匀,入锅摊成圆饼状,煎至微黄。

Fried sea cucumbers, shrimp, and eggs. Shandong dish. Scalded sea cucumbers and quick-fried shrimp meat mixed with egg paste, then fried into a pancake.

摊黄瓜丝饼 (tān huángguāsībǐng)

湖南点心。将面粉与黄瓜丝、虾仁泥、鸡蛋浆、葱花、盐、水调和,烙熟。

Cucumber cakes. Hunan snack. Flour, shredded cucumbers, minced shrimp meat, egg paste, chopped scallion, salt, and water mixed, then pan fried into cakes.

坛肉炖干豆腐 (tánròu dùn gāndòufu)

东北菜。先将五花猪肉块过油,加汤及调料炖烂,然后与干豆腐稍焖。

Stewed pork and bean curd. Northeastern dish. Quick-boiled cubed streaky pork stewed in seasoned soup, then simmered with dried bean curd.

坛子肉 (tánziròu)

山东菜。将猪肋肉放入瓷罐,加酱油、冰糖、肉桂、葱、姜,煨至汤浓肉烂。

Stewed pork with rock sugar. Home-style dish in Shandong. Pork with ribs stewed with soy sauce, rock sugar, cinnamon, scallion, and ginger in a porcelain container until reduced.

坦洋功夫茶 (tǎnyáng gōngfuchá)

红茶。中国名茶,产于福建福安、拓

荣等地，外形细长匀整，带白毫，色泽乌黑。汤色艳黄，叶底红匀光滑。

Tanyang gongfucha, famous black tea produced in Fu'an and Tuorong, Fujian. Its dry leaves are slim, even, black, and hairy. The tea is yellow with infused leaves red and glossy.

汤爆双脆 (tāngbào shuāngcuì)

山东菜。先将鸡肫和猪肚水焯，加黄酒和盐拌匀，然后浇上用清汤、葱、姜、辣椒、黄酒等调制的热汁。

Sautéed chicken gizzard and hog tripe. Shandong dish. Chicken gizzard and hog tripe quick-boiled, flavored with yellow wine and salt, then marinated in hot sauce of clear soup, green onion, ginger, red pepper, and yellow wine.

汤川天笋 (tāngchuāntiānsǔn)

福建菜。将水发笋干煸炒，然后加高汤和调料烹煮，勾芡。

Braised bamboo shoots. Fujian dish. Soaked bamboo shoots lightly fried, boiled in seasoned broth, then thickened with starch.

汤大玉 (tāngdàyù)

江苏菜。大玉即大虾仁。用鸡汤煮新鲜大虾仁。

Prawn meat in chicken soup. Jiangsu dish. Fresh shelled prawns boiled in chicken soup.

塘芹桂林炒鸽松 (tángqín guìlín chǎo gēsōng)

广东菜。将鹌鹑肉、腊鸭肝、马蹄、香菇、西芹等切成丁，加盐、料酒、蒜茸翻炒。

Quail and duck liver with vegetable. Guangdong dish. Diced quail meat, smoked duck liver, water chestnuts, dried mushrooms, and American celery stir-fried with salt, cooking wine, and mashed garlic.

糖醋脆皮鸡 (tángcù cuìpíjī)

南方菜。先将嫩鸡汆过，炸至半熟，挂糖醋浆，风干，然后炸至棕红。

Crisp sweet and sour chicken. Southern dish. Quick-boiled young chicken half-fried, coated with sugar and vinegar sauce, air-dried, then deep-fried.

糖醋带鱼 (tángcù dàiyú)

江浙沪皖家常菜。带鱼盐腌后炸透，加糖、醋、黄酒、姜、葱等烹煮。

Sweet and sour hairtail fish. Home-style dish in Jiangsu, Zhejiang, Shanghai, and Anhui. Hairtail fish pickled with salt, deep-fried, then simmered in sugar, vinegar, yellow wine, ginger, and scallion.

糖醋咕噜肉 (tángcù gūlūròu)

浙江菜。先将五花肉片腌渍，裹上鸡蛋淀粉浆，炸至半熟，然后同笋块拌炒，用糖、醋、蒜、辣椒、葱、番茄酱等调味，勾芡。

Sweet and sour pork. Zhejiang dish. Pickled sliced streaky pork coated with egg and starch paste, half-fried, stir-fried with bamboo shoots, flavored with sugar, vinegar, garlic, pepper, scallion, and tomato sauce, then thickened with starch.

糖醋辣白菜 (tángcù làbáicài)

西南家常凉菜。先将大白菜丝腌渍,浇糖醋汁,然后加入姜和干辣椒,浇热花椒油。

Sweet and sour Chinese cabbage. Home-style cold dish in southwestern areas. Pickled shredded Chinese cabbage drenched with sugared vinegar, then flavored with ginger, chili pepper, and heated Chinese prickly ash oil.

糖醋里脊 (tángcù lǐji)

各地家常菜。先将猪里脊肉挂糊油炸,然后浇上用醋、糖、酱油等制成的卤汁。

Sweet and sour pork tenderloin. Home-style dish in many places. Pork tenderloin coated with egg and starch paste, deep-fried, then washed with sauce of vinegar, sugar, and soy sauce.

糖醋鲤鱼 (tángcù lǐyú)

各地家常菜。先将鲤鱼稍腌,裹上面粉糊炸透,然后浇上糖醋卤汁。

Fried sweet and sour carp. Home-style dish in many places. Pickled carp coated with flour paste, deep-fried, then marinated in sweet and sour sauce.

糖醋排骨 (tángcù páigǔ)

各地家常菜。先将排骨块腌渍,挂蛋浆芡粉糊油炸,然后浇上醋糖卤汁。

Sweet and sour pork ribs. Home-style dish in many places. Pickled chopped pork ribs coated with egg and starch paste, deep-fried, then drenched with vinegar and sugar sauce.

糖醋三丝 (tángcù sānsī)

山东凉菜。先将白菜心、梨、山楂切成丝,然后浇上用白糖、醋、清水调成的汁。

Cabbage, pears, and haw with sauce. Shandong cold dish. Shredded baby Chinese cabbage, pears, and haws washed with dressing of sugar, vinegar, and water.

糖醋酥豌豆 (tángcù sūwāndòu)

各地家常菜。先将豌豆炸酥,然后用糖、醋、麻油等调成汁,浇在豌豆上。

Fried sweet and sour peas. Home-style dish in many places. Peas fried crisp, then washed with dressing of sugar, vinegar, and sesame oil.

糖醋驼峰 (tángcù tuófēng)

内蒙菜。先将驼峰片用盐和料酒盐渍,再裹上鸡蛋淀粉糊油炸,然后用糖醋汁烹煮,勾芡,淋油。

Sweet and sour camel hump. Inner Mongolian dish. Sliced camel hump pickled with salt and cooking wine, coated with egg and starch paste, then deep-fried. Boiled in sugar and vinegar soup, thickened with starch, then washed with lard.

糖醋鲜藕 (tángcù xiānǒu)

浙江家常菜。先将藕条挂面糊炸至金黄,再与蔬菜汤、酱油、白糖、醋翻炒,勾芡,淋猪油。

Sweet and sour lotus roots. Home-style dish in Zhejiang. Slivered lotus roots coated with flour paste, deep-fried, sautéed with vegetable soup, soy sauce, sugar, and vinegar, then washed with starched soup and lard.

糖醋鱼 (tángcùyú)

各地家常菜。先把鱼两面炸黄，然后加入糖、醋、酱油、姜、葱等烹煮，勾芡。

Sweet and sour fish. Home-style dish in many places. Fish fried until golden on both sides, simmered with sugar, vinegar, soy sauce, ginger, and scallion, then thickened with starch.

糖酱鸡块 (tángjiàng jīkuài)

山东菜。先将鸡块腌渍，稍炸，然后加糖和甜酱拌匀，用小火煨熟。

Simmered chicken with sweet sauce. Shandong dish. Pickled chicken chunks lightly fried, mixed with sugar and sweet fermented flour sauce, then simmered.

糖熘土豆 (tángliū tǔdòu)

东北家常菜。先将土豆块炸至金黄，再将白糖稍炒，加水，勾芡，然后倒入土豆块翻匀。

Sugared fried potatoes.
Northeastern home-style dish. Deep-fried cubed potatoes stir-fried with syrup of lightly fried sugar, starch, and water.

糖熘鲤鱼 (tángliū lǐyú)

山东菜。先将鲤鱼挂芡炸至金黄，然后放入用酱油、糖、醋、葱、姜、蒜、芡粉等做成的汤烹煮，浇原汤汁和鸡油。

Braised sweet and sour carp. Shandong dish. Deep-fried starched carp braised in soup seasoned with soy sauce, sugar, vinegar, scallion, ginger, garlic, and starch, then washed with broth and chicken oil.

糖麻圆 (tángmáyuán)

贵州点心。将糯米粉加水、碱、糖调和，揉成圆球，温油炸至金黄壳脆，撒芝麻。

Fried sweet sticky rice balls. Guizhou snack. Sticky rice flour mixed with water, soda, and sugar, kneaded into balls, fried until golden and crisp, then showered with sesame.

糖窝丝 (tángwōsī)

黑龙江点心。将面皮包上糖，搓成条，盘成饼，烙熟。

Baked sweet cakes. Heilongjiang snack. Flour dough sheets made into rolls filled with sugar, kneaded into ropes, coiled into cakes, then baked.

糖粘羊尾 (tángzhān yángwěi)

四川菜。先将猪肥膘片焯水，裹蛋浆豆粉糊炸至金黄，然后放入熬至起丝的白糖浆迅速翻炒。

Sugarcoated fat pork. Sichuan dish. Scalded sliced fat pork coated with mung bean powder and egg paste, deep-fried until golden, then quickly mixed with heated melting sugar.

糖汁草莓 (tángzhī cǎoméi)

各地家常菜。将草莓用白糖拌匀，

把银耳放在草莓中间，把芹菜放在草莓四周，撒糖。

Strawberry with white wood ears and celery. Home-style cold dish in many places. Strawberries mixed with sugar, then put on a platter with white wood ears at the center. Garnished with celery and showered with white sugar.

烫干丝 (tànggānsī)

江苏凉菜。将豆腐干用开水烫过，浇卤汁、麻油，撒姜丝、虾米。

Bean curd with flavorings. Jiangsu cold dish. Scalded dried bean curd washed with sauce and sesame oil, then sprinkled with shredded ginger and dried shrimp meat.

烫韭菜 (tàngjiǔcài)

河南家常菜。将韭菜氽过，浇上酱油、醋、蒜末、麻油等调成的汁。

Boiled chive with garlic dressing. Home-style dish in Henan. Chinese chive quick-boiled, then flavored with dressing of soy sauce, vinegar, mashed garlic, and sesame oil.

桃仁烩口蘑 (táorén huì kǒumó)

东北菜。先将口蘑蒸熟切片，核桃仁炸酥，然后将口蘑放入鲜汤烧沸，加入核桃仁及调料，勾芡。

Mushrooms with walnuts. Northeastern dish. Sliced steamed Mongolian mushrooms boiled in seasoned soup with fried walnuts, then thickened with starch.

桃仁鸡丁 (táorén jīdīng)

四川菜。先将核桃仁和鸡丁过油，然后加入香菇、辣椒、酱油、芹菜等翻炒。

Stir-fried walnuts and chicken. Sichuan dish. Lightly fried walnuts and diced chicken stir-fried with dried mushrooms, chili pepper, soy sauce, and celery.

桃源铜锤鸡腿 (táoyuán tóngchuí jītuǐ)

湖南桃源菜。先将土鸡鸡腿腌渍，再裹上芡粉蛋糊油炸。

Fried drumsticks. Dish in Taoyuan, Hunan. Pickled drumsticks of local chicken coated with starched egg paste, then deep-fried.

桃子色拉 (táozi sèlā)

浙江菜。先将去皮鲜桃放入用糖、丁香花、小豆蔻制成的糖水浸泡，然后去核、切片，浇上特制的卤汁。

Peach salad. Zhejiang dish. Peeled peaches steeped in soup of sugar, clove buds, and cardamom, cored, sectioned, then washed with special sauce.

陶陶居上月 (táotáojū shàngyuè)

广东著名点心。用面粉和油制皮，包入烧鸭、火腿、花生仁、莲子、蛋黄、糖、面粉、麻油等制成的馅，烤熟。

Moon cake with meat and nut filling. Famous Guangdong snack. Flour and oil made into cake wrappers, filled with mixture of broiled duck, ham, peanuts, lotus seeds, yolks, sugar, flour, and sesame oil, then baked.

套四宝 (tàosìbǎo)

河南菜。鸭、鸡、鸽、鹌鹑四禽层层相套,加调料蒸熟。

Four kinds of steamed poultry. Henan dish. Duck stuffed with a whole chicken. The chicken stuffed with a pigeon over a quail. The quail stuffed with assorted ingredients. Seasoned, then steamed.

特色水豆腐 (tèsè shuǐdòufu)

东北家常菜。将嫩豆腐,青豆,草菇,素肉酱磨成粗末,蒸熟定型,撒姜丝,麻油。

Steamed special bean curd. Home-style dish in northeastern areas. Tender bean curd, green beans, meadow mushrooms, and soy chicken sauce ground, steamed, then flavored with shredded ginger and sesame oil.

滕王阁红酥肉 (téngwánggé hóngsūròu)

江西菜。先将五花肉茸与蛋浆、芡粉、盐、水调和,抹在带皮肥膘上,撒生粉,炸至金黄,然后入高汤加调料焖煮,配绿叶菜装盘。

Fried and braised pork. Jiangxi dish. Ground streaky pork mixed with egg paste, starch, salt, and water, spread on fat pork with skins, powdered with dry starch, then deep-fried golden. Braised in broth with seasonings, then garnished with green vegetable.

提丝发糕 (tísī fāgāo)

四川点心。先将面粉加糖水调和并发酵,加入猪油,搓成细条,蒸熟成糕丝,然后撒白糖、糖桂花、芝麻。

Steamed hairy cakes. Sichuan snack. Flour, sugar, and water mixed, then fermented. Mixed with lard, then kneaded into fine slivers. Steamed into hairy cakes, then showered with granulated sugar, sweetened osmanthus flower, and sesame.

蹄筋北菇扒菜胆 (tíjīn běigū pá càidǎn)

广东菜。先将蹄筋与冬菇爆炒,用黄酒、蚝油调味,然后放在氽过的菜胆上。

Pork tendons and mushrooms over cabbage. Guangdong dish. Pork tendons and winter mushrooms stir-fried, flavored with yellow wine and oyster sauce, then served on a bed of quick-boiled baby Chinese cabbage.

天荡薇菜鱼丝 (tiāndàng wēicài yúsī)

陕西菜。先将鱼肉丝腌渍,划油,然后与薇菜加调料翻炒,勾芡。

Sautéed fish and common vetch. Shaanxi dish. Pickled shredded fish meat lightly fried, sautéed with common vetch and flavorings, then starched.

天下第一菜 (tiānxià dì-yī cài)

江苏、浙江菜。先将锅巴炸脆,再将虾仁、鸡丝及调料入汤烧开,勾芡,然后浇在锅巴上。

Fried rice crusts in chicken soup. Jiangsu and Zhejiang dish. Sticky rice crusts fried crisp, then washed with boiling soup of shrimp meat, shredded chicken, seasonings, and starch.

田鸡饭煲 (tiánjī fànbāo)

广东食品。先将腌过的田鸡加入红枣、云耳、葱煮熟,倒在米饭上,然后加入姜汁、蚝油煲煮。

Braised frogs and rice. Guangdong food. Pickled frogs boiled with red dates, white fungus, and scallion, then braised with cooked rice, ginger juice, and oyster sauce.

田螺塞肉 (tiánluó sāi ròu)

广东菜。先将猪肉和田螺肉剁碎,加盐、葱、姜、黄酒等制成馅,然后塞入田螺壳,蒸熟。

Steamed snail meat and pork. Guangdong dish. Minced pork, field snails meat, salt, scallion, ginger, and yellow wine mixed into filling, put in snail shells, then steamed.

田园翠笋 (tiányuán cuìsǔn)

江浙皖菜。将滑熟的牛肉片与芦笋、甜椒快炒,用盐、葱、姜等调味,勾芡。

Sautéed beef and asparagus. Home-style dish in Jiangsu, Zhejiang, and Anhui. Lightly fried sliced beef sautéed with asparagus, sweet pepper, salt, scallion, and ginger, then thickened with starch.

甜豆炒肾球 (tiándòu chǎo shènqiú)

广东菜。将菜豆、鹅肾、甘笋花、花枝片等同炒,用蒜茸、姜、盐、料酒调味。

Stir-fried snap beans and goose kidney. Guangdong dish. Snap beans and goose kidney stir-fried with carrot cuts and sliced cuttlefish, then flavored with mashed garlic, ginger, salt, and cooking wine.

甜金瓜丸 (tiánjīnguāwán)

南方点心。先将南瓜泥与糕粉、糖、水和成面团,包豆沙馅,做成蛋形,炸至金黄,然后浇糖汁,撒芝麻和糖粉。

Fried stuffed pumpkin eggs. Southern snack. Mashed pumpkin mixed with cake powder, sugar, and water, made into egg-shape buns filled with bean paste, then fried golden. Coated with syrup, then sprinkled with sesame and sugar powder.

条头糕 (tiáotóugāo)

上海点心。用糯米粉做成筒状,包甜豆沙蒸熟。

Sticky rice cake with bean paste. Shanghai snack. Tube-like sticky rice cake filled with sweet red or black bean paste, then steamed.

跳竹蛏 (tiàozhúchēng)

江苏菜。先将蛏肉急炒,再与笋片拌炒,加盐、姜末、料酒、原汁,勾芡,撒白胡椒粉。

Razor clam with bamboo shoots. Jiangsu dish. Razor clam meat quick-fried, sautéed with prepared sliced bamboo shoots, flavored with salt, minced ginger, cooking wine, and broth, thickened with starch, then sprinkled with ground white pepper.

铁板海皇豆腐 (tiěbǎn hǎihuáng dòufu)

广东菜。先将豆腐裹上芡粉蛋浆油炸,然后浇上用虾仁、带子、鲜鱿、鲍汁做成的海鲜酱,装在热铁盘里上桌。

Fried bean curd with seafood sauce. Guangdong dish. Bean curd coated with egg and starch paste, then deep-fried. Covered with sauce of shrimp, shellfish, fresh squid, and abalone broth. Served on a hot iron platter.

铁板蚝油栗子鸡 (tiěbǎn háoyóu lìzi jī)

山东菜。先将栗子稍炸,然后和鸡块扒烂,用蚝油、糖、黄酒等调味,装在烧热的铁盘里上桌。

Braised chicken and chestnuts. Shandong dish. Lightly fried chestnuts braised with chopped chicken, flavored with oyster sauce, sugar, and yellow wine, then served on a hot iron platter.

铁板鲈鱼 (tiěbǎn lúyú)

南方菜。先将鲈鱼腌过,炸至金黄,放在加热的铁板上,然后将玫瑰酱、豆瓣酱、芡粉等制成卤汁,浇在鱼上。

Bass with savoury sauce. Southern dish. Pickled bass deep-fried, put on a hot iron platter, then washed with gravy of rose sauce, fermented broad bean sauce, and starch.

铁板牛柳 (tiěbǎn niúliǔ)

广东菜。先将牛柳腌渍,过油,然后和洋葱、青椒、红甜椒及调料翻炒,装在热铁盘里上桌。

Stir-fried beef with onion and pepper. Guangdong dish. Pickled beef lightly fried, stir-fried with scallion, green pepper, sweet red pepper, and other seasonings, then served on a hot iron platter.

铁板鲜鱼 (tiěbǎn xiānyú)

广东菜。将鲜鱼腌渍,用文火煎熟,浇上用柠檬、茴香、辣椒、番茄酱等制成的卤汁,装在热铁盘里,撒葱花和黄油屑。

Fish with spicy lemon sauce. Guangdong dish. Pickled fresh fish fried over low heat, then washed with dressing of lemon, fennel, hot pepper, and tomato sauce. Put on a hot iron platter, then sprinkled with chopped scallion and butter crumbs.

铁观音 (tiěguānyīn)

乌龙茶。中国名茶,产于福建安溪。外形肥壮紧结,质重,呈砂绿色。汤色浅金黄,香高久远。叶底肥厚柔软,镶红边。

Tieguanying, famous oolong tea produced in Anxi, Fujian. Its dry leaves are stout, tight, solid, and sandy green. The tea is light golden with a rich and enduring aroma. The infused leaves are thick and soft with red fringes.

铁观音炖子鸡 (tiěguānyīn dùn zǐjī)

福建、广东菜。将草鸡和枸杞子放

入铁观音茶汤,加盐炖熟。

Chicken with tea and herb. Fujian and Guangdong dish. Hen stewed with Chinese wolfberry in *tieguanyin* tea soup and seasoned with salt.

铁锅鱼头 (tiěguō yútóu)

湖北菜。先将草鱼头稍煎,烹入醋和酱油,然后加高汤和调料煮熟,放在玉米面做的贴饼上。

Braised carp head with corn cakes. Hubei dish. Grass carp head quick-fried, flavored with vinegar and soy sauce, then boiled in seasoned broth. Served over baked corn cakes.

铁盔将军鸭 (tiěkuī jiāngjūnyā)

湖南菜。将洞庭湖产的老水鸭加入特制调料,慢火煨炖。

Stewed duck. Hunan dish. Mature duck from Dongting Lake stewed with special seasonings.

铁素四色球 (tiěsù sìsèqiú)

山东菜。先将青笋、胡萝卜、南荠用鸡汤煨熟,然后配蒸熟的鲜蘑装盘,浇原汤。

Four vegetables in chicken soup. Shandong dish. Stem lettuce, carrots, and water chestnuts simmered in chicken soup, combined with steamed fresh mushrooms, then marinated in broth.

桐叶粑粑 (tóngyè bāba)

湖南点心。用糯米粉和籼米粉加香蒿汁和清水调和,做成圆块,内包红豆或腊肉丁,外面用桐叶包好,蒸熟。

Steamed rice cake in tung leaves. Hunan snack. Sticky rice flour and non-sticky rice flour mixed with sweet wormwood juice and water, made into flat round cakes filled with red beans or diced smoked cured pork, wrapped in Chinese tung leaves, then steamed.

筒仔米糕 (tǒngzǎi mǐgāo)

台湾点心。先将五花肉、红葱头、香菇、虾米等炒熟,再和糯米、咸蛋黄一道放入杯中蒸熟。

Sticky rice and pork cakes. Taiwan snack. Stir-fried streaky pork, red onion heads, dried mushrooms, and shrimp meat steamed with sticky rice and salted egg yolks in mugs.

筒子肉 (tǒngziròu)

北方菜。将猪肉末加调料做成馅,用猪网油裹馅成卷,挂鸡蛋芡粉浆,先略蒸,然后煎至金黄。

Fried meat rolls. Northern dish. Seasoned ground pork rolled in web lard, coated with egg and starch paste, quick-steamed, then fried golden.

头粿 (tóukè)

福建点心。先将米浆、萝卜丝、虾米、盐拌匀,倒入抹油的粿盘拉平,然后蒸熟切块,炸至金黄。

Fried turnip and shrimp cakes. Fujian snack. Rice paste mixed with shredded turnips, shrimp meat, and salt, put in a larded flat container, then steamed into a lump cake. Cut into cubes, then deep-fried.

土豆饼 (tǔdòubǐng)

福建点心。将土豆泥、面粉、黄油、

蜂蜜、白糖、水等拌和，做成小饼，煎熟。

Fried potato cakes. Fujian snack. Mashed potato, flour, butter, honey, sugar, and water mixed, made into small cakes, then deep-fried.

土豆炒羊头 (tǔdòu chǎo yángtóu)

西藏菜。将羊头肉与土豆、红辣椒翻炒。

Stir-fried potatoes, mutton and red pepper. Tibetan dish. Lamb head meat stir-fried with potatoes and red pepper.

土豆罐焖鸡 (tǔdòu guànmèn jī)

湖南菜。先将炸过的鸡腿和香菇略炒，然后加入红枣、土豆及调料，小火焖煮。

Braised potatoes and drumsticks. Hunan dish. Deep-fried drumsticks stir-fried with dried mushrooms, then braised with red dates, potatoes, and flavorings.

土豆烧牛肉 (tǔdòu shāo niúròu)

东北家常菜。先将土豆炸至金黄，再将牛肉块炝黄酒、酱油稍炒，然后将牛肉和土豆入汤慢炖。

Stewed beef and potatoes. Home-style dish in northeastern areas. Cubed beef stir-fried with yellow wine and soy sauce, then stewed with deep-fried potatoes.

土鸡板栗罐 (tǔjī bǎnlì guàn)

湖南菜。先将土鸡块腌渍，然后加入板栗、色拉油、胡椒粉、辣椒酱，装入瓦罐焖煮。

Braised chicken and chestnuts. Hunan dish. Pickled range chicken mixed with chestnuts, vegetable oil, ground black pepper, and chili sauce, then braised in a clay pot.

土笋冻 (tǔsǔndòng)

福建泉州点心。土笋即星虫加水熬煮，冷却成冻，配海蜇、芫荽、萝卜丝、番茄片、香辣酱。

Sea worm jelly. Special snack in Quanzhou, Fujian. Sea worms boiled thoroughly, chilled into jelly, then served with jelly fish, coriander, shredded radish, sliced tomatoes, and spicy sauce.

氽金银丝 (tǔn jīnyínsī)

朝鲜族菜。先将狗肉丝和鸡肉丝挂浆，滑油，然后与冬笋、黄瓜及调料煸炒。

Stir-fried dog and chicken. Korean ethnic minority dish. Shredded dog meat and chicken starched, lightly fried, then stir-fried with winter bamboo shoots, cucumbers, and flavorings.

脱壳鳜鱼 (tuōqiào guìyú)

江苏菜。先将鳜鱼用盐和香糟汁腌渍，然后扑上淀粉，抹上猪油，裹上蛋糊，炸至皮脆，淋麻油，撒胡椒粉和椒盐。

Fried Chinese perch with spicy salt. Jiangsu dish. Chinese perch pickled with salt and fermented sticky rice juice, dusted with dry starch, larded, coated with egg paste, then fried crisp. Sprinkled with sesame oil, ground black pepper, and spicy salt.

脱内鳜鱼 (tuōnèi guìyú)

山东菜。先将鳜鱼切为头、身两段，再将鱼头炸透，鱼身去骨炸至金黄，然后一并装盘，浇上用香菇、鸡汤、油等调制的卤汁。

Fried Chinese perch with sauce. Shan-dong dish. Chinese perch cut into 2 sections of head and body. Body section boned and fried golden. Head section deep-fried. Both sections marinated in sauce of dried mushrooms, chicken soup, lard, and other condiments.

沱牌曲酒 (tuópáiqūjiǔ)

白酒，产于四川射洪。以高粱、糯米为原料，用小麦、大麦制曲发酵，酒精含量为 38-54 度，属浓香型酒。

Tuopaiqujiu, liquor produced in Shehong, Sichuan. White spirit made with sorghum, glutinous rice, and yeast of wheat and barley. Contains 38-54% alcohol, and has a thick aroma.

W

蛙式黄鱼 (wāshì huángyú)

江苏菜。先将黄鱼用黄酒和酱油浸渍,切成两瓣蛙形片,扑上干淀粉,炸至深黄,然后配青菜叶装盘,浇卤汁。

Fried yellow croaker with sauce. Jiangsu dish. Yellow croaker pickled with yellow wine and soy sauce, cut into 2 frog-shape pieces, dusted with dry starch, then deep-fried. Garnished with green vegetable leaves, then washed with dressing.

瓦缸田螺土鸡汤 (wǎgāng tiánluó tǔjī tāng)

湖南菜。在高汤里放入土鸡和田螺,烧开后改用小火煨煮。

Chicken and field snail soup. Hunan dish. Range chicken and field snails boiled in broth, then simmered.

豌豆粉蒸肉 (wāndòu fěnzhēngròu)

南方家常菜。将大块肉片与红糖、姜、葱、酱油、醪糟汁、米粉拌和,面上铺豌豆,蒸至肉烂。

Steamed pork steaks with peas. Home-style dish in southern areas. Pork steaks mixed with brown sugar, ginger, scallion, soy sauce, sweet fermented sticky rice juice, and ground rice, covered with peas, then steamed.

豌豆黄 (wāndòuhuáng)

北京点心。将豌豆煮成糊,放入白糖和糖桂花,凝固后切成三角形,缀蜜饯。

Pea cakes. Beijing snack. Peas boiled into paste, mixed with sugar and sweet osmanthus flower, chilled into lump cake, cut into triangle pieces, then garnished with candied fruits.

豌豆鸡丝 (wāndòu jīsī)

福建菜。先将鸡丝挂浆,划油,再与焯过的豌豆拌炒,加调料,勾芡。

Sautéed chicken and green peas. Fujian dish. Shredded chicken coated with wet starch, lightly fried, sautéed with scalded peas, then thickened with starch.

万峦猪脚 (wànluánzhūjiǎo)

台湾万峦名菜。先将猪脚稍煮,速冻,然后放入烧热的卤汁煮透。

Marinated hog trotters. Famous dish in Wanluan, Taiwan. Hog trotters quick-boiled, quickly chilled, then boiled in gravy.

王朝半干白葡萄酒 (wángcháo bàngān báipútáojiǔ)

葡萄酒,产于天津,用麝香葡萄为原

料酿成，酒精度 12 度。

Dynasty half-dry white wine produced in Tianjin, China. Made with muscat grapes. Contains 12% alcohol.

威化肉粒 (wēihuà ròulì)

广东菜。用猪肉丁、香菇、笋丝、甜椒丝、葱丝、盐等做成馅，用威化纸包成条状炸熟。

Fried pork and mushroom package. Guangdong dish. Diced pork, dried mushrooms, shredded bamboo shoots, sweet pepper, scallion, and salt mixed into stuffing, wrapped into tiny pillows with edible rice paper, then deep-fried.

威化鱼丝卷 (wēihuà yúsījuǎn)

广东菜。用鳕鱼、香菇、笋丝、甜椒丝及调料做成馅，用威化纸包成条状炸熟。

Fried codfish and mushroom package. Guangdong dish. Codfish, dried mushrooms, shredded bamboo shoots, sweet pepper, and seasonings mixed into stuffing, wrapped into tiny pillow-shape packages with edible rice paper, then deep-fried.

威化纸包鸡 (wēihuàzhǐ bāo jī)

广东菜。先将鸡肉片用胡椒粉、豉汁等料浸腌，再用威化纸包成小方块炸熟。

Fried chicken with fermented soybean juice. Guangdong dish. Sliced chicken pickled with ground black pepper and fermented soybean juice, packed into small cubes with edible rice paper, then deep-fried.

微山湖荷香鸭 (wēishānhú héxiāngyā)

山东微山湖菜。将本地鸭腌渍，与毛桃、白芷、香叶等置荷叶上蒸熟。

Steamed duck on lotus leaves. Specialty in Weishanhu Lake area, Shandong. Pickled local duck steamed on a lotus leaf with wild peaches, angelica roots, and bay leaves.

煨炒猴头蘑 (wēichǎo hóutóumó)

朝鲜族菜。将猴头蘑加调料入鸡汤煮熟，勾芡，装盘，配炒熟的牛肉丝。

Boiled mushrooms with fried beef. Korean ethnic minority dish. Monkey-head mushrooms boiled in seasoned chicken soup and thickened with starch. Served with stir-fried shredded beef.

煨豆腐 (wēidòufu)

福建家常菜。将豆腐块、目鱼、猪脚、排骨等加入调料，用炭火煨炖，佐酱油、蒜、冰糖等做成的卤汁。

Braised bean curd with meat varieties. Home-style dish in Fujian. Bean curd braised with milk fish, hog trotters, spareribs, and seasonings over charcoal. Served with dressing of soy sauce, garlic, and rock sugar.

煨烤鳜鱼 (wēikǎo guìyú)

广东菜。先将鳜鱼腌渍，再将炒熟的肉丝填入鳜鱼腹，用鲜荷叶将鱼包好，然后用炭火烤熟。

Grilled Chinese perch. Guangdong

dish. Pickled Chinese perch stuffed with stir-fried shredded pork, wrapped in fresh lotus leaves, then grilled over charcoal.

潍坊一晶锅 (wéifāng yījīngguō)

山东潍坊菜。先将猪蹄、肘子、四喜丸子、海参、鱼肚、鱼翅、干贝等蒸透,再放入清汤加调料煮沸。

Boiled pork varieties and seafood. Specialty in Weifang, Shandong. Hog trotters, knuckles, meatballs, sea cucumbers, fish maw, shark fin, and dried scallops steamed, then boiled in seasoned clear soup.

味噌鸡腿 (wèicēng jītuǐ)

福建、台湾菜。将鸡腿去骨,用味噌酱即豆面酱腌渍,烤熟,放在生菜叶上。

Broiled chicken legs. Fujian and Taiwan dish. Boned drumsticks pickled with miso sauce, broiled, then served on fresh lettuce.

温拌腰片 (wēnbàn yāopiàn)

东北菜。先将猪腰片、兰片、青椒片分别焯水,然后加盐、酱油、醋、辣酱、麻油拌匀。

Hog kidney with vegetables.

Northeastern dish. Hog kidney, soaked bamboo shoots, and sliced green pepper scalded separately, then mixed with salt, soy sauce, vinegar, chili sauce, and sesame oil.

温州黄汤 (wēnzhōuhuángtāng)

黄茶,又称平阳黄汤。产于浙江温州,外形条索细紧,色泽黄绿多毫。汤色橙黄透明,叶底嫩匀。

Wenzhouhuangtang, also called *Pingyanghuangtang*, yellow tea produced in southern areas of Zhejiang. Its dry leaves are slim, tight, yellowish green, and hairy. The tea is crystal yellowish with infused leaves tender and even.

文昌按粑 (wénchāng'ànbā)

海南文昌点心。先将糯米粉加水、油和成面团,做成小圆饼,用清水煮熟,然后裹上椰茸、红糖、炒花生粉、炒白芝麻粉。

Sticky rice cake with tasty powder. Snack in Wenchang, Hainan. Sticky rice flour, water, and oil mixed, made into small cakes, then boiled. Coated with mixture of mashed coconut meat, brown sugar, ground baked peanuts, and ground white sesame.

文昌空心煎堆 (wénchāng kōngxīn jiānduī)

海南文昌点心。先将糯米团煮熟,再加入生面粉和糖做成空心球,经反复油炸至膨胀酥脆。

Fried sticky rice balls. Special snack in Wenchang, Hainan. Sticky rice flour balls boiled, kneaded with flour and sugar, made into hollow balls, then fried several times until puffy and crisp.

文楼涨蛋 (wénlóuzhàngdàn)

江苏菜。将鸡蛋、肥猪肉粒、面包屑、虾仁、鸡汤及调料搅拌成糊,煎至膨胀,加盖略焖。

Simmered eggs with pork and

shrimp. Jiangsu dish. Egg paste, diced fat pork, bread crumbs, shrimp meat, and clear chicken soup mixed into paste, fried until spongy, then simmered briefly in a covered pot.

文思豆腐羹 (wénsī dòufugēng)

江苏菜。香菇丝、竹笋丝、胡萝卜丝、豆腐条、紫菜丝、肉丝等先炒后煮,加鸡汤,勾芡。

Bean curd and vegetables in chicken soup. Jiangsu dish. Shredded dried mushrooms, bamboo shoots, carrots, bean curd, laver, and pork sautéed, boiled in chicken soup, then thickened with starch.

纹露美鲍 (wénlù měibào)

广东菜。先将鲍鱼片略炒,然后加高汤、冬菇、胡萝卜及调料煮熟。

Abalone with mushrooms and carrots. Guangdong Dish. Sliced abalone lightly fried, then boiled in broth with winter mushrooms, carrots, and flavorings.

炆牛腩 (wènniúnǎn)

广东菜。将牛腩用酱油、蚝油等浸腌,加入水和煸炒过的花椒、八角等用小火炖熟。

Stewed beef flank. Guangdong dish. Beef flank pickled in soy sauce and oyster sauce, then stewed in water with quick-fried Chinese prickly ash and aniseed.

乌发汤 (wūfàtāng)

各地药膳。将何首乌与羊骨、黑豆等炖汤。

Lamb bone and black bean soup. Folk-medicinal dish in many places. Fleeceflower root stewed with lamb bones and black beans.

乌狼鲞烤肉 (wūlángxiǎng kǎoròu)

浙江菜。乌狼鲞即河豚鱼干。先将猪肉和笋块煮熟,然后放入河豚鱼干焖煮,用葱、姜、黄酒等调味。

Globefish with pork and bamboo shoots. Zhejiang dish. Dried salty globefish braised with boiled pork and bamboo shoots. Seasoned with scallion, ginger, and yellow wine.

乌龙 (wūlóng)

乌龙茶。产于福建安溪,属安溪色种。外形瘦小质轻,色泽褐绿。汤色青黄,带焦糖香。叶底叶张薄,叶脉浮现。

Wulong or oolong, one of the oolong varieties from Anxi, Fujian. Its dry leaves are light, small, and dark green. The tea is crystal yellowish green with a caramel aroma. The infused leaves are thin and veiny.

乌龙茶 (wūlóngchá)

半发酵茶。鲜茶叶经萎凋、做青、炒青、揉捻、干燥等工序制成。干茶叶呈黑褐色,茶汤清澈金黄,兼有红茶和绿茶的品质特征。

Oolong tea or semi-fermented tea. It is processed by withering, shaking, light-baking, rubbing-twisting, and drying. Its dry leaves are black-brown. The tea is clear and golden, having the characteristics of both

black tea and green tea.

无花果海底椰生鱼煲 (wúhuāguǒ hǎidǐyē shēngyú bāo)

香港菜。将黑鱼块煎至金黄,加入无花果和海底椰炖熟。

Fish stewed with figs and sea coconuts. Hong Kong dish. Deep-fried snakehead fish stewed with dried figs and sea coconuts.

无为板鸭 (wúwéibǎnyā)

安徽无为菜。先将鸭用盐腌渍,开水烫过,再入熏锅用木屑焖熏,然后加调料焖煮。

Smoked duck. Specialty in Wuwei, Anhui. Duck pickled with salt, scalded with boiling water, smoked with sawdust in a covered wok, then braised with seasonings.

吴家熏肉 (wújiāxūnròu)

山西菜。先将鸡、猪头、猪心、猪肝、猪肠、猪肚、猪蹄、猪舌等放入老汤卤熟,然后用秘制熏料熏制。据传由吴姓家族所创。

Marinated and smoked meat varieties. Shanxi dish. Chicken and hog offal, such as head, heart, liver, large intestine, tripe, trotter, and tongue, marinated in aged gravy, then smoked. It was created try a Wu family.

芜糕 (wúgāo)

福建点心。糯米粉加水和糖调成浆,放入铺有薄荷草的竹笼蒸熟。

Sticky rice cake with peppermint. Fujian snack. Sticky rice flour mixed with water and sugar, steamed in a bamboo steamer with peppermint leaves at the bottom.

梧州纸包鸡 (wúzhōu zhǐbāojī)

广西梧州菜。先将鸡用汾酒、五香粉、盐、胡椒粉腌渍,然后用油浸过的玉扣纸包好,油炸。

Fried chicken in bamboo paper. Specialty in Wuzhou, Guangxi. Chicken pickled with Fenjiu liquor, five-spice powder, salt, and ground black pepper, wrapped in oiled bamboo paper, then deep-fried.

五彩拼盘 (wǔcǎi pīnpán)

东北菜。将胡萝卜、白菜、海带、黄瓜、干豆腐、粉皮切丝摆盘,佐特制的卤汁。

Assorted vegetable plate. Northeastern dish. Carrots, Chinese cabbage, kelp, cucumbers, dried bean curd, and bean jelly shredded, arranged on the same plate, then served with special dressing.

五花肉烧冻豆腐 (wǔhuāròu shāo dòngdòufu)

山东家常菜。先将辣椒、蒜、葱、姜爆香,再放入五花肉爆炒,然后加入高汤、冻豆腐块炖煮。

Streaky pork with frozen bean curd. Home-style dish in Shandong. Streaky pork stir-fried with quick-fried chili pepper, garlic, and ginger, then stewed in broth with frozen bean curd.

五粮液 (wǔliángyè)

白酒,产于四川宜宾,用小麦、大米、糯米、高粱、玉米五种谷物酿造,并因此得名。酒精含量在 39-68 度之间。属浓香型酒。

Wuliangye liquor, white spirit produced in Yibin, Sichuan. It is a high alcohol drink made from wheat, rice, glutinous rice, sorghum, and corn. It contains 39-68% alcohol, and has a thick aroma.

五柳青鱼 (wǔliǔ qīngyú)

江苏菜。将青鱼微火焖熟,浇用辣椒、酱姜、酱乳瓜、冬笋、冬菇等做成的卤汁。

Simmered carp with sauce. Jiangsu dish. Black carp simmered, then marinated in sauce of shredded hot pepper, pickled ginger, pickled baby cucumbers, winter bamboo shoots, and winter mushrooms.

五沫烧海参 (wǔmò shāo hǎishēn)

北方菜。先将肉末与切碎的葱、姜、蒜、干辣椒煸炒,然后与海参入鸡汤煨煮,加鸽精。

Sea cucumbers with flavorings. Northern dish. Sea cucumbers stewed in chicken soup with quick-fried ground pork, chopped scallion, ginger, garlic, and chili pepper, then flavored with pigeon essence.

五仁包 (wǔrénbāo)

山东点心。先将炒熟的花生、芝麻、核桃、杏仁、瓜子碾碎,同红枣泥、桂花酱、白糖调成馅,然后用面皮包馅做成包子,蒸熟。

Steamed five-nut buns. Shandong snack. Flour dough made into buns filled with mixture of baked and ground peanuts, sesame, walnuts, almonds, sunflower seeds, red date paste, osmanthus flower sauce, and sugar, then steamed.

五仁金瓜煲 (wǔrén jīnguā bāo)

浙江菜。将炸熟的南瓜、碎核桃仁、松子、腰果、花生、芝麻等一起烹煮。

Pumpkin with five nuts. Zhejiang dish. Fried pumpkin braised with crushed walnuts, pine nuts, cashew nuts, peanuts, and sesame.

五味香卤鸡 (wǔwèi xiānglǔjī)

广东菜。把鸡放入用盐、料酒、酱油、沙姜、草果、甘草、桂皮、八角等制成的卤水中煮熟,浸泡。

Marinated five-taste chicken. Guangdong dish. Boiled chicken marinated in gravy of salt, cooking wine, soy sauce, kaempferia rootstock, tsaoko amomum fruit, licorice root, cinnamon, and aniseed.

五香茶鸡蛋 (wǔxiāng chájīdàn)

各地家常菜。鸡蛋煮八成熟,敲破壳,放入茶叶、盐、花椒、八角、酱油,文火煨煮。

Five-spice eggs. Home-style dish in many places. Eggs half-boiled, cracked, then simmered with tea leaves, salt, Chinese prickly ash, aniseed, soy sauce, and other seasonings.

五香豆腐干 (wǔxiāng dòufugān)

家常菜。将豆腐干放入用茶叶、花椒、八角、茴香、酱油等做成的五香卤汤慢火煨煮。

Five-spice dried bean curd. Home-

style dish in many places. Dried bean curd braised in five-spice gravy of tea leaves, Chinese prickly ash, aniseed, fennel, and soy sauce.

五香狗肉 (wǔxiāng gǒuròu)

贵州菜。先将狗肉块微炸,然后加入盐、白酒、桂皮、八角、花椒、姜、葱等炖煮。

Stewed five-spice dog. Guizhou dish. Lightly fried cubed dog meat seasoned with salt, Chinese liquor, cinnamon, aniseed, Chinese prickly ash, ginger, and scallion, then stewed.

五香花生 (wǔxiāng huāshēng)

四川菜。先将花生油炸,然后迅速裹上炒化的砂糖和五香粉。

Fried five-spice peanuts. Sichuan dish. Fried peanuts quickly coated with heated melting sugar and five-spice powder.

五香酱牛肉 (wǔxiāng jiàngniúròu)

南方家常菜。将新鲜牛肉加丁香、大茴香、桂皮、酱油、料酒等用文火煮熟,置凉后切片。

Five-spice beef. Home-style dish in southern areas. Beef braised with a variety of condiments, such as clove, aniseed, cinnamon, soy sauce, and cooking wine. Cooled, then sliced.

五香酱鸭 (wǔxiāng jiàngyā)

东北菜。先将鸭子用盐、葱、姜等腌渍,然后用酱油、黄酒、胡椒粉等调成的卤汤煮熟。

Five-spice duck. Northeastern dish. Duck pickled with salt, scallion, and ginger, then boiled in soup seasoned with soy sauce, yellow wine, and ground black pepper.

五香烧鸡 (wǔxiāng shāojī)

东北菜。先将鸡用盐、葱、姜等腌渍,炸至金黄,然后浇上用酱油、黄酒、胡椒粉等调成的卤汁,蒸烂。

Five-spice chicken. Northeastern dish. Chicken pickled with salt, scallion, and ginger, then deep-fried. Washed with dressing of soy sauce, yellow wine, and ground black pepper, then steamed.

五香熏鱼 (wǔxiāng xūnyú)

南方菜。先将鱼块腌渍,炸至金黄,再用葱、姜、桂皮、花椒、八角、酱油、糖等制成五香卤汁,然后将鱼放入汁中浸泡,最后用糖烟熏。

Five-taste smoked fish. Southern dish. Fish chunks pickled, deep-fried, marinated in gravy of scallion, ginger, cinnamon, Chinese prickly ash, aniseed, soy sauce, and sugar, then sugar smoked.

五元神仙鸡 (wǔyuán shénxiānjī)

湖南菜。将鸡肉、桂圆、荔枝、红枣、莲子放入瓦钵,加入冰糖、盐、枸杞子、清水,炖酥。

Chicken with five dried fruits. Hunan dish. Chicken stewed in water with logan pulp, litchi, red dates, lotus seeds, rock sugar, salt, and Chinese wolfberry.

武汉肉枣 (wǔhànròuzǎo)

湖北武汉菜。先把腌渍过的猪瘦肉

绞碎，与配料拌和，灌制成枣状肉肠，然后稍晾，熏烤。

Mini pork sausage. Specialty in Wuhan, Hubei. Pickled ground lean pork mixed with flavorings, made into date-size sausage, slightly air-dried, smoked, then broiled.

武陵酒 (wǔlíngjiǔ)

白酒，产于湖南常德。以高粱为主料，用小麦制曲酿造，酒精含量为48-53度，属酱香型酒。

Wulingjiu, liquor produced in Changde, Hunan. White spirit made with sorghum and wheat yeast. Contains 48-53% alcohol, and has a soy aroma.

武夷岩茶 (wǔyíyánchá)

乌龙茶。产于福建武夷山的乌龙茶通称为武夷岩茶。其外形肥壮匀整，色泽青褐润亮。汤色橙黄，叶底软亮，显红边。

Wuyi yancha, a general name for all oolong teas produced in Mount Wuyi, Fujian. Its dry leaves are stout, even, and glitteringly black brown. The tea is orange yellow with infused leaves soft yellow and red-fringed.

雾熘鳜鱼 (wùliū guìyú)

山东菜。先将鳜鱼打斜刀、氽水、盐渍，然后蒸熟，浇上用火腿、玉兰片、香菇等做成的卤汁。

Steamed Chinese perch with sauce. Shandong dish. Chinese perch carved on the back, scalded, pickled with salt, and steamed. Marinated in dressing of ham, soaked bamboo shoots, and dried mushrooms.

X

西凤酒 (xīfèngjiǔ)

白酒,中国传统四大名酒之一,产于陕西凤翔柳林镇。用高粱为主料,以大麦、豌豆制曲酿造,酒精含量为45-65度,属兼香型酒。

Hsifengchiew, or *xifengjiu* liquor, one of the top 4 traditional alcohol drinks in China, originated in Liulin Town of Fengxiang, Shaanxi. White spirit made with sorghum and yeast of barley and peas. Contains 45-65% alcohol, and has a mixed aroma.

西瓜盅 (xīguāzhōng)

福建点心。先把西瓜切成瓜盖和瓜盅,掏出瓜瓤,将冰糖水倒入瓜盅,放入冰箱至冷,然后把备好的西瓜瓤、苹果、雪梨、荔枝、菠萝等混装入瓜盅,盖上瓜盖,放入冰箱至冰凉。

Chilled watermelon soup. Fujian snack. Watermelon cut into a lid and a keg-like container, then hollowed. Poured in rock sugar soup, then chilled. Filled with watermelon pulp, apple, pear, litchi, and pineapple, covered with the lid, then chilled again.

西红柿炒蛋 (xīhóngshì chǎo dàn)

北方家常菜。先将蛋浆煎熟,再将西红柿片加盐稍炒,然后将鸡蛋与西红柿翻匀。

Fried eggs with tomatoes. Home-style dish in northern areas. Fried egg paste sautéed with tomatoes and seasoned with salt.

西红柿松鼠鱼 (xīhóngshì sōngshǔyú)

南方家常菜。先将鱼身切菱形刀纹腌渍,再挂芡粉蛋浆油炸,然后浇上用葱、姜、蒜、番茄、糖、醋、酱油等做成的卤汁。

Fried fish with tomato sauce. Home-style dish in southern areas. Fish diamond-carved on the body, coated with egg and flour paste, then deep-fried. Marinated in dressing of scallion, ginger, garlic, tomatoes, sugar, vinegar, and soy sauce.

西湖莼菜汤 (xīhú chúncàitāng)

浙江杭州名菜。杭州西湖莼菜、鸡脯丝、火腿丝入鸡汤合氽。

Water shield soup. Famous dish in Hangzhou, Zhejiang. Local water shield, ham, and shredded chicken quick-boiled in chicken soup.

西湖醋鱼 (xīhúcùyú)

浙江杭州名菜。将草鱼剖成两半,用滚水烫熟装盘,然后用鱼汤加入糖、淀粉、酱油及大量醋做成浓卤汁,浇在鱼上。

Sweet and sour carp. Famous dish in Hangzhou, Zhejiang. Grass carp cut into halves, quick-boiled, put on a big plate, then marinated in thick gravy of fish broth, sugar, starch, soy sauce, and lots of vinegar.

西湖龙井 (xīhúlóngjǐng)

中国著名传统绿茶,产于浙江杭州龙井一带。外形光扁平直,色翠略黄。汤色碧绿,叶底细嫩成朵。

Xihulongjing, famous traditional green tea in China produced in Longjing area of Hangzhou, Zhejiang. Its dry leaves are flat, straight, and yellowish green. The tea is crystal green with infused leaves like flowers.

西陵特曲 (xīlíngtèqū)

白酒,产于湖北宜昌西陵。以红高粱为主料,用小麦制曲酿造,酒精含量为38-55度,属兼香型白酒。

Xilingtequ, liquor produced in Xiling of Yichang, Hubei. White spirit made with red sorghum and wheat yeast. Contains 38-55% alcohol, and has a mixed aroma.

西米芒果冻糕 (xīmǐ mángguǒ dònggāo)

广东点心。先将西米、果胶、白糖、鲜奶、芒果丁熬煮,然后倒入模具,冷却定型。

Sago and mango jelly. Guangdong snack. Sago, pectin, white sugar, milk, and mangoes boiled, put in moulds to take shape, then chilled.

西柠百花鸡卷 (xīníng bǎihuā jījuǎn)

香港菜。用鸡肉片裹上柠檬、香菇、火腿,蒸熟。

Chicken with lemon and ham. Hong Kong dish. Sliced chicken made into rolls filled with mixture of lemons, dried mushrooms, and ham, then steamed.

西柠蜜糖乳鸽 (xīníng mìtáng rǔgē)

广东菜。先将腌过的乳鸽浸入热油泡熟,再加入柠檬和蜜糖焖煮,然后浇上用原汤做的芡汁。

Simmered pigeon with honey. Guangdong dish. Pickled young pigeon steeped in heated oil, simmered with lemon and honey, then washed with gravy of broth and starch.

西柠蒸乌头鱼 (xīníng zhēng wūtóuyú)

广东菜。先将乌头鱼腌渍,配柠檬片蒸熟,然后撒芹菜、香菜,淋猪油。

Steamed mullet with lemon. Guangdong dish. Pickled mullet steamed with sliced lemon. Flavored with celery, coriander, and lard.

西芹百合炒圣果 (xīqín bǎihé chǎo shèngguǒ)

东北菜。圣果即小枣形西红柿。先将西芹片、百合、圣果分别焯水,然后加调料炒熟。

Sautéed celery, lily and grape tomatoes. Northeastern dish. Sliced American celery, lily bulbs, and grape tomatoes scalded, then

sautéed with flavorings.

西芹炒牛蛙 (xīqín chǎo niúwā)

湖南菜。先把干辣椒、花椒煸炒，再放入用料酒、盐、胡椒粉等腌过的牛蛙肉，加笋片、西芹翻炒。

Stir-fried bull frogs with celery. Hunan dish. Bull frogs pickled with cooking wine, salt, and ground black pepper, stir-fried with sliced bamboo shoots and American celery, then flavored with quick-fried chili pepper and Chinese prickly ash.

西芹柠檬鸡 (xīqín níngméng jī)

广东凉菜。将蒸熟的鸡腿肉切片，加入西芹，用柠檬汁、橄榄油、糖、盐等调成的卤汁拌和。

Chicken with celery and lemon. Guangdong cold dish. Sliced steamed drumsticks mixed with American celery, then flavored with dressing of lemon juice, olive oil, sugar, and salt.

西施带子 (xīshīdàizi)

广东菜。将干贝、腰果、蛋清、鲜奶、香菇、葱花、香菜、盐等翻炒。

Sautéed scallops and cashew nuts. Guangdong dish. Scallops and cashew nuts sautéed with egg white, milk, dried mushrooms, chopped scallion, coriander, and salt.

西汁虾仁 (xīzhī xiārén)

广东菜。先将虾仁裹上蛋清淀粉浆油炸，然后用番茄汁、黄酒、麻油、盐、上汤等调成的卤汁翻炒。

Sautéed shrimp meat. Guangdong dish. Shelled shrimp coated with starched egg white, fried, then sautéed with sauce of tomato juice, yellow wine, sesame oil, salt, and broth.

希尔曼馕 (xīěrmànnáng)

新疆食品。将发酵面团做成馕坯，抹上沙枣泥，入馕坑烤熟。

Flour and oleaster *nang*. Xinjiang food. Fermented flour dough made into *nang*, covered with oleaster paste, then baked in a special oven.

锡伯饼 (xībóbǐng)

新疆锡伯族食品。将发面烙成饼，佐辣酱。

Pancake with chili sauce. Daily food of Xibo ethnic minority in Xinjiang. Fermented flour dough made into cakes and baked. Served with chili sauce.

锡纸八宝鮰鱼 (xīzhǐ bābǎo huíyú)

江苏菜。将鮰鱼片裹上用黄油、盐、面粉、水调成的浆，放上葱、姜、西芹等配料，用锡纸包好烤熟。

Roasted longsnout catfish. Jiangsu dish. Longsnout catfish coated with mixture of butter, salt, flour, and water. Garnished with scallion, ginger, American celery, and other condiments, wrapped in foil, then roasted.

习水大曲 (xíshuǐdàqū)

白酒，产于贵州习水。以当地红高粱为主料，用小麦制曲酿造，酒精含量为 38-52 度，属浓香型酒。

Xishuidaqu, liquor produced in Xishui, Guizhou. White spirit made with local red sorghum and wheat yeast. Contains 38-52% alcohol, and has a thick aroma.

虾干扒白菜 (xiāgān pá báicài)

东北菜。将白菜和虾干加调料煮熟,勾芡,淋猪油。

Braised shrimp meat and cabbage. Northeastern dish. Chinese cabbage braised with dried shrimp meat, thickened with starch, then washed with lard.

虾干粉丝节瓜煲 (xiāgān fěnsī jiéguā bāo)

广东菜。干虾仁、节瓜丝、粉丝配葱、姜、腐乳煲煮。

Dried shrimp meat with hairy gourd. Guangdong dish. Dried shrimp meat braised with hairy gourd, starch noodles, scallion, ginger, and fermented bean curd.

虾胶瓤鱼肚 (xiājiāo ráng yúdǔ)

广东菜。先把鱼肚用沸水氽过,再与葱、姜、盐、黄酒煨煮,然后铺上裹蛋清淀粉糊的虾胶,蒸熟。

Steamed fish maw and shrimp glue. Guangdong dish. Scalded fish maw simmered with scallion, ginger, salt, and yellow wine, topped with shrimp gruel coated with starched egg white, then steamed.

虾饺 (xiājiǎo)

广东点心。将精面粉和生粉做成饺皮,用鲜虾仁、肥肉丁、鸡蛋清、笋丝做成馅,包成饺子,蒸熟。

Shrimp *jiaozi*. Guangdong snack. *Jiaozi* made with flour and potato starch, stuffed with shrimp meat, fat pork, egg white, and shredded bamboo shoots, then steamed.

虾龙糊 (xiālónghú)

江苏家常菜。先将青豆仁和虾仁放入加调料的汤煮熟,然后将鸡蛋直接打入汤内煮开。

Beans and shrimp in egg soup. Home-style dish in Jiangsu. Shelled fresh shrimp and green beans boiled in seasoned soup, then eggs added to poach.

虾米拌圆白菜 (xiāmǐ bàn yuánbáicài)

东北家常菜。先将圆白菜丝焯水,然后加入熟虾米和调料拌匀。

Dried shrimp meat with cabbage. Northeastern home-style dish. Scalded shredded Chinese cabbage mixed with prepared dry shrimp meat and seasonings.

虾米炒黄瓜 (xiāmǐ chǎo huángguā)

江苏菜。先将虾米和葱花、姜末轻炒,再与黄瓜加盐快炒。

Sautéed shrimp meat and cucumbers. Jiangsu dish. Dried shrimp meat quick-fried with chopped scallion and minced ginger, sautéed with cucumbers, then seasoned with salt.

虾仁白菜 (xiārén báicài)

东北家常菜。将腌过的鲜虾仁迅速炒熟,放在氽过的白菜丝上,浇勾芡的卤汁。

Shrimp meat and Chinese cabbage with sauce. Home-style dish in

northeastern areas. Pickled fresh shrimp meat quick-fried, put on a bed of quick-boiled shredded Chinese cabbage, then washed with starched soy sauce soup.

虾仁莼菜汤 (xiārén chúncài tāng)

江苏菜。将虾仁、火腿片、笋片等用鸡汤煮熟,然后倒入汆莼菜铺底的汤碗中。

Shrimp meat and water shield soup. Jiangsu dish. Shrimp meat, ham, and sliced bamboo shoots boiled in chicken soup, then poured over scalded water shield.

虾仁熘豆腐 (xiārén liū dòufu)

东北家常菜。先将豆腐块炸至金黄,然后加虾仁稍煮,勾芡,淋猪油。

Fried bean curd with shrimp meat. Northeastern home-style dish. Deep-fried bean curd quick-boiled with shrimp meat, thickened with starch, then washed with lard.

虾仁酥 (xiārénsū)

南方家常菜。先将虾仁腌渍,然后挂蛋浆芡糊炸酥。

Fried shrimp meat. Home-style dish in southern areas. Pickled shelled shrimp coated with egg starch paste, then deep-fried.

虾仁玉子豆腐 (xiārén yùzǐ dòufu)

南方家常菜。先将虾仁与蒜、葱、草菇、红椒等翻炒,然后放入玉子豆腐加调料焖煮。

Shrimp meat and tender bean curd. Home-style dish in southern areas. Fresh shrimp meat stir-fried with garlic, scallion, meadow mushrooms, and red pepper, then simmered with tender bean curd and flavorings.

虾肉粉果 (xiāròu fěnguǒ)

福建点心。先将澄粉、盐、猪油等加开水和成面团,做成饺皮,再将猪肉、虾肉、冬菇、冬笋等烹炒、勾芡成馅,配胡萝卜花和香菜叶,包成饺子,蒸熟。

Steamed shrimp and pork *jiaozi*. Fujian snack. Wheat starch, salt, lard, and boiling water mixed, made into *jiaozi* stuffed with mixture of prepared pork, shrimp meat, dried mushrooms, winter bamboo shoots, starch, carrot cuts, and coriander leaves, then steamed.

虾肉鸳鸯菜 (xiāròu yuānyāngcài)

广东菜。将过油的白菜心和焯过的芥菜心放入奶汤煨熟,配蒸熟的虾仁装盘,浇芡汁。

Cabbage and mustard with shrimp. Guangdong dish. Lightly fried baby Chinese cabbage and quick-boiled baby mustard greens simmered in milk soup. Arranged on a platter with steamed shrimp meat, then washed with starched soup.

虾蟹菠萝 (xiāxiè bōluó)

江苏菜。将虾茸与芡粉、蛋清、盐调和,做成大虾球,中间塞入蟹黄,外面嵌上干咸面包丁成菠萝状,油煎。

Fried shrimp and crab meatballs. Jiangsu dish. Shrimp meat, starch, egg white, and salt made into large balls stuffed with crab roe. Dried diced salty bread stuck to the balls to make them look like pineapples, then deep-fried.

虾籽豆腐 (xiāzǐ dòufu)

广东菜。先用油炒香虾米和姜末,倒入用开水焯过的豆腐丁及调料炒匀,然后浇麻油。

Stir-fried shrimp meat and bean curd. Guangdong dish. Scalded diced bean curd stir-fried with prepared shrimp meat, minced ginger, and seasonings, then washed with sesame oil.

虾籽管廷 (xiāzǐ guǎntíng)

安徽菜。先将管廷入汤煮至近熟,再油炸,然后与河虾籽、竹笋、冬菇片、姜、葱等翻炒,加入鸡汤,勾芡。

Stir-fried hog gullets with shrimp roe. Anhui dish. Hog gullets half-boiled in soup, then fried. Stir-fried with fresh water shrimp roe, bamboo shoots, winter mushrooms, ginger, and scallion. Flavored with chicken soup, then thickened with starch.

虾子海参 (xiāzǐ hǎishēn)

广东菜。将汆过的海参条同虾仁翻炒,用胡椒粉、盐、料酒调味,用鸡汤勾芡。

Sautéed sea cucumbers and shrimp meat. Guangdong dish. Scalded sea cucumber strips sautéed with shrimp meat, flavored with ground black pepper, salt, and cooking wine, then washed with starched chicken soup.

夏巴吐 (xiàbātǔ)

西藏菜。先将面粉做成面皮,包肉末做成丸子,然后与萝卜丝、奶渣等入骨头汤煮熟。

Meat dumplings in bone soup. Tibetan dish. Flour wrappers filled with ground meat, made into balls, then boiled with shredded turnips and dairy dregs in bone soup.

夏果露笋炒带子 (xiàguǒ lùsǔn chǎo dàizi)

广东菜。将鲜贝、露笋、夏威夷果、甘笋翻炒,用盐、蒜茸、黄酒调味。

Sautéed scallops, asparagus, macadamia, and carrots. Guangdong dish. Fresh scallops sautéed with asparagus, macadamia nuts, and carrots, then flavored with salt, minced garlic and yellow wine.

夏果西芹炒百合 (xiàguǒ xīqín chǎo bǎihé)

南方菜。先将西芹煸炒,然后加入百合、夏果、胡萝卜片及调料翻炒。

Sautéed macadamia, celery, and lily bulb. Southern dish. American celery lightly fried, then sautéed with lily bulb cloves, macadamia nuts, sliced carrots, and seasonings.

夏河蹄筋 (xiàhétíjīn)

甘肃菜。先将夏河县产的羊蹄筋过油,煮透,再入鸡汤蒸烂,然后浇上用木耳、黄花、玉兰片、淀粉做的卤汁,淋麻油。

Steamed lamb tendons. Gansu dish.

Lightly fried and boiled lamb tendons from Xiahe County steamed in chicken soup, marinated in dressing of wood ears, day-lily buds, soaked bamboo shoots, and starch, then sprinkled with sesame oil.

仙草冰 (xiāncǎobīng)

福建点心，亦称仙冻。先将仙草煮水，加入曲粉，冷却后切块，然后加入冷开水和白糖。

Mesona herb jelly. Fujian snack. Chinese mesona herb boiled, then added red yeast rice powder. Chilled into jelly, cut into small cubes, then served with cool water and sugar.

仙菇竹荪 (xiāngū zhúsūn)

四川菜。将鲜蘑菇、竹荪、豆腐、竹笋、肉片加调料炖煮。

Stewed mushrooms and long net stinkhorn. Sichuan dish. Fresh mushrooms stewed with long net stinkhorn, bean curd, bamboo shoots, sliced pork, and flavorings.

仙桃蒸三元 (xiāntáo zhēng sānyuán)

湖北菜。将肉茸与豆腐茸加调料做成丸子，裹上浸过水的糯米，蒸熟。

Steamed pork and bean curd balls. Hubei dish. Ground pork and minced bean curd mixed with seasonings, made into balls, coated with soaked sticky rice, then steamed.

鲜贝冬瓜球 (xiānbèi dōngguāqiú)

山东菜。先将鲜贝用热油滑熟，再与冬瓜球入高汤煨煮，勾芡。

Shellfish and wax gourd soup. Shandong dish. Fresh shellfish lightly fried, simmered with wax gourd balls in broth, then thickened with starch.

鲜贝虾茸圆子 (xiānbèi xiāróng yuánzi)

南方家常菜。将虾仁、蛋清、盐、味精、葱、姜、鲜贝等制成虾糊，做成丸子，炸至金黄。

Fried shrimp and scallop balls. Home-style dish in southern areas. Fresh shrimp meat, egg white, salt, MSG, scallion, ginger, and scallops mixed into gruel, made into balls, then deep-fried.

鲜茶蒸水鱼 (xiānchá zhēng shuǐyú)

湖南菜。将水鱼即甲鱼用盐、味精、黄酒、葱、干淀粉腌渍，下垫茶叶蒸熟。

Steamed turtle with tea leaves. Hunan dish. Turtle meat pickled with salt, MSG, yellow wine, scallion, and dry starch, put on a bed of fresh tea leaves, then steamed.

鲜菇扒菜胆 (xiāngū pá càidǎn)

东北家常菜。先将油菜烫熟，再用鲜菇加调料炒熟，浇在油菜上。

Sautéed mushrooms and rape greens. Home-style dish in northeastern areas. Rape greens scalded, then covered with sautéed fresh mushrooms.

鲜菇扒鸭掌 (xiāngū pá yāzhǎng)

广东菜。将氽过去皮的鸭掌与鲜菇同炒,用黄酒、蚝油、老抽、糖、盐、胡椒粉等调味,用上汤勾芡。

Stir-fried duck feet and mushrooms. Guangdong dish. Skinned scalded duck feet stir-fried with fresh mushrooms, seasoned with yellow wine, oyster sauce, black soy sauce, sugar, salt, and ground black pepper, then washed with starched broth.

鲜菇鱼片 (xiāngū yúpiàn)

浙江菜。先将鱼片用蛋清芡粉浆腌渍,滑油,然后同焯水的鲜菇片翻炒,用盐、姜、葱、酱油、料酒等调味,勾芡。

Sautéed fish and mushrooms. Zhejiang dish. Fish fillets pickled with egg white and starch, lightly fried, sautéed with scalded sliced fresh mushrooms, seasoned with salt, ginger, scallion, soy sauce, and cooking wine, then thickened with starch.

鲜瓜炖四喜 (xiānguā dùn sìxǐ)

台湾菜。先将香菇、白果、素火腿、白木耳等煮熟,再倒入哈密瓜盅内微蒸。

Four vegetables in muskmelon. Taiwan dish. Boiled dried mushrooms, ginkgoes, soy ham, and white wood ears steamed briefly in a hollowed muskmelon.

鲜果银耳 (xiānguǒ yíněr)

广东点心。先将银耳蒸烂,再与鲜水果同煮,加糖,勾芡,撒糖桂花。

White wood ear and fruit soup. Guangdong snack. White wood ears steamed, boiled with fresh fruits, sugared, starched, then showered with sweet osmanthus flower.

鲜蚝煎蛋饼 (xiānháo jiān dànbǐng)

广东菜。鲜蚝肉用姜、葱、盐等腌渍,裹上蛋糊,煎至金黄。

Oyster and egg cakes. Guangdong dish. Fresh oysters pickled with ginger, scallion, and salt, coated with egg paste, then deep-fried.

鲜荷豉味鸡 (xiānhé chǐwèijī)

广东菜。先将鸡脯肉、冬菇、火腿等用豆豉、蒜茸、黄酒等腌渍,再用鲜荷叶包好蒸熟。

Pickled chicken in lotus leaves. Guangdong dish. Chicken breast, winter mushrooms, and ham pickled with fermented soybean, mashed garlic, and yellow wine, wrapped in fresh lotus leaves, then steamed.

鲜荷叶瓜粒汤 (xiānhéyè guālì tāng)

香港菜。先将瘦肉和鲜虾仁用糖、盐、姜汁、黄酒腌渍,然后放入滚水,加入冬瓜粒、鲜荷叶、姜片等煮熟。

White gourd and lotus leaf soup. Hong Kong dish. Lean pork and fresh shrimp meat pickled with sugar, salt, ginger juice, and yellow wine, then boiled with diced wax gourd, fresh lotus leaves, and sliced ginger.

鲜荷叶鸡蒸粉 (xiānhéyè jī zhēng fěn)

浙江菜。先把鸡脯肉和肥猪肉用

盐、黄酒、白糖、甜面酱等腌渍，然后裹上蛋清和炒米粉，用荷叶包好蒸熟。

Chicken with rice crumbs in lotus leaves. Zhejiang dish. Chicken breast and fat pork pickled with salt, yellow wine, sugar, and sweet fermented flour sauce. Coated with egg white and ground baked rice, wrapped in lotus leaves, then steamed.

鲜荷叶焗盐水鱼 (xiānhéyè júyán shuǐyú)

广东菜。先将姜、陈皮、香菇、桂圆肉、枸杞子和甲鱼块拌匀，用鲜荷叶包好，然后埋入热粗盐焖熏至熟。

Salt-baked turtle. Guangdong dish. Turtle chunks mixed with ginger, processed tangerine peels, dried mushrooms, longans, and Chinese wolfberry, wrapped in fresh lotus leaves, then buried in heated coarse salt for baking.

鲜莲炒子鸡 (xiānlián chǎo zǐjī)

浙江菜。鸡肉、新鲜莲子、香菇拌炒，烹入黄酒、酱油、白糖、醋、味精，勾芡，淋麻油。

Stir-fried chicken and lotus seeds. Zhejiang dish. Young chicken, fresh lotus seeds, and mushrooms sautéed, seasoned with yellow wine, soy sauce, sugar, vinegar, and MSG, thickeneed with starch, then washed with sesame oil.

鲜熘肉片 (xiānliū ròupiàn)

四川菜。先将肉片用芡粉、盐、料酒腌渍，然后与红椒、笋片等熘炒。

Sautéed pork and vegetables. Sichuan dish. Sliced pork pickled with starch, salt, and cooking wine, then sautéed with red pepper and bamboo shoots over low heat.

鲜蘑豆腐 (xiānmó dòufu)

浙江菜。将汆过的豆腐丁和鲜蘑菇丁拌炒，烹入黄酒、酱油、白糖、醋、味精，勾芡，淋麻油。

Sautéed mushrooms and bean curd. Zhejiang dish. Quick-boiled diced bean curd sautéed with diced fresh mushrooms, seasoned with yellow wine, soy sauce, sugar, vinegar, and MSG, thickened with starch, then washed with sesame oil.

鲜蘑鸳鸯菜花 (xiānmó yuānyāng càihuā)

东北家常菜。先将口蘑、菜花、西蓝花分别焯水，然后拌入特制的调料。

Mushrooms with cauliflower and broccoli. Home-style dish in northeastern areas. Mongolian mushrooms, cauliflower, and broccoli quick-boiled separately, then mixed with special seasonings.

鲜奶芒果露 (xiānnǎi mángguǒ lù)

广东点心。将芒果浆、椰汁、鲜奶调和，蒸熟。

Mango and milk soup. Guangdong snack. Mashed mango meat mixed with coconut juice and milk, then steamed.

鲜柠脆虾球 (xiānníng cuìxiāqiú)

广东菜。先将虾胶做成虾球，裹上

淀粉浆油炸，然后加柠檬汁、上汤、白醋、糖、黄酒等翻炒，勾芡。

Shrimp balls with lemon sauce. Guangdong dish. Shrimp gruel made into balls, coated with starch paste, then fried. Stir-fried with lemon juice, broth, vinegar, sugar, and yellow wine, then thickened with starch.

鲜人参炖竹丝鸡 (xiānrénshēn dùn zhúsījī)

广东菜。将乌骨鸡、鲜人参、火腿用高汤炖煮。

Stewed chicken with ginseng. Guangdong dish. Black-bone chicken stewed in broth with fresh ginseng and ham.

鲜肉小酥饼 (xiānròu xiǎosūbǐng)

广东点心。将面粉、水、油做成饼皮，包入用五花肉、香菇、虾仁及调料制成的馅，做成小圆饼，烤熟。

Baked meat cakes. Guangdong snack. Flour, water, and oil made into small cakes stuffed with mixture of streaky pork, dried mushrooms, shrimp meat, and seasonings, then baked.

鲜笋炒肉片 (xiānsǔn chǎo ròupiàn)

广东菜。将鲜笋片、肉片与姜、葱、盐翻炒。

Sautéed bamboo shoots and pork. Guangdong dish. Sliced pork sautéed with sliced fresh bamboo shoots, ginger, scallion, and salt.

鲜笋红烧肉 (xiānsǔn hóngshāoròu)

南方家常菜。将汆过的五花肉与红枣和笋块炖煮，加入葱、姜、酱油、糖，勾芡。

Stewed streaky pork and bamboo shoots. Home-style dish in southern areas. Scalded streaky pork stewed with red dates and cubed bamboo shoots, seasoned with scallion, ginger, soy sauce, and sugar, then thickened with starch.

鲜笋火腿 (xiānsǔn huǒtuǐ)

江浙皖菜。在火腿片上放生菜、鲜笋、乳酪，撒黑胡椒，裹成卷烤熟。

Roasted ham with bamboo shoots. Jiangsu, Zhejiang, and Anhui dish. Ham rolls filled with lettuce, fresh bamboo shoots, and cheese, sprinkled with black pepper, then grilled.

鲜虾冬瓜蓉 (xiānxiā dōngguāróng)

广东菜。先将鲜虾仁稍炸，然后与冬瓜球、蘑菇、丝瓜入上汤加盐、味精、麻油煮熟。

Shrimp meat with gourd. Guangdong dish. Fresh shrimp meat lightly fried, boiled in broth with wax gourd balls, mushrooms, and luffa, then seasoned with salt, MSG, and sesame oil.

鲜虾粉丝煲 (xiānxiā fěnsī bāo)

广东菜。将鲜虾仁、粉丝、五花肉、香菇、洋葱炖煮，用葱、蒜、黄酒、盐等调味。

Shrimp meat with starch noodles. Guangdong dish. Fresh shrimp meat and starch noodles braised with streaky pork, dried mushrooms, and

onion, then flavored with scallion, garlic, yellow wine, and salt.

鲜虾蟹柳色拉 (xiānxiā xièliǔ sèlā)

广东凉菜。将虾仁煮熟,放在用熟蟹肉、黄瓜、甘蓝叶、蛋黄酱拌成的沙拉上。

Shrimp and crab salad. Guangdong cold dish. Boiled shelled shrimp served over salad of prepared crab meat, cucumbers, Chinese broccoli leaves, and mayonnaise.

鲜鱿鲛鱼茄汁煲 (xiānyóu jiāoyú qiézhī bāo)

浙江菜。先将鲛鱼块、鱿鱼圈、洋葱、蒜茸翻炒,然后加入大头菜和番茄,添上汤焖煮,用盐、胡椒粉调味。

Mackerel and squid with tomato sauce. Zhejiang dish. Mackerel chunks and squid rings stir-fried with onion and garlic, braised in broth with mustard turnips and tomatoes, then seasoned with salt and ground black pepper.

鲜鱼生菜汤 (xiānyú shēngcài tāng)

湖南菜。将草鱼片铺在生菜上,倒入煮沸的高汤,加入鸡精、盐调味。

Fish and lettuce soup. Hunan dish. Grass carp fillets put on a bed of romaine lettuce, scalded with boiling broth, then flavored with chicken essence and salt.

鲜炸鱿鱼圈 (xiānzhá yóuyúquān)

东北家常菜。先将鱿鱼腌渍,再炸至金黄,然后切成圈状,撒椒盐。

Fried squid. Home-style dish in northeast areas. Pickled squid deep-fried, then cut into rings. Sprinkled with spicy salt.

鲜竹牛肉 (xiānzhú niúròu)

广东菜。牛肉末加入淀粉、盐、酱油、料酒、马蹄茸、陈皮丝等做成肉丸,放在炸过的腐竹上蒸熟。

Steamed beef meatballs. Guangdong dish. Ground beef mixed with starch, salt, soy sauce, cooking wine, minced water chestnuts, and shredded processed tangerine peels, made into balls, then steamed over fried bean curd skins.

咸菜肉丝毛豆 (xiáncài ròusī máodòu)

江浙沪家常菜。先将肉丝腌渍,过油,再与咸菜翻炒,然后放入煸炒过的毛豆拌匀。

Pork with preserved vegetable and soybean. Home-style dish in Jiangsu, Zhejiang, and Shanghai. Pickled shredded pork lightly fried, sautéed with preserved salty vegetable, then mixed with stir-fried green soybean.

咸蛋黄茶树菇 (xiándànhuáng cháshùgū)

广东菜。先将汆过的茶树菇裹上蛋清淀粉浆油炸,然后与咸蛋黄拌炒,加盐和糖。

Sautéed mushrooms and egg yolks. Guangdong dish. Tea bush mushrooms quick-boiled, coated with starched egg white, deep-fried, sautéed with salty duck egg yolks, then seasoned with salt and sugar.

咸蛋黄炒南瓜 (xiándànhuáng chǎo nánguā)

河南家常菜。先将咸鸭蛋黄加黄酒蒸熟，捣为糊状，然后与南瓜拌炒，加盐和鸡精。

Sautéed pumpkin with salty egg yolks. Home-style dish in Henan. Salty duck egg yolks steamed in yellow wine, mashed into paste, sautéed with pumpkin, then seasoned with salt and chicken essence.

咸蛋黄酥鱼条 (xiándànhuáng sūyú-tiáo)

山东家常菜。先将蒸熟的咸蛋黄压碎，再将炸好的鱼放在咸蛋黄上，佐椒盐、辣酱。

Fried fish with salty egg yolks. Home-style dish in Shandong. Fish coated with starch, fried, put on mashed steamed salty egg yolks, then served with spicy salt and chili sauce.

咸蛋黄玉米 (xiándànhuáng yùmǐ)

湖南菜。先将咸蛋黄蒸熟，搓碎，然后与裹上淀粉的玉米粒翻炒。

Sautéed salty egg yolks and corn. Hunan dish. Smashed salty egg yolks sautéed with starched corn.

咸蛋荔茸扇贝球 (xiándàn lìróng shànbèi qiú)

广东点心。将扇贝裹上咸蛋黄，包入芋头面团，做成球状，炸至金黄。

Salty egg yolks with taros. Guangdong snack. Scallops coated with salty egg yolks and mashed taros, made into balls, then deep-fried.

咸蛋酥 (xiándànsū)

广东点心。用面粉和蛋浆制皮，用花生仁、核桃仁、芝麻、盐等制馅，做成小饼，烤焙而成。

Salty egg and nut cakes. Guangdong snack. Flour and egg paste made into cake wrappers stuffed with mixture of peanuts, walnuts, sesame, and salt, then baked.

咸马蹄糕 (xiánmǎtígāo)

广东点心。将广式香肠、叉烧、香菇、虾米、马蹄浆调和，蒸熟成糕。

Salty water chestnut cakes. Guangdong snack. Water chestnut paste mixed with Guangdong style sausages, barbecued pork, dried mushrooms, and dried shrimp meat, then steamed.

咸鱼 (xiányú)

各地家常菜。先将鱼用重盐腌，然后或煎或蒸。

Salty fish. Home-style dish in many places. Fish pickled with lots of salt, then deep-fried or steamed.

咸鱼鸡丁豆腐煲 (xiányú jīdīng dòu-fu bāo)

广东菜。先将咸鱼与鸡丁、葱花、姜末翻炒，然后与豆腐丁煲煮。

Salty fish with chicken and bean curd. Guangdong dish. Dried salty fish sautéed with diced chicken, chopped scallion, and ginger, then braised with diced bean curd.

咸鱼茄子煲 (xiányú qiézi bāo)

南方家常菜。将鸡丁炒熟，加入炸

过的茄条和咸鱼茸,焖煮并勾芡。

Braised salty fish and eggplants. Home-style dish in southern areas. Quick-fried diced chicken braised with deep-fried eggplants and mashed salted fish, then thickened with starch.

咸鱼芋艿 (xiányú yùnǎi)

江浙沪皖家常菜。将煸炒过的咸鱼同芋艿放入黄酒和鸡汤焖煮,勾芡,淋葱油。

Salty fish with taros. Home-style dish in Jiangsu, Zhejiang, Shanghai, and Anhui. Quick-fried salty fish and taros braised in yellow wine and chicken soup, thickened with starch, then sprinkled with scallion oil.

咸鱼蒸豆腐 (xiányú zhēng dòufu)

浙江菜。将豆腐、咸鱼、五花肉丝、辣椒丝、姜丝蒸熟。

Steamed bean curd and salty fish. Zhejiang dish. Bean curd steamed with dried salty fish, shredded streaky pork, hot pepper, and ginger.

咸鱼蒸肉饼 (xiányú zhēng ròubǐng)

东北家常菜。将咸鱼粒和肉末加调料拌匀,加姜丝,蒸熟。

Steamed salty fish and pork. Home-style dish in northeastern areas. Salty fish granules and ground pork mixed with seasonings, garnished with shredded ginger, then steamed.

陷心馒头 (xiànxīn mántou)

福建点心。先将糯米蒸熟,然后与红糖、麻油拌匀,做成拳头大小的团子,外面裹上熟粳米。

Steamed sticky rice buns. Fujian snack. Steamed sticky rice mixed with brown sugar and sesame oil, made into fist-size round buns, then coated with prepared round grain rice.

相思肉 (xiāngsīròu)

东北家常菜。先将五花肉煮熟,在肉皮上涂醋和蜂蜜油炸,再加汤略煮,然后将肉片与豆腐干码入碗内,香菇作底,加调料蒸烂。

Steamed and fried pork with bean curd. Northeastern home-style dish. Boiled streaky pork coated with vinegar and honey on the skin side, deep-fried, re-boiled briefly, then steamed with dried bean curd and seasonings over dried mushrooms.

香菜饺 (xiāngcàijiǎo)

北方食品。用面粉制皮,香菜、虾仁、盐、味精制馅,包成饺子,蒸煮均可,配酱油、陈醋。

Coriander and shrimp *jiaozi*. Northern food. Flour dough made into *jiaozi* stuffed with mixture of shrimp meat, coriander, salt, MSG, then steamed or boiled. Served with soy sauce and vinegar.

香菜猪血煲 (xiāngcài zhūxuè bāo)

湖南菜。将猪血块加麻油、盐等调料入沙锅煲煮,加芫荽。

Stewed hog blood with coriander. Hunan dish. Hog blood lumps stewed in a clay pot, then seasoned with sesame oil, salt, and coriander.

香肠炒油菜 (xiāngcháng chǎo yóucài)

东北家常菜。将香肠和油菜炒熟，勾芡，淋猪油。

Stir-fried sausage with rape greens. Home-style dish in northeastern areas. Sliced sausage stir-fried with rape greens, thickened with starch, then washed with lard.

香肠烧菜花 (xiāngcháng shāo càihuā)

广东菜。将花菜和香肠翻炒，加黄酒、葱、鸡精、盐，勾芡。

Stir-fried sausage and cauliflower. Guangdong dish. Cauliflower and sausage stir-fried, flavored with yellow wine, scallion, chicken essence, and salt, then thickened with starch.

香肠蒸蛋 (xiāngcháng zhēng dàn)

江苏家常菜。将蛋浆、水、盐、鱼豉油、麻油调匀，蒸至凝固，撒香肠末。

Egg custard with sausage. Home-style dish in Jiangsu. Egg paste mixed with water, salt, soy sauce for seafood, and sesame oil, steamed into custard, then garnished with chopped sausage.

香炒螺片 (xiāngchǎo luópiàn)

江苏、浙江菜。将青螺片、红椒、木耳加调料翻炒，勾芡。

Stir-fried sea snails meat. Jiangsu and Zhejiang dish. Sliced dark green sea snails stir-fried with red pepper, wood ears, and seasonings, then thickened with starch.

香橙糕 (xiāngchénggāo)

浙江点心。先将蛋黄和糖橙汁加热，放入吉利丁片，加入鲜奶油打发，然后放入巧克力调和，置冷成糕。

Orange custard. Zhejiang snack. Egg yolks and sugared orange juice warmed, whipped with gelatin and fresh cream, then mixed with chocolate. Chilled into custard.

香橙鸡翅 (xiāngchéng jīchì)

东北家常菜。先将鸡翅煎至金黄，再加调料烹煮，然后加橙汁和橙皮丝略煮。

Orange-flavored chicken wings. Home-style dish in northeastern areas. Chicken wings fried golden, braised with seasonings, then flavored with orange juice and shredded orange peels.

香橙鸡盏 (xiāngchéng jīzhǎn)

江苏菜。将鸡丁、胡萝卜、豌豆拌炒，加入橙汁、柠檬汁和糖，然后装入掏空的橙子中。

Chicken and vegetables in orange cups. Jiangsu dish. Diced chicken, carrots, and peas sautéed, flavored with orange juice, lemon juice, and sugar, then put into hollowed oranges.

香橙妙龄鸽 (xiāngchéng miàolínggē)

广东菜。先将乳鸽腹内填入特制的馅料，再用橙汁浸腌，然后炸至金黄。

Fried young pigeon. Guangdong dish. Young pigeon stuffed with

special filling, pickled with orange juice, then deep-fried.

香橙牛肝 (xiāngchéng niúgān)

台湾菜。先将牛肝煎熟,再加香橙酒微炒,然后与橙瓣、橙皮茸等拌匀。

Ox liver with oranges. Taiwan dish. Fried ox liver light-sautéed with orange wine, then mixed with orange sections and mashed orange skins.

香橙玉米鱼 (xiāngchéng yùmǐyú)

广东菜。先将草鱼去骨,切成两半,在肉面剞十字刀,再裹上加调料的淀粉糊,炸至金黄,然后浇上用橙汁、糖、盐、生粉等做成的卤汁。

Fried carp with orange sauce. Guangdong dish. Boned grass carp halved, cross-carved, coated with seasoned starch paste, deep-fried until golden, then marinated in dressing of orange juice, sugar, salt, and starch.

香椿拌豆腐 (xiāngchūn bàn dòufu)

各地家常凉菜。将水焯香椿切碎,与豆腐丁、煮黄豆、酱油、麻油等拌匀。

Chinese toon leaves with bean curd. Home-style cold dish in many places. Scalded chopped tender Chinese toon leaves mixed with diced bean curd, boiled soybean, soy sauce, sesame oil, and other flavorings.

香椿豆腐鱼 (xiāngchūn dòufuyú)

河南菜。将豆腐皮作鱼皮,豆腐、香椿、肥猪膘及调料作鱼肉,水发腐竹作鱼骨,做成鱼状,先蒸后炸,浇特制的卤汁。

Soy fish. Home-style dish in Henan. Soy fish made with dried bean curd skins as fish skin. Bean curd, tender Chinese toon leaves, and fat pork as fish meat. Soaked bean curd skin rolls as fish bones. Steamed, deep-fried, then washed with special sauce.

香脆冻豆腐 (xiāngcuì dòngdòufu)

东北家常凉菜。将冻豆腐丁和青笋丁加葱油、味精、白糖、麻油、盐拌匀。

Stem lettuce with frozen bean curd. Home-style cold dish in northeastern areas. Diced bean curd and stem lettuce mixed with scallion oil, MSG, sugar, sesame oil, and salt.

香脆金针菇 (xiāngcuì jīnzhēngū)

广东菜。先将金针菇用紫菜包卷成束,用面糊粘牢,然后炸至皮脆,撒上胡椒粉。

Fried mushrooms in laver. Guangdong dish. Golden lily mushrooms bound into bouquets with laver, sealed with flour paste, fried crisp, then sprinkled with ground black pepper.

香脆苹果虾仁 (xiāngcuì píngguǒ xiārén)

广东菜。先将腌渍过的虾仁裹上蛋清淀粉糊,轻炒至半熟,再加入用柠檬汁泡过的苹果丁拌炒,勾芡。

Sautéed apples and shrimp meat. Guangdong dish. Pickled shelled shrimp coated with starched egg white, half-fried over low heat, sautéed with diced apples prepared with lemon juice, then starched.

香榧焖鸡脯 (xiāngfěi mèn jīpú)

江苏、浙江名菜。先将鸡脯肉炸至金黄,然后与炸过的冬笋和香榧仁及调料入鸡汤焖煮,勾芡。

Braised chicken with Chinese torreya. Famous dish in Jiangsu and Zhejiang. Chicken breast deep-fried, braised in seasoned chicken soup with quick-fried winter bamboo shoots and Chinese torreya kernels, then thickened with starch.

香干炒肉 (xiānggān chǎo ròu)

各地家常菜。先将猪肉用黄酒、蛋清、盐、生粉等腌渍,冷冻,然后与熏香干爆炒。

Stir-fried pork and smoked bean curd. Home-style dish in many places. Pork pickled with yellow wine, egg white, salt, and starch, chilled, then stir-fried with smoked bean curd.

香干炒肉丝 (xiānggān chǎo ròusī)

江浙沪菜。将五香豆腐干丝和肉丝先煮熟,然后加入调料爆炒。

Stir-fried dried bean curd and pork. Jiangsu, Zhejiang, and Shanghai dish. Shredded dried bean curd and pork boiled, then stir-fried with seasonings.

香干马兰头 (xiānggān mǎlántóu)

江浙沪家常小菜。马兰为当地野菜。先将马兰用开水烫过,剁碎,然后与豆腐干丁凉拌,用盐、麻油调味。

Bean curd with wild herb. Home-style side dish in Jiangsu, Zhejiang, and Shanghai. Local edible herb *malan* scalded, chopped, mixed with diced dried bean curd, then flavored with salt and sesame oil.

香梗目鱼蛋 (xiānggěng mùyúdàn)

广东菜。先将咸目鱼蛋蒸熟,再与香菜梗拌匀,加麻油、味精调味。

Steamed cuttlefish eggs with coriander. Guangdong dish. Salted cuttlefish eggs steamed, mixed with coriander stems, then flavored with sesame oil and MSG.

香菇炒腐皮 (xiānggū chǎo fǔpí)

各地家常菜。将香菇丝、笋丝、豆腐皮丝加盐、酱油等炒熟。

Sautéed mushrooms and bean curd skins. Home-style dish in many places. Shredded dried mushrooms sautéed with shredded bamboo shoots and shredded dried bean curd skins, then seasoned with salt and soy sauce.

香菇炒青菜 (xiānggū chǎo qīngcài)

浙江菜。青菜与香菇同炒,用盐和味精调味。

Sautéed mushrooms and greens. Zhejiang dish. Green vegetable sautéed with dried mushrooms, then seasoned with salt and MSG.

香菇炖鸡 (xiānggū dùn jī)

河南、山东家常菜。先将鸡块氽过,然后与香菇、笋片及调料炖煮。

Stewed chicken, mushrooms, and bamboo shoots. Home-style dish in

Henan and Shandong. Scalded chopped chicken stewed with dried mushrooms, sliced bamboo shoots, and seasonings.

香菇腐竹煲 (xiānggū fǔzhú bāo)

南方家常菜。先将五花肉翻炒，再与香菇、腐竹及调料烹煮，然后放入胡萝卜焖熟。

Braised mushrooms and bean curd skins. Home-style dish in Southern areas. Stir-fried streaky pork braised with dried mushrooms, bean curd skins, and seasonings, then simmered with carrots.

香菇干贝烩豆腐 (xiānggū gānbèi huì dòufu)

浙江菜。先将嫩豆腐、蛋清、牛奶搅拌成糊状，蒸熟后切成菱形方块，再将蒸过的干贝、火腿、豌豆、香菇片等入鸡汤煮开，然后放入豆腐块，勾芡，浇麻油。

Bean curd with scallops and mushrooms. Zhejiang dish. Tender bean curd, egg white, and milk mixed, then steamed into curd. Cut into diamond pieces, boiled in chicken soup with steamed dried scallops, ham, peas, and mushrooms, thickened with starch, then sprinkled with sesame oil.

香菇鸡丝凉拌 (xiānggū jīsī liángbàn)

东北家常凉菜。将熟香菇丝、熟鸡腿肉丝、熟豌豆荚丝加酱油、醋、麻油、辣油等拌匀。

Mushrooms, chicken, and peapods with dressing. Home-style cold dish in northeastern areas. Cooked shredded mushrooms, drumsticks, and peapods mixed, then flavored with soy sauce, vinegar, sesame oil, and chili oil.

香菇茭白 (xiānggū jiāobái)

浙江菜。先将茭白与香菇加调料同炒，然后焖煮，勾芡，淋麻油。

Sautéed wild rice stems and mushrooms. Zhejiang dish. Wild rice stems sautéed with dried mushrooms and seasonings, simmered, thickened with starch, then sprinkled with sesame oil.

香菇里脊 (xiānggū lǐji)

浙江菜。先将猪里脊肉片上浆，过油，然后与香菇、姜、葱等翻炒，用上汤勾芡。

Stir-fried pork tenderloin and mushrooms. Zhejiang dish. Sliced pork tenderloin pickled with wet starch, lightly fried, stir-fried with mushrooms, ginger, and scallion, then thickened with starch.

香菇焖鸡 (xiānggū mèn jī)

山东家常菜。将鸡块与煸过的葱姜爆炒，加入香菇、料酒、糖、酱油等焖煮。

Stir-fried chicken with mushrooms. Home-style dish in Shandong. Cubed chicken stir-fried with quick-fried scallion and ginger, then braised with dried mushrooms, cooking wine, sugar, soy sauce, and other condiments.

香菇焖田鸡腿 (xiānggū mèn tiánjītuǐ)

南方家常菜。将汆过的田鸡腿与香菇、葱、姜、辣椒翻炒,加入高汤、料酒、蚝油焖熟。

Braised frogs with mushrooms. Home-style dish in southern areas. Scalded frog legs stir-fried with mushrooms, scallion, ginger, and pepper. Braised in broth, cooking wine, and oyster oil.

香菇面筋 (xiānggū miànjin)

山东菜。先将油面筋、香菇、冬笋片等翻炒,再入鲜汤烧沸,然后转小火焖煮,勾芡。

Stir-fried gluten with mushrooms. Shandong dish. Fried gluten, dried mushrooms, and sliced winter bamboo shoots stir-fried, boiled in seasoned soup, simmered, then thickened with starch.

香菇牛肉丝 (xiānggū niúròusī)

各地家常菜。将牛肉丝、香菇丝、辣椒丝加盐、味精炒熟,勾芡。

Stir-fried beef with mushrooms. Home-style dish in many places. Shredded beef stir-fried with shredded dried mushrooms, green pepper, salt, and MSG, then thickened with starch.

香菇啤酒鸡翅 (xiānggū píjiǔ jīchì)

山东家常菜。先将鸡翅轻炒,然后加入啤酒、香菇及调料煨煮。

Chicken wings and mushrooms with beer. Home-style dish in Shandong. Chicken wings quick-fried, then simmered with beer, dried mushrooms, and seasonings.

香菇肉片 (xiānggū ròupiàn)

各地家常菜。将猪肉片和香菇片加调料炒熟,放盐、姜、味精。

Pork with mushrooms. Home-style dish in many places. Sliced pork stir-fried with sliced dried mushrooms, then seasoned with salt, ginger, and MSG.

香菇烧菜心 (xiānggū shāo càixīn)

浙江菜。先将菜心过油,与香菇入鲜汤加调料煮沸,然后用小火略焖。

Simmered mushrooms and baby cabbage. Zhejiang dish. Baby Chinese cabbage quick-fried, boiled with mushrooms in seasoned soup, then simmered.

香菇油菜 (xiānggū yóucài)

各地家常菜。将水发香菇与油菜加盐、葱、酱油等翻炒。

Sautéed mushrooms and rape greens. Home-style dish in many places. Soaked dried mushrooms sautéed with rape greens, then seasoned with salt, scallion, and soy sauce.

香瓜炒素鸡 (xiāngguā chǎo sùjī)

江苏家常菜。先将素鸡片拌入调料,煎至两面焦黄,然后与香瓜略炒。

Stir-fried soy chicken and muskmelon. Home-style dish in Jiangsu. Seasoned soy chicken fried golden brown on both sides, then quickly sautéed with muskmelon.

香滑鸡球 (xiānghuá jīqiú)

广东菜。先将鸡肉配红椒、青椒、香菇做成鸡球,再与姜片、高汤、调料翻炒。

Sautéed chicken balls. Guangdong dish. Chicken meat mixed with red and green pepper and dried mushrooms, shaped into balls, then sautéed with ginger, broth, and seasonings.

香滑鲈鱼球 (xiānghuá lúyúqiú)

广东名菜。先将鲈鱼肉茸配红椒、青椒、香菇做成鱼丸,再与姜片、高汤同炒。

Sautéed perch balls. Famous Guangdong dish. Mashed perch meat mixed with red and green pepper and dried mushrooms, made into balls, then sautéed with sliced ginger and broth.

香煎白鳝 (xiāngjiān báishàn)

广东菜。将白鳝片用蒜、酱油、糯米粉腌渍,然后炸至金黄。

Fried white eel. Guangdong dish. Sliced white eel pickled with garlic, soy sauce, and sticky rice flour, then deep-fried.

香煎菜脯蛋 (xiāngjiān càipúdàn)

台湾菜。菜脯即萝卜干。先将油渣和菜脯及调料炒香,再与葱花、蛋液等调匀,煎至金黄。

Fried preserved turnips and eggs. Taiwan dish. Chopped preserved turnips and lard dregs sautéed with seasonings, mixed with chopped scallion and egg paste, then fried.

香煎鲫鱼 (xiāngjiān jìyú)

各地家常菜。先将鲫鱼用调料腌渍,然后煎熟。

Fried crucian. Home-style dish in many places. Crucian pickled with seasonings, then deep-fried.

香煎金枪鱼 (xiāngjiān jīnqiāngyú)

广东菜。先将新鲜金枪鱼片用盐略腌,然后炸至金黄。

Fried tuna. Guangdong dish. Fresh tuna fillets pickled with salt, then deep-fried.

香煎牛小排 (xiāngjiān niúxiǎopái)

广东菜。先将牛排用料酒、酱油、苏打等腌渍,稍煎,然后放入用蚝油、番茄酱、辣椒油、糖调成的卤汁烹煮,撒香菜。

Beef steaks in spicy gravy. Guangdong dish. Beef steaks pickled with cooking wine, soy sauce, and soda, then lightly fried. Braised in gravy of oyster sauce, tomato sauce, chili oil, and sugar, then sprinkled with coriander.

香煎苹果薯饼 (xiāngjiān píngguǒ shǔbǐng)

广东点心。将炒脆的碎烟肉、洋葱、苹果丝、番薯、葱、盐、胡椒粉拌匀,用小火煎至两面金黄。

Fried apple and sweet potato cakes. Guangdong snack. Chopped bacon and onion fried until crisp, mixed with shredded apples, sweet potatoes, chopped scallion, salt, and ground black pepper, then pan fried golden on both sides.

香煎翘鱼 (xiāngjiān qiàoyú)

湖南菜。先将腌渍过的翘鱼煎至金黄，然后与青椒、红椒、野山椒等翻炒。

Fried topmouth culter. Hunan dish. Pickled topmouth culter deep-fried, then stir-fried with green, red, and wild peppers.

香酱炒卤肉 (xiāngjiàng chǎo lǔròu)

山东传统菜。将卤熟的五花肉同姜、蒜、葱等爆炒，加酱油和猪油。

Stir-fried marinated pork. Traditional dish in Shandong. Marinated streaky pork stir-fried with ginger, garlic, and scallion, then flavored with soy sauce and lard.

香蕉蛋糕 (xiāngjiāo dàngāo)

上海点心。将香蕉泥、砂糖、蛋浆、黄油、牛奶、葡萄酒、面粉、发粉等调和，烘烤成糕。

Banana cakes. Shanghai snack. Mashed bananas mixed with sugar, egg paste, butter, milk, wine, flour, and baking powder, then baked.

香蕉果炸 (xiāngjiāo guǒzhá)

江苏点心。先将面粉、淀粉、蛋浆加清水调和，慢火煮熟成厚糊，然后放入香蕉精、葡萄干、糖冬瓜丁等搅匀，冷却后切块，裹干淀粉油炸，撒白糖。

Fried fruit cakes. Jiangsu snack. Flour, starch, egg paste, and water mixed, boiled into gruel over low heat, mixed with banana essence, raisins, and diced candied wax gourd, then chilled. Cut into pieces, coated with dry starch, fried, then dusted with granulated sugar.

香蕉肉脯 (xiāngjiāo ròupú)

浙江菜。先将肉茸、蛋浆、淀粉、盐等调和，做成小肉饼，蒸熟，油炸，然后加入黄酒、高汤、盐、糖、茄汁等焖煮，配炸香蕉丁装盘。

Meat cake with fried bananas. Zhejiang dish. Ground pork mixed with egg paste, starch, and salt, made into small cakes, then steamed. Deep-fried, then simmered with yellow wine, broth, salt, sugar, and tomato sauce. Served with fried diced bananas.

香蕉虾卷 (xiāngjiāo xiā juǎn)

南方点心。用春卷皮包入虾仁、香蕉、色拉酱，裹成卷状，炸至金黄，配香醋。

Fried banana and shrimp rolls. Southern snack. Flour sheets made into rolls filled with shrimp meat, bananas, and salad sauce, then deep-fried. Served with vinegar.

香辣脆肠 (xiānglà cuìcháng)

湖南菜。先将猪大肠腌渍，然后加入姜、蒜、豆瓣酱、干辣椒等爆炒。

Stir-fried hog large intestine. Hunan dish. Pickled hog large intestine stir-fried with ginger, garlic, fermented broad bean sauce, and chili pepper.

香辣脆皮明虾 (xiānglà cuìpí míngxiā)

广东菜。先将去壳的明虾用葱、姜、

盐、料酒、胡椒粉腌渍，裹上淀粉糊炸至香脆，再放入红油、葱、姜、蒜、香辣酱翻炒，烹入料酒、麻油、白糖。

Spicy crisp prawns. Guangdong dish. Shelled prawns pickled with scallion, ginger, salt, yellow wine, and ground black pepper, coated with starch paste, then fried crisp. Stir-fried with chili oil, scallion, ginger, garlic, and spicy sauce, then flavored with cooking wine, sesame oil, and sugar.

香辣豆角香干 (xiānglà dòujiǎo xiānggān)

东北传统菜。将猪肉、豆腐干、豆角与煸炒过的蒜茸和辣椒及调料翻炒。

Stir-fried pork, bean curd, and cowpeas. Traditional dish in northeastern areas. Pork, dried bean curd, and cowpeas stir-fried with quick-fried garlic, chili pepper, and seasonings.

香辣肚尖 (xiānglà dǔjiān)

湖南菜。先将酸泡菜和鲜红椒煸炒，然后与汆过的猪肚尖和青蒜熘炒。

Sautéed hog tripe. Hunan dish. Scalded sliced hog tripe and garlic shoots sautéed with quick-fried sour pickled vegetables and red pepper.

香辣蜂蛹 (xiānglà fēngyǒng)

湖南菜。先将蜂蛹煎熟，然后加入秘制香辣调料煸炒。

Spicy bee pupas. Hunan dish. Fried bee pupas sautéed with special spicy condiments.

香辣狗肉 (xiānglà gǒuròu)

湖南菜。先将狗肉腌渍，煮熟，用清水漂过，再加入葱、姜、干辣椒、黄酒炖煮，然后与辣椒、花椒爆炒，加红辣椒、姜、蒜等炖烂。

Braised spicy dog. Hunan dish. Pickled and boiled dog meat rinsed thoroughly with water, then stewed with scallion, ginger, chili pepper, and yellow wine. Stir-fried with chili pepper and Chinese prickly ash, then braised with red pepper, ginger, and garlic.

香辣麻仁鱼条 (xiānglà márén yútiáo)

湖南菜。将草鱼切成条，裹上淀粉、盐、辣椒粉、花椒粉等油炸。

Fried spicy carp. Hunan dish. Slivered grass carp coated with starch, salt, chili powder, and ground Chinese prickly ash, then deep-fried.

香辣免治猪肉 (xiānglà miǎnzhì-zhūròu)

香港菜。五花肉与炒米、柠檬汁、鱼露、干辣椒翻炒。

Stir-fried streaky pork with baked rice. Hong Kong dish. Streaky pork stir-fried with baked rice, lemon juice, fish sauce, and chili pepper.

香辣牛蒡 (xiānglà niúbàng)

四川菜。先将牛蒡煮熟，切成薄片，然后加入煸炒过的豆豉辣酱、醋、辣油、麻油等拌匀。

Spicy burdock. Sichuan dish. Boiled

sliced burdock roots mixed with quick-fried hot fermented soybean sauce, vinegar, chili oil, and sesame oil.

香辣双丝 (xiānglà shuāngsī)

东北家常菜。将鱼丝和鸡肉丝与红辣椒丝、酱油、盐、料酒等炒熟。

Spicy fish and chicken. Home-style dish in northeastern areas. Shredded fish and chicken stir-fried with shredded red pepper, soy sauce, salt, and cooking wine.

香辣小龙虾 (xiānglà xiǎolóngxiā)

湖南菜。将炸透的小龙虾与煸过的辣椒、蒜、姜等焖煮。

Spicy red swamp crayfish. Hunan dish. Red swamp crayfish deep-fried, then simmered with quick-fried hot pepper, garlic, and ginger.

香辣蟹 (xiānglàxiè)

四川菜。先将螃蟹用料酒、盐、味精腌渍,油炸,再加花椒、姜、葱炒香,烹入高汤和料酒。

Spicy crab. Sichuan dish. Crab pickled with cooking wine, salt, and MSG, then deep-fried. Stir-fried with Chinese prickly ash, ginger, and scallion, then flavored with broth and cooking wine.

香露炖花菇 (xiānglù dùn huāgū)

东北家常菜。将花菇加猪油拌匀稍煮,然后入清汤炖煮,加鸡油、黄酒、白糖、味精、盐,撇掉浮油。

Braised mushrooms. Home-style dish in northeastern areas. Quick-boiled larded grained mushrooms braised in clear soup with chicken oil, yellow wine, sugar, MSG, and salt. Degreased before serving.

香露全鸡 (xiānglù quánjī)

广东菜。将鸡、火腿片、香菇加入鸡汤、丁香、料酒、盐等蒸烂。

Chicken with ham and mushrooms. Guangdong dish. Chicken, ham, and dried mushrooms steamed in chicken soup, then seasoned with clove, cooking wine, salt, and other condiments.

香满脆环虾 (xiāngmǎn cuìhuánxiā)

福建菜。虾仁腌渍,黄瓜掏空切成环状,嵌入虾仁,滑炒,调味。

Sautéed shrimp meat and cucumbers. Fujian dish. Pickled shelled shrimp filled in hollowed cucumber sections, sautéed, then seasoned.

香芒炒虾球 (xiāngmáng chǎo xiāqiú)

广东菜。将腌渍过的虾仁与芒果肉、红椒翻炒,炝葡萄酒。

Stir-fried shrimp meat and mangoes. Guangdong dish. Pickled shrimp meat stir-fried with mango meat and red pepper, then flavored with wine.

香芒冻奶酪 (xiāngmáng dòngnǎilào)

广东点心。先将鲜奶与备好的吉利丁调匀,煮沸,然后冷冻成奶酪状,加入芒果汁。

Milk curd with mango juice. Guangdong snack. Fresh milk boiled with prepared gelatine, chilled into curd, then immersed in mango juice.

香芒鸡柳盏 (xiāngmáng jīliǔ zhǎn)

东南方家常菜。先将鸡丁用鸡精、酱油、麻油、胡椒粉腌渍,然后与胡萝卜、芒果爆炒,装入芒果皮做的盘。

Stir-fried chicken with carrots and mangoes. Home-style dish in southeastern areas. Diced chicken pickled with chicken essence, soy sauce, sesame oil, and ground black pepper, stir-fried with carrots and mangoes, then served in a hollowed mango.

香茅浸鸡 (xiāngmáo jìnjī)

广东凉菜。将鸡加青柠檬汁、蒜茸、香茅、姜、芫荽、辣椒酱煮熟,迅速冷却,涂麻油,切块,佐酱油、香醋、蒜茸做成的卤汁。

Citronella chicken with sauce. Guangdong cold dish. Chicken boiled in gravy of green lemon juice, minced garlic, citronella, ginger, coriander, and chili sauce, then quickly chilled. Coated with sesame oil, chopped, then served with dressing of soy sauce, vinegar, and minced garlic.

香嫩芝麻鸡块 (xiāngnèn zhīma jīkuài)

各地家常菜。先将鸡块裹芡粉蛋浆煎熟,然后浇用葱、姜、蘑菇、芡粉等做成的卤汁,撒芝麻。

Fried chicken with sesame sauce. Home-style dish in many places. Chicken coated with egg paste and starch, deep-fried, washed with starched sauce of scallion, ginger, and mushrooms, then sprinkled with sesame.

香柠斑块 (xiāngníng bānkuài)

台湾菜。将腌过的石斑鱼块炸至金黄,浇上用柠檬汁、糖、粟粉等调制的卤汁。

Fried grouper with lemon sauce. Taiwan dish. Pickled grouper chunks deep-fried, then marinated in dressing of lemon juice, sugar, and corn powder.

香柠佛手鱼 (xiāngníng fóshǒuyú)

浙江菜。先将新鲜比目鱼头用盐和黄酒腌渍,裹上生粉炸成佛手形,然后与青豆和甜红椒拌炒,用柠檬汁勾芡。

Flatfish head with vegetables. Zhejiang dish. Fresh flatfish heads pickled with salt and yellow wine, coated with starch, then deep-fried. Sautéed with green soy beans and sweet red pepper, then dressed with starch and lemon soup.

香馓黄金贝 (xiāngsǎn huángjīnbèi)

江苏菜。扇贝肉焯水后蒸熟,浇上咸鸭蛋黄液,撒上青红椒末,配麻油馓装盘。

Steamed scallops with fried dough. Jiangsu dish. Scallops quick-boiled, steamed, then washed with salted duck egg yolks. Sprinkled with chopped green and red pepper, then served with fine dough twists fried in sesame oil.

香酥菜肉饼 (xiāngsū càiròubǐng)

广东点心。用发酵面制皮,包入五花肉和青菜及调料制成的馅,做成饼，煎至金黄。

Pork and vegetable cakes. Guangdong snack. Flour dough made into cakes filled with streaky pork and green vegetable, then deep-fried until golden.

香酥朝饼 (xiāngsū cháobǐng)

台湾点心。用豆皮包上猪肉、桔饼、冬瓜糖、蛋浆等调成的馅，稍蒸，炸至金黄。

Fried bean curd cakes. Taiwan dish. Mixture of pork, candied tangerines, candied wax melon, and egg paste packed in bean curd skins, half-steamed, then deep-fried.

香酥带子 (xiāngsū dàizi)

广东菜。先将带子腌渍，裹上生粉，炸至金黄，再与花椒、葱花翻炒。

Crisp and savoury scallops. Guangdong dish. Pickled scallops coated with potato starch, deep-fried, then stir-fried with Chinese prickly ash and chopped scallion.

香酥豆腐 (xiāngsū dòufu)

山东家常菜。将豆腐蒸熟，淋酱油、撒芹菜末和炒过的豆酥。

Steamed bean curd with fermented bean floss. Home-style dish in Shandong. Bean curd steamed, then flavored with soy sauce, chopped celery, and stir-fried fermented soybean floss.

香酥鹅掌 (xiāngsū ézhǎng)

四川菜。将鹅掌汆过，挂淀粉蛋浆，炸至金黄，入卤汤煮烂。

Marinated goose feet. Sichuan dish. Scalded goose feet coated with starch and egg paste, deep-fried, then marinated in gravy.

香酥凤腿 (xiāngsū fèngtuǐ)

广东菜。先将鸡腿用料酒、酱油、胡椒粉、糖腌渍，再裹上蛋黄淀粉糊，炸至金黄。

Fried drumsticks. Guangdong dish. Drumsticks pickled with cooking wine, soy sauce, ground black pepper, and sugar, coated with egg yolk and starch paste, then deep-fried.

香酥糕 (xiāngsūgāo)

南方点心。将面团分成小块，抹上用猪油、面粉、盐调成的糊，做成小圆饼，烙至金黄皮脆。

Fried salty cakes. Southern snack. Flour dough cut into small pieces, brushed with paste of lard, flour, and salt, made into small round cakes, then baked until golden and crisp.

香酥狗肉 (xiāngsū gǒuròu)

东北菜。先将狗肉腌渍，蒸熟，然后裹上淀粉，炸至金黄。

Fried dog meat. Northeastern dish. Pickled dog meat steamed, coated with starch, then deep-fried.

香酥猴头菇 (xiāngsū hóutóugū)

河南传统菜。先将猴头菇入葱姜汤煮熟，再裹上干淀粉炸至金黄，然后与葱、蒜、辣椒、盐、酱油等拌炒。

Sautéed mushrooms. Traditional dish in Henan. Monkey-head mushrooms boiled in scallion and ginger

soup, coated with dry starch, then deep-fried. Sautéed with scallion, garlic, hot pepper, salt, and soy sauce.

香酥鸡 (xiāngsūjī)

山东菜。先将腌渍过的鸡块蒸熟，再抹上米酒炸至色黄皮脆，佐甜面酱和麻油。

Fried chicken with sweet sauce. Shandong dish. Pickled chicken steamed, coated with sweet fermented sticky rice juice, then deep-fried until crisp. Served with fermented sweet flour sauce and sesame oil.

香酥鲤鱼 (xiāngsū lǐyú)

北方家常菜。先将鲤鱼块挂糊，炸至金黄，然后加姜、葱、盐、料酒等焖熟。

Fried and braised carp. Home-style dish in northern areas. Carp chunks coated with starch paste, deep-fried, then braised with ginger, scallion, salt, and cooking wine.

香酥牛肉 (xiāngsū niúròu)

浙江菜。先把牛里脊肉酱与蛋清、清汤、芡粉、油、盐等调和，蒸熟切片，然后裹上蛋黄浆和面包屑炸至金黄，佐番茄酱。

Fried beef with tomato sauce. Zhejiang dish. Ground beef tenderloin mixed with egg white, clear soup, starch, oil, and salt, then steamed. Sliced, coated with egg yolk paste and bread crumbs, deep-fried, then served with tomato sauce.

香酥山药 (xiāngsū shānyào)

四川菜。先将山药蒸熟，炸至金黄，然后加糖和水焖煮，勾芡，淋猪油。

Braised sweet Chinese yam. Sichuan Dish. Chinese yam steamed, deep-fried, braised with sugar and water, starched, then washed with lard.

香酥丸子 (xiāngsū wánzi)

东北家常菜。将猪肉末加淀粉及调料做成丸子，炸熟。

Fried pork meatballs. Northeastern home-style dish. Minced pork mixed with starch and seasonings, made into balls, then deep-fried.

香酥咸烧白 (xiāngsū xiánshāobái)

四川菜。先将五花肉氽过，炸至半熟，再切成夹板片，嵌入芋头片，加调料蒸熟，然后裹上加调料的面糊，炸至金黄发脆。

Fried streaky pork and taros. Sichuan dish. Half-fried scalded streaky pork cut into clamping pieces, filled with sliced taros, then steamed with flavorings. Coated with seasoned flour paste, then fried golden and crisp.

香酥响铃 (xiāngsū xiǎnglíng)

江苏、浙江菜。将切碎的金针菇和胡萝卜加盐、麻油等炒熟成馅，用豆皮包馅成三角形，炸至金黄发脆。

Fried vegetable package. Jiangsu and Zhejiang dish. Chopped golden mushrooms and carrots sautéed with salt, sesame oil, and other ingredients, packed into triangle buns with bean curd skins, then deep-fried until golden and crisp.

香酥小刁 (xiāngsū xiǎodiāo)

湖北菜。将刁子鱼腌渍,裹上芡粉,炸熟。

Fried small carp. Hubei dish. Pickled local *diaozi* carp coated with starch, then deep-fried.

香酥小河虾 (xiāngsū xiǎohéxiā)

家常菜。先将小河虾炸酥,然后加青椒丁、洋葱丁及调料速炒。

Stir-fried river shrimp. Home-style dish. Small river shrimp deep-fried, then quick-fried with diced green pepper, onion, and seasonings.

香酥鸭 (xiāngsūyā)

四川、湖南菜。先将鸭用盐腌过,再涂上黄酒、酱油、花椒、八角、甜面酱,蒸熟,然后炸至金黄。

Crisp and spicy duck. Sichuan and Hunan dish. Duck pickled with salt, then coated with mixture of yellow wine, soy sauce, Chinese prickly ash, aniseed, and sweet fermented flour sauce. Steamed, then deep-fried.

香酥鱼 (xiāngsūyú)

江浙沪家常菜。先将腌过的草鱼油炸,再加入姜、葱、酱油、料酒、糖焖煮收汁,撒葱花。

Simmered and fried grass carp. Home-style dish in Jiangsu, Zhejiang, and Shanghai. Deep-fried pickled grass carp simmered with ginger, scallion, soy sauce, cooking wine, and sugar, reduced, then sprinkled with chopped scallion.

香蒜咖喱炒蟹 (xiāngsuàn gālí chǎo xiè)

广东菜。先将花蟹块裹生粉入热油浸熟,再与葱、蒜、红椒、咖喱酱烹炒,然后入上汤加椰酱烹煮。

Garlic and curry crab. Guangdong dish. Crab chunks coated with starch, then immersed in heated oil. Stir-fried with garlic, scallion, red pepper, and curry, then simmered in broth and coconut sauce.

香荽鱼松酿银萝 (xiāngsuī yúsōng niàng yínluó)

广东菜。先将白萝卜厚片掏空,酿入鲮鱼胶蒸熟,然后浇火腿虾米汤,撒芫荽叶。

Steamed turnips with mud carp. Guangdong dish. Turnips sectioned into thick pieces, hollowed, filled with mud carp paste, then steamed. Marinated in ham and dried shrimp meat soup, then sprinkled with coriander.

香桃鹌鹑蛋 (xiāngtáo ānchúndàn)

湖南菜。先将去壳熟鹌鹑蛋、虾仁及火腿末贴在桃型面包片上,然后油炸。

Fried quail eggs on bread. Hunan dish. Shelled boiled quail eggs, shrimp meat, and minced ham stuck to heart-shape bread, then deep-fried.

香油石鳞 (xiāngyóu shílín)

福建菜。将石鳞鱼去骨腌渍,挂干淀粉炸熟,配青椒花和红绿樱桃

装盘。

Fried codfish. Fujian dish. Boned codfish pickled, coated with dry starch, then deep-fried. Served with green pepper cuts and preserved red and green cherries.

香鱼白菜 (xiāngyú báicài)

浙江菜。将白菜嫩杆与辣豆瓣酱翻炒,加入用酱油、醋、糖、淀粉、黄酒等调成的芡汁。

Savoury Chinese cabbage stems. Zhejiang dish. Young Chinese cabbage stems sautéed with hot fermented broad bean sauce, then flavored with gravy of soy sauce, vinegar, sugar, starch, and yellow wine.

香芋扣肉 (xiāngyù kòuròu)

南方菜。先将五花肉和芋头油炸,裹上佐料酱,蒸烂,然后配焯过的青菜装盘,浇芡汁。

Steamed taros and pork. Southern dish. Streaky pork and taros deep-fried, coated with special sauce, steamed, garnished with scalded green cabbage, then washed with starched soup.

香芋烧竹鸡 (xiāngyù shāo zhújī)

江苏菜。将竹鸡块与香芋用鸡汤炖煮,加入盐、黄酒、酱油、姜、葱,勾芡。

Stewed bamboo partridge and taros. Jiangsu dish. Bamboo partridge chunks stewed with taros, seasoned with salt, yellow wine, soy sauce, ginger, and scallion, then thickened with starch.

香芋汶鸭 (xiāngyù wèn yā)

香港菜。先将鸭块加调料爆炒,加水煮开,然后放入芋头焖煮,用鸭酱调味。

Braised duck and taros. Hong Kong dish. Duck chunks quick-fried with seasonings, boiled, braised with taros, then seasoned with duck sauce.

香炸金片 (xiāngzhá jīnpiàn)

江苏菜。将冬瓜片裹上鸡蛋淀粉糊和面包屑,煎至金黄。

Fried wax gourd. Jiangsu dish. Sliced wax gourd coated with egg and starch paste and bread crumbs, then deep-fried.

香炸龙凤蛋 (xiāngzhá lóngfèngdàn)

广东点心。将虾茸丸嵌入咸蛋黄,裹上蛋清淀粉浆和面包屑,炸至金黄。

Fried shrimp balls with yolks. Guangdong snack. Shrimp balls filled with salty duck egg yolks, coated with starched egg white and bread crumbs, then deep-fried.

香炸螺丝卷 (xiāngzhá luósījuǎn)

广东点心。用面粉、猪油、水制成面皮,包莲蓉馅,捏成螺状,油炸。

Fried stuffed buns. Guangdong snack. Flour, lard, and water mixed, made into snail-shape buns filled with sweet mashed lotus seeds, then deep-fried.

香寨 (xiāngzhài)

西藏菜。将羊肉块用酥油烹炒,加

入土豆焖熟,用咖喱、藏蔻调味。

Mutton simmered with potatoes. Tibetan dish. Cubed mutton stir-fried with Tibetan butter, simmered with potatoes, then flavored with curry and Tibetan myristica.

香汁茄子 (xiāngzhī qiézi)

各地家常菜。先将茄子煎至熟软,然后浇上用胡萝卜、西芹、芡粉及调料做的卤汁。

Fried eggplants with sauce. Home-style dish in many places. Eggplants deep-fried soft, then washed with dressing of carrots, American celery, seasonings, and starch.

香煮油脾 (xiāngzhǔ yóupí)

西藏菜。用羊脾和羊油煮制而成。

Boiled lamb spleen. Tibetan dish. Lamb spleen boiled in suet oil.

湘江抱盐鱼 (xiāngjiāng bàoyányú)

湖南菜。先将湘江草鱼盐腌后油煎,再加水焖煮,然后淋红油和麻油,撒葱花。

Salty grass carp. Hunan dish. Salted grass carp deep-fried, simmered in water, sprinkled with chili oil and sesame oil, then garnished with chopped scallion.

湘南选籽鱼 (xiāngnán xuǎnzǐyú)

湖南菜。先将腊刁子鱼油炸,再加入干豆角及调料先焖后蒸,配香菜装盘。

Diaozi carp with dried cowpeas. Hunan dish. Smoked Hunan *diaozi* carp deep-fried, simmered with dried cowpeas, steamed, then garnished with coriander.

湘泉酒 (xiāngquánjiǔ)

白酒,产于湖南吉首。以当地糯高粱和香糯米为主料,用小麦制曲酿造,酒精含量为 38-54 度,属兼香型酒。

Xiangquanjiu, liquor produced in Jishou, Hunan. White spirit made with local sticky sorghum, aromatic glutinous rice, and wheat yeast. It contains 38-54% alcohol, and has a mixed aroma.

湘式火培鱼 (xiāngshì huǒpéiyú)

湖南菜。将嫩子鱼加入用小红辣椒、紫苏、葱、姜、蒜、盐等调成的卤汁焙熟。

Baked spicy small fish. Hunan dish. Small fresh water fish seasoned with sauce of red pepper, purple basil, scallion, ginger, garlic, and salt, then baked over low heat.

湘味辣子鸡 (xiāngwèi làzijī)

四川、湖南菜。将腌渍过的鸡块与煸炒过的姜、蒜、干辣椒和花椒等爆炒,加入葱段、酱油、黄酒、熟芝麻。

Spicy chicken. Sichuan and Hunan dish. Pickled cubed chicken stir-fried with quick-fried ginger, garlic, chili pepper, and Chinese prickly ash, then flavored with scallion stems, soy sauce, yellow wine, and prepared sesame.

湘西土匪鸭 (xiāngxī tǔfěiyā)

湖南菜。将当地肥鸭切成条,与多种调料烹煮而成。传说鸭子糟蹋庄稼,被比喻作土匪,就被做成了

菜肴。

Braised tasty duck. Hunan dish. Slivered fat duck braised with a variety of spices. It is a local legend that ducks were called bandits because they damaged crops, so they were cooked.

镶豆腐 (xiāngdòufu)

家常菜。用豆腐片夹肉馅,挂蛋糊,炸至金黄。

Fried stuffed bean curd. Home-style dish. Minced pork sandwiched between bean curd pieces, coated with egg and starch paste, then deep-fried.

祥云凤展翅 (xiángyún fèngzhǎnchì)

山东菜。先将鸡翅煮熟,拆骨,再填入蒸好的鱼翅和火腿丝,加汤、盐、味精、糖蒸透,原汤勾芡。

Steamed chicken wings and shark fin. Shandong dish. Boiled and boned chicken wings filled with steamed shark fin and shredded ham, steamed in soup flavored with salt, MSG, and sugar, then washed with starched broth.

响铃鸭子 (xiǎnglíng yāzi)

山东菜。先将鸡肉小馄饨炸脆,再倒入鸭汤。

Fried small chicken wontons in duck soup. Shandong dish. Small chicken wontons fried until crisp, then poured into duck soup.

响萝卜丝 (xiǎngluóbosī)

湖南菜。先将萝卜丝用盐腌,然后用熟猪油、干辣椒末、大蒜丝轻炒,勾芡。

Sautéed shredded turnips. Hunan dish. Shredded turnips pickled with salt, sautéed with lard, chili powder, and garlic shreds, then thickened with starch.

像生雪梨果 (xiàngshēng xuělíguǒ)

广东点心。用熟薯、咸蛋黄等料制成皮,用猪肉、鸡肉、冬菇、笋等制成馅,包成梨状,炸至金黄。

Fried sweet potato and meat buns. Guangdong snack. Cooked sweet potatoes and salted egg yolks made into pear-shape buns filled with mixture of pork, chicken, winter mushrooms, and diced bamboo shoots, then deep-fried.

潇湘五元龟 (xiāoxiāng wǔyuánguī)

湖南菜。先将煸过的土龟蒸至半熟,再加入桂圆、荔枝、黑枣、莲仁、枸杞子蒸烂,加盐、胡椒粉。

Steamed turtle with five treasures. Hunan dish. Quick-fried turtle meat half-steamed, combined with longan pulp, litchi, black dates, lotus seeds, and Chinese wolfberry, then steamed again. Seasoned with salt and ground black pepper.

小肠肺汤 (xiǎocháng fèi tāng)

湖南菜。先将猪肺、小肠、萝卜、红枣及调料用旺火煮滚,再用小火熬炖。

Hog lung and small intestine soup. Hunan dish. Hog lung, small intestine, turnips, and red dates boiled with seasonings, then stewed.

小炒羊杂 (xiǎochǎo yángzá)

湖南菜。将盐渍过的羊腰、羊肚、羊肝、羊心、羊肠加入调料爆炒。

Stir-fried lamb offal. Hunan dish. Lamb offal, such as kidney, tripe, liver, heart, and intestine, pickled with salt, then stir-fried with seasonings.

小葱拌豆腐 (xiǎocōng bàn dòufu)

北方家常凉菜。将豆腐丁和葱花、盐、酱油、麻油等拌匀。

Bean curd with scallion. Home-style cold dish in northern areas. Diced bean curd flavored with chopped scallion, salt, soy sauce, and sesame oil.

小黄瓜鸡肫 (xiǎohuángguā jīzhūn)

江浙沪皖家常凉菜。将煮熟的鸡肫和生小黄瓜切片,浇上用蒜、红辣椒、姜、酱油、麻油等调制的卤汁。

Boiled chicken gizzard with cucumbers. Home-style cold dish in Jiangsu, Zhejiang, Shanghai, and Anhui. Sliced boiled chicken gizzard and fresh baby cucumbers seasoned with dressing of garlic, red pepper, ginger, soy sauce, and sesame oil.

小黄鱼炖豆腐 (xiǎohuángyú dùn dòufu)

江浙沪家常菜。先将小黄鱼裹淀粉油炸,豆腐水汆,再将鱼和豆腐放入加调料的汤炖煮。

Yellow croaker with bean curd. Home-style dish in Jiangsu, Zhejiang, and Shanghai. Small yellow croaker coated with starch, deep-fried, then braised with quick-boiled bean curd in seasoned soup.

小蓟焖田螺 (xiǎojì mèn tiánluó)

浙江菜。将盐腌过的田螺配小蓟、姜丝、料酒煸炒,再加酱油、料酒等焖煮。

Braised field snails. Zhejiang dish. Field snails pickled with salt, quick-fried with thistle, shredded ginger, and cooking wine, then braised with soy sauce, cooking wine, and other seasonings.

小煎仔鸡 (xiǎojiān zǐjī)

南方菜。先将嫩鸡脯肉用料酒和盐腌渍,轻炒至半熟,然后与青笋、西芹、姜、葱、泡椒爆炒,浇上用酱油、糖、醋等做成的芡汁。

Stir-fried young chicken. Southern dish. Young chicken breast pickled with cooking wine and salt, half-fried, stir-fried with lettuce stems, American celery, ginger, scallion, and pickled red pepper, then washed with soup of soy sauce, sugar, vinegar, and starch.

小笼蒸羊排 (xiǎolóng zhēng yángpái)

北方菜。先将羊排腌渍,然后裹上米粉蒸熟。

Steamed lamb steaks. Northern dish. Lamb steaks pickled, coated with ground rice, then steamed.

小米粉蒸鸡 (xiǎomǐfěn zhēng jī)

四川、贵州菜。先将鸡块腌渍,再与小米、五香粉、盐等拌和,蒸熟。

Steamed millet and chicken. Sichuan and Guizhou dish. Pickled chopped chicken mixed with millet, five-spice powder, and salt, then steamed.

小米扣肉 (xiǎomǐ kòuròu)

四川、贵州菜。将小米与腌过的五花肉、五香粉、盐等拌匀,蒸熟。

Steamed millet and streaky pork. Sichuan and Guizhou dish. Millet mixed with pickled streaky pork, five-spice powder, and salt, then steamed.

小米蒸排骨 (xiǎomǐ zhēng páigǔ)

四川、贵州菜。先将排骨腌渍,然后同小米、花椒粉、五香粉等拌匀,蒸熟。

Steamed millet and pork ribs.

Sichuan and Guizhou dish. Pickled pork ribs mixed with millet, ground Chinese prickly ash, and five-spice powder, then steamed.

小绍兴鸡粥 (xiǎoshàoxīng jīzhōu)

上海点心。用鸡汤煮粳米粥,放盐,撒鸡丝、姜丝、葱花。

Rice and chicken porridge. Shanghai snack. Round grain rice porridge cooked in chicken soup, then flavored with salt, shredded chicken, ginger, and scallion.

小棠菜煮双丸 (xiǎotángcài zhǔ shuāngwán)

上海菜。用鸡汤煮鱼丸、牛肉丸和小棠菜。

Shanghai greens with beef and fish balls. Shanghai dish. Fish balls, beef balls, and baby Shanghai greens boiled in chicken soup.

小桃园瓦罐鸡汤 (xiǎotáoyuán wǎguàn jītāng)

湖北武汉名菜。先将嫩母鸡块爆炒,然后放入瓦罐,加入调料煨煮。

Chicken soup in earthen jar.

Specialty in Wuhan, Hubei. Quick-boiled chopped young hen put in an earthen pot, then stewed with seasonings.

小叶苦丁茶 (xiǎoyè kǔdīngchá)

南方传统清凉饮料,产于华南和西南地区,常用作保健茶。将嫩苦丁茶叶经过萎凋、杀青、揉捻、干燥四个过程制成。外形条索粗壮,无茸毫,汤色透明深绿,叶底摊张,呈深绿,味先微苦后甘,耐冲泡。

Small-leaf *kuding* tea. Traditional summer beverage in southern areas, also used as a tonic drink. Made with tender leaves of broadleaf holly trees through withering, gentle-baking, rubbing-twisting, and drying. Its dry leaves are stout, tight, and hairless. The tea is crystal dark green with a soft bitter taste then a sweet aftertaste. The infused leaves are flat. Good for repeated brewing.

小鱿鱼酿肉 (xiǎoyóuyú niàng ròu)

浙江菜。先将猪肉、芹菜、冬菇、洋葱炒成馅,填进鱿鱼肚,用面粉蛋浆封口,油煎,然后加高汤和调料焖煮收汁。

Braised stuffed squid. Zhejiang dish. Small squid stuffed with cooked mixture of pork, celery, winter mushrooms, and onion, sealed with egg and flour paste, deep-fried, then braised in broth with seasonings until reduced.

小鱼花生 (xiǎoyú huāshēng)

山东菜。先将小鱼干煎熟，再放入干辣椒、黄酒、酱油、糖等焖煮收汁，然后加入油炸花生拌匀。

Spicy fingerlings with peanuts. Shandong dish. Dried fingerlings fried, simmered with chili pepper, yellow wine, soy sauce, and sugar until reduced, then mixed with fried peanuts.

笑口枣 (xiàokǒuzǎo)

南方点心。将面粉、发粉、苏打粉、猪油、糖、水拌和，做成小条，裹上芝麻，炸至黄脆。

Flour and sesame cookie. Southern snack. Flour, baking powder, and soda mixed with lard, sugar, and water, made into small strips, coated with sesame, then deep-fried golden and crisp.

蟹扒菜胆 (xiè pá càidǎn)

广东菜。用鲜蟹肉、香菇、芡粉及调料制卤，淋在焯过的菜心上。

Baby greens with crab sauce. Guangdong dish. Quick-boiled baby greens marinated in gravy of crab meat, dried mushrooms, starch, and seasonings.

蟹粉豆腐 (xièfěn dòufu)

上海菜。先把豆腐用开水烫过，再加入蟹肉和蟹黄翻炒，烹入鲜汤、姜末、葱花，勾芡。

Sautéed bean curd with crab meat. Shanghai dish. Scalded bean curd sautéed with crab meat and roe, seasoned soup, minced ginger, and chopped scallion, then thickened with starch.

蟹黄灌汤包 (xièhuáng guàntāngbāo)

浙江点心。将猪肉茸、蟹肉、蟹黄、酱油、料酒、肉皮冻等调成馅，用面皮包上，蒸熟。

Steamed pork and crab buns. Zhejiang snack. Flour buns filled with mixture of ground pork, crab meat and roe, soy sauce, cooking wine, and pork skin jelly, then steamed.

蟹黄海参 (xièhuáng hǎishēn)

山东菜。先将海参氽过，加清汤、黄酒、白糖、盐等煮透，然后将蟹黄汤烧沸，浇在海参上。

Sea cucumbers in crab soup. Shandong dish. Scalded sea cucumbers boiled in clear soup flavored with yellow wine, sugar, and salt, then marinated in crab roe soup.

蟹黄蹄筋 (xièhuáng tíjīn)

山东菜。先将油发蹄筋入清汤炖煮，用白糖、盐、黄酒调味，然后把蟹黄炒匀，配蹄筋装盘。

Stewed pork tendons with crab roe. Shandong dish. Oil-soaked pork tendons stewed in clear soup, flavored with sugar, salt, and yellow wine, then served with stir-fried crab roe.

蟹壳黄 (xièkéhuáng)

江浙沪点心。把发酵面团做成饼，用

炒过的咸菜肥肉丁或者糖为馅，撒上芝麻，烤熟。其形如蟹壳，色如蟹黄。

Crab-shape cakes. Snack in Jiangsu, Zhejiang, and Shanghai. Fermented flour dough made into cakes, filled with sugar or sautéed salty preserved vegetable and diced fat pork, topped with sesame, then baked. It has a yellowish colour, and looks like a crab's back.

蟹酿橙 (xièniàngchéng)

浙江点心。先将鲜橙切成杯状，挖出橙瓤，装入煸炒过的蟹肉、橙肉、橙汁，加入特制卤汁和白菊花，盖上橙盖，蒸熟。

Steamed crab and orange cup. Zhejiang snack. Oranges cut into cups, hollowed, filled with sautéed crab meat, orange pulp, orange juice, special sauce, and white chrysanthemum petals, covered with orange lids, then steamed.

蟹肉扒翅 (xièròu pá chì)

江苏菜。先将蟹肉加调料爆炒，再与鱼翅烹煮，勾芡。

Crab meat with shark fin. Jiangsu dish. Crab meat stir-fried with seasonings, braised with shark fin, then thickened with starch.

蟹肉豆腐 (xièròu dòufu)

安徽菜。先将氽过的蟹肉裹上面糊炸酥，再与油炸豆腐条拌炒，加姜、葱、盐、料酒。

Sautéed crab meat and bean curd. Anhui dish. Quick-boiled crab meat coated with flour paste, deep-fried, sautéed with deep-fried bean curd strips, then seasoned with ginger, scallion, salt, and cooking wine.

蟹肉卷 (xièròujuǎn)

福建菜。将煎蛋皮裹上氽过的蟹肉成卷，用牙签固定，切块装盘。

Crab meat rolls. Fujian dish. Fried egg sheets made into rolls filled with scalded crab meat, fastened with toothpicks, then cut into segments.

蟹肉小笼包 (xièròu xiǎolóngbāo)

南方沿海一带点心。先用面粉加猪油和水制成面团，发酵，再制成面皮，包入蟹肉馅，蒸熟，用酱油、醋、麻油佐食。

Steamed crab meat buns with sauce. Snack in southern coastal areas. Flour, lard, and water mixed, fermented, made into small buns filled with crab meat, then steamed. Served with soy sauce, vinegar, and sesame oil.

蟹子鸡肉卷 (xièzǐ jīròu juǎn)

江苏菜。先将鸡肉铺在紫菜上，放上蟹黄，裹成卷，然后炸至表面酥脆，切斜片装盘。

Crab roe and laver rolls. Jiangsu dish. Crab meat rolled in sliced chicken breast, wrapped in laver, then fried until crisp. Cut into elliptical pieces before serving.

新安烫面饺 (xīn'ān tàngmiànjiǎo)

河南新安食品。用烫面擀成饺皮，包入用猪肉、大葱、韭黄、白菜心、姜、盐等做成的馅，蒸熟。

Steamed pork and vegetable *jiaozi*.

Specialty in Xin'an, Henan. Half-boiled flour dough kneaded into wrappers, made into *jiaozi* filled with mixture of pork, scallion, yellow Chinese chive, baby Chinese cabbage, ginger, and salt, then steamed.

新疆拌面 (xīnjiāngbànmiàn)

新疆食品。将拉面煮熟,放入炒好的羊肉或牛肉,用辣椒酱、蒜泥等调味。

Noodles and meat with sauce. Xinjiang food. Hand-made noodles boiled, topped with cooked mutton or beef, then flavored with chili sauce and mashed garlic.

新疆凉面 (xīnjiāngliángmiàn)

新疆食品。细拉面煮熟,过凉水,淋油,配葫芦丝、辣椒酱等。

Noodles with gourd and chili sauce. Xinjiang food. Fine hand-made noodles boiled, cooled in cold water, then sprinkled with oil. Served with shredded gourd and chili sauce.

信阳毛尖 (xìnyángmáojiān)

绿茶。产于河南信阳大别山。外形细直,色泽翠绿,白毫显露。汤色碧透,叶底嫩绿匀整。

Xinyangmaojian, green tea produced in Mount Dabie in Xinyang, Henan. Its dry leaves are slim, straight, jade green, and hairy. The tea is crystal green with infused leaves tender green and even.

杏花村虾肉 (xìnghuācūn xiāròu)

山西菜,源于山西杏花村。将虾肉和山药用特制杏汁煨至汤浓,撒香菜。

Shrimp meat with apricot sauce. Shanxi specialty originated in Xinghua Village—Apricot Blossoming Village, Shanxi. Shrimp meat and Chinese yam braised with apricot sauce, then garnished with co-riander.

杏花村竹叶青 (xìnghuācūn zhúyèqīng)

以汾酒作底酒,用公丁香、竹叶、陈皮、砂仁、当归、零陵香、紫檀香等药材和冰糖、白砂糖浸泡制成,酒精含量为38-45度,属浓香型酒。

Xinghuacunzhuyeqing, white spirit made from *fenjiu*, male cloves, bamboo leaves, processed tangerine peels, villous amomum, Chinese angelica, basil, purple sandal wood, rock sugar, and granulated sugar. Contains 38-45% alcohol, and has a thick aroma.

杏片香虾筒 (xìngpiàn xiāngxiā tǒng)

广东菜。用糯米纸卷凤梨片和虾仁成筒状,涂上蛋黄液,沾满杏仁片,油炸。

Fried pineapple and almond rolls. Guangdong dish. Sliced pineapple and shelled shrimp rolled in edible sticky rice paper, coated with egg yolks and sliced almonds, then fried.

杏仁冻 (xìngréndòng)

山东点心。先将洋菜煮溶,加糖,然后同鲜奶和杏仁露调匀,置凉成冻。

Agar and almond jelly. Shandong snack. Boiled and sugared agar

mixed with fresh milk and almond milk, then chilled into jelly.

杏仁豆腐 (xìngrén dòufu)

各地家常点心。将杏仁浆、鲜奶、米浆、白糖等调和,蒸熟,冷却成糕状,切成小块,加入糖水、山楂糕、黄瓜片等。

Almond curd. Home-style snack in many places. Mixture of almond paste, milk, rice paste, and sugar steamed, chilled into curd, then cut into small pieces. Served in sugar soup with haw candies and sliced cucumbers.

杏仁葛粉包 (xìngrén gěfěn bāo)

江苏点心。先将熟花生末、熟猪肥膘、白糖等拌和,做成小圆粒,裹上葛粉,入沸水烫熟,然后放入牛奶杏仁露汤。

Arrowroot balls in almond soup. Jiangsu snack. Mixture of ground roasted peanuts, fat pork, and sugar made into pearl-size balls, coated with arrowroot powder, then boiled. Served in milk and almond soup.

杏仁煨梨 (xìngrén wēi lí)

台湾点心。将梨和杏仁用砂锅煨熟。

Simmered almonds and pears. Taiwan snack. Pears and almonds simmered in a casserole.

杏仁猪肺汤 (xìngrén zhūfèi tāng)

南方家常菜。先将猪肺略炒,再放入杏仁及调料,用小火煨炖。

Almond and hog lung soup. Home-style dish in southern areas. Hog lung quick-fried, braised with almonds and seasonings.

杏香鱼块 (xìngxiāng yúkuài)

台湾菜。先将鱼块裹蛋汁和杏仁片,炸至金黄,然后用洋葱、青椒、盐、黄酒、芡粉等做成卤汁,浇在鱼块上。

Fried fish and almonds with sauce. Taiwan dish. Fish chunks coated with starch and egg paste and sliced almonds, fried golden, then marinated in sauce of onion, green pepper, salt, yellow wine, and starch.

杏元凤爪炖水鱼 (xìngyuán fèngzhǎo dùn shuǐyú)

广东菜。将水鱼加杏肉、桂圆肉、鸡爪熬炖,用盐、姜、料酒等调味。

Stewed turtle with chicken feet and fruit. Guangdong dish. Soft-shell turtle stewed with apricots, longan pulp, and chicken feet, then seasoned with salt, ginger, and cooking wine.

杏圆炖甲鱼 (xìngyuán dùn jiǎyú)

福建菜。先将甲鱼加调料爆炒,然后与氽过的猪肉、杏仁、桂圆肉等加清汤炖至酥烂。

Stewed turtle with almonds and longan pulp. Fujian dish. Turtle stir-fried with seasonings, then stewed in clear soup with quick-boiled pork, almonds, and dried longan pulp.

熊掌豆腐 (xióngzhǎng dòufu)

东北传统菜。将煎过的豆腐和煸炒过的肉片加入盐、豆瓣酱、姜、葱、蒜、肉汤等焖煮,浇芡汁,淋猪油。

Soy bear paw with pork.

Northeastern traditional dish. Fried bean curd and quick-fried pork braised in broth with salt, fermented broad bean sauce, ginger, scallion, and garlic, then washed with starched soup and lard.

秀珍菇炒虾仁 (xiùzhēngū chǎo xiārén)

江苏家常菜。先将秀珍菇氽熟,虾仁过油,然后加入盐、黄酒、味精等拌炒。

Sautéed mushrooms and shrimp meat. Home-style dish in Jiangsu. Quick-boiled oyster mushrooms and lightly fried shrimp meat sautéed with salt, yellow wine, MSG, and other seasonings.

绣球干贝 (xiùqiú gānbèi)

山东菜。先将大虾肉和猪肉加蛋清及调料做成丸子,裹上干贝丝、火腿丝、香菇丝等成球状,然后蒸熟,勾芡。

Steamed prawn and pork balls. Shandong dish. Prawn meat and minced pork mixed with egg white and seasonings, made into balls, coated with shredded dried scallops, ham, and dried mushrooms, steamed, then washed with starched soup.

绣球鱼翅 (xiùqiú yúchì)

沿海一带名菜。将肉末、虾茸、荸荠末、盐、芡粉、水等调和,做成丸子,裹上鱼翅和香菇丝,蒸熟,浇芡汁。

Pork and shrimp balls with shark fin. Famous dish along coastal areas. Ground pork, minced shrimp meat, mashed water chestnuts, salt, starch, and water mixed, made into balls, coated with shark fin and shredded dried mushrooms, steamed, then washed with starched soup.

学果馍馍 (xuéguǒmómo)

西藏点心,即土豆包子。把土豆泥与面粉和成面团,擀成面皮,包入炒熟的肉馅,用酥油炸熟。

Fried stuffed potato buns. Tibetan snack. Mashed potatoes and flour mixed, made into buns stuffed with cooked meat, then deep-fried in Tibetan butter.

雪白奶馒头 (xuěbái nǎimántou)

广东点心。将糖面粉用干酵母发酵,做成小馒头,蒸熟,配炼奶。

Sweet steamed bread with condensed milk. Guangdong snack. Sugared flour fermented with dry yeast, made into small buns, and steamed. Served with condensed milk.

雪菜白鲳 (xuěcài báichāng)

上海、江苏菜。先把腌过的白鲳蒸熟,再把炒熟的雪菜和笋丁勾芡,浇在鱼上。

Steamed butterfish with preserved mustard. Shanghai and Jiangsu dish. Pickled white butterfish steamed, then covered with sautéed and starched preserved potherb mustard and diced bamboo shoots.

雪菜炒豆芽 (xuěcài chǎo dòuyá)

东北家常菜。腌雪里蕻与黄豆芽加盐、葱、姜炒熟。

Sautéed preserved mustard and soybean sprouts. Northeastern home-style dish. Preserved potherb mustard sautéed with soybean sprouts, then seasoned with salt, scallion, and ginger.

雪菜炒毛豆 (xuěcài chǎo máodòu)

江浙沪菜。毛豆与雪菜加调料同炒。

Sautéed soybean with preserved mustard. Jiangsu, Zhejiang, and Shanghai dish. Fresh soybean sautéed with preserved potherb mustard and seasonings.

雪菜大汤黄鱼 (xuěcài dàtāng huángyú)

浙江菜。先将大黄鱼稍煎,再用盐和黄酒腌渍,然后加水煮沸,转用小火炆煮至汤变白,最后放入雪菜和笋片。

Yellow croaker and preserved mustard soup. Zhejiang dish. Lightly fried big yellow croaker pickled with salt and yellow wine, then boiled. Braised until soup milky white, then combined with preserved potherb mustard and sliced bamboo shoots.

雪菜豆板汤 (xuěcài dòubǎn tāng)

上海家常菜。将水发去皮蚕豆与煸炒过的雪菜煮汤。

Broad bean and preserved mustard soup. Home-style dish in Shanghai. Peeled soaked broad beans boiled with quick-fried preserved potherb mustard.

雪菜发芽豆 (xuěcài fāyádòu)

上海家常菜。雪菜炒发芽蚕豆.

Preserved mustard with broad beans. Home-style dish in Shanghai. Preserved potherb mustard sautéed with sprouted broad beans.

雪菜黄鱼 (xuěcài huángyú)

江浙沪皖家常菜。先把调料、豆腐块、雪里蕻入汤烧开,然后放入炸过的黄鱼用小火焖煮,加香菜。

Yellow croaker with preserved mustard. Home-style dish in Jiangsu, Zhejiang, Shanghai, and Anhui. Yellow croaker deep-fried, simmered in seasoned soup with prepared bean curd and preserved potherb mustard, then garnished with coriander.

雪菜火鸭卷 (xuěcài huǒyā juǎn)

香港菜。用豆腐皮卷上雪里蕻和烤鸭肉,切块,配特制卤汁。

Mustard greens and broiled duck. Hong Kong dish. Bean curd skins made into rolls filled with potherb mustard greens and broiled duck meat, cut into segments, then served with special sauce.

雪菜苦瓜 (xuěcài kǔguā)

安徽菜。将煸炒过的苦瓜丝与雪菜速炒,烹入黄酒、酱油、葱花。

Sautéed bitter gourd and mustard. Anhui dish. Quick-fried shredded bitter gourds sautéed with preserved potherb mustard, then flavored with yellow wine, soy sauce, and chopped scallion.

雪菜目鱼 (xuěcài mùyú)

上海菜。将比目鱼片与雪菜炒熟。

Stir-fried flatfish with potherb mustard. Shanghai dish. Sliced flatfish stir-fried with preserved potherb mustard.

雪菜肉丝 (xuěcài ròusī)

浙江菜。将猪肉丝与雪菜翻炒,烹入豆瓣酱、盐、味精、黄酒、酱油,勾芡。

Sautéed shredded pork and preserved mustard. Zhejiang dish. Shredded pork sautéed with preserved potherb mustard, then seasoned with fermented broad bean sauce, salt, MSG, yellow wine, soy sauce, and starched soup.

雪菜素腿炒莲子 (xuěcài sùtuǐ chǎo liánzǐ)

山东菜。将雪里蕻、素火腿、莲子等用高汤煮熟,勾芡。

Preserved mustard with lotus seeds. Shandong dish. Preserved potherb mustard, soy ham, and lotus seeds boiled in broth, then thickened with starch.

雪耳蟹黄虾仁 (xuěěr xièhuáng xiārén)

广东菜。将虾仁和蟹黄用高汤和麻油翻炒,撒胡椒粉,配熟雪耳装盘。

Stir-fried shrimp meat with crab roe and white wood ears. Guangdong dish. Shrimp meat and crab roe stir-fried with broth and sesame oil, then sprinkled with ground black pepper. Served with prepared white wood ears.

雪梗珍珠羹 (xuěgěng zhēnzhū gēng)

浙江菜。在酱油汤中投入虾仁和咸菜梗煮沸,勾芡,撒上火腿末和香菜末。

Shrimp meat and preserved mustard soup. Zhejiang dish. Shrimp meat and preserved potherb mustard stalks boiled in soy sauce soup, thickened with starch, then sprinkled with minced ham and chopped coriander.

雪花八卦鸡 (xuěhuā bāguàjī)

江苏菜。用鸡脯肉作主料,用鸡蛋清、火腿、紫菜、蛋糕作配料,摆成八卦图状,蒸熟。

Steamed chicken breast. Jiangsu dish. Chicken breast, egg white, ham, laver, and cake arranged like Eight Trigrams, then steamed.

雪花菜 (xuěhuācài)

东北家常菜。将雪里蕻末加豆腐渣及调料炒熟。

Sautéed mustard greens with bean dregs. Northeastern home-style dish. Chopped potherb mustard greens fried with bean dregs and seasonings.

雪花豆腐 (xuěhuā dòufu)

东北家常菜。先将豆腐丁氽过,然后同蘑菇丁入鸡汤烧沸,加调料,勾芡,撒入炒熟的虾仁。

Bean curd with mushrooms. Northeastern home-style dish. Scalded diced bean curd and diced

mushrooms boiled in chicken soup, seasoned, thickened with starch, then topped with sautéed shrimp meat.

雪花豆腐羹 (xuěhuā dòufugēng)

浙江菜。先将嫩豆腐丁用开水烫过,再将香菇、蘑菇、火腿等入高汤煮沸,加调料并勾芡,做成卤汁,然后将卤汁浇在豆腐上,撒上过油的松子和虾米。

Tender bean curd with flavorings. Zhejiang dish. Scalded diced tender bean curd washed with dressing of dried and fresh mushrooms, ham, seasonings, and starch. Garnished with lightly fried pine nuts and dried shrimp meat.

雪花鸡 (xuěhuājī)

四川菜。将鸡脯肉茸与蛋清、干豆粉、盐、胡椒粉、清汤等调和,入油锅低温炒熟,撒火腿末。

Sautéed chicken with egg white. Sichuan dish. Ground chicken breast mixed with egg white, mung bean powder, salt, ground black pepper, and clear soup, sautéed over low heat, then sprinkled with ham crumbs.

雪花梨片 (xuěhuā lípiàn)

北方点心。将鸭梨去皮,切薄片装盘,撒白糖。

Pears with sugar. Fruit plate in northern areas. Pears peeled, sliced, then flaked with white sugar.

雪花蟹斗 (xuěhuā xièdòu)

江苏苏州名菜。以蟹壳为容器,内放蟹粉,面上浇蛋清泡浆,撒火腿末,蒸熟。

Steamed crab meat under egg white. Famous dish in Suzhou, Jiangsu. Crab shells filled with minced crab meat, covered with whipped egg white, garnished with minced ham, then steamed.

雪花鱼翅 (xuěhuā yúchì)

广东菜。先将鱼翅放入鸡汤加调料蒸烂,再与鸡肉茸和蛋清调成的糊翻炒,然后淋鸡油,撒火腿末。

Sautéed shark fin with egg white. Guangdong dish. Shark fin steamed in seasoned chicken soup, sautéed with mixture of minced chicken and whipped egg white, then flavored with chicken oil and minced ham.

雪花鱼肚 (xuěhuā yúdǔ)

广东菜。将腌渍好的鱼肚茸与蛋清调和,用中火翻炒,配焯过的油菜心装盘。

Sautéed fish maw with egg white. Guangdong dish. Pickled mashed fish maw mixed with egg white, stir-fried over medium fire, then served with quick-boiled baby rape greens.

雪花鱼糕 (xuěhuā yúgāo)

福建菜。先把鱼肉、猪肥肉、地瓜粉、水、蛋清等加调料调成糊,蒸熟成糕,然后把糕切成菱形,把蛋清涂在糕面上,回笼稍蒸。

Steamed fish and pork cakes. Fujian dish. Fish, fat pork, sweet potato starch, water, egg white, and seasonings mixed, then steamed into cake. Cut into diamond cubes, coated with egg white, then re-steamed briefly.

雪梨凤脯 (xuělí fèngpú)

浙江菜。先将鸡片腌渍,再与雪梨片和青红椒片翻炒,然后浇上用姜、黄酒、盐、味精等调制的卤汁。

Sautéed chicken with pears and pepper. Zhejiang dish. Pickled sliced chicken sautéed with sliced snow pear, green and red pepper, then flavored with sauce of ginger, yellow wine, salt, and MSG.

雪梨肘棒 (xuělí zhǒubàng)

广东菜。先将肘棒加调料炖烂,再加入梨块稍煮,浇原汤浓汁。

Stewed hog joint with pears. Guangdong dish. Hog joint stewed with seasonings, then quick-boiled with pear sections. Washed with thick broth.

雪里藏珠 (xuělǐcángzhū)

山东菜。先将煮熟去壳的鸽蛋挂蛋清面粉糊炸至浅黄,然后加米醋和汤烹煮,浇原汤浓汁。

Pigeon eggs in vinegar soup. Shandong dish. Boiled and shelled pigeon eggs coated with flour and egg white paste, fried light yellow, simmered in vinegar soup, then marinated in its thick broth.

雪里豆干 (xuělǐ dòugān)

台湾菜。将肉丝、雪里蕻、豆腐干加辣椒丝、白酱油等翻炒。

Stir-fried dried bean curd and mustard. Taiwan dish. Shredded pork, potherb mustard, and dried bean curd stir-fried, then seasoned with shredded pepper and white soy sauce.

雪里蕻炒肉末 (xuělǐhóng chǎo ròumò)

东北家常菜。将雪里蕻末与肉末及调料炒熟。

Mustard greens and ground pork. Northeastern home-style dish. Chopped potherb mustard greens sautéed with minced pork and seasonings.

雪丽蛋泊肉 (xuělì dànbóròu)

江苏、浙江菜。先将肥肉丸裹蛋清生粉糊炸至金黄,然后放入炒化出丝的白糖翻匀。

Sugarcoated fat pork balls. Jiangsu and Zhejiang dish. Fat pork meatballs coated with egg white and potato starch paste, deep-fried, then quick-fried with heated melting sugar.

雪莲美蹄 (xuělián měití)

西藏菜。将羊蹄与雪莲花、西蓝花煮熟。

Lamb hooves with snow lotus flower. Tibetan dish. Lamb hooves boiled with snow lotus flower and broccoli.

雪山潭虾 (xuěshān tánxiā)

福建菜。先将虾仁挂蛋清、菱粉爆熟,再将虾仁放在蛋清和菱粉打成的蛋白上,上面覆盖蛋白,撒上火腿末和香菜叶,然后用文火炸至

微黄。

Fried shrimp meat with egg white. Fujian dish. Shelled shrimp coated with egg white and water caltrop starch, then stir-fried. Placed on a bed of whipped stiff egg white and starch, covered with whipped stiff egg white, topped with diced ham and coriander, then fried light yellow.

雪笋平鱼 (xuě sǔn píngyú)

山东菜。先将平鱼用盐和黄酒腌渍,微蒸,然后加入炒熟的雪里蕻和笋丝,蒸透。

Steamed butterfish with mustard. Shandong dish. Butterfish pickled with salt and yellow wine, then half-steamed. Covered with sautéed preserved potherb mustard and shredded bamboo shoots, then steamed again.

雪衣银鱼 (xuěyī yínyú)

东北家常菜。银鱼挂糊后油炸,蘸椒盐。

Fried silver fish. Home-style dish in northern areas. Silver fish coated with starch paste, deep-fried, then served with spicy salt.

雪域明珠 (xuěyù míngzhū)

西藏菜。将冬虫夏草插在黄菇上,蒸熟,浇芡汁。

Caterpillar fungi on mushrooms. Tibetan dish. Chinese caterpillar fungi stuck onto yellow mushrooms, steamed, then washed with starched soup.

雪域雄鹰 (xuěyù xióngyīng)

西藏菜。用卤牛肉、卤猪舌、川味香肠、藏式香肠、熟蛋黄、熟蛋白、黄瓜等在盘里摆成雄鹰状。

Assorted meat plate. Tibetan dish. Marinated beef, marinated hog tongue, spicy sausage, Tibetan sausage, cooked egg yolks, cooked egg white, and cucumbers arranged on a platter in the shape of an eagle.

雪域羊头 (xuěyù yángtóu)

西藏菜。先将糍粑放在焖熟的羊头周围,再撒上辣椒粉、葱花、姜末。

Lamb head with sticky rice cakes. Tibetan dish. Simmered lamb head served with sticky rice cakes. Sprinkled with chili powder, chopped leek, and minced ginger.

血旺牛柳煲 (xuèwàng niúliǔ bāo)

四川菜。先将生猪血汆过,然后与牛肉焖煮,加盐、花椒、辣椒、姜、葱、酱油。

Braised spicy hog blood and beef. Sichuan dish. Scalded hog blood braised with beef. Seasoned with salt, Chinese prickly ash, chili pepper, ginger, scallion, and soy sauce.

熏干炒蒜苗 (xūngān chǎo suànmiáo)

西南家常菜。将蒜苗与烟熏豆腐干翻炒,加入盐、酱油、辣椒。

Stir-fried garlic sprouts with smoked bean curd. Southwestern dish. Garlic sprouts stir-fried with smoked dried bean curd. Seasoned with salt, soy sauce, and hot pepper.

熏干豆腐卷 (xūn gāndòufujuǎn)

东北菜。先将猪肉末加调料做成馅,再用薄豆腐干裹馅成卷,蒸熟,然后入熏锅用糖熏3分钟,最后刷上麻油。

Smoked bean curd rolls. Northeastern dish. Ground pork mixed with condiments, rolled in dried bean curd sheets, then steamed. Smoked with sugar in a covered wok for 3 minutes, then sprinkled with sesame oil.

熏烤河鳗 (xūnkǎo hémán)

广东菜。将河鳗去骨,用多种调料腌渍,抹上鳗鱼骨做的卤汁烤熟,配香菜和酸萝卜。

Roasted eel. Guangdong dish. Boned river eel pickled with seasonings, coated with eel-bone sauce, then roasted. Served with coriander and sour pickled turnips.

熏马肠 (xūnmǎcháng)

新疆菜。将加了调料的碎马肉灌入马肠,再用果木烟熏。

Smoked horse meat sausage. Xinjiang dish. Horse intestine filled with seasoned chopped horse meat, then smoked with the wood of fruit trees.

熏马肉 (xūnmǎròu)

新疆菜。将马肉用盐、白糖等腌渍,再用松柴烟熏。

Smoked horse meat. Xinjiang dish. Horse meat pickled with salt and sugar, then smoked with pine wood.

熏肉 (xūnròu)

东北菜。将大块五花猪肉加调料煮熟,然后熏烤。

Smoked streaky pork. Northeastern dish. Streaky pork chunks boiled with seasonings and then smoked.

熏沙半鸡 (xūnshābànjī)

四川菜。沙半鸡即山鹑。先将山鹑汆过,入酱汤煮熟,然后入熏锅,用茶叶、白糖熏至枣红色,涂麻油。

Marinated and smoked daurian partridge. Sichuan dish. Scalded daurian partridge marinated in gravy, smoked with tea leaves and sugar in a covered wok, then sprinkled with sesame oil.

熏兔肉 (xūntùròu)

东北菜。先将兔肉汆烫,再入鸡汤煮熟,然后熏烤。

Smoked rabbit. Northeastern dish. Scalded rabbit meat boiled in chicken soup, then smoked.

熏鱼 (xūnyú)

江浙沪菜。先把草鱼块用酱油、盐、黄酒、葱汁等腌渍,再油炸,然后加高汤、茴香、桂皮、姜、葱等焖煮收汁。

Smoked grass carp. Jiangsu, Zhejiang, and Shanghai dish. Grass carp chunks pickled with mixture of soy sauce, salt, yellow wine, and scallion juice, deep-fried, then simmered in broth with aniseed, cinnamon, scallion, and ginger till reduced.

鸭蛋豆腐 (yādàn dòufu)

各地家常菜。将豆腐丁与碾碎的咸鸭蛋黄、姜、葱、酱油等炒熟。

Duck egg yolks with bean curd. Home-style dish in many places. Diced bean curd sautéed with mashed salted duck egg yolks, ginger, scallion, and soy sauce.

鸭肝首乌汤 (yāgān shǒuwū tāng)

南方药膳。先将何首乌和鸭肝焯烫,切片,再与香菇、黑木耳等加盐炖汤。

Duck liver and herb soup. Southern folk-medicinal dish. Scalded and sliced fleeceflower root and duck liver stewed with dried mushrooms and wood ears, then seasoned with salt.

鸭子萝卜汤 (yāzi luóbo tāng)

湖北菜。先用黄酒将鸭块腌过,稍炒,再与白萝卜煮开,加调料入沙罐烹炖。

Duck and turnip soup. Hubei dish. Duck pickled with yellow wine, quick-fried, boiled with white turnips, then simmered with seasonings in an earthen pot.

牙签牛肉 (yáqiān niúròu)

湖南菜。先将牛肉丁用牙签串好,再用辣椒粉,花椒油,五香粉,料酒,蒜茸,淀粉等腌渍,然后油炸。

Fried spicy beef. Hunan dish. Diced beef strung on toothpicks, pickled with chili powder, Chinese prickly ash oil, five-spice powder, cooking wine, mashed garlic, and starch, then deep-fried.

牙签田螺 (yáqiān tiánluó)

湖南菜。先将田螺汆过,然后与辣椒、姜、花椒、盐、酱油等爆炒,用牙签挑出螺肉食用。

Stir-fried field snails. Hunan dish. Field snails scalded, then stir-fried with hot pepper, ginger, Chinese prickly ash, salt, and soy sauce. Served with toothpicks to remove meat from shells.

牙签羊肉 (yáqiān yángròu)

东北菜。先将羊肉丁腌渍,然后串在牙签上,炸至金黄。

Fried mutton. Northeastern dish. Pickled diced mutton strung with toothpicks, then deep-fried.

芽菜扣肉 (yácài kòuròu)

河南传统菜。将五花肉煮熟,在皮的一面抹酱油和糖,炸至金黄,然后将肉块皮朝下放入碗中,盖上芽菜,加酱油、糖、姜、葱等蒸至酥烂。

Steamed streaky pork with preserved cabbage. Traditional dish in Henan. Boiled streaky pork painted with soy sauce and sugar on the skin side, then deep-fried. Put in a bowl with skins down, covered with preserved cabbage, flavored with soy sauce, sugar, ginger, and scallion, then steamed.

烟鲳鱼 (yānchāngyú)

上海菜。先将鲳鱼腌渍,烤熟,再用茶叶和赤砂糖熏烤,然后涂上熟花生油,配色拉酱。

Broiled and smoked pomfret. Shanghai dish. Pickled silvery pomfret broiled, smoked with tea leaves and brown sugar, then brushed with prepared peanut oil. Served with salad dressing.

烟笋腊肉 (yānsǔn làròu)

湖南菜。将烟熏楠竹笋和腊肉装入土钵焖煮。

Smoked cured pork and bamboo shoots. Hunan dish. Smoked cured pork stewed with smoked bamboo shoots in a clay pot.

烟笋烧鸭 (yānsǔn shāoyā)

四川菜。将鸭块与烟笋爆炒,加入葱、姜、辣椒,加少量汤焖煮收汁。

Stir-fried bamboo shoots and duck. Sichuan dish. Chopped duck stir-fried with smoked bamboo shoots, scallion, ginger, and hot pepper, then simmered in a little soup until reduced.

烟台红葡萄酒 (yāntái hóngpútáojiǔ)

葡萄酒。产于山东烟台,以当地玫瑰香、玛瑙红、解百纳等优质葡萄为原料酿制而成,酒精含量为 7-12 度。

Red wine produced in Yantai, Shandong. It is made with local grapes such as muscat, agate red, and cabernet. Its alcohol content is 7-12%.

烟熏豆腐 (yānxūn dòufu)

四川、贵州菜。先将豆腐干用松柏树枝或谷壳薰过,再与肉片、芹菜、蒜苗及调料爆炒。

Stir-fried smoked bean curd. Sichuan and Guizhou dish. Dried bean curd smoked with rice husk or pine and cypress leaves. Stir-fried with pork, celery, garlic sprouts, and seasonings.

腌笃鲜 (yāndǔxiān)

上海菜。腌猪肉、鲜猪肉、春笋长时间煨炖。

Stewed pork and bamboo shoots. Shanghai dish. Preserved pork, fresh pork, and spring bamboo shoots stewed for several hours.

腌三样 (yānsānyàng)

湖南菜。先将茄子、豆角、辣椒分别用沸水烫过,晾干,腌渍,然后加调料煸炒。

Sautéed three vegetables. Hunan Dish. Scalded eggplants, string beans, and hot peppers air-dried, pickled, then sautéed with seasonings.

芫爆肚丝 (yánbào dǔsī)

山东菜。先将猪肚用开水烫过，再煮熟，然后切成丝，加醋、黄酒、酱油、香菜等翻炒。

Stir-fried hog tripe with coriander. Shandong dish. Scalded shredded hog tripe boiled, then stir-fried with vinegar, yellow wine, soy sauce, and coriander.

芫爆墨鱼 (yánbào mòyú)

山东菜。先将墨鱼焯水，然后加香菜翻炒，烹入用黄酒、醋、酱油、芡粉等调成的卤汁。

Stir-fried cuttlefish with coriander. Shandong dish. Scalded cuttlefish stir-fried with coriander, then flavored with sauce of yellow wine, vinegar, soy sauce, and starch.

芫爆牛百叶 (yánbào niúbǎiyè)

广东菜。先将牛百叶用水汆过，再与香菜、蒜苗、葱等爆炒，然后加盐、料酒，淋麻油。

Stir-fried ox tripe. Guangdong dish. Scalded ox tripe stir-fried with coriander, garlic sprouts, and scallion, seasoned with salt and cooking wine, then washed with sesame oil.

芫爆鱿鱼卷 (yánbào yóuyújuǎn)

山东菜。先将鱿鱼片焯水，滑熟，再加入香菜、黄酒、醋、盐等爆炒。

Stir-fried squid with coriander. Shandong dish. Lightly fried scalded squid stir-fried with coriander. Seasoned with yellow wine, vinegar, and salt.

胭脂鹅脯 (yānzhī épú)

曹雪芹《红楼梦》菜谱。先将鹅剖为两半，汆过，再加水、盐、黄酒、葱、姜、桂叶、苹果等煮至出骨，然后加入清汤、白糖、蜂蜜、盐、红曲粉炖至汁浓，淋少许香油，用蓑衣王瓜片饰盘。

Braised goose with red yeast rice. Menu from *A Dream of Red Mansions* by Cao Xueqin. Scalded halved goose boiled with salt, yellow wine, scallion, ginger, bay leaves, and apples until meat falls off bones. Braised in clear soup with sugar, honey, salt, and red yeast rice powder until reduced. Sprinkled with sesame oil, then garnished with sliced cucumbers.

盐焗鸡 (yánjújī)

广东菜。先将整鸡或鸡块用盐、味精等腌渍，再用锡纸包裹，埋入热粗盐焖熟。

Salt-baked chicken. Guangdong dish. Whole chicken or chicken chunks pickled with salt and MSG, wrapped in foil, buried in heated coarse salt, then baked.

盐焗蟹 (yánjúxiè)

广东菜。先将蟹肉用盐、味精等腌渍，再用锡纸包裹，埋入热粗盐焖烤。

Salt-baked crab. Guangdong dish. Crab meat pickled with salt and MSG, wrapped in foil, buried in heated coarse salt, then baked.

盐水豆腐 (yánshuǐ dòufu)

四川家常菜。将豆腐块放入盐水花椒汤中焖煮，出锅沥干水分，佐香辣油食用。

Salty bean curd with chili oil. Home-style dish in Sichuan. Cubed bean curd boiled in salt and Chinese prickly ash soup, drained, then served with chili oil.

盐水鸡 (yánshuǐjī)

江苏家常菜。先将鸡身内外抹盐，再用布包住，蒸熟。

Steamed salty chicken. Home-style dish in Jiangsu. Whole chicken pickled with salt, wrapped in cloth, then steamed.

盐水口条 (yánshuǐ kǒutiáo)

东北家常凉菜。将开水烫过的猪舌配姜、葱、料酒、花椒、茴香、盐等煮熟，切片。

Boiled hog tongue. Home-style cold dish in northeastern areas. Scalded hog tongue boiled with ginger, scallion, cooking wine, Chinese prickly ash, aniseed, salt, and other seasonings. Sliced to serve.

盐水虾 (yánshuǐxiā)

山东菜。先将对虾用清水加花椒、大料、盐等煮熟，然后用原汤浸泡。

Prawns in salty soup. Shandon dish. Prawns boiled and marinated in soup of Chinese prickly ash, aniseed, and salt.

盐水鸭 (yánshuǐyā)

江苏菜。先将鸭用热盐、五香粉、清卤水腌渍，然后入清汤焖煮，用姜、葱、八角、香醋调味，置凉后切块。

Braised salty duck. Jiangsu dish. Duck pickled with heated salt, five-spice powder, and light gravy, then braised in clear soup with ginger, scallion, aniseed, and vinegar. Chopped after cooled.

盐水羊肉 (yánshuǐ yángròu)

西北家常菜。先将焯过的羊肉配花椒、葱、姜、盐煮熟，然后入原汤卤泡。

Salty mutton. Northwestern home-style dish. Scalded mutton boiled with Chinese prickly ash, scallion, ginger, and salt, then marinated in broth.

燕皮 (yànpí)

福建食品。将猪肉茸与甘薯粉混合，制成薄片，切成丝状煮熟，配卤汁。也可与其他原料搭配，制成不同菜肴。

Pork and sweet potato powder sheets. Fujian food. Ground pork and sweet potato powder made into sheets, shredded, boiled, then served with special sauce. It may be combined with other ingredients to make different dishes.

燕窝炖雪梨 (yànwō dùn xuělí)

广东点心。将雪梨和燕窝加入冰糖和水，炖煮。

Pear and bird's nest soup. Guangdong snack. Pears and edible bird's nests stewed in water and sweetened with rock sugar.

燕窝鸽汤蛋 (yànwō gē tāng dàn)

广东菜。燕窝用开水焖过，放入煮熟去壳的鸽蛋，冲入鸡汤，加入胡椒粉、盐、味精。

Bird's nest and pigeon egg soup.

Guangdong dish. Edible bird's nests simmered in water. Boiled shelled pigeon eggs and chicken soup added. Seasoned with ground black pepper, salt, and MSG.

燕窝水晶球 (yànwō shuǐjīngqiú)

广东名菜。先将腌好的鲍鱼丁裹虾胶、碎燕窝蒸熟，然后淋上用鸡汤、蟹粉、芡粉、麻油等制成的卤汁。

Abalone with bird's nests. Famous Guangdong dish. Pickled diced abalone coated with minced shrimp meat and crumbs of edible bird's nests, then steamed. Washed with gravy of chicken soup, crab meat, starch, and sesame oil.

燕窝四大件 (yànwō sìdàjiàn)

台湾菜。将鸡脯肉末、鳜鱼肉末、肥肉膘泥、蛋清、芡粉、盐、水等调成糊，面上划出“万寿无疆”字样，蒸熟，在空隙处摆上熟燕窝。

Steamed bird's nest with assorted meat. Taiwan dish. Ground chicken breast, Chinese perch, and fat pork mixed with egg white, starch, salt, and water. Four Chinese characters of longevity, “万寿无疆”, carved on the surface of the paste. Steamed into cake, then garnished with prepared edible bird's nests.

燕窝椰子炖鸡 (yànwō yēzi dùn jī)

广东菜。将嫩鸡肉、燕窝、淮山、枸杞子、生姜、红枣加盐入椰汁炖煮。

Bird's nest with chicken and herb. Guangdong dish. Young chicken stewed in coconut milk with edible bird's nests and herbs, such as Chinese yam, Chinese wolfberry, ginger, and red date, then seasoned with salt.

扬州炒饭 (yángzhōuchǎofàn)

江苏扬州食品。先将鸡蛋炒熟，然后加入米饭、虾仁、豌豆、熟肉、香肠、胡萝卜等翻炒，用盐、葱等调味。

Stir-fried rice. Specialty of Yangzhou, Jiangsu. Fried eggs stir-fried with cooked rice, shrimp meat, peas, pork, ham, sausage, and carrots. Seasoned with salt and scallion.

羊背子 (yángbèizi)

内蒙菜。将全羊卸成七大件，带尾入锅，加盐煮熟。

Boiled salted mutton. Inner Mongolian dish. Whole dressed lamb boiled, seasoned with salt, then served in seven large sections including tail.

羊肚菌炖鸡 (yángdǔjūn dùn jī)

西北菜。先将乌鸡入高汤烧开，然后加入羊肚菌、料酒、盐、鸡精、葱、姜、枸杞、红枣，炖烂。

Stewed chicken and morchella mushrooms. Northwestern dish. Black-bone chicken boiled in broth, then stewed with morchella mushrooms, cooking wine, salt, chicken essence, scallion, ginger, Chinese wolfberry, and red dates.

羊方藏鱼 (yángfāng cáng yú)

西北菜。将大块羊肉用花椒、盐、黄酒、葱、姜腌渍，再氽烫，然后从中间剖开,藏入鲫鱼,加调料炖酥。

Stewed fish in mutton. Northwestern dish. Mutton chunks pickled with Chinese prickly ash, salt, yellow wine, scallion, and ginger. Quick-boiled, cut open, filled with a crucian carp, then stewed with seasonings.

羊羔肉 (yánggāoròu)

新疆菜。羊羔肉加盐、洋葱、胡椒等煮熟。

Boiled seasoned lamb. Xinjiang dish. Lamb boiled with salt, onion, and black pepper.

羊肉串 (yángròuchuàn)

新疆点心。将羊肉片腌渍,用铁钎串好,撒上盐、辣椒粉、孜然粉等烤熟。

Grilled spicy mutton. Xinjiang snack. Pickled sliced mutton strung on iron bars, sprinkled with salt, chili powder, and ground cumin, then grilled.

羊肉氽面 (yángròu cuānmiàn)

新疆食品。将面条用水煮熟,放入熟羊肉,加香菜等佐料。

Noodles with mutton. Xinjiang food. Noodles boiled, combined with cooked mutton, then flavored with coriander and other seasonings.

羊肉炖茄子 (yángròu dùn qiézi)

东北菜。将羊肉片和茄子块加五香粉、盐、葱、姜、黄酒等用小火炖熟。

Stewed mutton and eggplants. Northeastern dish. Sliced mutton stewed with cubed eggplants, then seasoned with five-spice powder, salt, leek, ginger, and yellow wine.

羊肉火锅 (yángròu huǒguō)

东北菜。先将海米和酸菜丝加入火锅汤煮沸,然后放入羊肉片、粉丝等涮熟。

Mutton in fire pot. Northeastern dish. Sliced mutton and starch vermicelli quick-boiled in fire pot soup flavored with dried shrimp meat and shredded sour pickled Chinese cabbage.

羊肉泡馍 (yángròu pàomó)

陕西点心。将羊肉和羊骨加调料炖煮,在汤里放入烙饼或馍。

Steamed bread in mutton soup. Shaanxi snack. Mutton and lamb bones stewed with seasonings, then served in broth with pancakes or steamed bread.

羊肉抓饭 (yángròu zhuāfàn)

新疆维吾尔族食品。先将羊肉块用羊油炸熟,再同胡萝卜丝翻炒,然后放入米饭焖煮,用盐、孜然粉等调味,用手抓食。

Rice with mutton and carrots. Uygur delicacy in Xinjiang. Mutton deep-fried in suet oil, then stir-fried with shredded carrots. Simmered with cooked rice, seasoned with salt and ground cumin, then served as finger food.

羊蹄子 (yángtízi)

新疆菜。先将羊蹄腌渍,再淋上羊油辣椒,蒸熟。

Steamed spicy lamb hooves.

Xinjiang dish. Pickled lamb hooves washed with chili suet oil, then steamed.

杨梅武昌鱼 (yángméi wǔchāngyú)

湖北菜。先将鱼头、鱼尾腌渍,上浆炸酥,再将鱼肉做成丸子,炸透,然后浇用杨梅汁、麻油等调成的卤汁。

Fried carp with waxberry dressing. Hubei dish. Wuchang carp head and tail pickled, coated with starch, then deep-fried. Carp meat made into balls, then deep-fried. Fried fish head, tail, and balls marinated in dressing of Chinese waxberry juice and sesame oil.

杨梅虾球 (yángméi xiāqiú)

南方菜。将虾仁、猪肥肉、葱姜水、芡粉、盐、蛋清调和,做成丸子,裹火腿屑油炸。

Shrimp meat and fat pork balls. Southern dish. Mixture of shrimp meat, ground fat pork, scallion and ginger soup, starch, salt, and egg white made into balls, coated with ham crumbs, then deep-fried.

杨枝甘露 (yángzhī gānlù)

香港点心。先将西米煮至透明,再加入冰糖水煮沸,置凉后加入牛奶、椰浆、柚肉、芒果肉,冷冻。

Chilled milky sagos and fruits. Hongkong snack. Sagos quick-boiled translucent, re-boiled in rock sugar soup, then cooled. Combined with milk, coconut milk, pomelo meat, and mango meat. Served chilled.

洋参枸杞薏米汤 (yángshēn gǒuqǐ yìmǐ tāng)

广东点心。将薏米、姜片、西洋参、枸杞子、红枣、冰糖熬炖。

Ginseng, wolfberry, and seeds of Job's tears soup. Guangdong snack. Seeds of Job's tears stewed with sliced ginger, American ginseng, Chinese wolfberry, red dates, and rock sugar.

洋参雪耳炖燕窝 (yángshēn xuěěr dùn yànwō)

广东点心。燕窝、雪耳、西洋参炖煮,加冰糖。

Ginseng, white wood ear, and bird's nest soup. Guangdong snack. Edible bird's nests stewed with white wood ears and American ginseng, then sweetened with rock sugar.

洋葱炒蛋 (yángcōng chǎo dàn)

各地家常菜。先将鸡蛋煎熟,再将洋葱炒熟,然后将鸡蛋和洋葱及调料炒匀。

Fried eggs and onion. Home-style dish in many places. Quick-fried onion sautéed with fried eggs and seasonings.

洋葱脆皮鲩鱼 (yángcōng cuìpíhuànyú)

广东菜。先将鲩鱼片用盐、黄酒、姜浸腌,炸熟,再用洋葱、麻油、芡粉、盐等制卤,浇在鱼片上。

Deep-fried grass carp with onion sauce. Guangdong dish. Grass carp fillets pickled with salt, yellow wine, and ginger, then fried. Washed with sauce of onion, sesame oil, starch, and salt.

洋葱酱 (yángcōngjiàng)

浙江菜。先将洋葱煸炒,再倒入醋、红葡萄酒、糖、石榴汁炆煮。常用作凉拌菜的调味料。

Onion sauce. Zhejiang dish. Sautéed onion simmered with vinegar, red wine, sugar, and pomegranate juice. It is often used as sauce for cold dishes.

洋河大曲 (yánghédàqū)

白酒,中国八大名白酒之一,产于江苏泗洋洋河镇。用高粱为主料,以小麦、大麦、豌豆制曲酿造,酒精含量为38-55度,属浓香型酒。

Yanghedaqu, liquor produced in Yanghe Town of Siyang, Jiangsu, one of the top eight white spirits in China. White spirit made with sorghum and yeast of wheat, barley, and peas. Contains 38-55% alcohol, and has a thick aroma.

洋烧白鳗 (yángshāo báimán)

台湾菜。先将白鳗炸至金黄,然后加酱油、糖、黄酒等翻炒,配焯过的菜心装盘。

White eel with baby greens. Taiwan dish. Deep-fried white eel stir-fried with soy sauce, sugar, and yellow wine. Served with quick-boiled baby greens.

洋烧排骨 (yángshāo páigǔ)

台湾菜。先将排骨炸至金黄,然后加番茄酱、糖、白醋等拌炒。

Pork spareribs with tomato sauce. Taiwan dish. Deep-fried pork spareribs stir-fried with tomato sauce, sugar, and white vinegar.

腰丁虾仁 (yāodīng xiārén)

山东菜。先将虾仁滑油,再与猪腰丁、南荠、青豆等翻炒,烹入盐、清汤、黄酒。

Stir-fried pork kidney and shrimp meat. Shandong dish. Lightly fried shrimp meat stir-fried with diced hog kidney, water chestnuts, and green beans, then flavored with salt, clear soup, and yellow wine.

腰果鸡丁 (yāoguǒ jīdīng)

广东菜。将炸好的腰果与鸡丁、芹菜段、鲜百合及调料同炒。

Stir-fried chicken and cashew. Guangdong dish. Fried cashew nuts stir-fried with diced chicken, sectioned celery, and fresh lily bulb cloves.

肴肉 (yáoròu)

福建凉菜。先将蹄髈腌过,再用老卤将肉煮熟,然后用木板加石块把肉压紧,冷透后切片。

Marinated hog knuckle. Fujian cold dish. Pickled hog knuckle boiled in aged gravy, then pressed tightly with a plank and a stone. Cooled and sliced.

瑶柱扒瓜脯 (yáozhù pá guāpú)

广东菜。瑶柱即元贝。将去皮黄瓜条与元贝丝、胡萝卜片蒸熟,然后淋用高汤、鱼露、麻油、蚝油等制成的卤汁。

Cucumbers with scallops and carrots. Guangdong dish. Peeled cucumber slivers steamed with shredded scallops and sliced carrots. Washed with gravy of broth, fish sauce, sesame oil, and oyster sauce.

瑶柱金钱鱼盒 (yáozhù jīnqiányú-hé)

广东菜。将元贝与鱼胶制成丸子,包上豆腐皮,炸至金黄。

Fried scallop and fish cakes. Guangdong dish. Scallops mixed with fish gelatin, made into cakes, wrapped in bean curd skins, then deep-fried.

瑶柱田鸡节瓜盅 (yáozhù tiánjī jié-guāzhōng)

广东菜。以焯水毛瓜为容器,放入田鸡、冬菇、元贝、高汤,蒸熟,撒火腿丝。

Steamed frogs in hairy gourd. Guangdong dish. Scalded hairy gourd filled with frogs, winter mush-rooms, dried scallops, and broth. Steamed, then garnished with shredded ham.

药酒 (yàojiǔ)

配制酒。以治疗疾病为目的,在酿酒过程中或在酒中加入中草药,可内服或外用。一般使用高度酒泡制。

Folk-medicinal liquor. Compound drink of liquor and herbs with therapeutic properties. Herbs are added in the process of liquor making or after fermentation. It may be used as a health drink or topical medicine. It is usually made with liquor of high alcohol content.

椰酱煎班戟 (yējiàng jiān bānjǐ)

广东点心。先用椰酱、面粉、鸡蛋、牛油调好班戟糊,再慢火煎至微黄。

Coconut pancakes. Guangdong snack. Coconut jam blended with flour, eggs, and butter, then fried into pancakes over low heat.

椰蓉鲜贝串 (yēróng xiānbèi chuàn)

山东菜。先将鲜贝用芡粉蛋浆腌渍,再用竹签串好,裹上椰蓉,炸至金黄。

Fried scallops with coconut crumbs. Shandong dish. Fresh scallops pickled with starch and egg white, strung with bamboo sticks, coated with coconut crumbs, then deep-fried.

椰丝灌汤虾球 (yēsī guàntāng xiā-qiú)

福建菜。把虾茸和肥肉茸加调料做成丸子,炸熟,配椰丝装盘。

Fried shrimp and pork balls. Fujian dish. Minced shrimp meat and fat pork seasoned, made into balls, then deep-fried. Served with shredded coconut meat.

椰香鸡 (yēxiāngjī)

海南菜。将鸡块装入椰子盅,加椰汁、红枣、盐,密封蒸熟。

Steamed chicken in coconut milk. Hainan dish. Cubed chicken put in a coconut pot, seasoned with coconut milk, red dates, and salt, sealed, then steamed.

椰香咖喱鸡 (yēxiāng gālíjī)

江苏菜。先将鸡块加咖喱、茴香籽等翻炒，再与土豆、胡萝卜炖熟，然后加入柿子椒和椰汁稍煮。

Curry chicken with vegetables. Jiangsu dish. Cubed chicken stir-fried with curry and fennel seeds, then stewed with potatoes and carrots. Flavored with sweet pepper and coconut milk.

椰香辣牛肉 (yēxiāng làniúròu)

广东菜。先将牛柳用辣椒、茴香、姜、酱油等腌渍，再与椰酱、椰茸同炒。

Sautéed beef with coconuts. Guangdong dish. Beef strips pickled with chili, fennel, ginger, and soy sauce, then stir-fried with coconut jam and mashed coconut meat.

椰汁糕 (yēzhīgāo)

海南点心。将糯米浆、磨碎的板兰叶、椰汁、生粉、白糖、炼奶调和，蒸熟。

Sticky rice and coconut cakes. Hainan snack. Sticky rice paste mixed with ground isatis leaves, coconut juice, potato starch, sugar, and condensed milk, then steamed.

椰汁黑糯米粥 (yēzhī hēinuòmǐ zhōu)

海南点心。将黑糯米加糖煮成粥，调入椰汁。

Sticky rice and coconut porridge. Hainan snack. Black sticky rice and sugar boiled into porridge, then mixed with coconut milk.

椰汁水晶冻 (yēzhī shuǐjīng dòng)

南方点心。将椰汁、洋菜液、牛奶煮开，倒入模具，冷却成冻。

Coconut, agar, and milk jelly. Southern snack. Coconut milk boiled with agar soup and milk, poured into a mould to take shape, then chilled into jelly.

椰汁豌豆糕 (yēzhī wāndòu gāo)

海南点心。将牛奶、椰浆、豌豆煮成浓浆，冷却后切块。

Coconut and pea cakes. Hainan snack. Milk, coconut milk, and peas boiled into thick paste, chilled into cake, then cut to serve.

椰汁西米露 (yēzhī xīmǐ lù)

海南点心。将椰肉和西米倒入烧开的椰汁和牛奶，加入白糖，煮沸，置凉。

Coconut and sago soup. Hainan snack. Coconut meat and sagos combined with boiling coconut juice and milk, flavored with white sugar, boiled, then chilled.

椰汁芋头鸡 (yēzhī yùtou jī)

海南菜。将氽过的鸡块和芋头加椰汁焖熟，用盐调味。

Stewed chicken and taros. Hainan dish. Quick-boiled chicken and taros stewed in coconut milk, then seasoned with salt.

椰子荔枝老鸽汤 (yēzi lìzhī lǎogē tāng)

广东菜。将老鸽、鸡脚、猪骨、荔枝放入椰汁熬炖。

Pigeon, coconut, and litchi soup.

Guangdong soup. Mature pigeon braised with chicken feet, hog bones, and litchi in coconut milk.

野鸡瓜子 (yějīguāzǐ)

曹雪芹《红楼梦》菜谱。先将野鸡肉丁用盐、黄酒、蛋清、淀粉腌渍,入油锅滑散,放入酱瓜和青椒,再与煸香的姜、蒜翻炒,然后快速烹入用酱油、黄酒、盐、鸡精、胡椒粉、水淀粉、清汤做成的卤汁。

Sautéed pheasant with pickled cucumbers. Menu from *A Dream of Red Mansions* by Cao Xueqin. Diced pheasant pickled with salt, yellow wine, egg white, and starch. Lightly fried with pickled cucumbers and green pepper. Sautéed with quick-fried ginger and garlic, then quickly washed with gravy of soy sauce, yellow wine, salt, chicken essence, pepper, wet starch, and clear soup.

野鸡崽子汤 (yějīzǎizitāng)

曹雪芹《红楼梦》菜谱。先将野仔鸡块过油,加入黄酒、盐、葱、姜,然后用高汤炖烂。

Young pheasant soup. Menu from *A Dream of Red Mansions* Dream by Cao Xueqin. Chopped young pheasant lightly fried, then braised in broth with yellow wine, salt, scallion, and ginger.

野山椒牛肉丝 (yěshānjiāo niúròusī)

湖南菜。将牛肉丝用蛋清及调料腌过,再与野山椒用猛火翻炒。

Stir-fried beef and wild pepper. Hunan dish. Shredded beef pickled with egg white and seasonings, then stir-fried with wild hot pepper.

夜香冬瓜盅 (yèxiāng dōngguāzhōng)

广东菜。先将冬瓜切成瓜盖和瓜盅,再把鸡骨、猪骨、鸭肉、火腿等入汤煮沸,然后连汤倒入瓜盅,加入调料,盖上瓜盖,蒸熟。

Bone soup in wax gourd. Guangdong dish. Wax gourd cut into a container and a lid. Boiled chicken bones, hog bones, duck, ham, and broth poured in the container. Seasoned, covered with the lid, then steamed.

一卵孵双鸡 (yī luǎn fū shuāngjī)

山东菜。将两只鸡炖烂,装入冬瓜盅内,加入海参、银耳、冬菇、鸡汤等蒸熟。

Steamed chickens in wax gourd. Shandong dish. Two stewed chickens put in a hollowed wax gourd. Sea cucumbers, white wood ears, winter mushrooms, and chicken soup added, then steamed.

一品豆腐 (yīpǐn dòufu)

山东曲阜孔府菜。先将豆腐块挖空,填入肉馅,用豆腐片封好,然后与肘片一同放入沙锅,加高汤及调料慢火炖煮,勾芡。

Stuffed bean curd and ham. Specialty of the Kong Family in Qufu, Shandong. Large bean curd cubes hollowed, stuffed with meat, then sealed with sliced bean curd.

Braised in seasoned broth with sliced ham in an earthen pot, then thickened with starch.

一品寿桃 (yīpǐn shòutáo)

山东曲阜孔府点心。将红枣泥和山药泥做成巨形寿桃,用青梅摆成叶,用山楂条在桃身上摆成寿字,蒸熟,浇清芡。

Steamed red date and Chinese yam cakes. Birthday snack created by the Kong Family in Qufu, Shandong. Mashed red dates and Chinese yam kneaded into a big peach. Decorated with candied green plums as leaves, and a character of longevity "寿" with haw cake strips. Steamed, then washed with lightly starched clear soup.

一品燕菜 (yīpǐn yàncài)

山东菜。将水发燕菜洗净,入清汤煮沸,加糖或盐。

Bird's nest soup. Shandong dish. Edible bird's nests soaked in water and cleaned, then boiled in clear soup. Sugar or salt added.

一蛇三吃 (yī shé sān chī)

湖南菜。由拌蛇皮、油炸蛇段、白酒冲蛇胆和蛇血三道菜组成。

Three dishes made with one snake. Hunan dish. Three mini courses made with one snake. Seasoned snake skins, fried snake sections, and snake gallbladder and blood with Chinese liquor.

伊府面 (yīfǔmiàn)

福建点心。先将面条煮熟后油炸,再与肉丝、笋丝、白菜丝翻炒,然后添汤稍焖。

Stir-fried noodles with meat and vegetables. Fujian snack. Boiled noodles deep-fried, sautéed with shredded pork, stem lettuce, and cabbage, then simmered briefly with a little soup.

夷陵春卷 (yílíngchūnjuǎn)

湖北宜昌点心。宜昌古称夷陵。将面皮卷入韭菜、腊肉、豆腐干、冬菇等炒成的馅,油炸。

Smoked pork and vegetable rolls. Snack in Yichang, Hubei. Yichang was named Yiling in ancient times. Flour sheets made into rolls filled with quick-fried Chinese chive, smoked cured pork, dried bean curd, and dried mushrooms, then deep-fried.

宜红功夫茶 (yíhóng gōngfuchá)

红茶。产于鄂西宜县、恩施两地山区。外形紧细有金毫,色泽乌润。汤色红亮,叶底红亮柔软。

Yihong gongfucha, black tea produced in the mountainous areas of Yixian and Enshi Counties in western Hubei. Its dry leaves are slim, tight, golden-hairy, and off-black. The tea is bright red with infused leaves red and tender.

宜兰鸭赏 (yílányāshǎng)

台湾家常菜,源于宜兰。鸭皮、鸭胗切片,同葱、姜、蒜爆炒,烹入黄酒、酱油。

Stir-fried duck skins and gizzard.

Home-style dish originated in Yilan, Taiwan. Sliced duck skins and duck gizzard stir-fried with scallion, ginger, and garlic, then seasoned with yellow wine and soy sauce.

弋阳醋鸡 (yìyángcùjī)

江西弋阳菜,即酸辣鸡。先把鸡用盐腌渍两天,然后切成块,与醋、酱油、辣椒、米酒、芝麻油等拌和,罐封15天。

Sour and spicy chicken. Specialty in Yiyang, Jiangxi. Chicken pickled in salt for 2 days, chopped, then mixed with vinegar, soy sauce, chili pepper, sweet fermented sticky rice juice, and sesame oil. Sealed in an earthen jar for 15 days before serving.

银川烩羊杂碎 (yínchuān huìyángzásuì)

宁夏菜。先将羊肺灌入面糊,煮熟,然后与煮熟的羊内脏、羊头等入原汤烩煮,放入盐、葱、姜、蒜、辣椒油、酱油、香菜。

Braised lamb offal. Ningxia dish. Lamb lung filled with flour paste, boiled, then braised in broth with prepared lamb intestine and head. Seasoned with salt, leek, ginger, garlic, chili oil, soy sauce, and coriander.

银耳炒鸡片 (yíněr chǎo jīpiàn)

山东菜。将白木耳、鸡脯肉、青椒、葱、红辣椒拌炒,加盐、料酒,勾芡。

Sautéed white wood ears and chicken. Shandong dish. White wood ears and sliced chicken breast sautéed with green pepper, scallion, and red pepper, then flavored with salt, cooking wine, and starched soup.

银耳陈皮炖乳鸽 (yíněr chénpí dùn rǔgē)

广东菜。将乳鸽与白木耳汆过,与陈皮入高汤蒸熟,用盐调味。

Pigeon with white wood ears and tangerine peels. Guangdong dish. Young pigeon and white wood ears quick-boiled, steamed with processed tangerine peels in broth, then seasoned with salt.

银耳山药羹 (yíněr shānyào gēng)

广东点心。将山药和银耳同锅烹煮,加入黑砂糖。

Chinese yam and white fungus soup. Guangdong snack. Chinese yam boiled with white wood ears, then flavored with black sugar.

银粉牛肉丝 (yínfěn niúròusī)

江苏菜。先将腌渍过的牛肉丝轻炒,再加入韭菜及卤汁快炒,然后把炒好的韭菜牛肉丝倒在粉丝上。

Quick-fried beef and Chinese chive. Jiangsu dish. Pickled shredded beef lightly fried, then quick-fried with Chinese chive and sauce. Served over vermicelli.

银球滚绣球 (yínqiú gǔn xiùqiú)

湖南菜。先将肉末、鱼茸、生粉及调料拌和,做成小球,裹上干贝丝、蛋皮丝、红椒丝、香菇丝,然后与冬瓜球一道入高汤蒸熟,配菜心装盘。

Gourd balls with meatballs. Hunan dish. Ground pork, minced fish, starch, and seasonings mixed, made into meatballs coated with shredded scallops, egg sheets, red pepper, and dried mushrooms, then steamed with wax gourd balls in broth. Served with baby cabbage.

银丝拌白菜 (yínsī bàn báicài)

东北家常凉菜。先将绿豆芽焯水，然后与白菜丝和粉丝拌匀，加酱油、醋、麻油。

Mung bean sprouts with cabbage. Home-style cold dish in northeastern areas. Mung bean sprouts scalded, mixed with shredded Chinese cabbage and starch vermicelli, then seasoned with soy sauce, vinegar, and sesame oil.

银丝长鱼 (yínsī chángyú)

江苏菜。先将鳝鱼丝上浆，划油，再与笋丝及佐料煸炒，撒白胡椒粉。

Sautéed eel and bamboo shoots. Jiangsu dish. Pickled shredded field eel lightly fried. Sautéed with shredded bamboo shoots and flavorings, then flavored with ground white pepper.

银杏豆腐花 (yínxìng dòufuhuā)

广东点心。将杏仁汁煮沸，加入鲜奶、豆奶、糖等搅匀，冷却成豆腐花状。

Almond custard. Guangdong snack. Almond milk boiled, mixed with sugar, milk, and soybean milk, then chilled into custard.

银杏天山雪莲粥 (yínxìng tiānshān xuělián zhōu)

广东点心。将白果、天山雪莲子、龙眼、红枣熬煮。

Gingko and snow lotus seed soup. Guangdong snack. Gingkoes boiled with snow lotus seeds, longan pulp, and red dates.

银杏芋泥 (yínxìng yùní)

福建点心。先将银杏加冰糖煮熟，再将芋头蒸熟，捣成芋泥，然后用葱油加糖水把芋泥炒松软，装碗，面上摆银杏。

Sweet taro paste with ginkgoes. Fujian snack. Mashed steamed taros cooked with scallion oil and syrup until soft, then topped with gingkoes boiled with rock sugar.

银杏蒸鸭 (yínxìng zhēng yā)

四川菜。先将鸭去骨切块，加入煮熟去皮去芯的银杏，然后加调料蒸熟。

Steamed duck and ginkgoes. Sichuan dish. Boned duck steamed with peeled cored gingkoes and seasonings.

银芽韭菜花 (yínyá jiǔcàihuā)

浙江菜。先将绿豆芽氽烫，再与轻炒过的韭菜花拌炒，用盐、葱、姜等调味。

Sautéed bean sprouts and Chinese chive. Zhejiang dish. Scalded mung bean sprouts sautéed with quick-fried Chinese chive blossoms, then seasoned with salt, scallion, and ginger.

银芽炒牛肉 (yínyá chǎo niúròu)

浙江菜。先将牛肉片用盐、淀粉、料酒、蛋清腌渍，再用小火炒散，然后与豆芽、韭菜、春笋、香菇、洋葱煸炒。

Sautéed beef and vegetables.

Zhejiang dish. Sliced beef pickled with salt, starch, cooking wine, and egg white, lightly fried, then sautéed with bean sprouts, Chinese chive, spring bamboo shoots, dried mush-rooms, and onion.

银芽鲩鱼片 (yínyá huànyúpiàn)

湖南菜。鲩鱼片加入豆芽、红辣椒丝、灯笼椒丝及调料翻炒。

Stir-fried grass carp with bean sprouts. Hunan dish. Shredded grass carp stir-fried with bean sprouts, shredded hot and sweet peppers, and seasonings.

银芽鸡丝 (yínyá jīsī)

山东菜。先将红辣椒和豆芽煸炒，然后与煸炒过的鸡肉丝加盐、酱油、味精翻炒。

Sautéed chicken and bean sprouts. Shandong dish. Lightly fried sliced chicken sautéed with quick-fried red pepper and bean sprouts, then seasoned with salt, soy sauce, and MSG.

银鱼肉丝 (yínyú ròusī)

江西菜。先将银鱼丝和肉丝用盐和芡粉腌渍，过油，然后加入姜葱丝、料酒等翻炒，撒胡椒粉。

Stir-fried silver fish and pork. Jiangxi Dish. Shredded silver fish and pork pickled with salt and starch, then lightly fried. Stir-fried with shredded ginger, scallion, and cooking wine, then sprinkled with ground black pepper.

银鱼稀卤豆花 (yínyú xīlǔ dòuhuā)

湖南菜。将豆花加入银鱼、鸡蛋羹及调料煨熟。

Icefish with jellied bean curd. Hunan dish. Jellied bean curd simmered with lake icefish, egg custard, and seasonings.

银针炒鸡丝 (yínzhēn chǎo jīsī)

广东菜。鸡肉丝、菇丝与摘掉豆瓣和根的黄豆芽及调料同炒。

Sautéed chicken and bean sprout stems. Guangdong dish. Shredded chicken sautéed with shredded mush-rooms and soybean sprout stems, then flavored with seasonings.

银针海蛏 (yínzhēn hǎichēng)

山东菜。先将汆过的掐菜稍炒，然后与汆腌过的海蛏肉快炒，烹入醋、麻油。

Sautéed razor clam and bean sprout stems. Shandong dish. Scalded and pickled razor clam meat quickly sautéed with scalded bean sprout stems, then flavored with vinegar and sesame oil.

银针鸡汁鱼片 (yínzhēn jīzhī yúpiàn)

湖南菜。先将汆好的鳜鱼、冬笋、口蘑、菜苞放入鸡汤煮熟，再添入银针茶。

Fish and tea leaves in chicken soup.

Hunan dish. Scalded Chinese perch, winter bamboo shoots, Mongolian mushrooms, and baby cabbage boiled in chicken soup, then flavored with Junshan Yinzhen tea leaves.

银针土鱿丝 (yínzhēn tǔyóusī)

广东菜。将鱿鱼丝、摘掉豆瓣和根的黄豆芽与虾干、姜片及调料同炒。

Sautéed squid and soybean sprout stems. Guangdong dish. Shredded squid sautéed with soybean sprout stems, dried shrimp, and sliced ginger.

应城扒肉 (yìngchéngpáròu)

湖北应城菜。先将蹄髈氽过,刷上糖浆,油煎至发红,再放入酱油、糖、盐、料酒、八角、桂皮、姜片、葱结等蒸烂,然后将浓汤勾芡,浇在蹄髈上。

Steamed hog knuckle with sauce. Dish in Yingcheng, Hubei. Quick-boiled hog knuckle coated with syrup, then fried red. Steamed with soy sauce, sugar, salt, yellow wine, aniseed, cinnamon, ginger, and scallion, then washed with starched thick broth.

樱花虾糙米粥 (yīnghuāxiā cāomǐ zhōu)

广东点心。将糙米、樱花虾、芹菜用小火熬成粥,加盐和麻油。

Rice and shrimp porridge. Guangdong snack. Brown rice boiled into porridge with spotted shrimp and celery, then flavored with salt and sesame oil.

樱橘蛤士蟆 (yīngjú háshìmá)

安徽菜。先将泡发好的蛤士蟆加清水、料酒、葱、姜等蒸熟,再用清水浸泡,然后与橘瓣入冰糖水烧开,缀上红樱桃。

Forest frog and orange soup. Anhui dish. Soaked Chinese forest frog oviduct steamed with water, cooking wine, scallion, and ginger, then immersed in water. Boiled with orange sections in rock sugar soup, then decorated with red cherries.

樱桃肉 (yīngtáoròu)

浙江菜。先将红薯切樱桃大小,油炸,然后入浓冰糖水煮开。

Fried sweet potatoes in sweet soup. Zhejiang dish. Sweet potatoes cut into cherry-size pieces, deep-fried, then boiled in thick rock sugar soup.

邕州鱼角 (yōngzhōuyújiǎo)

广西邕州菜。将鲮鱼肉茸加芡粉搅拌成鱼胶,包入用冬菇、虾米、荸荠等制成的馅,做成扇形饺,蒸熟。

Steamed carp *jiaozi*. Specialty in Yongzhou, Guangxi. Minced mud carp meat and starch mixed into gelatin, made into fan-shape *jiaozi* filled with mixture of winter mushrooms, dried shrimp meat, and water chestnuts, then steamed.

永定牛肉丸 (yǒngdìng niúròuwán)

福建永定菜。牛肉末加少许薯粉、盐、水等制成肉丸,放入加目鱼、香菇的汤煮熟,用胡椒、葱花等调味。

Beef meatball soup. Specialty of Yongding, Fujian. Ground beef,

sweet potato starch, salt, and water made into balls, boiled in milk fish and dried mushroom soup, then flavored with ground black pepper and chopped scallion.

尤溪卜鸭 (yóuxībǔyā)

福建尤溪菜。先将鸭煮熟，抹盐，用烟熏透，然后入加调料的热鸭汤浸泡。

Marinated smoked duck. Specialty in Youxi, Fujian. Boiled duck pickled with salt, smoked, then marinated in heated seasoned duck soup.

油爆双脆 (yóubào shuāngcuì)

山东菜。先将肚头和鸡肫刻花，用盐和湿淀粉腌渍，然后爆炒，勾芡。

Stir-fried hog tripe and chicken gizzard. Shandong dish. Carved hog tripe and chicken gizzard pickled with salt and wet starch, stir-fried, then thickened with starch.

油爆双鲜 (yóubào shuāngxiān)

山东菜。先将赤贝肉和海螺片汆过，然后与冬笋片、冬菇及调料爆炒，勾芡。

Stir-fried red shellfish and whelks. Shandong dish. Scalded red shellfish and sliced whelks stir-fried with sliced winter bamboo shoots, winter mushrooms, and seasonings, then thickened with starch.

油爆乌鱼花 (yóubào wūyúhuā)

山东菜。先将鲜乌贼鱼片用开水烫过，然后同黄瓜片和笋片及调料翻炒，勾芡。

Stir-fried cuttlefish with vegetables. Shandong dish. Quick-boiled sliced cuttlefish stir-fried with sliced cucumbers, bamboo shoots, and seasonings, then thickened with starch.

油爆虾 (yóubàoxiā)

上海菜。先将虾高温油炸，然后与葱、酱油、糖翻炒。

Deep-fried shrimp with scallion. Shanghai dish. Shrimp fried at high temperature and then stir-fried with scallion, soy sauce, and sugar.

油爆鱿鱼卷 (yóubào yóuyújuǎn)

浙江菜。先将鱿鱼和竹笋焯水，过油，然后加调料翻炒，勾芡，淋花椒油。

Stir-fried squid. Zhejiang dish. Lightly fried scalded squid and bamboo shoots stir-fried with condiments, thickened with starch, then sprinkled with Chinese prickly ash oil.

油爆鱼芹 (yóubào yú qín)

山东菜。先将草鱼块裹上用肥肉泥、鸡茸、芹菜末、蛋清、料酒、盐调成的糊，炸至金黄，然后放入特制的卤汁翻炒。

Fried carp with sauce. Shandong dish. Grass carp chunks coated with paste of ground fat pork, minced chicken, chopped celery, egg white, cooking wine, and salt, deep-fried until golden, then quick-fried with special sauce.

油茶 (yóuchá)

湖北点心。先将面粉炒香，然后放

入羊油、胡椒粉、豆豉、葱、姜、盐，加水煮成糊。

Suet oil and flour chowder. Hubei snack. Half-baked flour boiled into chowder with suet oil, ground black pepper, fermented soybean, scallion, ginger, salt, and water.

油脆 (yóucuì)

南方家常点心。将蒸熟的糯米饭舂茸，做成糍粑，包入白糖、芝麻或豆沙馅，炸至金黄。

Fried stuffed sticky rice cakes. Home-style snack in southern areas. Sticky rice steamed, pestled in a mortar, made into cakes filled with white sugar, sesame, or bean paste, then deep-fried.

油豆腐粉丝鸡汤 (yóudòufu fěnsī jītāng)

上海家常菜。将鸡肉丁、油豆腐、粉丝用鸡汤烹煮，加入盐、味精、姜、葱、猪油或麻油。

Bean curd and starch noodle soup. Home-style dish in Shanghai. Diced chicken, fried bean curd, and starch noodles boiled in chicken soup. Seasoned with salt, MSG, ginger, scallion, and lard or sesame oil.

油豆腐镶肉 (yóudòufu xiāng ròu)

南方菜。在油豆腐中塞入加调料的肉馅，入上汤煮透，淋麻油。

Fried bean curd with pork. Southern dish. Fried cubed bean curd filled with seasoned minced pork, boiled in broth, then washed with sesame oil.

油豆角炖排骨 (yóudòujiǎo dùn páigǔ)

东北菜。将排骨段与油豆角炖熟，放入辣椒、盐、面酱、胡椒粉。

Stewed pork spareribs and snap beans. Northeastern dish. Chopped pork spareribs stewed with snap beans. Seasoned with chili pepper, salt, fermented flour sauce, and ground black pepper.

油豆泡 (yóudòupào)

山西菜。将豆泡切开口，填入加调料的肉馅，用面糊封口，蒸熟。

Steamed stuffed bean curd puff. Shanxi dish. Deep-fried bean curd puffs cut open, stuffed with seasoned ground meat, sealed with flour paste, then steamed.

油墩子 (yóudūnzi)

上海点心。把白萝卜丝同调料拌匀，装入平底金属勺，浇上面粉浆，炸熟。

Fried turnip cakes. Shanghai snack. White shredded turnips mixed with seasonings, put in a flat-bottom metal ladle, drenched with flour paste, then deep-fried.

油浸鸡 (yóujìnjī)

浙江菜。先将鸡用盐腌渍，再往鸡腹内放入葱、姜、白酒、八角等调料，在鸡皮上涂满糖浆，然后把鸡放入烧滚的花生油浸熟。

Chicken marinated in oil. Zhejiang dish. Salt-pickled chicken stuffed with scallion, ginger, Chinese liquor, and aniseed, coated with syrup, then immersed in heated peanut oil until ready.

油卷 (yóujuǎn)

福建点心。用面粉加水擀成面皮,用猪肉、豆腐、豆酱、葱等制成馅,用皮裹馅成卷,切成段,先蒸熟,然后油煎至皮脆。

Fried pork and bean curd rolls. Fujian snack. Flour and water made into sheets, made into rolls filled with mixture of prepared pork, bean curd, bean sauce, and scallion. Rolls cut into sections, steamed, then fried crisp.

油淋草鱼 (yóulín cǎoyú)

东北菜。先将草鱼腌渍,加调料蒸熟,然后浇热麻油。

Steamed carp with sesame oil. Northeastern dish. Pickled grass carp steamed with seasonings, then washed with hot sesame oil.

油淋鸡 (yóulínjī)

各地家常菜。将全鸡腌渍,炸至金黄,切块,浇特制油卤汁。

Fried chicken with oil sauce. Home-style dish in many places. Pickled whole chicken fried until golden, chopped, then washed with special oil sauce.

油焖大虾 (yóumèn dàxiā)

沿海一带家常菜。将大虾配姜丝用葱油煸炒,加入料酒、高汤焖煮至汁干。

Simmered prawns in scallion oil. Home-style dish in coastal areas. Prawns and shredded ginger quick-fried in scallion oil, then simmered in cooking wine and broth until reduced.

油焖凤尾鱼 (yóumèn fèngwěiyú)

沿海一带家常菜。先将凤尾鱼炸至金黄,再放入料汤烧开,然后用小火焖煮至汤浓。

Braised anchovy. Home-style dish in coastal areas. Tapertail anchovy deep-fried, boiled in seasoned soup, then simmered until reduced.

油焖笋 (yóumènsǔn)

江浙沪菜。先将竹笋爆炒,然后入高汤焖煮,用黄酒、酱油、白糖、味精等调味,淋麻油。

Braised bamboo shoots. Zhejiang dish. Stir-fried bamboo shoots braised in broth, seasoned with yellow wine, soy sauce, sugar, and MSG, then washed with sesame oil.

油焖虾 (yóumènxiā)

南方家常菜。先将虾配葱、姜爆炒,然后加入料酒、酱油、糖、盐,鸡精焖煮。

Simmered shrimp. Home-style dish in southern areas. Shrimp stir-fried with scallion and ginger, then simmered with cooking wine, soy sauce, sugar, salt, and chicken essence.

油面筋塞肉 (yóumiànjīn sāi ròu)

上海菜。先往油面筋里塞满用肉末、蛋清、盐等拌成的馅,然后入高汤煮熟,加酱油、味精。

Boiled stuffed gluten. Shanghai dish. Fried gluten puffs stuffed with mixture of minced meat, egg white, and salt, boiled in broth, then seasoned with soy sauce and MSG.

油泡金银鱿 (yóupào jīnyínyóu)

广东菜。将红鱿、白鱿、干笋花加姜、葱、黄酒、盐等爆炒，勾芡。

Stir-fried red and white squid. Guangdong dish. Red squid and white squid stir-fried with sliced dried bamboo shoots, ginger, scallion, yellow wine, and salt, then thickened with starch.

油泡青鱼丸 (yóupào qīngyúwán)

广东菜。先将青鱼茸加蛋清、淀粉、盐做成丸子，煮熟，然后用猪油、芡汁翻炒，加调料，淋麻油。

Stir-fried black carp balls. Guangdong dish. Minced black carp meat made into balls and boiled. Stir-fried with lard and starched soup, seasoned, then sprinkled with sesame oil.

油泡肾球 (yóupào shènqiú)

广东菜。将鹅肾、姜花、笋花加料酒、盐、酱油等爆炒，勾芡。

Stir-fried goose kidney with bamboo shoots. Guangdong dish. Goose kidney stir-fried with sliced ginger, sliced soaked bamboo shoots, cooking wine, salt, and soy sauce, then thickened with starch.

油泡爽肚 (yóupào shuǎngdǔ)

广东菜。将猪肚片、姜花、笋花加盐、酱油、黄酒爆炒，勾芡。

Stir-fried hog tripe and bamboo shoots. Guangdong dish. Sliced hog tripe stir-fried with sliced ginger, sliced soaked bamboo shoots, salt, soy sauce, and yellow wine, then thickened with starch.

油泡虾仁 (yóupào xiārén)

山东菜。先将腌过的虾仁用热油泡至半熟，然后与葱、姜翻炒，浇芡汁。

Sautéed shrimp meat. Shandong dish. Pickled shrimp meat marinated in hot oil. Sautéed with scallion and ginger, then washed with starched soup.

油泡腰花 (yóupào yāohuā)

浙江菜。先将猪腰片稍腌，滑透，然后加芹菜梗、泡椒、姜、葱、料酒等翻炒。

Stir-fried hog kidney. Zhejiang dish. Pickled sliced hog kidney lightly fried, then stir-fried with celery stems, pickled red pepper, ginger, scallion, and cooking wine.

油泡鸳鸯虾 (yóupào yuānyāngxiā)

广东菜。将虾、姜花、干笋花加黄酒及调料爆炒，勾芡。

Stir-fried shrimp and bamboo shoots. Guangdong dish. Shrimp stir-fried with carved ginger, sliced soaked bamboo shoots, yellow wine, and other seasonings, then thickened with starch.

油泼鸡 (yóupōjī)

山东菜。先将鸡块用黄酒、酱油、姜汁腌渍，炸至深红，然后将清汤和腌鸡的卤汁烧开，放入香菜、味精、麻油，浇在鸡上。

Fried chicken with gravy. Shandong

dish. Chicken pickled with yellow wine, soy sauce, and ginger juice, deep-fried until dark red, then marinated in gravy of clear soup, pickling sauce, coriander, MSG, and sesame oil.

油泼青鱼 (yóupō qīngyú)

山东菜。将青鱼蒸熟,浇上滚烫的植物油和酱油。

Steamed black carp in oil. Shandong dish. Black carp steamed, then washed with burning vegetable oil and soy sauce.

油泼羊肉 (yóupō yángròu)

南方菜。将羊肉煮熟,切片,撒上葱丝和香菜,浇烧热的素油。

Boiled mutton in oil. Southern dish. Sliced boiled mutton sprinkled with shredded scallion and coriander, then washed with heated vegetable oil.

油塔子 (yóutǎzi)

新疆点心。先将面团擀成面皮,抹上羊尾油,裹成卷,切成段,拧成塔状,然后蒸熟。

Steamed oiled buns. Xinjiang snack. Flour dough sheets coated with lamb tail oil, rolled up, cut into segments, hand-kneaded into small towers, then steamed.

油条 (yóutiáo)

各地传统点心。将发酵面团切成条,炸至金黄泡脆,配甜或咸豆浆。

Fried flour dough sticks. Traditional snack in many places. Fermented flour dough cut into slivers, then deep-fried golden, puffy, and crisp. Served with sweet or salty soybean milk.

油条烩鳝鱼 (yóutiáo huì shànyú)

江西菜。先将鳝鱼去骨,过油,再与玉兰片和韭菜略炒,然后与油条翻炒,加调料,勾芡。

Eel with fried dough sticks. Jiangxi Dish. Lightly fried boned eel quick-fried with soaked bamboo shoots and Chinese chive. Sautéed with deep-fried flour dough sticks and seasonings, then thickened with starch.

油条西舌 (yóutiáo xīshé)

福建菜。先将油条炸酥脆,再将西施舌片汤炒、加调料、勾芡,然后倒在油条上。

Fried dough sticks with clam sauce. Fujian dish. Clam meat quickly fried in soup with seasonings, thickened with starch, then served over deep-fried flour dough sticks.

油盐炒枸杞芽儿 (yóuyán chǎo gǒuqǐyáer)

曹雪芹《红楼梦》菜谱。枸杞嫩叶用油盐清炒。

Sautéed tender Chinese wolfberry leaves. Menu from *A Dream of Red Mansions* by Cao Xueqin. Tender Chinese wolfberry leaves sautéed with oil and salt.

柚子花菜 (yòuzihuācài)

朝鲜族点心。先将柚子丝用糖腌渍,再与梨丝成扇形对称装碗,然后把石榴粒放在中心,加入凉糖水。

Pomelo and pear soup. Korean

ethnic minority snack. Shredded pomelos pickled with sugar, then arranged like an open fan on one side of a bowl. Shredded pears put on the other side of the bowl. Garnished with pomegranate seeds at the center of the bowl. Cold sugar soup added.

莜面搓鱼子 (yóumiàn cuō yúzǐ)

四川菜。将莜麦面加水搓成小鱼条子,蒸熟,浇上辣味卤汁。

Steamed bell wheat pieces with sauce. Sichuan dish. Bell wheat flour and water mixed, kneaded into small fish-shape pieces, steamed, then served with spicy dressing.

莜面栲栳 (yóumiàn kǎolǎo)

山西点心。将莜面用开水合成面团,碾成薄皮,卷成中空小卷,急火蒸熟,佐肉卤食用。

Steamed oat rolls with sauce. Shanxi snack. Oat flour and boiling water made into dough, pressed into thin sheets, then made into small hollow rolls. Quick-steamed, then served with meat sauce.

鱿鱼肉丝 (yóuyú ròusī)

湖南菜。先将猪肉丝挂芡,过油,然后与鱿鱼、柿子椒、冬笋及调料翻炒。

Stir-fried squid and pork. Hunan dish. Squid and lightly fried shredded pork stir-fried with sweet pepper, winter bamboo shoots, and seasonings.

鱿鱼汤 (yóuyútāng)

南方菜。将氽透的鱿鱼入清汤煮沸,加盐、葱、姜、黄酒、猪油。

Squid soup. Southern dish. Scalded squid boiled in clear soup seasoned with salt, scallion, ginger, yellow wine, and lard.

油炸凤尾鱼 (yóuzhá fèngwěiyú)

沿海一带家常菜。先将凤尾鱼用酱油浸透,然后炸脆。

Deep-fried anchovy. Home-style dish in coastal areas. Tapertail anchovy pickled in soy sauce, then fried crisp.

鱼包三经 (yú bāo sānjīng)

山东菜。将冬菇、冬笋、火腿、熟鸡肉及调料拌成馅,用鱼片裹馅挂浆,炸至金黄。

Fried fish rolls. Shandong dish. Fish fillets made into rolls filled with seasoned mixture of winter mushrooms, winter bamboo shoots, ham, and prepared chicken, coated with starch and egg paste, then deep-fried.

鱼肠蒸蛋 (yúcháng zhēng dàn)

山东菜。将鸡蛋浆同鱼肝和鱼肠调匀,上置油条,蒸熟。

Steamed eggs and fish intestine. Shandong dish. Egg paste mixed with fish liver and intestine, topped with deep-fried flour dough sticks, then steamed.

鱼翅羹 (yúchìgēng)

广东菜。将鱼翅和香菇、竹笋、虾仁加调料同炖,用生粉勾芡,撒胡椒粉。

Shark fin soup. Guangdong dish. Shark fin stewed with dried mushrooms, bamboo shoots, shrimp meat, and seasonings. Thickened

with starch, then flavored with ground black pepper.

鱼翅饺 (yúchìjiǎo)

广东点心。用面粉和太白粉制皮，用鱼翅、发菜、北海道青虾及调料制馅，包成饺子，蒸熟。

Shark fin *jiaozi*. Guangdong snack. Flour and potato starch made into *jiaozi* filled with seasoned mixture of shark fin, black moss and Hokkaido shrimp, then steamed.

鱼翅四丝 (yúchì sìsī)

山东菜。先将鸡脯肉、海参、蹄筋、冬笋等切丝，加调料翻炒，勾芡，然后将鱼翅蒸烂，焖至汁浓，勾芡，盖在四丝上。

Shark fin over meat and bamboo shoots. Shandong dish. Shredded chicken breast, sea cucumbers, pork tendons, and winter bamboo shoots stir-fried with seasonings, thickened with starch, then covered with steamed shark fin and gravy.

鱼浆腐皮肉卷 (yújiāng fǔpí ròu juǎn)

广东菜。用猪里脊肉、鱼茸、洋葱、香菇及调料制馅，用豆腐皮裹馅成卷，蒸炸均可。

Pork and fish rolls. Guangdong dish. Bean curd skins made into rolls filled with seasoned mixture of pork tenderloin, minced fish, onion, and dried mushrooms. Steamed or fried.

鱼烙 (yúlào)

广东菜。将鲜鱼片用盐、酱油、葱、蛋清、胡椒粉腌渍，煎至金黄。

Fried fish fillets. Guangdong dish. Fresh fish fillets pickled with salt, soy sauce, scallion, egg white, and ground black pepper, then fried golden.

鱼柳烩杂海鲜 (yúliǔ huìzá hǎixiān)

广东菜。将鲈鱼块焗熟，浇上用虾仁、青口、番茄、青椒、洋葱、香菜、盐、蚝油、黄酒等炒制的卤汁。

Roasted perch with seafood sauce. Guangdong dish. Perch chunks roasted, then marinated in dressing of shrimp meat, mussel meat, tomatoes, green pepper, onion, coriander, salt, oyster sauce, and yellow wine.

鱼露乳鸽 (yúlù rǔgē)

广东菜。先将乳鸽焯水，然后用加了鱼露和白糖的汤卤熟。

Marinated young pigeon. Guangdong dish. Quick-boiled young pigeon marinated in gravy flavored with fish sauce and sugar.

鱼腩煲 (yúnǎnbāo)

福建菜。先将鱼脯烫熟，再与姜、葱、蒜、南豆腐乳、豆豉酱等爆炒，然后煲煮，加入黄酒，勾芡，撒葱花和香菜。

Braised savoury fish fillets. Fujian dish. Scalded fish fillets stir-fried with ginger, scallion, garlic, southern-style fermented bean curd, and fermented soybean sauce. Braised in yellow wine and starched soup, then sprinkled with chopped scallion and coriander.

鱼匹子 (yúpǐzi)

内蒙菜。先将鱼身刻成片状,撒上盐发酵,再晾晒,然后洗净,切段,蒸熟。

Steamed pickled fish. Inner Mongolian dish. Fish carved into connected pieces, pickled with salt, fermented, then air-dried. Cleaned, sectioned, then steamed.

鱼茸蹄筋 (yúróng tíjīn)

河南菜。将鱼茸和肥猪肉末加调料拌成馅,用烫过的牛蹄筋片裹馅,入开水煮熟,沥干,然后放入清汤,加盐、料酒、味精、葱、姜烩煮,勾芡。

Beef tendons with fish and pork. Henan dish. Scalded sliced beef tendons made into rolls filled with mixture of mashed fish, fat pork, and seasonings. Boiled, drained, simmered in clear soup with salt, cooking wine, MSG, scallion, and ginger, then thickened with starch.

鱼生 (yúshēng)

福建菜。生草鱼片洒上麻油,蘸酱油、醋、芥酱等佐料。

Raw carp fillets with sauce. Fujian dish. Sliced boned grass carp sprinkled with sesame oil, then served with soy sauce, vinegar, and mustard sauce.

鱼松豆芽 (yúsōng dòuyá)

南方菜。先将鱼肉做成饼状煎熟,再与鱼松和黄豆芽拌炒,加盐和葱姜丝。

Fish cake with bean sprouts. Southern dish. Deep-fried fish cakes sautéed with fish floss and soybean sprouts, then seasoned with salt and shredded scallion and ginger.

鱼头冻豆腐 (yútóu dòngdòufu)

江苏菜。将胖鱼头两面煎黄,再与冻豆腐焖煮,用盐、姜、葱调味。

Fried carp head with bean curd. Jiangsu dish. Bighead carp head fried golden on both sides, boiled with frozen bean curd, then flavored with salt, ginger, and scallion.

鱼头砂锅 (yútóu shāguō)

湖北菜。先将豆腐和腌渍过的鱼头略煎,再将煸过的冬菇和肉片加调料略煮,然后全部放入砂锅,加粉皮焖煮。

Braised fish head and bean curd. Hubei dish. Pickled fish head and bean curd lightly fried, then braised in an earthen pot with quick-fried winter mushrooms, sliced pork, starch noodles, and seasonings.

鱼丸 (yúwán)

福建点心。将鱼茸与淀粉和调料做成丸子,包入猪肉、虾仁、葱、姜等制成的馅,入清汤煮熟。

Stuffed fish ball soup. Fujian snack. Minced fish meat mixed with starch and seasonings, made into balls filled with mixture of pork, shrimp meat, scallion, and ginger, then boiled in clear soup.

鱼香菜心 (yúxiāng càixīn)

南方菜。将油菜心用豆瓣酱、葱、姜、蒜等煸炒,烹入用糖、醋、酱油、味精、盐、淀粉等调成的卤汁。

Sautéed rape greens with fish flavorings. Southern dish. Baby rape greens sautéed with fermented broad bean sauce, scallion, ginger, and garlic. Washed with sauce of sugar, vinegar, soy sauce, MSG, salt, and starch.

鱼香豆腐 (yúxiāng dòufu)

南方菜。先将豆腐炸至金黄,再与豆瓣酱、葱、姜、蒜等翻炒,烹入用糖、醋、酱油、味精、盐、淀粉等调成的卤汁。

Sautéed bean curd with fish flavorings. Southern dish. Deep-fried bean curd sautéed with fermented broad bean sauce, scallion, ginger, and garlic, then washed with sauce of sugar, vinegar, soy sauce, MSG, salt, and starch.

鱼香苦瓜 (yúxiāng kǔguā)

浙江菜。先将苦瓜丝和红辣椒丝焯水,再将葱、姜、蒜、豆瓣酱炒成调味汁,淋在苦瓜上。

Bitter gourd with fish flavorings. Zhejiang dish. Shredded bitter gourd and red pepper scalded, then washed with sauce of stir-fried scallion, ginger, garlic, and fermented broad bean sauce.

鱼香茄子 (yúxiāng qiézi)

各地家常菜。先将茄子炸软,然后加泡红椒、豆瓣酱等同炒,勾芡,淋麻油。

Eggplants with fish flavorings. Home-style dish in many places. Deep-fried eggplants stir-fried with pickled red pepper and fermented broad bean sauce, thickened with starch, then washed with sesame oil.

鱼香茄子煲 (yúxiāng qiézibāo)

湖南菜。先将炸透的茄子、碎猪肉及蒸熟的咸鱼加入鱼香佐料煮开,然后装入瓦煲炖煮。

Braised eggplants, pork, and salty fish. Hunan dish. Deep-fried eggplants braised with chopped pork and steamed salty fish in a clay pot, then seasoned with fish flavorings.

鱼香肉丝 (yúxiāng ròusī)

四川菜。先将肉丝滑散,再与泡椒、木耳、胡萝卜急炒,烹入黄酒、豆瓣酱、酱油、醋、糖、葱、蒜。

Stir-fried pork with fish flavorings. Sichuan dish. Lightly fried shredded pork stir-fried with pickled pepper, wood ears, and carrots. Flavored with yellow wine, fermented broad bean sauce, soy sauce, vinegar, sugar, scallion, and garlic.

鱼香腰花 (yúxiāng yāohuā)

山东菜。先将猪腰条用油溜过,再与冬菇和笋片翻炒,烹入黄酒、豆瓣酱、酱油、醋、糖、葱、蒜。

Hog kidney with fish flavorings. Shandong dish. Lightly fried slivered hog kidney stir-fried with winter mushrooms and bamboo shoots. Seasoned with yellow wine, fermented broad bean sauce, soy sauce, vinegar, sugar, scallion, and garlic.

鱼香蒸蛋 (yúxiāng zhēngdàn)

湖南菜。将肉馅、荸荠粒、木耳、辣豆瓣酱、糖、清汤、葱等制成香卤,浇在蒸好的鸡蛋羹上。

Steamed egg custard with sauce. Hunan dish. Scrambled eggs steamed into custard, then covered with sauce of minced pork, chopped water chestnuts, wood ears, spicy fermented broad bean sauce, sugar, clear soup, and scallion.

鱼咬羊 (yúyǎoyáng)

湖南菜。先将鳜鱼腹内填入熟羊腰窝肉,油煎,然后加入清汤、羊汤、调料,用小火焖煮。

Stewed fish and mutton. Hunan dish. Chinese perch stuffed with prepared mutton loin, deep-fried, then stewed in gravy of seasoned clear soup, mutton soup, and seasonings.

榆钱糕 (yúqiángāo)

北京点心。将发面和白糖、榆叶调和,蒸熟。

Elm leaf cakes. Beijing snack. Fermented flour dough mixed with sugar and elm leaves, then steamed.

雨花茶 (yǔhuāchá)

绿茶。产于江苏南京雨花台。外形紧直,茸毫隐露,色呈墨绿。汤色绿而清澈,叶底匀嫩。

Yuhuacha, green tea produced in Yuhuatai area in Nanjing, Jiangsu. Its dry leaves are tight, straight, deep green, and hairy. The tea is crystal green with infused leaves even and tender green.

雨花石汤圆 (yǔhuāshí tāngyuán)

广东点心。糯米粉与可可粉加水揉成雨花石纹的汤圆皮,包入豆沙、芋泥等馅,煮熟,加入甜汤。

Stuffed sticky rice and cocoa *tangyuan*. Guangdong snack. Sticky rice flour, cocoa powder, and water mixed, kneaded into dumpling wrappers, filled with mashed red beans or taros, then boiled. Served in sugar soup.

玉参水鱼汤 (yù shēn shuǐyú tāng)

广东菜。先将水鱼块、玉竹、沙参、桂圆肉、红枣、陈皮等用武火煮沸,然后加调料用文火煲煮。

Turtle and herb soup. Guangdong dish. Softshell turtle chunks boiled with solomonseal root, glehnia root, longan pulp, red dates, and processed tangerine peels, then stewed.

玉带才鱼卷 (yùdài cáiyújuǎn)

湖北菜。先用才鱼片裹火腿丝、冬菇丝、玉兰片丝、姜丝、葱丝等成卷,蒸熟,然后浇上勾芡的鸡汤,撒胡椒粉。

Steamed dark sleeper rolls. Hubei dish. Dark sleeper fillets made into rolls filled with shredded ham, winter mushrooms, soaked bamboo shoots, ginger, and scallion. Steamed, dressed with starched chicken soup, then sprinkled with ground black pepper.

玉凤还朝 (yùfèng huáncháo)

陕西菜。先将鸭加入葱、姜、料酒等

煮沸，再用小火炖烂，然后把炒好的嫩豌豆糊浇在鸭上。

Stewed duck with green peas. Shaanxi dish. Duck boiled with scallion, ginger, and yellow wine, then stewed. Covered with prepared green pea paste.

玉灌肺 (yùguànfèi)

河南传统点心。将面粉、油饼、芝麻、松子、胡桃、茴香拌和，裹成卷蒸熟，切成肺状块，佐枣汁。

Steamed nut cakes. Traditional snack in Henan. Mixture of flour, fried cakes, sesame, pine nuts, walnuts, and fennel made into rolls, then steamed. Cut into lung-shape pieces, then served with red date sauce.

玉林牛肉丸 (yùlín niúròuwán)

广西玉林菜。先将牛肉末加调料做成肉丸，用水煮熟，再油炸，然后入高汤加葱、姜、红枣等烹煮。

Braised beef balls. Specialty in Yulin, Guangxi. Ground beef mixed with seasonings, made into balls, boiled, deep-fried, then braised in broth with scallion, ginger, and red dates.

玉麟香腰 (yùlín xiāngyāo)

湖南菜，又称宝塔香腰。此菜由七种点心重叠组成，由下至上分别为：红枣虎皮蛋、滑肉、锅烧丸、黄雀肉、鱼丸、蛋卷、腰花。

Snack pagoda. Hunan dish. It is also called pagoda dish. Seven kinds of snacks piled up: red dates with tiger-skin eggs; stir-fried pork; braised meatballs; fried siskin; fish balls; egg rolls; pork kidney cuts.

玉鸟银丝 (yùniǎo yínsī)

南方菜。先将鱼茸调入盐味，放在小白菜叶上成鸟状，蒸熟装盘，再将大白菜丝煸炒后围放在鸟周围。

Steamed fish with Chinese cabbage. Southern dish. Minced fish salted, arranged like birds on small Chinese cabbage leaves, then steamed. Surrounded by shredded quick-fried Chinese cabbage on a platter.

玉树麒麟生鱼 (yùshù qílín shēngyú)

广东菜。先将生鱼块、火腿片、香菇等在碗里码好，蒸熟，然后围上汆过的菜心，浇高汤做的芡汁。

Steamed snakehead fish with sauce. Guangdong dish. Snakehead fish chunks, sliced ham, and dried mushrooms piled in a bowl, then steamed. Surrounded by scalded baby greens, then washed with starched broth.

玉树乌参 (yùshù wūshēn)

广东菜。先将乌参与葱、姜稍炒，然后放入清汤焖煮，加料酒、酱油、盐、姜汁，勾芡。

Braised sea cucumbers. Guangdong dish. Sea cucumbers quick-fried with scallion and ginger, braised in clear soup seasoned with cooking wine, soy sauce, salt, and ginger juice, then thickened with starch.

玉珠双珍 (yùzhū shuāngzhēn)

河南菜。将煮熟去壳的鹌鹑蛋置于

猴头菇片上，再将鸡蛋浆抹在鹌鹑蛋上，然后用香菜叶和火腿片点缀，加入羊肚菌蒸透，浇调味芡汁。

Steamed quail eggs with mushrooms and ham. Henan dish. Shelled boiled quail eggs coated with egg paste, set on sliced monkey-head mushrooms, then decorated with coriander and sliced ham. Steamed with morchella mushrooms, then washed with seasoned and starched soup.

玉竹炖鹧鸪 (yùzhú dùn zhègū)

广东菜。将鹧鸪、玉竹、红枣、生姜炖熟，用盐调味。

Stewed partridge with herbs. Guangdong dish. Partridge stewed with herbs, such as solomonseal root, red date, and ginger, then seasoned with salt.

芋艿鸡骨酱 (yùnǎi jīgǔ jiàng)

上海菜。先将鸡块略腌、过油，然后加入调料和芋艿焖煮。

Braised chicken and taros. Shang-hai Dish. Pickled chopped chicken quick-fried, then braised with taros and seasonings.

芋头包 (yùtoubāo)

福建点心。先将芋头煮熟，捣成泥，然后与地瓜粉擀成面皮，做成各种馅料的包子，蒸熟。

Steamed stuffed taro buns. Fujian snack. Mashed boiled taros mixed with sweet potato powder, made into buns, stuffed with a variety of fillings, then steamed.

芋头糕 (yùtougāo)

广东点心。先将米粉、面粉、水调成浆，加入爆香的虾米、冬菇、芋头调匀，然后蒸熟，撒芝麻，置凉，切块，油煎。

Taro cakes. Guangdong snack. Rice flour, wheat flour, and water made into paste, combined with quick-fried dried shrimp meat, winter mushrooms, and taros, then steamed. Sprinkled with sesame, cooled, cut into cubes, then fried.

芋头蒸鲫鱼 (yùtou zhēng jìyú)

山东菜。先将芋头稍煮，然后同腌渍过的小鲫鱼一道蒸熟。

Steamed taro and small crucian. Shandong dish. Quick-boiled taros steamed with pickled small crucian.

芋屯 (yùtún)

台湾点心。将芋泥包入豆沙馅，蒸熟，勾芡，用果脯装饰。

Taros with sweet bean paste. Taiwan snack. Minced taros made into buns stuffed with sweet bean paste, then steamed. Washed with starched soup, then decorated with candied fruits.

芋味糯米脆皮鸡 (yùwèi nuòmǐ cuìpíjī)

广东菜。将腊味鸡胸肉、芋头、糯米等用荷叶包好蒸熟，切片装盘。

Steamed chicken breast, taros, and sticky rice. Guangdong dish. Smoked chicken breast, taros, and sticky rice wrapped in lotus leaves, then steamed. Cut into pieces to serve.

御笔猴头 (yùbǐ hóutóu)

山东菜。先将猴头蘑蒸烂，刻成12个毛笔头样，摆成扇形装盘，中间置书形黄蛋糕，上用冬菇摆出“御笔猴头”字样，再稍蒸。

Steamed monkey-head mushrooms with cakes. Shandong dish. Steamed monkey-head mushrooms carved like 12 Chinese brush pens, then arranged like a fan around a book-shape yellow cake at the center of a platter. Shredded winter mushrooms arranged into 4 Chinese characters, “御笔猴头”, meaning “the emperor’s pen and the monkey’s head”, on the cakes. Re-steamed briefly before serving.

御福翅泡饭 (yùfú chìpàofàn)

湖北菜。先用老鸡、老鸭、瘦猪肉煮成高汤，再将鱼翅入高汤炖煮，然后添入鸡汤、火腿、鸡肉，盐、糖等烧开，配白米饭。

Shark fin soup with rice. Hubei dish. Shark fin stewed in broth of mature chicken, mature duck, and lean pork. Boiled in chicken soup, enriched with ham, chicken, salt, and sugar, then served with rice.

豫章酥鸭 (yùzhāngsūyā)

江西菜。江西古称豫章。先将鸭腌渍，蒸至半熟，再抹上酱油和料酒晾干，炸至皮酥，然后去骨切块，与姜、香菇、干椒、冬笋、目鱼等同蒸，浇原汤芡汁、红油，撒胡椒粉。

Fried and steamed duck. Jiangxi dish. Jiangxi was named Yuzhang in ancient times. Pickled duck half-steamed, coated with soy sauce and yellow wine, drained, then fried crisp. Boned and chopped, then steamed again with ginger, dried mushrooms, chili pepper, bamboo shoots, and flatfish. Washed with starched broth and chili oil, then dusted with chili powder.

鸳鸯戏水 (yuānyāng xìshuǐ)

药膳，又名参附鸽。先将雌雄乳鸽与党参、附片、冰糖等隔水蒸熟，然后去掉药渣，用盐、米酒调味。

Steamed pigeons and herbs. Folk-medicinal dish. Young male and female pigeons steamed with pilose asiabell root, processed kusnezoff monkshood baby root, and rock sugar. Herb dregs removed. Seasoned with salt and sweet fermented sticky rice juice.

鸳鸯雪花卷 (yuānyāng xuěhuā juǎn)

江苏菜。先将猪肥膘煮熟，切成大片，卷入鸡茸和火腿末，裹上芡粉和芝麻，炸透，然后配熟菜心装盘。

Fried fat pork and chicken rolls. Jiangsu dish. Sliced boiled fat pork made into rolls filled with minced chicken and ham, coated with starch and sesame, then deep-fried. Served with prepared baby greens.

鸳鸯鱼扇 (yuānyāng yúshàn)

山东菜。先将一扇猴子鱼分别用清汤、料酒、糖、酱油、盐、辣酱等烧成红色咸甜香辣味，再将另一扇用白油、清汤、料酒、盐烧成白色咸鲜香嫩味，然后合装一盘。

Two-taste monkey fish. Shandong dish. One half of a monkey fish cooked with clear soup, cooking wine, sugar, soy sauce, salt, and hot sauce. Another half cooked with lard, clear soup, cooking wine, and salt. Two halves served on the same platter.

鸳鸯珍珠汤 (yuānyāng zhēnzhū tāng)

山东菜。先将鸡蛋打成泡状，做成鸳鸯形，蒸熟，然后放入煮熟的偏口鱼丸汤中，用盐、酱油、香菜梗、芝麻油等调味。

Flatfish and egg white soup. Shandong dish. Egg white whipped, shaped like 2 mandarin ducks, then steamed into cakes. Cakes put in boiled flatfish ball soup, then flavored with salt, soy sauce, coriander stalks, and sesame oil.

元宝牛蒡 (yuánbǎo niúbàng)

广东菜。先将去皮牛蒡用滚水汆过，然后切块，与排骨、腰果同炖。

Burdock roots with spareribs and cashew. Guangdong dish. Quick-boiled peeled burdock roots chopped, then braised with pork spareribs and cashew nuts.

元宝酥 (yuánbǎosū)

浙江点心。用面粉和猪油擀成面皮，包入豆沙，做成元宝形，油炸。

Fried buns filled with mashed beans. Zhejiang snack. Flour and lard made into wrappers filled with bean paste, arranged in the shape of Chinese ingots, then deep-fried.

元旦焗虾 (yuándàn jú xiā)

台湾菜。将虾、鸡蛋、肉酱、咸鲢鱼等同置一碗，入烤箱烤熟。

Roasted shrimp and eggs. Taiwan dish. Shrimp, eggs, meat sauce, and preserved silver carp put in a bowl, then roasted in oven.

元红酒 (yuánhóngjiǔ)

又称状元红，绍兴黄酒的代表品种。色泽橙黄清亮，酒精含量在15-18度之间。

Yuanhong also called *zhangyuanhong*, the top of Shaoxing yellow wine varieties. It has a clear orange color and an alcohol content of 15-18%.

元蹄炖鸡 (yuántí dùn jī)

浙江菜。母鸡和猪蹄配火腿片、冬菇、姜、盐等蒸熟。

Steamed hen and hog trotters. Zhejiang dish. Hen and hog trotters steamed with sliced ham, dried mushrooms, ginger, salt, and other seasonings.

元蹄焖乌参 (yuántí mèn wūshēn)

江苏菜。先将猪蹄煮熟切块，加调料炖煮，然后加入乌参稍焖，勾芡。

Simmered hog trotters and sea cucumbers. Jiangsu dish. Chopped boiled hog trotters stewed with seasonings. Simmered with black sea cucumbers, then thickened with starch.

原汁武陵水鱼 (yuánzhī Wǔlíngshuǐ yú)

湖南武陵菜。先将武陵水鱼腌渍，

然后与五花肉、火腿、口蘑煨煮，用盐、黄酒等调味。

Stewed soft-shell turtle. Specialty of Wuling, Hunan. Local soft-shell turtle stewed with streaky pork, ham, and Mongolian mushrooms. Seasoned with salt, yellow wine, and other condiments.

原盅淮杞炖山瑞 (yuánzhōng huáiqǐ dùn shānruì)

香港菜。将山瑞即海南鳖、猪肉、鸡脚、枸杞子、当归等用砂锅熬炖，用盐、黄酒调味。

Stewed turtle with herbs. Hong Kong dish. Hainan turtle stewed in a clay pot with pork, chicken feet, Chinese wolfberry, and Chinese angelica. Seasoned with salt and yellow wine.

约拉 (yuēlā)

西藏食品，即酥油麦糊。将西藏本地小麦磨成粉，煮成糊，加入酥油。

Wheat chowder with Tibetan butter. Tibetan food. Ground Tibetan wheat boiled into chowder. Tibetan butter added.

粤式腊肠卷 (yuèshì làchángjuǎn)

广东点心。将发酵面团搓成长条，绕在腊肠上，蒸熟。

Steamed bread ropes around sausage. Guangdong snack. Fermented flour dough made into ropes, wound around sausages, then steamed.

云岭咖啡 (yúnlǐngkāfēi)

云南咖啡。以云南小粒咖啡豆为原料焙制。口味浓而不苦，香而不烈，略带果味。

Yunling Coffee. Yunnan coffee made with roasted coffee beans grown in southern Yunnan. It tastes heavy but not bitter, and has a light fruit aroma.

云梦鱼面 (yúnmèngyúmiàn)

湖北点心。将鱼茸、面粉、盐、苏打、清水拌匀，压成薄面皮，做成面条，煮、炒均可。

Noodles of fish meat and flour. Hubei snack. Minced fish meat mixed with flour, salt, soda, and water, made into noodles, then boiled or fried.

云片猴头 (yúnpiàn hóutóu)

山东菜。先将猴头蘑、鸡骨架、猪肉、火腿、冬菇、冬笋、鸡脯肉等加清汤蒸熟，然后勾芡。

Steamed monkey-head mushrooms with ham. Shandong dish. Monkey-head mushrooms, chicken bones, pork, ham, winter mushrooms, winter bamboo shoots, and chicken breast steamed in clear soup, then thickened with starch.

云片鹿角菜 (yúnpiàn lùjiǎocài)

山东菜。将鹿角菜、火腿、玉兰片、黄瓜皮加调料拌匀，旺火蒸熟，浇芡汁，淋麻油。

Carrageen with ham and bamboo shoots. Shandong dish. Sliced carrageen, ham, soaked bamboo shoots, and cucumber peels mixed with seasonings, then steamed.

Washed with starched soup and sesame oil.

云腿鸽子汤 (yúntuǐ gēzi tāng)

江苏菜。将腌渍过的鸡肉茸填入仔鸽腹,加入云腿片及调料蒸烂。

Steamed pigeon with ham and chicken. Jiangsu dish. Young pigeon stuffed with seasoned ground chicken, then steamed with sliced Yunnan ham and flavorings.

云托八鲜 (yúntuō bāxiān)

湖南菜。先将鱿鱼、海参、鱼皮、香菇、虾肉、鲜贝、蟹肉、黄瓜焯水,然后放在嫩豆腐上,用鸡汤焖煮。

Assorted seafood on bean curd. Hunan dish. Squid, sea cucumbers, fish skins, dried mushrooms, shrimps, scallops, crab, and cucumbers scalded, placed on tender bean curd, then braised in chicken soup.

云香绿豆沙 (yúnxiāng lǜdòushā)

广东点心。先将绿豆与臭草、红糖煮烂收汁,然后捣成泥。

Scented sweet mung bean paste. Guangdong snack. Mung beans boiled with common rue herb and brown sugar, reduced, then mashed.

芸豆焖肉片 (yúndòu mèn ròupiàn)

山东菜。先将肉片煸炒,然后加入芸豆和甜面酱,用小火焖熟。

Simmered wax beans and pork. Shandong dish. Stir-fried sliced pork simmered with wax beans and sweet fermented flour sauce.

Z

糌粑 (zānbā)

藏族传统主食。先将青稞炒熟,连皮磨成细面,然后加入酥油茶,搅匀,捏成团,即成糌粑。

Tibetan barley buns, or *zanba*. Baked ground Tibetan barley mixed with Tibetan butter tea, then kneaded into round buns.

砸鱼汤 (záyútāng)

山东菜。将鱼头和鱼尾砸碎,入清汤煮沸,加盐、猪油,撒香菜末、胡椒粉。

Fish head and tail soup. Shandong dish. Fish head and tail pounded, then boiled in clear soup. Flavored with salt and lard, then sprinkled with coriander and ground black pepper.

糌粑蜕 (zānbātuì)

西藏传统点心。将糌粑与酥油、奶渣、糖等调和,做成各种形状。

Tibetan barley and butter cakes. Tibetan traditional snack. *Zanba* mixed with Tibetan butter, dairy dregs, and sugar, then made into cakes of different shapes.

藏餐吉祥羊头 (zàngcān jíxiáng yángtóu)

西藏菜。由羊头、嫩羊肉煮成。藏民过新年须吃羊头,表示吉祥。

Boiled lamb head. Tibetan dish. Lamb head and lamb meat boiled. Traditionally, the Tibetans eat lamb head to celebrate Tibetan New Year.

藏餐羊血肠 (zàngcān yángxuècháng)

西藏菜。将羊血、糌粑粉、羊肉末等加调料混合,灌入羊肠,扎成小段,煮熟。

Boiled mutton sausage. Tibetan dish. Lamb intestine stuffed with mixture of lamb blood, ground baked Tibetan barley, ground mutton, and seasonings, bound into small segments, then boiled.

藏红花羊排 (zànghónghuā yángpái)

西藏菜。先将仔羊排用高压锅煮熟,切块,然后加入盐和藏红花煨煮,勾芡。

Lamb ribs with saffron. Tibetan dish. Young lamb ribs boiled in a pressure cooker, chopped, simmered with salt and Tibetan saffron, then thickened with starch.

藏式烤全羊 (zàngshì kǎoquányáng)

西藏菜。先将藏北肥羊用香料和藏药腌渍,煮熟,然后用柴火烤至皮脆。

Broiled whole lamb. Tibetan dish. Northern Tibetan whole sheep pickled with spices and Tibetan herbs, boiled, then broiled over wood fire until skin crisp.

糟钵头 (zāobōtóu)

上海菜。将猪下水如肠段、肚条、肺块、猪爪、猪心片、猪肝片等与油豆腐、笋片一同放在砂锅里,加入猪骨汤和糟卤焖煮。

Braised hog offal. Shanghai dish. Hog offal, such as intestine, tripe, lung, trotter, heart, and liver, braised with deep-fried bean curd and sliced bamboo shoots in hog bone soup flavored with gravy made from sweet fermented sticky rice, salt, yellow wine, and broth.

糟蛋 (zāodàn)

各地家常点心。将醪糟加水煮开,打入鸡蛋煮熟。

Eggs in sweet fermented sticky rice soup. Home-style snack in many places. Eggs poached in sweet fermented sticky rice soup.

糟鹅掌 (zāoézhǎng)

江浙沪菜。鹅掌煮熟后用糟卤泡渍。

Marinated goose feet. Jiangsu, Zhejiang, and Shanghai dish. Goose feet boiled, then marinated in gravy made from sweet fermented sticky rice, salt, yellow wine, and broth.

糟鹅掌鸭信 (zāo ézhǎng yāxìn)

曹雪芹《红楼梦》菜谱。将鹅掌及鸭舌煮熟,剔骨,用鸡汤加盐复煮,佐香糟汁或糟油。

Goose feet and duck tongue. Menu from *A Dream of Red Mansions* by Cao Xueqin. Boiled and boned goose feet and duck tongue simmered in salted chicken soup. Served with fermented sticky rice juice, or sauce made with aged sweet fermented sticky rice juice and a variety of seasonings.

糟鸡 (zāojī)

上海菜。鸡块煮熟后用糟卤泡渍。

Marinated chicken. Shanghai dish. Chicken chunks boiled, then marinated in gravy made with sweet fermented sticky rice, salt, yellow wine, and broth.

糟熘鱼片 (zāoliū yúpiàn)

山东菜。先将鲮鱼片挂浆,下油锅滑散,再放入鸡汤烧沸,然后加香糟,勾芡。

Dace in fermented sticky rice soup. Shandong dish. Sliced dace coated with starch and egg white, lightly fried, then boiled in chicken soup. Flavored with sweet fermented sticky rice juice and starched soup.

糟田螺 (zāotiánluó)

上海菜。田螺、茴香、桂皮、糟卤同煮。

Braised field snails. Shanghai dish. Field snails braised with aniseed, cinnamon, and gravy made with sweet fermented sticky rice, salt, yellow wine, and broth.

糟煨冬笋 (zāowēi dōngsǔn)

山东菜。先将冬笋氽过,再加奶汤、香糟酒等烧开,然后用小火煨煮,勾芡。

Bamboo shoots in fermented sticky rice soup. Shandong dish. Scalded winter bamboo shoots simmered in milk and sweet fermented sticky rice soup, then thickened with starch.

糟煨鳜鱼 (zāowēi guìyú)

山东菜。先将煎过的鳜鱼与葱、姜、蒜等煸炒,加入盐和香糟汤煨煮,然后加入南荠、冬菇、玉兰片,勾芡,淋猪油。

Simmered Chinese perch with sauce. Shandong dish. Lightly fried Chinese perch fish quick-fried with scallion, ginger, and garlic, then simmered in salted sweet fermented sticky rice soup with water chestnuts, winter mushrooms, and bamboo shoots. Enriched with starched broth and lard.

糟煨蒲菜 (zāowēi púcài)

山东菜。先将蒲菜加奶汤和黄酒略煨,再加入香糟汤烧开,加盐,勾芡。

Cattail roots in fermented sticky rice soup. Shandong dish. Cattail roots simmered with milk and yellow wine, boiled in sweet fermented sticky rice soup, flavored with salt, then thickened with starch.

糟油青鱼划水 (zāoyóu qīngyúhuá-shuǐ)

安徽菜。先将青鱼尾稍煎,加调料煮沸,然后小火焖至汁浓,加入糟汁和其他调料烧沸,勾芡。

Braised carp tails with sauce. Anhui dish. Black carp tails quick-fried, boiled with seasonings, then braised with sweet fermented sticky rice juice and other condiments until soup thick. Thickened with starch.

糟汁汆海蚌 (zāozhī cuān hǎibàng)

福建菜。先将汆过的海蚌肉用黄酒稍腌,再用姜末、香糟、鸡汤等做成卤汁,然后将卤汁过滤,浇在蚌肉上。

Clam meat with sauce. Fujian dish. Quick-boiled clam meat pickled with yellow wine, then washed with filtered gravy made with chicken soup, ginger, and sweet fermented rice juice.

糟猪蹄 (zāozhūtí)

各地家常菜。先将猪蹄煮熟切开,放入卤汁腌渍,然后切块,淋糟酒。

Marinated hog feet. Home-style dish in many places. Boiled hog feet cut into halves, then marinated in gravy. Chopped, then served with dressing of yellow wine and sweet fermented sticky rice juice.

早红桔酪鸡 (zǎohóngjúlàojī)

江苏菜。先将仔鸡油炸,然后浸入江苏洞庭出产的早红桔酪酱蒸熟。

Chicken in mandarin orange jam. Jiangsu dish. Fried young chicken steamed in orange jam made with autumn mandarin oranges from Dongting, Jiangsu.

枣莲炖雪蛤 (zǎo lián dùn xuěhá)

东北点心。先将雪蛤汆过,加入清水、冰糖、红枣、莲子煮沸,然后蒸烂。

Forest frogs with dates and lotus seeds. Northeastern snack. Scalded Chinese forest frog oviduct boiled with rock sugar, red dates, and lotus seeds, then steamed.

枣泥夹沙肉 (zǎoní jiāshāròu)

湖南菜。将猪五花肉煮熟,切成夹片,嵌入枣泥,挂鸡蛋面糊,炸至金黄,撒糖,淋熟猪油。

Fried stuffed streaky pork. Hunan dish. Boiled streaky pork cut into clamping slices, filled with red date paste, coated with egg and flour paste, then deep-fried. Sprinkled with sugar and prepared lard.

枣泥酥条 (zǎoní sūtiáo)

北方点心。把面团和油酥糅和,擀成长片,包入枣泥,扭成麻花,涂上蛋黄,烤至黄脆。

Baked flour and date twists.

Northern snack. Flour dough and pastry sheets made into rolls filled with mashed red dates, then kneaded into twisted strips. Coated with egg yolks, then baked crisp.

枣香丝糕 (zǎoxiāng sīgāo)

江苏点心。用玉米粉、面粉、鸡蛋、蜂蜜和泡红枣的水、泡打粉等调成糊,嵌入泡过的红枣,蒸熟。

Corn cake with red dates. Jiangsu snack. Corn flour, wheat flour, egg white, honey, soaking water of red dates, and baking powder mixed into paste, decorated with soaked dates, then steamed.

扎西东嘎 (zāxīdōnggā)

西藏菜,又名吉祥海螺。先将蛋清、蛋黄分别与奶调和,分别蒸熟成白色和黄色的糕,切成片或丝,再与胡萝卜片、心里美片、黄瓜片等摆成海螺状。

Dairy cake with vegetables. Tibetan dish. Eggs separated, mixed with milk, then steamed into white and yellow cakes separately. Cakes sliced or shredded, then arranged in the shape of a conch with sliced carrots, pink turnips, and cucumbers.

炸八块 (zhábākuài)

湖南菜。先将仔鸡剁为八块,腌渍,然后挂糊油炸。

Fried chicken chunks. Hunan dish. Young chicken cut into eight chunks, pickled with seasonings, coated with starch paste, then deep-fried.

炸扳指 (zhábānzhi)

四川菜。先将肥肠头氽过,加调料蒸熟,然后炸至金黄,切成短节,淋麻油,配辣味卤汁。

Fried pork large intestine with sauce. Sichuan dish. Quick-boiled hog large intestine steamed with seasonings, then deep-fried. Sectioned, washed with sesame oil, then served with spicy dressing.

炸比目鱼 (zhábǐmùyú)

南方菜。将比目鱼块用盐等调料腌制,挂淀粉蛋浆,炸至金黄。

Fried flatfish. Southern dish. Flatfish chunks pickled with salt and other seasonings, coated with starch and egg paste, then deep-fried.

炸蚕蛹鸡 (zhácányǒngjī)

山东菜。将鸡脯肉末与芡粉、蛋清、

盐等拌成馅,用猪网油片裹馅成蚕蛹状,炸至金黄。

Fried chicken in web lard. Shandong dish. Web lard pieces made into silkworm-shape rolls filled with mixture of ground chicken, starch, egg white, and salt, then deep-fried.

炸赤鳞鱼 (zháchìlínyú)

山东菜。将腌渍过的泰山赤鳞鱼裹上面粉,用微火炸至淡黄。

Fried red squama. Shandong dish. Red squama fish from Mount Taishan pickled, coated with flour, then fried over low heat until light yellow.

炸大扁 (zhádàbiǎn)

山东菜。将腌渍过的猪瘦肉丁、熟肥肉、海参、熟荸荠丁、冬笋丁、薄菜头制成馅,放在蛋浆上,然后浇上蛋浆,用低温烙成夹饼,炸至金黄。

Eggs, meat, and vegetable cakes. Shandong dish. Mixture of pickled diced lean pork, prepared fat pork, sea cucumbers, water chestnuts, bamboo shoots, and radish put on a bed of egg paste, then covered with egg paste. Pan fried into sandwich cakes over low heat, then fried golden.

炸豆腐盒 (zhádòufuhé)

东北家常菜。先将豆腐块中间挖空,灌入加调料的肉馅,然后炸至金黄。

Fried stuffed bean curd. Northeastern home-style dish. Cubed dried bean curd hollowed, stuffed with seasoned ground pork, then deep-fried.

炸豆腐丸子 (zhá dòufuwánzi)

东北家常菜。将豆腐、面粉、盐、海米等加水做成丸子,炸至金黄。

Fried bean curd balls. Northeastern home-style dish. Bean curd mixed with flour, minced dried shrimp meat, salt, and water, made into balls, then deep-fried.

炸二丝卷 (zháèrsījuǎn)

湖南菜。用猪网油将鸡肉丝和火腿丝包裹成条,裹芡粉蛋浆油炸,佐甜面酱食用。

Fried chicken and ham in lard. Hunan dish. Shredded chicken breast and ham wrapped in web lard, coated with starch and egg paste, then deep-fried. Served with sweet fermented flour sauce.

炸佛手通脊 (zhá fóshǒu tōngjǐ)

山东菜。将腌过的肉片逐层裹上干粉、蛋液、面包屑,做成佛手状,油炸。

Fried palm-shape pork. Shandong dish. Pickled sliced pork coated with flour, egg white, and bread crumbs, arranged like a palm, then deep-fried.

炸灌肺 (zháguànfèi)

西藏菜,又称灌肺。先把面粉、五香粉、盐、酥油、水等调成糊,灌满羊肺,再把羊肺放入开水煮熟,冷却后切片,油煎。

Fried stuffed lamb lung. Tibetan

dish. Lamb lung filled with mixture of flour, five-spice powder, salt, Tibetan butter, and water, then boiled. Cooled, sliced, then fried.

炸灌汤丸子 (zhá guàntāngwánzi)

山东菜。先将猪肉末加蛋清和芡粉做成肉丸,包入高汤冻,然后裹上面包屑,炸至棕黄。

Fried meatballs stuffed with broth jelly. Shandong dish. Ground pork mixed with egg white and starch, made into meatballs stuffed with broth jelly, coated with bread crumbs, then deep-fried until yellow brown.

炸海蟹 (zháhǎixiè)

山东菜。先将海蟹切成两半,用辣椒粉、盐、黄酒等腌渍,然后裹上淀粉,炸至金黄。

Fried spicy crab. Shandong dish. Crab cut into halves, pickled with ground black pepper, salt, and yellow wine, coated with starch, then deep-fried.

炸荷包鲜鱿 (zhá hébāo xiānyóu)

广东菜。将糯米饭、叉烧、香菇、虾米、肥肉、莲子等加调料制成八宝饭,填进鲜鱿筒,先蒸后炸。

Fried stuffed squid. Guangdong dish. Fresh squid stuffed with seasoned mixture of cooked sticky rice, barbecued pork, dried mushrooms, shelled shrimp, fat pork, and lotus seeds, steamed, then deep-fried.

炸荷花 (zháhéhuā)

山东济南名菜。先将豆沙涂在荷花瓣上,再将荷花瓣对折,沾上鸡蛋面粉糊,用中火炸熟,然后撒上糖桂花。

Fried lotus flower. Famous dish in Jinan, Shandong. Lotus flower petals layered with sweet bean paste, folded in the middle, coated with flour and egg paste, fried over medium heat, then sprinkled with preserved sweet osmanthus flower.

炸红糟鲤 (zháhóngzāolǐ)

广东菜。将鲤鱼条用盐、葱、姜、黄酒略腌,裹上用鸡蛋、面粉、红糟调成的糊,炸至金黄。

Fried savoury carp. Guangdong dish. Carp chunks pickled with salt, scallion, ginger, and yellow wine, coated with paste of eggs, flour, and sweet fermented sticky rice juice, then deep-fried golden.

炸藿香 (zháhuòxiāng)

山东菜。将藿香嫩尖裹上用面粉、盐、发酵面和素油调成的糊,炸至金黄。

Fried wrinkled gianthyssop herb. Shandong dish. Tender wrinkled gianthyssop herb coated with paste of flour, salt, fermented dough, and vegetable oil, then fried golden.

炸酱鸡 (zhájiàngjī)

浙江菜。先将鸡块稍腌,再与煸炒过的葱、蒜、番茄、香茅、羌花、海鲜酱、豆瓣酱等焖煮,然后撒上辣椒粉和胡椒粉。

Braised spicy chicken. Zhejiang

dish. Pickled chicken chunks braised with quick-fried scallion, garlic, tomatoes, lemongrass, notopterygium flower, seafood sauce, and fermented broad bean sauce. Sprinkled with chili powder and ground black pepper.

炸酱面 (zhájiàngmiàn)

各地传统食品。先将面条煮熟，再浇上用炸肉丁、炸豆腐、炸花生、咸萝卜、甜面酱等做成的浇头，配油、盐、酱油、醋、辣椒酱等食用。

Noodles with stir-fried varieties. Traditional food in many places. Boiled noodles covered with dressing of fried diced pork, bean curd, peanuts, preserved radish, and sweet fermented flour sauce. Served with oil, salt, soy sauce, vinegar, and chili sauce.

炸金钩棒 (zhájīngōubàng)

山东菜。猪肉片卷入用海米、南荠、火腿及调料制成的馅，裹上蛋浆和面包末，炸至金黄。

Fried stuffed meat rolls. Shandong dish. Sliced pork made into rolls filled with mixture of dried shrimp meat, water chestnuts, ham, and seasonings, coated with egg paste and bread crumbs, then deep-fried.

炸菊花虾包 (zhá júhuā xiābāo)

山东菜。用猪网油片包入用虾仁、海参、荸荠、蛋清等做成的馅，炸至金黄，码成菊花形装盘。

Fried lard and sea food rolls. Shandong dish. Web lard package stuffed with mixture of shrimp meat, sea cucumbers, water chestnuts, and egg white, deep-fried, then arranged like a chrysanthemum flower on a platter.

炸蛎黄 (zhálìhuáng)

山东菜。将牡蛎黄裹上面粉浆，炸熟。

Fried oyster roe. Shandong dish. Oyster roe coated with flour paste and deep-fried.

炸熘素肉丸 (zháliū sùròuwán)

东北家常菜。将豆腐末与熟红薯泥拌和，做成丸子，炸至金黄，再加盐、姜、葱、味精等熘炒。

Sautéed vegetable balls. Northeastern home-style dish. Mashed bean curd and boiled potatoes mixed, made into balls, then deep-fried. Sautéed with salt, ginger, scallion, and MSG.

炸龙肠 (zhálóngcháng)

山东菜。先将对虾、鸡肉、海参、火腿、荸荠加调料做成馅，再将猪网油膜包馅成卷，蒸熟，然后裹蛋浆和芝麻，油炸至芝麻呈黄色。

Fried rolls of meat varieties. Shandong dish. Prawn meat, chicken, sea cucumbers, ham, water chestnuts, and seasonings mixed, rolled in web lard, then steamed. Rolls coated with egg paste and sesame, then fried yellow.

炸牡蛎 (zhámǔlì)

沿海一带传统菜。牡蛎肉裹生粉炸至金黄，佐椒盐食用。

Fried oysters with spicy salt. Traditional dish along coastal areas. Oyster meat coated with corn starch, deep-fried, then served with spicy salt.

炸烹铁雀 (zhápēng tiěquè)

北方、南方菜,亦称炒禾花雀。先将铁雀腌渍,再挂糊油炸,然后用特制卤汁烹炒。

Sautéed fried buntings. Northern and southern dish. Pickled yellow-breast buntings coated with starch paste, fried, then sautéed with special sauce.

炸烹银鱼 (zhápēng yínyú)

东北家常菜。先将银鱼腌渍,再挂糊油炸,然后用卤汁烹炒。

Fried silver fish. Home-style dish in northeastern areas. Pickled silver fish coated with starch paste, deep-fried, then sautéed with special sauce.

炸烹子蟹 (zhápēng zǐxiè)

各地家常菜。先将子蟹切开,炸熟,然后用特制卤汁烹炒。

Fried crab. Home-style dish in many places. Young crab cut into halves, deep-fried, then stir-fried with special sauce.

炸千子 (zháqiānzi)

东北家常菜。先将鸡蛋摊成皮,裹上猪肉馅成卷,然后切段,炸至金黄。

Fried egg rolls. Northeastern home-style dish. Seasoned ground pork rolled in egg sheets, cut into small sections, then fried golden.

炸肉卷 (zháròujuǎn)

江苏菜。将猪网油铺上一层淀粉,把猪肉荸荠馅摊在上面,裹成卷,用蛋浆封口,油炸。

Pork and water chestnut rolls. Jiangsu dish. Hog web lard dusted with dry starch, made into rolls filled with minced pork and water chestnuts, sealed with egg paste, then deep-fried.

炸田鸡腿 (zhátiánjītuǐ)

东北菜。先将田鸡腿用盐、黄酒,味精,花椒水腌渍,然后裹面粉,炸至金黄。

Fried frog legs. Northeastern dish. Frog legs pickled with salt, yellow wine, MSG, and Chinese prickly ash soup, coated with flour, then deep-fried.

炸土豆肉饼 (zhá tǔdòuròubǐng)

东北家常菜。将土豆泥包上猪肉馅,做成饼,炸至金黄。

Fried potato cakes. Northeastern home-style dish. Mashed potatoes made into cakes stuffed with seasoned ground pork, then deep-fried.

炸五香 (zháwǔxiāng)

福建菜。将里脊肉、肥猪肉、虾仁、笋、葱白加调料做成馅,用豆腐皮裹馅成卷,用面糊粘牢,油炸。

Fried meat and vegetable package. Fujian dish. Pork tenderloin, fat pork, shrimp meat, bamboo shoots, scallion stems, and seasonings made into stuffing, rolled in bean curd skins, sealed with flour paste, then deep-fried.

炸羊尾 (zháyángwěi)

西北清真菜。先将羊尾片卷入白糖、京糕、麻仁、面粉等做成的馅,然后裹上鸡蛋面粉浆,炸至金黄,撒上白糖。

Fried lamb tail rolls. Muslim dish in northwestern areas. Sliced lamb tail made into rolls filled with mixture of white sugar, Beijing rice cake, hemp seeds, and flour, coated with flour and egg paste, then deep-fried. Sprinkled with white sugar.

炸鸳鸯嘎渣 (zhá yuānyānggāzhā)

山东、河南菜。将鸡蛋、淀粉、面粉、桂花酱搅匀,分成两碗,一碗加入碎山楂糕。两碗浆分别炒熟,摊成红黄两块嘎渣。两块嘎渣叠放切条,裹上面粉油炸。

Fried egg and haw cakes. Shandong and Henan dish. Eggs, starch, flour, and osmanthus flower sauce mixed into paste, then put in 2 bowls. Haw cake crumbs added to one of them. Baked separately into red and yellow pancakes or *gazha*. Red and yellow *gazha* piled, cut into strips, coated with flour paste, then deep-fried.

炸脂盖 (zházhīgài)

山东菜。先将半肥瘦羊肉腌渍,加调料蒸熟,再裹上淀粉蛋糊,炸至金黄,配甜面酱、芝麻油、葱段、蒜瓣。

Fried streaky mutton. Shandong dish. Pickled streaky mutton steamed with seasonings, coated with starch and egg paste, then deep-fried. Served with sweet fermented flour sauce, sesame oil, leek sections, and garlic cloves.

炸紫酥肉 (zházǐsūròu)

河南菜。先将腌渍过的猪肋条肉蒸熟,油炸,然后在皮上涂醋,反复炸三次至柿黄色,佐大葱和甜面酱。

Fried streaky pork with leek and sauce. Henan dish. Pickled streaky pork steamed, fried, brushed on the skins with vinegar, then re-fried 3 times until yellow red. Served with leek and sweet fermented flour sauce.

榨菜肉丝汤 (zhàcài ròusī tāng)

各地家常菜。先将猪瘦肉丝和榨菜丝烫熟,再加入清汤、葱花、姜丝、黄酒、酱油。

Mustard tuber and pork soup. Home-style dish in many places. Shredded lean pork and preserved mustard tubers quick-boiled in clear soup, then flavored with chopped scallion, shredded ginger, yellow wine, and soy sauce.

诈马宴 (zhàmǎyàn)

蒙古族宴席,即整牛席或整羊席。先将去毛的牛或羊去除内脏,洗净,再把盐和五香调料放进腹腔,将开膛处缝好,然后放入封闭的烤炉烤熟。

Roasted whole ox or lamb. Mongolian beef or mutton banquet. Whole ox or lamb dressed, filled with salt and five-spice seasonings, then sealed. Roasted in a closed oven.

斩鱼圆 (zhǎnyúyuán)

浙江菜。用草鱼茸、黄酒、猪油、火腿末等做成鱼丸,煮熟。

Grass carp balls. Zhejiang dish. Mashed grass carp meat mixed with yellow wine, lard, and minced ham, made into balls, then boiled.

张弓酒 (zhānggōngjiǔ)

白酒,产于河南宁陵县张弓镇。用当地高粱为主料,以小麦、大麦制曲酿造,酒精含量为 28-54 度,属浓香型酒。

Zhanggongjiu, liquor produced in Zhanggong Town of Ningling County, Henan. White spirit made with local sorghum and yeast of barley and wheat. Contains 28-54% alcohol, and has a thick aroma.

张记烧鸡 (zhāngjìshāojī)

河南洛阳菜。将土鸡配丁香、茴香、良姜、花椒、党参、桂皮、陈皮、沙仁及调料烹煮。

Chicken with herbs. Specialty in Luoyang, Henan. Local chicken cooked with sea-sonings and herbs such as clove, aniseed, lesser galangal, Chinese prickly ash, pilose asiabell root, cinnamon, processed tangerine peel, and fructus amomi.

张裕解百纳干红葡萄酒 (zhāngyù jiěbáinà gānhóng pútáojiǔ)

世界名酒,产于山东烟台。以当地红色或紫色葡萄为主要原料制成,酒精含量为 12 度。

Zhangyu, or *Changyu*, cabernet, world famous dry red wine produced in Yantai, Shandong. It is made with local red or purple grapes, and contains 12% alcohol.

张裕金奖白兰地 (zhāngyù jīnjiǎng báilándì)

产于山东烟台。以当地葡萄为主要原料制成,酒精含量 38-40 度。

Zhangyu, or *Changyu*, Special fine brandy produced in Yantai, Shandong. It is made with local grapes, and contains 38-40% alcohol.

张裕雷司令干白葡萄酒 (zhāngyù léisīlìng gānbái pútáojiǔ)

中国名酒,产于山东烟台。以当地白葡萄为主要原料制成,酒精含量为 12 度。

Zhangyu, or *Changyu*, riesling dry white wine produced in Yantai, Shandong. It is made with local white grapes, and contains 12% alcohol.

章鱼莲藕汤 (zhāngyú liánǒu tāng)

广东菜。将章鱼、莲藕、猪蹄、红枣熬炖。

Octopus and lotus root soup. Guangdong dish. Octopus stewed with lotus roots, hog trotters, and red dates.

彰化肉圆 (zhānghuàròuyuán)

台湾彰化菜。先用米粉和番薯粉做成丸子,包入用香菇、肉片等炒制的馅,蒸熟,然后浇上用酱油膏、海山酱等调成的卤汁。

Steamed stuffed balls with sauce.

Specialty in Zhanghua, Taiwan. Rice flour and sweet potato powder made into balls, stuffed with prepared dried mushrooms and pork, then steamed. Washed with dressing of soy sauce and seafood sauce.

樟茶鸭 (zhāngcháyā)

四川菜。先将鸭用料酒、醪糟汁、胡椒粉、花椒、盐腌渍,用沸水烫过,再用樟树叶、花茶叶等薰至棕黄,然后蒸熟,炸至棕红。

Fried smoked duck. Sichuan dish. Duck pickled with cooking wine, sweet fermented sticky rice juice, ground black pepper, Chinese prickly ash, and salt. Scalded, then smoked with camphor leaves and scented tea leaves. Steamed, then fried red brown.

掌上明珠 (zhǎngshàng-míngzhū)

安徽菜。先将鸭掌煮熟去骨,用盐和料酒稍腌,再扑上面粉,抹上肉泥,置入鹌鹑蛋,配上火腿末和香菜末蒸熟,然后浇鸡汤芡汁。

Duck feet with quail eggs. Anhui dish. Boned boiled duck feet pickled with salt and cooking wine, dusted with flour, spread with minced pork, inlaid with a quail egg in each, garnished with ham crumbs and chopped coriander, then steamed. Washed with starched chicken soup.

朝阳鸡蛋豆腐 (zhāoyáng jīdàn dòufu)

浙江菜。先将嫩豆腐蒸熟,再将爆炒过的蛋白、瘦肉、虾仁等烹煮,浇在豆腐上,洒上碎熟蛋黄和葱花。

Bean curd with eggs and meat. Zhejiang dish. Tender bean curd steamed, covered with stir-fried mixture of egg white, lean pork, and shrimp meat, then garnished with mashed prepared egg yolks and chopped scallion.

招远蒸丸 (zhāoyuǎnzhēng wán)

山东招远菜。将五花肉、鹿角菜、海米加蛋清、盐、芡粉做成小肉丸,蒸熟。

Steamed small meatballs. Specialty in Zhaoyuan, Shandong. Mixture of streaky pork, hart-horn vegetable, dried shrimp meat, egg white, salt, and starch made into small balls, then steamed.

折当果折 (zhédāngguǒzhé)

西藏点心。先将面团切条,炸熟,然后挂熬化的红糖浆,沥干。

Fried sugarcoated flour strips. Tibetan snack. Dough cut into strips, deep-fried, coated with melted brown sugar, then air-dried.

哲色莫古 (zhésèmògǔ)

西藏点心,即人参果饭。将大米和人参果煮熟,然后加入融化的酥油、红糖、白糖、葡萄干、盐。

Rice with silverweed roots. Tibetan snack. Rice and silverweed roots boiled, then flavored with melted Tibetan butter, brown sugar, white sugar, raisins, and salt.

柘荣牛肉丸 (zhèróng niúròuwán)

福建柘荣菜。先将牛肉末、淀粉、酱油、辣椒粉、醋、黄酒等拌和，做成丸子，然后煮熟，放入麻油、香菜、香葱等调味。

Spicy beef meatballs. Specialty in Zherong, Fujian. Ground beef, starch, soy sauce, dried chili powder, vinegar, and yellow wine mixed, made into balls, then boiled. Flavored with sesame oil, coriander, and scallion.

针菇鸡丝 (zhēngū jīsī)

湖北菜。先把鸡肉丝用蛋清、盐、芡粉等腌渍，再和氽过的金针菇煸炒，加入姜、葱等配料，然后加入卤汁烧开。

Sautéed chicken and mushrooms. Hubei dish. Shredded chicken pickled with egg white, salt, and starch, sautéed with scalded golden mushrooms, ginger, and scallion, then boiled in special sauce.

珍味鸭舌 (zhēnwèi yāshé)

台湾菜。先将鸭舌氽过，再加入甘草、八角、人参须、辣椒、酱油、料酒等煮熟，然后在原汤里卤泡。

Marinated duck tongue. Taiwan dish. Scalded duck tongue boiled with licorice root, aniseed, ginseng tail, chili pepper, soy sauce, and yellow wine, then marinated in broth.

珍珠豆腐 (zhēnzhū dòufu)

福建菜。先把豆腐做成珍珠状颗粒，用清水煮沸，然后放入用母鸡和干贝炖出的清汤，用盐调味。

Bean curd in chicken and scallop soup. Fujian dish. Bean curd made into pearl-size balls, boiled, put in clear chicken and scallop soup, then flavored with salt.

珍珠饺 (zhēnzhūjiǎo)

南方点心。将饺皮包上用鱼肉、干贝、冬菇等制成的馅，捏成圆形小饺子，入鸡汤煮熟。

Mini *jiaozi* in chicken soup. Southern snack. Flour wrappers packed with mixture of fish meat, dried scallops and dried winter mushrooms, made into mini round *jiaozi*, then boiled in chicken soup.

珍珠鲤鱼 (zhēnzhū lǐyú)

广东菜。将鲤鱼茸加入蛋清、盐、味精搅匀，做成珍珠状小丸子，入上汤煮熟，放在烹煮好的鱼头和鱼尾中间，浇上用清汤做的芡汁。

Mini fish ball soup. Guangdong dish. Minced carp meat mixed with egg white, salt, and MSG., made into pearl-size balls, then boiled in broth. Balls placed between cooked fish head and tail, then washed with starched clear soup.

珍珠南瓜 (zhēnzhū nánguā)

广东菜。珍珠指鹌鹑蛋。南瓜片和煮熟去壳的鹌鹑蛋加青椒拌炒，用盐、姜、葱、白糖调味，勾芡。

Sautéed quail eggs and pumpkin. Guangdong dish. Sliced pumpkin sautéed with shelled boiled quail eggs and green pepper, seasoned with salt, ginger, scallion, and sugar, then thickened with starch.

珍珠酥皮鸡 (zhēnzhū sūpíjī)

四川菜。先将鸡块腌渍,然后裹上面包屑和蛋浆,炸至金黄。

Fried chicken. Sichuan dish. Pickled chopped chicken coated with bread crumbs and egg paste, then deep-fried.

珍珠咸水饺 (zhēnzhū xiánshuǐjiǎo)

广东点心。用糯米粉制皮,猪肉、冬菇、虾米、盐等制馅,做成饺子,炸至金黄。

Fried meat and mushroom *jiaozi*. Guangdong snack. Sticky rice flour dough made into *jiaozi* filled with mixture of pork, winter mushrooms, shrimp meat, and salt, then deep-fried.

珍珠鸭蛋卷 (zhēnzhū yādàn juǎn)

江苏菜。豆腐皮切成方块,卷入鸭蛋黄酱,外面裹上浸泡过的糯米,蒸熟。

Steamed duck yolk rolls. Jiangsu dish. Dried bean curd skins cut into squares, made into rolls filled with duck yolk sauce, coated with soaked sticky rice, then steamed.

珍珠鱼丸 (zhēnzhū yúwán)

浙江杭州家常菜。将鳙鱼茸与淀粉、葱、姜、盐、水等调和,做成小丸子,入清汤煮熟,放入黄瓜片。

Bighead fish ball soup. Home-style dish in Hangzhou, Zhejiang. Mixture of minced bighead carp, starch, scallion, ginger, salt, and water made into small balls, boiled in clear soup, then garnished with sliced cucumbers.

珍珠圆子 (zhēnzhū yuánzi)

四川点心。将糯米粉加水揉成面团,用白糖、芝麻、猪油为馅,做成圆子,裹上湿糯米蒸熟。

Steamed stuffed sticky rice balls. Sichuan snack. Sticky rice flour mixed with water, made into round dumplings filled with mixture of sugar, sesame, and lard, coated with soaked sticky rice, then steamed.

蒸扒三丝干贝 (zhēngpá sānsī gānbèi)

山东菜。先将笋丝、冬菇丝、火腿丝煸炒,再加汤和调料煮沸,勾芡,然后放在干贝上蒸透,浇原汤和鸡油。

Steamed ham, winter mushrooms, and bamboo shoots. Shandong dish. Quick-fried sliced bamboo shoots, winter mushrooms, and ham boiled in seasoned and starched soup, placed on prepared scallops, then steamed. Washed with broth and chicken oil.

蒸糕 (zhēnggāo)

四川点心。先将鸡蛋搅打起泡,再加入面粉、水、糖、葡萄干调和,蒸熟。

Steamed egg cake with raisins. Sichuan snack. Whipped eggs mixed with flour, water, sugar, and raisins, then steamed.

蒸牛舌 (zhēngniúshé)

西藏菜。将牛舌煮熟,去掉表皮,切片,加调料蒸透。

Steamed ox tongue. Tibetan dish. Boiled and skinned ox tongue sliced, then steamed with seasonings.

蒸麒麟鱼 (zhēngqílínyú)

广东菜。用鲈鱼片夹上腌渍过的火腿、香菇、猪肥肉等蒸熟,浇芡汁。

Steamed perch. Guangdong dish. Perch fillets layered with prepared ham, mushrooms, and fat pork, steamed, then washed with starched soup.

蒸人参鸡 (zhēngrénshēnjī)

朝鲜族菜。先将鸡用盐腌渍,再往鸡肚里填入人参和糯米,蒸熟,然后将鸡切块,配蒸熟的糯米和人参装盘。

Steamed chicken with ginseng. Korean ethnic minority dish. Whole chicken pickled with salt, stuffed with ginseng and sticky rice, then steamed. Cooked chicken chopped, then served with cooked sticky rice and ginseng.

蒸山水豆腐 (zhēng shānshuǐdòufu)

东北家常菜。山水豆腐是用山泉水和当地黄豆制成的嫩豆腐。先将豆腐焯水,切片,撒上虾仁和调料,然后蒸熟。

Steamed tender bean curd.

Northeastern home-style dish. Tender bean curd made with spring water and local soybean. Quick-boiled, sliced, sprinkled with shrimp meat and seasonings, then steamed.

整鱼两吃 (zhěngyú liǎng chī)

山东菜。先将整鱼剖成两半,留头去骨。将半条鱼刻十字花刀,腌渍,连同鱼头炸成金黄色,浇上番茄青豆卤汁。将另外半条鱼劈出相连的鱼片,裹入虾馅,抹上蛋清芡粉汁,蒸熟,浇清芡汁,撒香菜梗。

Two-taste fish. Shandong dish. Fish halved, then boned save the head. One half of the fish cross cut, pickled, deep-fried, then washed with sauce of tomatoes and green beans. The other half carved into connected slices, rolled with shrimp filling, coated with starch and egg paste, steamed, washed with starched clear soup, then garnished with coriander stems.

政和功夫茶 (zhènghé gōngfuchá)

红茶。主产于福建政和县。外形肥壮多毫,色泽乌润。汤色红浓,叶底肥壮尚红。

Zhenghe gongfucha, black tea produced in Zhenghe, Fujian. Its dry leaves are stout, hairy, and off-black. The tea is red and thick with infused leaves stout and reddish.

之江鲈莼羹 (zhījiāng lúchúngēng)

浙江菜。将鲈鱼丝上浆,过油,入清汤加调料煮沸,放入莼菜、火腿丝、鸡丝、葱丝,淋鸡油。

Perch and water shield soup.

Zhejiang dish. Shredded perch pickled with starch paste, lightly fried, boiled in seasoned clear soup with water shield, shredded ham, chicken, and scallion, then washed with chicken oil.

芝麻茶 (zhīmachá)

广东潮州点心。先将芝麻和花生炒熟,再和粳米加水磨成浆,然后把浆徐徐倒入沸水煮熟,搅匀,加糖。

Sesame tea. Snack in Chaozhou, Guangdong. Round grain rice, roasted sesame, and roasted peanuts ground into paste with water, tipped into boiling water, mixed, then sweetened with sugar.

芝麻糊 (zhīmahú)

广东点心。先将芝麻粉用沸水调匀,再用小火煮开,然后加糖。

Sesame gruel. Guangdong snack. Sesame powder and boiling water mixed into paste, boiled over low heat, then sweetened with sugar.

芝麻球 (zhīmaqiú)

西南点心。先将糯米粉与水调和,揉成圆球,放入加糖浆的热油煎至金黄壳脆,然后撒上芝麻。

Fried sticky rice balls with sesame. Southwestern snack. Sticky rice flour and water mixed, kneaded into balls, fried in oil and syrup until golden and crisp, then showered with sesame.

芝麻条 (zhīmatiáo)

江苏点心。用面粉、鸡蛋、芝麻、白糖做成面团,切成条,油炸。

Fried sweet sesame strips. Jiangsu snack. Dough made of flour, eggs, sesame, and white sugar, cut into strips, then deep-fried.

枝江大曲 (zhījiāngdàqū)

白酒,产于湖北宜昌枝江。用红高粱、大米、小麦等为原料酿造,酒精含量为 33-54 度,属浓香型酒。

Zhijiangdaqu, liquor produced in Zhijiang of Yichang, Hubei. White spirit made with red sorghum, rice, and wheat. Contains 33-54% alcohol, and has a thick aroma.

直沽烧 (zhígūshāo)

白酒,产于天津直沽。以高粱为原料,酒精含量为 45-50 度,属清香型酒。

Zhigushao, liquor produced in Zhigu, Tianjin. White spirit made with sor-ghum. Contains 45-50% alcohol, and has a delicate aroma.

植物扒四宝 (zhíwù pá sìbǎo)

浙江菜。先把笋尖、冬菇、蘑菇、竹笋过油,再下入鸡汤烩煮,加入鸡油、蚝油、味精、糖,勾芡。

Four vegetables in chicken soup. Zhejiang dish. Quick-fried bamboo shoot tips, winter mushrooms, fresh mushrooms, and bamboo shoots braised in chicken soup, flavored with chicken oil, oyster sauce, MSG, and sugar, then thickened with starch.

纸包柠檬烤鱼 (zhǐbāo níngméng kǎo yú)

江苏菜。把鸡腿菇和金针菇与用盐、柠檬汁、白葡萄酒腌过的鳜鱼用锡纸包好,烤熟。

Broiled Chinese perch with lemon and mushrooms. Jiangsu dish. Drumstick mushrooms and golden lily mushrooms wrapped in foil with Chinese perch pickled with salt, lemon juice, and white wine, then broiled.

纸包牛掌 (zhǐbāo niúzhǎng)

西藏菜。先将卤牛掌丁与香菇、鸡脯、土豆泥等拌匀，再用威化纸包好，裹上鸡蛋糊，沾上面包屑，油煎。

Fried ox feet in wafer paper. Tibetan dish. Diced marinated ox feet mixed with dried mushrooms, chicken breast, and mashed potatoes, wrapped in wafer paper, coated with egg paste and bread crumbs, then deep-fried.

纸包秀珍菇 (zhǐbāo xiùzhēngū)

山东菜。将氽过的秀珍菇片与盐、姜末等调料拌匀，铺在放有香菜叶和红辣椒的米纸上，裹成卷，油炸至脆。

Fried mushroom rolls. Shandong dish. Scalded sliced oyster mushrooms mixed with salt and ginger, placed on rice paper with coriander and red pepper, made into rolls, then fried crisp.

中和汤 (zhōnghétāng)

安徽菜。先将豆腐用开水焯过，再与虾米、香菇、笋尖等入鸡清汤煮沸，然后加调料和猪油，慢火炆煮。

Bean curd in chicken soup. Anhui dish. Scalded bean curd boiled in clear chicken soup with dried shrimp meat, dried mushrooms, and bamboo shoot tips, then simmered with flavorings and lard.

仲景羊肉汤 (zhòngjǐng yángròutāng)

河南菜。将氽过的羊肉配当归、生姜等炖煮。

Mutton and herb soup. Henan dish. Scalded mutton stewed with Chinese angelica, ginger, and other herbs.

粥 (zhōu)

各地家常食品。用大米或粟米等加水煮成粥，可加入各种配料。

Porridge. Home-style food in many places. Rice, corn, or other grain boiled into porridge with water. A variety of ingredients may be added.

周村烧饼 (zhōucūnshāobing)

山东点心。将面粉加水和白砂糖揉匀，做成薄饼，撒上芝麻，烘烤至熟脆。

Crisp flour and sesame cakes. Shandong snack. Flour, water, and sugar made into thin cakes, sprinkled with sesame, then baked crisp.

绉纱鸽蛋 (zhòushā gēdàn)

广东菜。先将鸽蛋煮熟，去壳，裹淀粉炸至皮绉，然后与冬笋、冬菇、菜心等加调料烩煮，勾芡。

Pigeon eggs with vegetables. Guangdong dish. Shelled boiled pigeon eggs coated with starch, then fried creasy. Braised with winter bamboo shoots, winter mushrooms, baby greens, and seasonings, then thickened with starch.

猪肺百合杏仁汤 (zhūfèi bǎihé xìngrén tāng)

药膳。猪肺片、百合、杏仁、蜜枣加入盐、料酒炖煮。

Hog lung and herb soup. Folk-medicinal dish. Sliced hog lung stewed with lily bulb cloves, almonds, and honey dates, then seasoned with salt and cooking wine.

猪肉黄豆炖豆腐 (zhūròu huángdòu dùn dòufu)

东北菜。将猪肉丝、泡涨的黄豆、豆腐块加雪里蕻炖熟。

Pork with soybean and bean curd. Northeastern dish. Shredded pork stewed with soaked soybean, bean curd, and preserved potherb mustard.

猪肉苦芝 (zhūròu kǔzhī)

福建菜。苦芝是一种当地野菜。用晒干的苦芝和猪肉加调料炖煮。

Stewed pork with wild vegetable. Fujian dish. Dried *kuzhi*, a local wild vegetable, stewed with pork.

猪油饽饽 (zhūyóu bōbo)

湖北点心。用发酵面团做面皮，用猪油和麻油拌面粉做油酥。把大葱嫩蕊切碎，与生猪油和盐拌成馅。用面皮包油酥，在油酥里塞入馅，煎至金黄。

Fried lard cakes. Hubei snack. Fermented flour dough made into wrappers. Flour, lard, and sesame oil made into pastry. Chopped tender leek, lard, and salt mixed into stuffing. Wrappers made into cakes filled with pastry over stuffing. Fried golden.

竹筏紫菜卷 (zhúfá zǐcàijuǎn)

广东菜。先将湿豆腐衣铺上蛋皮、苔菜、目鱼胶、黄瓜条、胡萝卜条，然后裹成卷，撒上面包屑，蒸熟，炸黄，配芥末酱或辣酱。

Laver, fish, and vegetable rolls. Guangdong dish. Wet bean curd skins made into rolls filled with egg sheets, laver, flatfish paste, cucumber strips, and carrot cuts. Coated with bread crumbs, steamed, then fried golden. Served with mustard sauce or chili sauce.

竹节鸡盅 (zhújié jīzhōng)

湖南菜。将鸡肉、瘦猪肉、干贝、荸荠剁碎加调料拌匀，装入竹筒，蒸熟。

Chicken and pork in bamboo tubes. Hunan dish. Diced chicken, lean pork, dried scallops, and water chestnuts mixed with seasonings, then steamed in bamboo tubes.

竹篱牛肉 (zhúlí niúròu)

湖北菜。先将牛肉用多种调料腌好，切片，置竹篱上稍晾，然后蒸熟。

Steamed spicy beef. Hubei dish. Beef pickled with seasonings, sliced, drained on bamboo drippers, then steamed.

竹篱飘香肉 (zhúlí piāoxiāngròu)

湖南菜。先将五花肉片挂蛋黄、吉士粉、米酒、淀粉调成的浆油炸，再加干辣椒煸炒，然后装入竹篱上桌。

Fried spicy streaky pork. Hunan

dish. Sliced streaky pork coated with mixture of egg yolks, custard powder, sweet fermented sticky rice juice, and starch, deep-fried, then stir-fried with chili pepper. Served in a bamboo sieve.

竹丝鸡煲鱼翅 (zhúsījī bāo yúchì)

广东菜。将竹丝鸡和鱼翅加入高汤炖煮。

Stewed chicken and shark fin. Guangdong dish. Black-bone chicken stewed with shark fin in broth.

竹荪百花鲍脯 (zhúsūn bǎihuā-bàopú)

广东菜。先将鲍鱼面抹上虾胶,底部用竹荪包上,蒸熟,然后浇芡汁,配轻炒芦笋装盘。

Steamed abalone with asparagus. Guangdong dish. Abalone topped with minced shrimp meat, wrapped in long net stinkhorn at the bottom, then steamed. Washed with starched soup, then served with quick-fried asparagus.

竹荪鱼糕 (zhúsūn yúgāo)

北方家常菜。将鱼茸、淀粉,水、盐、料酒、姜葱汁等调成糊,蒸熟成糕,置凉后切片,然后将鱼糕片入竹荪汤煮沸。

Fish cake in mushroom soup. Home-style dish in northern areas. Minced fish mixed with starch, water, salt, cooking wine, and ginger-scallion juice, then steamed into cakes. Cooled, sliced, then boiled in long net stinkhorn soup.

竹笋红烧肉 (zhúsǔn hóngshāoròu)

各地家常菜。先将五花肉过油,然后放入竹笋,加酱油、糖、葱、姜等调料,炖至收汁。

Stewed pork and bamboo shoots. Home-style dish in many places. Lightly fried streaky pork stewed with bamboo shoots, soy sauce, sugar, scallion, and ginger until reduced.

竹笋鸡 (zhúsǔnjī)

南方家常菜。先将竹笋与鸡腿爆炒,加清汤炖煮,然后放入竹荪,加调料煮熟。

Bamboo shoots and chicken in mushroom soup. Home-style dish in southern areas. Stir-fried drumsticks and bamboo shoots stewed in clear soup, then boiled with long net stinkhorn and seasonings.

竹笋木耳炒肉片 (zhúsǔn mùěr chǎo ròupiàn)

南方家常菜。先将猪肉片炒熟,再加入竹笋片、木耳及调料翻炒,勾芡。

Pork with bamboo shoots and wood ears. Home-style dish in southern areas. Sliced quick-fried pork sautéed with sliced bamboo shoots, wood ears, and seasonings, then thickened with starch.

竹笋芋艿鲫鱼汤 (zhúsǔn yùnǎi jìyú tāng)

浙江菜。先将竹笋、芋艿、鲫鱼、瘦肉、蜜枣入沸水煮滚,再加入调料用文火炖煮。

Crucian, bamboo shoot, and taro

soup. Zhejiang dish. Bamboo shoots, taros, crucian carp, lean pork, and honey dates boiled, then braised with seasonings.

竹筒豆豉蒸排骨 (zhútǒng dòuchǐ zhēng páigǔ)

湖南菜。先将排骨略腌,然后裹上淀粉,与蒜茸、豆豉、辣椒一道装入竹筒,蒸熟。

Steamed pork ribs with fermented soybean. Hunan dish. Lightly pickled pork ribs coated with starch, put in a bamboo tube with mashed garlic, fermented soybean, and hot pepper, then steamed.

竹筒粉蒸肠 (zhútǒng fěnzhēngcháng)

江西菜。先将肥肠焯水,腌渍,然后与五香米粉拌和,装入竹筒蒸烂,浇热红油。

Steamed hog large intestine in bamboo tube. Jiangxi dish. Pickled scalded hog large intestine mixed with ground rice and five-spice powder, put in a bamboo tube, then steamed. Flavored with heated chili oil.

竹筒鸡 (zhútǒngjī)

云南菜。先将鸡肉、鸡肝、鸡肫加调料腌渍,然后与火腿、冬菇、玉兰片装入竹筒蒸熟。

Steamed chicken in bamboo tube. Yunnan dish. Chopped chicken, chicken liver, and chicken gizzard pickled, put in a bamboo tube with ham, winter mushrooms, and soaked bamboo shoots, then steamed.

竹筒排骨 (zhútǒng páigǔ)

湖南菜。将猪排裹上糯米粉和剁椒,装入竹筒,蒸熟。

Spicy pork ribs in bamboo tube. Hunan dish. Pork ribs coated with sticky rice flour and pickled red pepper, then steamed in a bamboo tube.

竹香鲫鱼 (zhúxiāng jìyú)

湖南菜。先将腌好的鲫鱼油炸,然后把孜然、葱、姜等炒香盖在鲫鱼上。

Spicy crucian carp. Hunan dish. Pickled crucian carp deep-fried, then covered with quick-fried cumin, scallion, and ginger.

竹影海参 (zhúyǐng hǎishēn)

南方菜。先将海参入高汤汆熟,加入做成竹节状的鸡肉丸子,然后放上清水汆过的豌豆苗,倒入煮沸的鲜汤。

Sea cucumbers and chicken balls. Southern dish. Sea cucumbers quick-boiled in broth, bamboo-shape chicken balls added, garnished with scalded pea sprouts, then washed with boiling seasoned soup.

竹蔗马蹄羊腩煲 (zhúzhè mǎtí yángnǎn bāo)

广东菜。将羊腩、马蹄、竹蔗、姜加盐、料酒炖煮。

Steamed mutton flank. Guangdong dish. Mutton flank stewed with water chestnuts, sugar cane, and ginger, then seasoned with salt and cooking wine.

煮糟青鱼 (zhǔzāoqīngyú)

江苏菜。将糟腌青鱼加葱、姜、黄酒等煮熟,再把火腿片、冬笋、香菇、菠菜等入鱼汤烹煮,然后把煮好的配料排放在鱼身上,淋鱼汤和熟猪油。

Black carp with vegetables. Jiangsu dish. Pickled black carp boiled with scallion, ginger, and yellow wine. Covered with sliced ham, winter bamboo shoots, dried mushrooms, and spinach boiled in fish soup, then washed with fish soup and prepared lard.

柱侯浸鸡 (zhùhóu jìnjī)

广东菜。将鸡浸入柱侯酱,加入猪油、麻油及调料焖煮。柱侯酱为佛山名产,由清代当地厨师梁柱侯首制。

Braised chicken with special sauce. Guangdong dish. Chicken coated with Zhuhou sauce, then braised with lard, sesame oil, and other condiments. Zhuhou sauce is a specialty of Foshan, Guangdong, created by a famous chef, Liang Zhuhou, in Qing Dynasty.

柱侯文山腩 (zhùhóu wénshānnǎn)

广东菜。先将羊腩汆水,过油,再把姜、蒜、腐乳、柱侯酱等调料炒香,然后放入羊腩炖烂。

Stewed mutton flank with sauce. Guangdong dish. Scalded mutton flank lightly fried, then stewed with quick-fried ginger, garlic, fermented bean curd, and Zhuhou sauce.

抓草鱼片 (zhuācǎoyúpiàn)

北京菜。先把草鱼片挂浆腌渍,炸至金黄,再与煸炒过的姜葱翻炒,勾芡,淋猪油。

Stir-fried carp fillets. Beijing dish. Grass carp fillets pickled with salt and starch paste, fried golden, then stir-fried with quick-fried ginger and scallion. Thickened with starch, then washed with lard.

抓炒掌中宝 (zhuāchǎo zhǎngzhōngbǎo)

广东菜。先将腌好的鸡爪心肉炸至金黄,再与青瓜、姜、蒜等快炒,淋麻油。

Stir-fried chicken feet meat. Guangdong dish. Pickled meat of chicken feet fried golden, then quick-fried with cucumbers, ginger, and garlic. Washed with sesame oil.

壮乡竹筒鱼 (zhuàngxiāng zhútǒngyú)

广西壮族菜。将鲶鱼茸、香菇、葱、姜加盐拌和,装入竹筒,用糯米饭封口,蒸熟。

Steamed catfish in bamboo tube. Specialty of Zhuang ethnic minority in Guangxi. Minced catfish, mushrooms, scallion, and ginger seasoned with salt, put into a bamboo tube, sealed with cooked sticky rice, then steamed.

状元糕 (zhuàngyuangāo)

福建泰宁点心。先将糯米用水泡透,蒸熟,再舂打成糕,做成小球,然后埋入加糖的炒黄豆粉、炒花生粉或炒芝麻粉,加香葱、猪油。

Sticky rice balls in baked gourmet

powder. Snack in Taining, Fujian. Sticky rice soaked thoroughly in water, steamed, pestled in a mortar, then made into small balls. Balls coated generously with sugared powder of baked soybean, peanuts, or sesame, then flavored with scallion and lard.

状元卤味 (zhuàngyuan lǔwèi)

浙江菜。先将姜、葱、蒜、辣椒、香菇等煸炒,加入水、酱油、糖、料酒、麻油等煮沸,然后放入鸡肝、豆腐干等卤熟。

Marinated chicken liver and bean curd. Zhejiang dish. Chicken liver and dried bean curd marinated in gravy of water, soy sauce, sugar, cooking wine, sesame oil, ginger, scallion, garlic, chili pepper, and dried mushrooms.

卓巴卡渣 (zhuóbākǎzhā)

西藏菜,即咖喱牛肚。将熟牛肚切片,加咖喱和茴香等拌匀。

Ox tripe with curry. Tibetan dish. Sliced boiled ox tripe seasoned with curry and fennel.

卓玛莫古 (zhuómǎmògǔ)

西藏点心。将蕨麻即人参果煮熟,加入融化的酥油和白糖。

Silverweed roots with butter and sugar. Tibetan snack. Boiled silverweed roots mixed with melted Tibetan butter and white sugar.

卓学 (zhuóxué)

西藏菜,即酸奶拌人参果。先把人参果煮熟,然后与酸奶拌和。

Silverweed roots with yogurt. Tibetan dish. Boiled silverweed roots mixed with yogurt.

孜然羊肉 (zīrán yángròu)

新疆菜。先将羊肉片腌渍,用油滑熟,然后加入孜然和干辣椒爆炒,撒香菜。

Stir-fried mutton with cumin. Xinjiang dish. Pickled sliced mutton lightly fried, stir-fried with cumin and chili pepper, then garnished with coriander.

滋补酒 (zībǔjiǔ)

配制酒。各地家常含酒精饮品。在酿酒过程中或在酒中加入中草药,以滋补养生健体为主。可内服或外用。

Tonic liquor. Home-style compound drink of liquor and tonic herbs or folk medicines around the country. Herbs or folk medicines are added in the process of or after fermentation. May be used as a health drink or applied topically.

姊妹团子 (zǐmèi tuánzi)

湖南点心。将糯米粉与水调和,做成甜咸两种团子,一种包入桂花糖红枣泥馅,另一种包入香菇猪肉馅,或蒸或炸。

Sweet and salty sticky rice balls. Hunan snack. Sticky rice flour and water made into 2 kinds of balls. One sweet kind filled with osmanthus-scented sugar and red date paste, the other salty kind stuffed with mushrooms and pork.

Both may be steamed or fried.

紫菜汤 (zǐcàitāng)

沿海一带家常菜。将水发海带、海藻、紫菜放入沙锅炖煮,加调料和麻油。

Sea vegetable soup. Home-style dish in coastal areas. Water-soaked kelp simmered with seaweed and laver in an earthen pot. Flavored with seasonings and sesame oil.

紫菜鸭卷 (zǐcài yājuǎn)

江苏菜。先将鸭去骨加调料腌渍,卷入香菜和紫菜,用线捆紧,下卤水锅焖煮,然后抹上芝麻油,切片。

Marinated duck with laver. Jiangsu dish. Pickled boned duck rolled with coriander and laver, bound with strings, then marinated in gravy. Brushed with sesame oil, then sliced.

紫菜芋枣 (zǐcài yùzǎo)

台湾点心。将面皮包上芋泥和咸蛋黄,外面裹上紫菜条,炸至金黄。

Fried taro and laver buns. Taiwan snack. Flour buns filled with minced taros and salty egg yolks, wrapped in laver, then fried golden.

紫盖肉 (zǐgàiròu)

福建菜。先把猪五花肉用清汤、酱油、糖和酒炆煮,再抹上用菱粉、蛋浆、面粉调成的糊,炸至紫红色。

Fried sweet and salty pork. Fujian dish. Streaky pork simmered with soy sauce, sugar, and yellow wine in clear soup. Coated with starch, egg paste, and flour, then deep-fried until purple red.

紫桂焖大排 (zǐguì mèn dàpái)

山东菜。先将猪大排炸至金黄,再入鸡汤焖煮,用煸炒过的葱、姜、紫肉桂、盐、料酒等调味,浇原汁。

Fried and simmered pork steaks. Shandong dish. Deep-fried pork steaks simmered in chicken soup, flavored with quick-fried scallion, ginger, purple cinnamon, salt, and cooking wine, then marinated in broth.

紫金骨 (zǐjīngǔ)

江苏菜。先将猪大排用酱油腌渍,炸至深红,然后加酱油、番茄酱、松仁、葡萄干、红茶等焖煮,用蜂蜜和白兰地调味。

Fried and braised pork steaks. Jiangsu dish. Pork steaks pickled with soy sauce, fried until deep red, braised with soy sauce, tomato sauce, pine nuts, raisins, and black tea, then flavored with honey and brandy.

紫龙脱袍 (zǐlóng tuōpáo)

湖南菜。先将腌渍过的鳝鱼丝过油,再与葱姜丝、冬笋丝、香菇丝、柿子椒丝等翻炒,撒香菜。

Stir-fried field eel. Hunan dish. Lightly fried pickled shredded field eel stir-fried with shredded ginger, scallion, winter bamboo shoots, and round sweet pepper, then garnished with coriander.

紫米花宵枣泥球 (zǐmǐ huāxiāo zǎoníqiú)

广东点心。糯米粉与糖水和面,包入枣泥,做成球状,外沾紫色糯米,蒸熟。

Sticky rice and date balls. Guangdong snack. Sticky rice flour, sugar, and water made into balls, stuffed with mashed red dates, coated with purple rice, then steamed.

紫米珍珠丸 (zǐmǐ zhēnzhūwán)

广东菜。将猪肉末与太白粉加调料拌和,做成珍珠状丸子,裹上紫色糯米,蒸熟。

Steamed meatballs with purple rice. Guangdong dish. Ground pork and corn starch mixed with seasonings, made into pearl-size meatballs, coated with purple sticky rice, then steamed.

紫苏炒田螺 (zǐsū chǎo tiánluó)

广东菜。先把紫苏叶用沙茶酱、豆豉、蒜茸等爆香,再与田螺爆炒,勾芡。

Stir-fried snails with perilla. Guangdong dish. Field snails stir-fried with purple perilla leaves prepared with barbecue sauce, fermented soybean, and minced garlic, then thickened with starch.

棕叶香菇鸡翅 (zōngyè xiānggū jīchì)

湖南菜。将红烧鸡翅、糯米饭、浸汁香菇和藕丁用棕叶包好,蒸熟。

Steamed chicken wings in palm leaves. Hunan dish. Prepared chicken wings, cooked sticky rice, seasoned dried mushrooms, and diced lotus roots wrapped in palm leaves, then steamed.

粽子 (zòngzi)

各地端午节点心,源于湖南汨罗县。将水发糯米与红枣、赤豆、绿豆、莲子、腊肉、咸蛋黄、火腿等用竹叶或芦苇叶包裹成圆锥形,扎牢,煮熟,佐白糖或蜂蜜。

Boiled *zongzi* or sticky rice buns. Dragon Boat Festival snack around the country, originated in Miluo, Hunan. Soaked sticky rice mixed with red dates, red beans, mung beans, lotus seeds, smoked cured pork, salty egg yolks, or ham, wrapped into pyramid-shape dumplings or *zongzi* with bamboo or reed leaves, fastened, then boiled. Served with white sugar or honey.

走油豆豉扣肉 (zǒuyóu dòuchǐ kòuròu)

湖南菜。先将带皮五花肉氽过,油炸,然后加入浏阳黑豆豉、甜酒汁、葱,蒸至酥烂。

Streaky pork with fermented soybean. Specialty in Liuyang, Hunan. Deep-fried scalded streaky pork steamed with Liuyang fermented soybean, sweet fermented sticky rice juice, and scallion.

走油肉 (zǒuyóuròu)

各地家常菜。先将猪五花肉油炸,再用水浸软,切厚片,皮朝下放入碗中,加酱油、酒、糖、葱、姜蒸烂。

Fried and steamed pork. Home-style dish in many places. Deep-fried streaky pork soaked in water until soft, cut into thick pieces, then put in a bowl with skins down. Steamed with soy sauce, yellow wine, sugar, scallion, and ginger.

走油蹄髈 (zǒuyóu típǎng)

上海菜。先将蹄髈蒸熟,再炸至皮脆,然后加入黄酒、酱油、糖等炖烂,收汁。

Sweet and salty hog knuckle. Shanghai dish. Hog knuckle steamed, then fried until skins crisp. Stewed with yellow wine, soy sauce, and sugar until reduced.

组庵玉结鱼翅 (zǔān yùjié yúchì)

湖南菜。用鱼翅、肥母鸡、猪肘肉、干贝等加调料煨制。

Shark fin with hen and pork. Hunan dish. Shark fin stewed with hen, pork knuckle, dried scallops, and seasonings.

醉蚌肉 (zuìbàngròu)

福建菜。先将活蚌去壳切块,用高粱酒腌渍,再用酱油、白糖、芝麻油、醋、酒等调制的卤汁翻炒,然后放在醉卤里浸泡。

Mussels with liquored gravy. Fujian dish. Live mussels shelled, chopped, pickled in sorghum liquor, stir-fried with gravy of soy sauce, sugar, sesame oil, vinegar, and liquor, then marinated in gravy.

醉鹅掌 (zuìézhǎng)

广东凉菜。先将鹅掌煮熟,再放入用上汤、黄酒、盐、糖、蒜瓣做成的卤汁浸泡,配皮蛋装盘。

Marinated goose feet with preserved eggs. Guangdong cold dish. Boiled goose feet marinated in gravy of broth, yellow wine, salt, sugar, and garlic. Served with preserved duck eggs.

醉活虾 (zuìhuóxiā)

浙江菜。先将葱、姜、盐、酱油、黄酒、味精、麻油放在碗内调匀,再倒入活河虾,盖住闷约5分钟,配麻油醋碟。

Fresh shrimp in liquored dressing. Zhejiang dish. Live fresh water shrimp drowned in dressing of ginger, scallion, salt, soy sauce, yellow wine, MSG, and sesame oil for about 5 minutes. Served with sesame oil and vinegar.

醉排骨 (zuìpáigǔ)

南方菜。先将排骨块挂芡粉、盐、糖、醋、蛋清等调成的糊,炸熟,然后将加黄酒的卤汁浇在排骨上。

Fried sweet and sour pork ribs. Southern dish. Chopped pork ribs coated with paste of starch, salt, sugar, vinegar, and egg white, then deep-fried. Washed with gravy made with yellow wine and other condiments.

醉翁软鸡 (zuìwēngruǎnjī)

台湾菜。先将鸡腿用桂皮、丁香、胡椒粉、盐等腌渍,蒸熟,然后放入用黄酒、虾油等调制的卤汁浸泡。

Marinated drumsticks. Taiwan dish. Drumsticks pickled with cinnamon, clove, ground black pepper, and salt, then steamed. Marinated in gravy of yellow wine and shrimp oil.

醉蟹 (zuìxiè)

江浙沪皖菜。将活蟹洗净,放入用盐、糖、花椒、黄酒、白酒、姜、葱等调制的醉卤浸腌五天左右,切块装盘,浇醉卤。

Crab in liquored gravy. Jiangsu, Zhejiang, Shanghai and Anhui dish. Live fresh water crab cleaned, then immersed in gravy of salt, sugar, Chinese prickly ash, yellow wine, Chinese liquor, ginger, and scallion for about 5 days. Chopped, then marinated in liquored gravy.

左公鸡 (zuǒgōngjī)

湖南菜。先将鸡块和干辣椒与煸过的葱、蒜爆炒,然后放入番茄、香菇、青豆、菠萝、醋、酱油、盐、糖等翻炒。

Spicy savoury chicken. Hunan dish. Cubed chicken and chili pepper stir-fried with quick-fried onion and garlic, then sautéed with tomatoes, dried mushrooms, green beans, pineapple, vinegar, soy sauce, salt, and sugar.

参考文献

北京市人民政府外事办公室，北京市旅游局．北京市菜单英文译法（讨论稿）[Z]．北京旅游信息网，www. bjta. gov. cn/document/20070823171158406596. doc. 2007. 12. 12-2008. 5. 31.

曹雪芹．红楼梦[M]．北京：人民文学出版社，2006. 12．周汝昌汇校．

陈丕琮．新英汉餐饮词典[M]．上海：上海译文出版社，1999.

陈忠明，屠志详，张叶琴．扬州私房菜[M]．北京：中国纺织出版社，2006.

冯志伟．2003 年的读书笔记：理论词和形式词[Z]．北京：教育部语言文字应用研究所．http://ling. cuc. edu. cn/htliu/feng/xingshici. pdf. 2009-01-15.

冯志伟．确定切词单位的某些非语法因素[J]．《中文信息学报》，2000.

冯志伟．学者新论：中文信息技术标准：汉字注音？拼音正词法？[Z]．人民网．http://www. people. com. cn/GB/guandian/30/20030416/972825. html. 2009. 02. 07.

各地餐饮行业协会网站．

各省、自治区、市、县旅游局网站．

国际标准化组织（ISO）．文献工作——中文罗马字母拼写法【7098】[S]. 1982. 8. 1.

国家标准 GB13715，信息处理用现代汉语分词规范[S]．北京：中国标准出版社，1992.

国家语言文字工作委员会．汉语拼音正词法基本规则（国家标准）[S]．北京：国家技术监督局，1996.

汉典．汉典工具[DB]. http://www. zdic. net/tools/. 2004. 03. 01－2008. 05. 31.

黄山烹饪协会．徽菜技术标准[DB]．http://www. czhzw. com/Article_Show. asp? ArticleID＝408. 2004. 03. 01－2008. 03. 15.

李春丽，张慜（主编）．肿瘤疾病食疗指南[M]．北京：农村读物出版社，2000.

联合国第三届地名标准化会议．关于中国地名拼法的决议[Z]．联合国，1977. 9.

林明金，霍金根(主编). 简明英美语言与文化词典[M]. 上海:上海外语教育出版社，2003.

陆丙甫. 汉字文本便读格式初探[A]. 南昌:汉字书写系统改进国际研讨会，2004. 06. 12.

陆谷孙(主编). 英汉大词典[M]. 上海:上海译文出版社，1993.

美食中国. 菜谱大全[DB]. http://www.meishichina.com/English/Cate/12591_3.html. 2004. 03. 01－2008. 05. 31.

潘富俊(文)，吕胜由(摄影). 诗经植物图鉴[M]. 上海:上海书店出版社，2003.

祁公任等. 药膳百方[M]. 福州:福建科学技术出版社，2001.

企业商标注册网. 中国商标查询[DB]. http://sbcx.saic.gov.cn/trade/index.jsp. 2008. 10. 15.

清华大学. 英汉技术词典[M]. 北京:国防工业出版社，1985.

全国人民代表大会. 汉语拼音方案[S]. 北京:新华社，1958. 2. 11.

食品伙伴网. 食品图库[DB]. http://www.foodmate.net/4images/. 2004. 03. 01-2008. 05. 31.

丝绸之旅国际互联网. 美食图库[DB]. http://www.travel-silkroad.com/chinese/dongfangmeishi/zgwm_index.htm. 2004. 03. 01-2008. 05. 31.

王军等(译). 英汉双解饭店旅游餐饮词典[M]. 北京:世界图书出版公司，1998. (*Dictionary of Hotels, Tourism and Catering Management*. Peter Collin Publishing, 1994.)

王同亿(主编). 英汉辞海[M]. 北京:国防工业出版社，1987.

王增. 食鱼治百病[M]. 上海:上海书店出版社，2000.

王忠民(主编). 家常川味菜[M]. 济南:山东科学技术出版社，2005.

夏金龙(主编). 好学易学做. 川菜1000样[M]. 长春:吉林科技出版社，2005.

潇雪(主编). 川味家常菜[M]. 广州:世界图书出版公司，2004.

亚麻油健康推广中心. 公司产品[EB]. www.yamayou.cn[EB]. 2004. 03. 012008. 03. 09.

杨任之. 中国典故辞典[M]. 北京:北京出版社，1993.

药典在线. 中国药典[DB]. http://www.newdruginfo.com/ypbz/cp2000/cp2c021.htm. 2004. 03. 01-2008. 04. 30.

药品资讯网. 中草药数据库[DB]. http://www.chemdrug.com/Herbal.asp. 2004. 03. 01-2008. 05. 31.

葉美利，湯志真，黃居仁，陳克健. 漢語的動詞名物化初探——漢語中帶論元的名物化派生詞[A]. 台北:第五屆計算語言學研討會論文集，1992, pp. 177－193.

英汉缩略语词典[M]. 西安:陕西人民出版社，1979.

张天生,陈茂春. 中国名菜营养配餐. 北京篇[M]. 上海:上海科技出版社,2006.
张天生,张海燕. 中国名菜营养配餐. 江苏篇[M]. 上海:上海科学技术出版社,2006.
章自福. 中华成语大辞典[M]. 海口:海南国际新闻出版中心,1996.
赵建芳. 时新杭州菜[M]. 北京:农村读物出版社,2003.
浙江萧山海润养殖有限公司. 产品介绍[DB]. http://www.cn-hryz.com/acy.htm.
郑云山,臧威霆(主编). 中外史地知识手册[M]. 上海:上海人民出版社,1984.
中国茶叶网. 茶叶种类[DB]. http://www.china-tea.org/List/List_33_1.Html. 2008.10.10.
中国大百科全书[CD]. 北京:中国大百科全书出版社,2000.
中国饭店协会. 中国美食节[EB]. http://www.chinahotel.org.cn/msj_difangmsj.asp. 2004.03.01-2008.05.31.
中国酒水在线. 产品大全[DB]. http://www.jiushui.net.cn/T_ProductList1.asp. 2009.01.01.
中国科普博览. 中国重要经济鱼类[DB]. http://kepu.itsc.cuhk.edu.hk/gb/lives/. 2004.03.01-2008.05.31.
中国烹饪百科全书[CD]. 北京:中国大百科全书出版社,1995 年修订版.
中国烹饪协会网站. 健康饮食[DB]. http://www.ccas.com.cn/[DB]. 2004.03.01-2008.05.31.
中国社会科学院语言研究所词典编辑室. 现代汉语词典[M]. 北京:商务印书馆,2005 年第 5 版.
中国社会科学院语言研究所词典编辑室. 现代汉语词典[M]. 北京:外语教学与研究出版社,2002.(汉英双语版)
中华美食网. 美食[DB]. http://www.zhms.cn/index.htm. 2004.03.01-2008.05.31.
中华全国供销合作总社. 茶叶资讯[DB]. http://www.chinacoop.gov.cn/Category_36/index.aspx. 2008.08.15.
钟肇恒. 英汉美术词典[M]. 上海:上海外语教育出版社,1984.
周范林(主编). 百姓家常菜 1000 例[M]. 北京:科学出版社,2004.
周锦(主编). 满汉全席[M]. 北京:农村读物出版社,2002.
周有光语文论集(第二卷)[M]. 上海:上海文化出版社,2002.
American Heritage Dictionary [M]. Boston: Houghton Mifflin, 1985, 2nd college edition.
Archambault, A. & Corbeil, J. *Macmillan Visual Dictionary*. U.S.A.: 1995.
Barnhart, C. L., & Barnhart, R. K. (eds.). *The World Book Dictionary*

[M]. Chicago: World Book-Childcraft International, Inc. 1981.

Berkowitz, R., & Doerfer, J. (eds.). *The Legal Sea Foods Cookbook* [M]. Boston: Main Street Books, 1988.

Dictionary of American Food and Drink, *A* [M]. New York: Hearst Books, 1994, Subsequent edition.

Dictionary of Food and Drink, *A* [M]. Essex, England: Longman, 1984, 2nd print.

English Duden, *the*. Bibliographisches Institut AG, Mannheim. Paris: Librairie Marcel Didier. 1960.

Fachredaktionen of the Bibliographisches Institut, and the Modern Languages Department of George G. Harrap & Company Ltd. *English Duden*, *the* [M]. Paris: Librairie Marcel Didier, 1960, revised edition.

Food-Image. Com. http://www. food-image. com/info/fruit-photos. html[DB]. 2004. 03. 01-2008. 05. 31.

Gernot Katzer's Spice Pages. All Spices[DB]. http://www. uni-graz. at/~katzer/engl/index. html. 2004. 03. 01-2008. 05. 31.

King, L. T. & Wexler, J. S. (eds.). *Martha's Vineyard Cooking Book*[M]. Guilford, Connecticut, U. S. A: The Globe Pequot Press. 2000.

Lite Delight[M]. Lincolnwood, IL: Publications International, 1990.

Mcmillan. *Food Lover's Encyclopedia*, *the*. NJ: Prentice Hall & IBD. 1998.

Microsoft Corporation. *Microsoft Encata Encyclopedia*[CD]. Redmond, U. S. A.: Author, 2006.

National Geographic. Fish [DB]. http://animals. nationalgeographic. com/animals/fish. html. 2004. 03. 01-2008. 05. 31.

Oxford English Dictionary. New York: Oxford University Press. 2007.

Parnwell, E. C. The New Oxford Picture Dictionary. New York: Oxford University Press. 1988.

Sea-Ex: Fish & Seafood Information[DB]. http://www. sea-ex. com/fish/. 2004. 03. 01-2008. 05. 31.

Seidemann, *J. World Spice Plants*. New York: Springer-. 2005.

Simpson, D. P. *Lantin English Dictionary*. UK: Cassell & Co Ltd, 1987.

Stock Photo and Stock Footage Library. Stock Photography and Stock Footage [DB]. http://www. fotosearch. com/photos-images/bitter-gourd. html. 2004. 03. 01-2008. 04. 30.

Tsao Hsueh-chin, Kao Hgo. A Dream of Red Mansions. Translated by Yang Hsien-Yi and Gladys Yang. Illustrated by Tai Tun-Pang. Beijing: Foreign

Language Press, 1999.

Unabridged Dictionary of Food & Drink[M]. LA: T. G. I. Friday, 1981.

Watson (ed.). *Longman Modern English Dictionary*[M]. London: Longman, 1976 edition.

Webster's Encyclopedic Unabridged Dictionary of the English Language[M]. Avenel, NJ: Gramercy Books. 1994, Deluxe Edition.

Webster's Gold Encyclopedia [CD]. U. S. A.: CounterTop Software & MultiEducator, Inc. 2000 Edition.